The Handbook of
Biomass Combustion and Co-firing

The Handbook of
Biomass Combustion and Co-firing

Edited by
Sjaak van Loo and Jaap Koppejan

London • Washington, DC

First published by Earthscan in the UK and USA in 2008
Published in paperback 2010

Copyright © Sjaak van Loo and Jaap Koppejan, 2008

All rights reserved

ISBN: 978-1-84971-104-3

Typeset by FiSH Books, Enfield
Cover design by Susanne Harris

For a full list of publications please contact:

Earthscan Ltd, Dunstan House, 14a St Cross Street, London EC1N 8XA, UK
Earthscan LLC, 1616 P Street, NW, Washington, DC 20036, USA
Earthscan publishes in association with the International Institute for Environment and Development

For more information on Earthscan publications, see www.earthscan.co.uk or write to earthinfo@earthscan.co.uk

A catalogue record for this book is available from the British Library

Library of Congress Cataloging-in-Publication Data

The handbook of biomass combustion and co-firing/edited by Sjaak van Loo and Jaap Koppejan.
    p. cm.
  Includes bibliographical references.
  ISBN-13: 978-1-84407-249-1 (hardback)
  ISBN-10: 1-84407-249-5 (hardback)
  1. Biomass—Combustion—Handbooks, manuals, etc. 2. Biomass energy—Handbooks, manuals, etc. I. Van Loo, Sjaak. II. Koppejan, Jaap.
  TP339.H36 2007
  662'.88—dc22
                                                                                    2007021163

At Earthscan we strive to minimize our environmental impacts and carbon footprint through reducing waste, recycling and offsetting our CO_2 emissions, including those created through publication of this book. For more details of our environmental policy, see www.earthscan.co.uk.

Printed and bound in the UK by CPI Antony Rowe.
The paper used is FSC certified.

Contents

List of figures and tables	xi
Preface	xx
List of contributors	xxi

1 Introduction — 1
1.1 Current status of bioenergy — 3
1.2 Combustion as main bioenergy technology — 4
1.3 This handbook — 5
1.4 References — 6

2 Biomass Fuel Properties and Basic Principles of Biomass Combustion — 7
2.1 Introduction — 7
2.2 The process of biomass combustion — 8
    2.2.1 Drying, pyrolysis, gasification and combustion — 9
    2.2.2 Operational and design variables affecting the combustion process — 11
    2.2.3 Principles of medium- to large-scale combustion applications — 22
    2.2.4 Thermodynamics and gas phase kinetics — 23
    2.2.5 Solid fuel kinetics and gas phase interaction — 23
    2.2.6 Basic engineering calculations — 23
    2.2.7 Basic emission conversion calculations — 29
    2.2.8 Batch versus continuous combustion — 31
    2.2.9 Modelling biomass combustion — 33
    2.2.10 Ash formation — 34
2.3 Physical and chemical characteristics of biomass fuels — 38
    2.3.1 Overview — 38
    2.3.2 Fuel characteristics and interaction with the combustion system — 38
    2.3.3 Databases on biomass fuel characteristics — 48
    2.3.4 Standardization of biomass fuels — 49
2.4 References — 50

3 Biomass Fuel Supply and Pre-treatment — 54
3.1 Quality-related influences along the supply chain of biomass fuels — 56
    3.1.1 Influences on the fuel quality during the growing phase — 57
    3.1.2 Influences on the fuel quality during the supply phase — 58
    3.1.3 Fuel quality control and price calculations — 59
3.2 Biomass production — 60
    3.2.1 Forest residues — 60
    3.2.2 Herbaceous biomass fuels — 61
    3.2.3 Short rotation woody crops — 62

	3.3	Fuel pre-treatment	64
		3.3.1 Comminution of woody biomass	64
		3.3.2 Pre-treatment of waste wood	69
		3.3.3 Baling and bundling of biomass fuels	72
		3.3.4 Pellets and briquettes	74
		3.3.5 Drying of biomass	78
	3.4	Storage, handling and transport systems	83
		3.4.1 Storage of biomass	83
		3.4.2 Fuel-feeding and handling systems	89
		3.4.3 Transport systems for biomass fuels	95
	3.5	System perspectives on bioenergy	96
		3.5.1 Case 1 Multifunctional bioenergy systems (MBS)	98
		3.5.2 Case 2 Co-firing opportunities in Poland	102
		3.5.3 Conclusions from the two cases	108
	3.6	References	108
4	**Domestic Wood-Burning Appliances**		**112**
	4.1	Introduction	112
	4.2	Design considerations	112
	4.3	Residential batch-fired wood-burning appliances	113
		4.3.1 Wood-stoves	113
		4.3.2 Fireplace inserts and zero clearance fireplaces	115
		4.3.3 Heat-storing stoves	116
		4.3.4 Wood log boilers	118
	4.4	Wood pellet appliances and burners	122
		4.4.1 Background	122
		4.4.2 Technical features	122
		4.4.3 Safety requirements regarding back-burning	123
		4.4.4 Emissions	123
		4.4.5 Efficiency	124
		4.4.6 Certification	124
		4.4.7 Pellet burners for central heating systems	125
		4.4.8 Wood pellet stoves	127
	4.5	Woodchip appliances	128
		4.5.1 Pre-ovens	128
		4.5.2 Under-fire boilers	129
		4.5.3 Stoker burners	129
	4.6	Certification and testing standards	129
		4.6.1 European standards	130
		4.6.2 North American standards	132
	4.7	References	133
5	**Combustion Technologies for Industrial and District Heating Systems**		**134**
	5.1	Introduction	134
	5.2	Fixed-bed combustion	135
		5.2.1 Grate furnaces	135
		5.2.2 Underfeed stokers	146

			Page
	5.3	Fluidized bed combustion	147
		5.3.1 Bubbling fluidized bed (BFB) combustion	148
		5.3.2 Circulating fluidized bed (CFB) combustion	149
	5.4	Pulverized fuel combustion	150
	5.5	Summary of combustion technologies	154
	5.6	Heat recovery systems and possibilities for increasing plant efficiency	155
	5.7	Process control systems for biomass combustion installations	161
		5.7.1 Control objectives	161
		5.7.2 Process dynamics	161
		5.7.3 State-of-the-art process control	163
		5.7.4 Advanced process control	164
	5.8	Techno-economic aspects of biomass combustion plant design	167
		5.8.1 Technical and economic standards for biomass combustion and district heating plants	168
		5.8.2 Plant dimensioning/boiler size	169
		5.8.3 Annual utilization rate of the biomass system	170
		5.8.4 Size of the fuel storage unit	170
		5.8.5 Construction and civil engineering costs	171
		5.8.6 Heat distribution network	171
		5.8.7 Heat generation costs and economic optimization	171
	5.9	References	173
6	**Power Generation and Co-generation**		**175**
	6.1	Overview of power generation processes	175
	6.2	Closed thermal cycles for power production	175
	6.3	Steam turbines	177
		6.3.1 Working principle	177
		6.3.2 Rankine cycle	179
		6.3.3 Economic aspects	182
	6.4	Steam piston engines	183
	6.5	Steam screw engines	187
	6.6	Organic Rankine cycle (ORC)	189
	6.7	Closed gas turbines	192
	6.8	Stirling engines	193
	6.9	Comparison of heat, power and CHP production	197
	6.10	Summary	200
	6.11	References	201
7	**Co-combustion**		**203**
	7.1	Introduction	203
	7.2	Operational experience	203
	7.3	Co-firing concepts	206
		7.3.1 Direct co-firing	207
		7.3.2 Indirect co-firing	207
		7.3.3 Parallel co-firing	207
	7.4	Examples of biomass co-firing in pulverized coal-fired boilers	208
		7.4.1 Direct co-firing of demolition wood waste with coal, Gelderland Power Station, Nijmegen, the Netherlands	208

	7.4.2	Direct co-firing of sawdust and woodchips with coal, Wallerawang Power Station, NSW, Australia	209
	7.4.3	Direct co-firing of straw with coal, Studstrup, Denmark	212
	7.4.4	Direct co-firing of wood fuels with coals, St Andrea, Austria	215
	7.4.5	Indirect co-firing of biomass fuel gas with coal, Zeltweg Power Plant, Austria	216
	7.4.6	Indirect co-firing of biomass fuel gas with coal, Amer Power Plant, the Netherlands	218
	7.4.7	Parallel co-firing of biomass and fossil fuels, Avedøre Power Plant, Denmark	219
	7.4.8	Summary of experience of biomass co-firing in pulverized coal-fired boilers	220
7.5	Fuel preparation, processing and handling issues		223
	7.5.1	Preliminary size reduction	224
	7.5.2	Bulk handling	225
	7.5.3	Long-term storage	226
	7.5.4	Drying	228
	7.5.5	Secondary size reduction	228
7.6	Operational and environmental issues		232
	7.6.1	General fuel characteristics of coals and biomass	232
	7.6.2	Particle size and residence time	233
	7.6.3	Boiler efficiency	235
	7.6.4	Rate of combustion and char burnout	235
	7.6.5	Flame stability	237
	7.6.6	System integration and control issues	237
	7.6.7	Ash deposition	237
	7.6.8	Gas-side corrosion of boiler components	239
	7.6.9	Impact on emissions	239
	7.6.10	Performance of NO_x and SO_x emissions abatement equipment	243
	7.6.11	Efficiency of particulate emissions abatement equipment	244
	7.6.12	Flyash utilization	245
7.7	References		247

8 Biomass Ash Characteristics and Behaviour in Combustion Systems   249
8.1	Introduction		249
8.2	Biomass ash characteristics		250
	8.2.1	Introduction	250
	8.2.2	Laboratory characterization techniques for biomass ashes	251
8.3	High temperature behaviour of inorganic constituents of biomass in combustion systems		258
	8.3.1	Introduction	258
	8.3.2	Grate-fired combustors	260
	8.3.3	Fluidized bed combustors	263
	8.3.4	Pulverized fuel combustion systems	264
8.4	Formation and nature of ash deposits on the surfaces of combustors and boilers firing or co-firing biomass materials		266
	8.4.1	Introduction to ash deposition	266
	8.4.2	Slag formation processes	267

		8.4.3	Convective section fouling processes	269

		8.4.3	Convective section fouling processes	269
		8.4.4	Deposit growth, shedding and online cleaning	271
	8.5	Impact of ash on the flue gas cleaning equipment in biomass-firing systems		272
		8.5.1	Introduction	272
		8.5.2	The impact of co-firing on electrostatic precipitators	273
		8.5.3	The impact of biomass co-firing on SCR catalysts	273
		8.5.4	The impact of biomass co-firing on FGD plants	275
	8.6	Impact of biomass ash on boiler tube corrosion, and on erosive and abrasive wear of boiler components		276
		8.6.1	Technical background to gas-side corrosion processes in boilers	276
		8.6.2	Corrosion mechanisms	277
		8.6.3	Plant experience and boiler probing trials with biomass firing and biomass–coal co-firing	278
		8.6.4	Preventive and remedial measures for fire side corrosion	283
		8.6.5	Erosion and abrasion of boiler components and other equipment	285
	8.7	Biomass ash utilization and disposal		286
	8.8	References		288
9	**Environmental Aspects of Biomass Combustion**			**291**
	9.1	Introduction		291
	9.2	Environmental impacts of biomass combustion		292
		9.2.1	Emission components and their main influencing factors	294
		9.2.2	Measuring emissions from biomass combustion	304
		9.2.3	Emissions data	304
		9.2.4	Primary emission reduction measures	309
		9.2.5	Secondary emission reduction measures	318
		9.2.6	Emission limits in selected IEA member countries	339
	9.3	Options for ash disposal and utilization		348
		9.3.1	Introduction: General, ecological and technological limitations	348
		9.3.2	Physical and chemical properties of biomass ashes	350
		9.3.3	Material fluxes of ash-forming elements during combustion of biomass	359
		9.3.4	Conclusions	361
	9.4	Treatment technologies and logistics for biomass ashes		362
		9.4.1	Combustion technology	362
		9.4.2	Downstream processes: Ash pre-treatment and utilization	363
		9.4.3	Examples of guidelines for the utilization of biomass ashes in Austria	365
		9.4.4	Recommended procedure and quantitative limits for biomass ash-to-soil recycling in Austria	365
	9.5	Waste-water handling from flue gas condensation		367
	9.6	References		372

10 Policies — 379
- 10.1 Introduction — 379
- 10.2 Global expansion of biomass combustion — 379
  - 10.2.1 Trends in selected OECD member countries — 381
  - 10.2.2 Trends in selected non-OECD member countries — 385
  - 10.2.3 Relevant policy issues — 385
- 10.3 Financial support instruments — 387
  - 10.3.1 Fixed feed-in tariffs and fixed premiums — 387
  - 10.3.2 Green certificate systems — 387
  - 10.3.3 Tendering — 387
  - 10.3.4 Investment subsidies — 387
  - 10.3.5 Tax deduction — 388
- 10.4 Other policies that influence the establishment of biomass combustion — 389
- 10.5 References — 390

11 Research and Development: Needs and Ongoing Activities — 391
- 11.1 Investigation of potentials of biomass resources — 391
- 11.2 Development of improved combustion technologies — 392
- 11.3 Gaseous (especially NO_x) reduction technologies — 392
- 11.4 Ash and aerosol-related problems during biomass combustion including dust (fine particulate) reduction technologies — 393
- 11.5 Innovative micro-, small- and medium-scale CHP technologies based on biomass combustion — 394
- 11.6 Technology development and unsolved problems concerning co-firing of biomass in large-scale power plants — 394
- 11.7 Fuel pre-treatment technologies — 395
- 11.8 CFD modelling and simulation of thermochemical processes — 395
- 11.9 References — 396

Annex 1 Mass Balance Equations and Emission Calculation — 408
Annex 2 Abbreviations — 413
Annex 3 European and National Standards or Guidelines for Solid Biofuels and Solid Biofuel Analysis — 417
Annex 4 Members of IEA Bioenergy Task 32: Biomass Combustion and Cofiring — 423

Index — 426

List of Figures, Tables and Boxes

Figures

1.1	Many countries have abundant resources of unused biomass readily available	1
1.2	Wood-stove commonly used in Cambodia	2
1.3	Influencing parameters for the optimal design of biomass combustion systems	4
1.4	Wood-fired heating plant used for district heating in Wilderswil, Switzerland	5
2.1	Thermochemical conversion technologies, products and potential end uses	7
2.2	The combustion of a small biomass particle proceeds in distinct stages	8
2.3	Thermogravimetric analysis of four wood samples	10
2.4	Adiabatic combustion temperature as a function of moisture content (w.b.) and excess air ratio for a continuous combustion process	12
2.5	GCV and NCV as a function of wt% C and H	13
2.6	NCV as a function of wt% moisture	13
2.7	Chemical composition of various solid fuels	14
2.8	Integral reaction flow analysis in a premixed stoichiometrical methane flame	16
2.9	Influence of residence time and temperature on CH_4 conversion	17
2.10	Main variables influencing emission levels and energy efficiency in wood-stoves and fireplaces	18
2.11	Adiabatic combustion temperature as a function of excess air ratio and inlet air temperature for a continuous combustion process	19
2.12	Time dependence of combustion processes in batch combustion applications	32
2.13	Modelling tools	33
2.14	Mechanisms involved in ash formation in biomass combustion	35
2.15	Images of coarse fly ash and aerosol particles from wood combustion in a grate furnace	36
2.16	Ash fractions produced in a biomass combustion plant	37
3.1	Fuel supply chain for woody biomass	54
3.2	Wood fuel flows from the forest to the end-user	55
3.3	Factors influencing the quality of solid biomass fuels	56
3.4	Content of Cl in straw as a function of chlorine supply through fertilizers	57
3.5	Leaching of barley straw by rainfall during storage on the field	58
3.6	Yield and moisture content at different harvesting dates for various cereals	59
3.7	Harvesting processes for herbaceous biomass fuels	62
3.8	Processes for harvesting short rotation coppice	63
3.9	Disc and drum chippers	64
3.10	Mobile chipper with a crane efficiently collecting roadside thinnings for fuel	66
3.11	Mobile chipper	66
3.12	Spiral chunker	67
3.13	Fine grinding mill	67

3.14	Hammer mill	68
3.15	Pilot plant for processing waste wood	70
3.16	Breaker for waste wood	71
3.17	Bundler	73
3.18	Bundles stored at the roadside	74
3.19	Truck and train transportation of bundles	75
3.20	Pelletizing plant	76
3.21	Ring die pelletizer and flat die pelletizer	77
3.22	Two briquetting technologies	78
3.23	Development of moisture content in the bark piles stored outdoors and indoors	80
3.24	Pulsating solar biomass drying process	81
3.25	Short-term biomass drying process based on pre-heated air from a flue gas condensation unit	82
3.26	Belt dryer for biomass	82
3.27	Temperature development and processes responsible for self-heating in stored biological material	84
3.28	Development of temperature in a bark pile	86
3.29	Sliding bar conveyor used for the discharge of bunkers	87
3.30	Walking floor used for the discharge of biomass fuels from a long-term storage hall	87
3.31	Inclined rotating screw used for the discharge of silos	88
3.32	Wheel loaders are often used for transporting sawdust and bark from the long-term storage to the feeding system of the biomass heating plant	89
3.33	Automatic crane for straw feeding	90
3.34	Cranes are often used for handling woodchips in the intermediate storage bunkers of a biomass heating plant	90
3.35	Tube-rubber belt conveyor	91
3.36	Chain trough conveyor	92
3.37	Screw conveyor	92
3.38	Bucket elevator	93
3.39	Pneumatic conveyor with low-pressure ventilators	94
3.40	Linking energy-forest cultivation with energy and sewage plants	99
3.41	Salix field irrigated with pre-treated municipal waste water in Enköping	100
3.42	Practical potential and production cost for multifunctional energy plantations on a Sweden-wide level	101
3.43	Power generation system of Poland by age and fuel, including hydro	103
3.44	Modelling methodology used in the co-firing example	105
3.45	Modelled electricity generation 2005–2020 in boilers less than 30 years old	106
3.46	Biomass co-firing potential in Polish power plants up to the year 2020	106
3.47	Illustrative modelling of the expansion of biomass use for co-firing in existing units in Poland over the period up to 2020	107
4.1	Classification of wood-stoves depending on primary air flow paths	114
4.2	Catalytic stove design	114
4.3	Fireplace insert	115
4.4	Contraflow system in Finnish heat-storing stoves	116
4.5	Heat-storing stove using down-draught combustion	117
4.6	Wood combustion with over-fire boiler	118

4.7	Under-fire boiler	119
4.8	Microprocessor-controlled down-draught boiler for wood (logs Fröling)	120
4.9	Two wood pellet boilers, used to heat a school in Denmark	123
4.10	Conversion of fuel nitrogen to NO_x in wood pellet combustion	124
4.11	Correlation between unburned compounds (CO and OGC) and NO_x	125
4.12	Wood pellet burners	126
4.13	Pellet stove	127
4.14	Woodchip pre-oven	128
4.15	Horizontal stoker burner	130
5.1	Principal combustion technologies for biomass	134
5.2	Combustion process in fixed fuel beds	136
5.3	Classification of grate combustion technologies	137
5.4	Technological principle of a travelling grate	138
5.5	Travelling grate furnace fed by spreader stokers	139
5.6	Operating principle of an inclined moving grate	140
5.7	Modern grate furnace with infrared control system section-separated grate and primary air control	141
5.8	Inclined moving grate furnaces	141
5.9	Diagram of a horizontally moving grate furnace	142
5.10	Horizontally moving grate furnace	143
5.11	Vibrating grate fed by spreader stokers	143
5.12	Cigar burner	144
5.13	Underfeed rotating grate	145
5.14	Underfeed stoker furnace	146
5.15	Post-combustion chamber with imposed vortex flow	147
5.16	Diagram of a BFB furnace	149
5.17	$25MW_e$ woodchip-fired power plant with BFB boiler in Cuijk, the Netherlands	150
5.18	CFB furnace	151
5.19	Pulverized fuel combustion plant (muffle furnace) in combination with water-tube steam boiler	152
5.20	Two-stage cyclone burner	153
5.21	Dust injection furnace	153
5.22	Influence of the oxygen content in the flue gas on plant efficiency	156
5.23	Influence of the oxygen content in the flue gas on the heat recoverable in flue gas condensation plants	157
5.24	Flue gas condensation unit for biomass combustion plants	158
5.25	Efficiency of biomass combustion plants with flue gas condensation units as a function of flue gas temperature	159
5.26	Flue gas condensation unit with an integrated air humidifier	160
5.27	Dynamic responses of a large incinerator installation to fuel dosage, grate speed, primary air and secondary air changes	162
5.28	Block diagram of combustion (CO/λ) and load control	164
5.29	Schematic representation of a model predictive process control system	166
5.30	Example of distribution between base load and peak load on the basis of the annual heat output line	170
5.31	Specific investment costs for biomass combustion plants in Austria and Denmark as a function of biomass boiler size	172

5.32	Specific capital costs for biomass combustion systems as a function of boiler capacity and boiler utilization	172
6.1	Single-stage radial flow steam turbine with gear shaft and generator used in a biomass-fired CHP plant of approximately $5MW_{th}$ and $0.7MW_e$	177
6.2	Axial flow steam turbine with five stages, typical for industrial applications of a few MW_e	178
6.3	Flow and T/S diagrams of back-pressure plants based on the Rankine cycle	180
6.4	Thermal efficiency of the Rankine cycle as a function of live steam parameters and back-pressure	181
6.5	Extraction condensing plant with use of steam at intermediate pressure (6B) for heat production and condensing operation for the non-utilized part of steam at lower presssure to drive the low-pressure section of the turbine	182
6.6	Steam engine (two cylinders) from Spillingwerk, Germany	184
6.7	Flow and T/s-diagrams for a steam cycle using saturated steam in a steam piston engine or a steam screw-type engine	185
6.8	Screw-type compressor and working principle of a steam screw engine	187
6.9	730kW$_e$ screw-type engine with high-pressure and low-pressure stage and generator in a demonstration plant in Hartberg, Austria	188
6.10	Possible processes with screw engines in the T/s-diagram	188
6.11	Principle of co-generation using a steam engine controlled by pressure-reducing valve and throttle valve	189
6.12	Flow and T/S diagrams for co-generation with an ORC process (above) and process in the T/s-diagram	190
6.13	ORC equipment	191
6.14	Flow and T/s diagrams of a closed gas turbine cycle with recuperation	193
6.15	V-shaped Stirling engine	194
6.16	T/s diagram of the Stirling cycle	195
6.17	CHP biomass combustion plant with Stirling engine	195
6.18	Stirling engines	196
6.19	Power-to-heat ratio α as a function of the plant size of biomass-fuelled CHP plants in Finland and Sweden with 1–20MW$_e$	197
6.20	Percentage of heat and electric power production in heating plants, CHP plants and power plants (qualitative figures)	198
6.21	Comparison of heat, CHP and power plant efficiencies by an exergetically weighted efficiency ($\eta_{ex} = \varepsilon\eta_e + \eta_h$)	199
7.1	Geographic distribution of power plants that have experience with co-firing biomass with coal, as of 2004	204
7.2	Distribution of firing systems with coal-fired power plants that have experience with co-firing biomass	204
7.3	Effects of biomass co-firing at a coal-fired power station	205
7.4	Wood co-firing system at the Gelderland Power Plant of Electrabel	210
7.5	Straw shredder at Studstrup power plant	213
7.6	Straw pre-processing equipment at Studstrup Unit 4	214
7.7	Biomass co-firing system at St Andrea, Austria	215
7.8	BIOCOCOMB system	217
7.9	Original layout of the AMERGAS biomass gasification plant at the Amer Power Plant in Geertruidenberg, the Netherlands	218

7.10	AMERGAS biomass gasification plant after modification	219
7.11	Avedøre 2 multifuel system	220
7.12	Biomass fuel handling facility, which directly meters biomass onto the coal conveyor belts, Wallerawang Power Station, Australia	223
7.13	Dumping woodchips on a coal conveyor before the mills	224
7.14	Comparison of flow function for coal and coal/biomass blends	227
7.15	Pilot-scale vertical spindle mill used in the CCSD study	230
7.16	Mill wear rate of pulverized samples	231
7.17	Product size distributions for 5wt% blends of biomass and Lithgow coal	232
7.18	Mill power consumption during size reduction	233
7.19	Typical ultimate analyses of biomass and coal	234
7.20	Typical ultimate analyses of ash in biomass and coal	234
7.21	Char burnout from CCSD pilot-scale co-firing studies	236
7.22	Ash deposition rate for various fuels	238
7.23	The molar ratio of sulphur to available alkali and chlorine is an indicator of the chlorine corrosion potential	240
7.24	Influence of co-firing various biofuels on SO_2 emissions	240
7.25	Effect on NO_x emissions when co-firing switchgrass and wood with coal	241
7.26	Influence of mixed combustion of wood and coal on N_2O emissions in a circulating fluidized bed boiler	242
7.27	Amount of aerating agent required to generate air entrainment within ASTM specifications for a variety of fly-ash compositions	246
7.28	Dependence of flexural strength on fly-ash composition	246
8.1	Calculated melting curves for salt mixtures with K/Na molar ratio 90/10, SO_4/CO_3 molar ratio 80/20 and Cl varying between 0 and 20 per cent of the total alkali	254
8.2	Basic chemical fractionation data for a range of solid fuels	256
8.3	Summary of the chemical fractionation data for a range of solid fuels	257
8.4	Fate of inorganic material in solid fuels during combustion processes	259
8.5	Key processes involved in fly-ash and aerosol release from the combustion of wood on a grate	261
8.6	Chemical analysis of key elements on an SCR catalyst surface before and after exposure to a flue gas from an alkali and alkaline earth metal-rich fuel	274
8.7	Principles of the proposed 'active oxidation' corrosion mechanism	279
8.8	Corrosion rates of boiler tube material specimens exposed to flue gases from straw combustion, plotted against the metal temperature	285
9.1	Fraction of fuel nitrogen converted to NO_x for various wood fuels in various wood combustion applications, as a function of fuel nitrogen content	296
9.2	Fuel-NO_x emission levels as a function of temperature and fuel type and comparison with thermal and prompt NO_x formation	297
9.3	CO emissions as a function of excess air ratio λ	299
9.4	CO and PAH emissions as a function of combustion temperature	300
9.5	TFN/Fuel-N ratio for NH_3, HCN, and a mixture of 50 per cent NH_3 and 50 per cent HCN as a function of primary excess air ratio for two ideal flow reactors	315
9.6	NO_x emission level as a function of primary excess air ratio for the 25kW test reactor	315

9.7	Three principles of combustion	316
9.8	Collection efficiencies for various particle control technologies	319
9.9	Settling chamber	321
9.10	Principle of a cyclone	322
9.11	Principle of a multicyclone	323
9.12	Electrostatic filters	325
9.13	Bag filters	328
9.14	Scrubbers	330
9.15	Panel bed filter	331
9.16	Rotating particle separator	332
9.17	Wood-firing system with NO_x reduction by SCR low dust process	335
9.18	Wood-firing system with NO_x reduction by SCR high dust process	335
9.19	Wood-firing system with NO_x reduction by SNCR	336
9.20	Comparison of NO_x reduction potential for various NO_x reduction measures versus fuel nitrogen	337
9.21	Flow diagram of a typical limestone wet scrubber	338
9.22	Simplified decision tree for emission limits in the Netherlands	343
9.23	Stable and unstable natural cycles of minerals in recycling of ash from biomass combustion	350
9.24	Average water-soluble amounts of elements in mixtures of bottom ash and cyclone fly ash of biomass fuels	359
9.25	Average distribution of heavy metals among the various ash fractions of bark and woodchip combustion plants	360
9.26	Average distribution of nutrients among the ash fractions of bark and woodchip combustion plants	360
9.27	Environmentally compatible cycle economy for ashes from biomass combustion plants with possibilities of ash-conditioning	362
9.28	Separation of condensate and sludge using a combination of sedimentation tank and sedimentation separator	369
9.29	Separation of condensate and sludge using a filter with wood dust	370
9.30	Separation of condensate and sludge using a combination of sedimentation tank and filter press	371
10.1	Expected growth in biomass power generation and installed capacity	380
10.2	Projected electricity production from biomass in the US	382

Tables

1.1	Primary energy consumption by energy source and region in 2006, PJ/year	3
2.1	Overview of the usefulness of various model types for certain purposes	34
2.2	Characteristics of solid biomass fuels and their most important effects	39
2.3	Moisture content, gross calorific value, net calorific value, bulk density and energy density of biomass fuels	40
2.4	C, H and O concentrations as well as amounts of volatile matter in biomass fuels	41
2.5	N, S and Cl concentrations in biomass fuels	41
2.6	Fuel-specific ash content of biomass fuels	43

2.7	Concentration ranges of major ash-forming elements in biomass fuels and ashes	44
2.8	Ash-melting behaviour of biomass fuels	45
2.9	Concentration ranges of minor ash-forming elements in biomass fuels	45
2.10	Guiding values and guiding ranges for elements in biomass fuels and ashes for unproblematic thermal utilization	47
3.1	Fertilizing plan of Salix plantations in Denmark	63
3.2	Characteristics of various types of chippers	65
3.3	Parameters affecting the combustion behaviour of waste wood, their typical ranges and recommended guiding values	70
3.4	Technical data for various baling technologies	72
3.5	Comparison of payload and energy content for truck transportation of forest residues in Sweden	73
3.6	Chemical and physical fuel properties in accordance with ÖNORM M7135	77
3.7	Typical energy consumption and specific processing costs for biomass fuel pre-treatment under Dutch conditions	94
3.8	Suitable fuel-feeding and combustion technologies according to shape and particle size of the biomass fuel	95
3.9	Comparison of transportation costs of different biomass fuel under Austrian framework conditions	97
4.1	Approved EU product standards for residential solid fuel-fired appliances	131
4.2	North American standards for residential solid fuel-fired appliances	132
5.1	Overview of advantages, disadvantages and fields of application of various biomass combustion technologies	154
5.2	Influence of various measures on the thermal efficiency of biomass combustion plants	156
5.3	Comparison of specific investment and fuel costs for biomass and fuel oil-fired combustion systems	169
6.1	Closed processes for power production by biomass combustion	176
6.2	Advantages and disadvantages of steam turbines for use in biomass combustion	183
6.3	Output power of a steam engine when using 10 t/h of dry and saturated steam	186
6.4	Specifications of steam piston engines from Spilling, Germany	186
6.5	Advantages and disadvantages of steam piston engines	186
6.6	Typical efficiencies for heating plants, CHP plants and power plants today and target values for the future	199
7.1	Specifications for chipping of wood for use at Gelderland Power Station	208
7.2	Specifications for powdered wood product for use at Gelderland Power Station	209
7.3	Fuel gas composition for the BIOCOCOMB process	217
8.1	Speciation of inorganic materials in higher plants	251
8.2	Typical ash elemental analysis data (major elements) for a number of important biomass materials	253
8.3	Melting/eutectic temperatures for pure compounds and binary mixtures relevant to chloride-rich, biomass ash deposits and associated corrosion reactions	278
8.4	Summary of the findings on corrosion and deposit morphologies on superheater surfaces in straw-fired boilers	280

9.1	Emissions from a typical 2000MW fossil fuel power station using coal, oil or natural gas	293
9.2	Pollutants from biomass combustion and their impacts on climate, environment and health	302
9.3	Arithmetic average emission levels from small-scale biomass combustion applications	305
9.4	Arithmetic average emission values from wood combustion applications	306
9.5	Emissions mainly influenced by combustion technology and process conditions	306
9.6	Emissions mainly influenced by fuel properties	307
9.7	Emissions from small industrial woodchip combustion applications in the Netherlands	307
9.8	Emissions from industrial wood-fired installations, using particle board, woodchips, MDF and bark	308
9.9	Emissions from straw-fired CHP plants	309
9.10	Effect of optimization on emissions and efficiency	313
9.11	Summary of typical sizes of particles removed by various particle control technologies	320
9.12	Summary of separation efficiency for different particle control technologies based on full-scale measurements	320
9.13	Characteristics of a settling chamber	321
9.14	Characteristics of a cyclone	323
9.15	Characteristics of an electrostatic filter	324
9.16	Temperature resistance of bag filter materials	327
9.17	Overview of LCP and WID	339
9.18	Emission limits for residential heating boilers fired with solid biomass fuels with a nominal heat output of up to 300kW	339
9.19	Emission limits for steam boilers and hot water boilers in industrial and commercial plants	340
9.20	Overview of most relevant emission limits in the Netherlands for solid biomass combustion	344
9.21	Distribution of ash fractions for biomass fuels	351
9.22	Average particle and bulk densities	352
9.23	Average concentrations of plant nutrients in ash fractions from bark, woodchip and sawdust combustion plants	352
9.24	Average concentrations of plant nutrients in ash fractions from straw and cereal combustion plants	353
9.25	Average concentrations of plant nutrients in ash fractions of combustion plants using wood residues and waste wood	353
9.26	Average concentrations of heavy metals in ash fractions of bark, woodchips and sawdust incinerators	354
9.27	Average concentrations of heavy metals in ash fractions of straw and cereal combustion plants	354
9.28	Average concentrations of heavy metals in ash fractions of residual and waste wood incinerators	355
9.29	Organic carbon, chlorine and organic contaminants in biomass ashes	356
9.30	pH value and electrical conductivity of biomass ashes	357

9.31	Si, Al, Fe, Mn and carbonate concentrations in biomass ashes by fuel type	358
9.32	Possibilities for influencing the composition and characteristics of ashes	363
9.33	Conditions for a controlled and ecologically friendly ash utilization	363
9.34	Logistics of a smooth-running closed-cycle economy for biomass ashes	364
9.35	Limiting values for concentrations of heavy metals in biomass ashes used on agricultural land and in forests and guiding values for soils according to existing Austrian regulations	367
9.36	Contents of Cd, Pb and Zn in the condensate after separating the sludge and condensate	368
10.1	Policies in selected EC countries for promoting bio-electricity production	388

Box

10.1	The dead koala RECs problem	389

Preface

Combustion technologies are commercially available throughout the world. They play a major role in energy production from biomass. For further implementation of biomass combustion, however, combustion technology needs to be optimized to meet demands for lower costs, increased fuel flexibility, lower emissions and increased efficiency.

IEA Bioenergy Task 32, Biomass Combustion and Cofiring, aims to accelerate the market introduction of improved combustion systems in its member countries by exchanging technical and non-technical information. This relates both to stand-alone small and medium-scale combustion plants as well as co-firing of biomass in existing coal power plants.

This second edition of the handbook again is the result of a collective effort of the members of Task 32. The first edition was produced in response to the many questions asked of us with respect to developments in biomass combustion technologies for both domestic and industrial use as well as for co-firing in coal power plants. It was spread around the world and translated into different languages.

Because of the speed of technology development in this field we believed it was necessary to update the handbook. Thanks to the joint efforts of all of our task members, this second edition is now available.

We feel that it represents a comprehensive overview of important issues and topics concerning biomass combustion and co-firing. It was carefully compiled on the basis of available literature sources, national information and experiences as well as suggestions and comments from equipment suppliers. The compilers especially benefited from the large number of international experts in this field that participated as authors. Although the figures and performance data will soon be outdated because of the speed of technological developments in this field, the general principles and combustion concepts explained will no doubt remain useful. Further, the reader is invited to regularly visit the internet site of our Task (www.ieabcc.nl) where updates and additional background information are provided.

We herewith express our deepest thanks to all who have contributed to this handbook. We trust this handbook will be a tangible contribution to the dissemination of valuable knowledge.

Sjaak van Loo/Jaap Koppejan
Task Leader/Co-Task Leader
IEA Bioenergy Task 32: Biomass Combustion and Cofiring
www.ieabcc.nl

List of Contributors

Contributions to this handbook were provided by:

- Göran Berndes, Chalmers University, Sweden
- Larry Baxter, Brigham Young University, USA
- Peter Coombes, Delta Electricity, Australia
- Jerome Delcarte, Département de Génie Rural, Belgium
- Anders Evald, Force Technology, Denmark
- Hans Hartmann, Technologie- und Forderzentrum, Germany
- Maarten Jansen, TNO, the Netherlands
- Jaap Koppejan, Procede Biomass BV, Netherlands (Task Secretary)
- William Livingston, Doosan Babcock Energy Limited, United Kingdom
- Sjaak van Loo, Procede Group BV, Netherlands (Task Leader)
- Sebnem Madrali, Department of Natural Resources, Canada
- Behdad Moghtaderi, University of Newcastle, Australia
- Erich Nägele, European Commission
- Thomas Nussbaumer, Verenum, Switzerland
- Ingwald Obernberger, TU Graz, Austria
- Heikki Oravainen, VTT, Finland
- Fernando Preto, Department of Natural Resources, Canada
- Øyvind Skreiberg, Norwegian University of Science and Technology, Norway
- Claes Tullin, SP Swedish National Testing and Research, Sweden
- Gerold Thek, BIOS BIOENERGIESYSTEME GmbH, Austria

Contact details can be found in Annex 4.

The compilation of individual chapters was co-ordinated by:

Anders Evald, Jaap Koppejan, William Livingston, Thomas Nussbaumer, Ingwald Obernberger and Øyvind Skreiberg.

1
Introduction

In a broad sense, *energy conversion* is the capacity to promote changes and/or actions (heating, motion, etc.), and *biomass* includes all kinds of materials that were directly or indirectly derived not too long ago from contemporary photosynthesis reactions, such as vegetal matter and its derivatives: wood fuel, wood-derived fuels, fuel crops, agricultural and agro-industrial by-products, and animal by-products. *Bioenergy* is the word used for energy associated to biomass, and *biofuel* is the bioenergy carrier, transporting solar energy stored as chemical energy. Biofuels can be considered a renewable source of energy as long as they are based on sustainable biomass production [1].

Worldwide, there is a growing interest in the use of solid, liquid and gaseous biofuels for energy purposes. There are various reasons for this, such as:

- political benefits (for instance, the reduction of the dependency on imported oil);
- employment creation – biomass fuels create up to 20 times more employment than coal and oil; and
- environmental benefits such as mitigation of greenhouse gas emissions, reduction of acid rain and soil improvements.

Figure 1.1 *Many countries have abundant resources of unused biomass readily available*

Figure 1.2 *Wood-stove commonly used in Cambodia*

Large amounts of wood and other solid biomass residues remain unused so far and could potentially be made available for use as a source of energy. In addition to this, wood and other biomass energy crops could be grown. There is, for instance, a policy debate on whether trees should be used to sequester carbon or to replace fossil fuels. Trees and other forms of biomass can act as carbon sinks but at the end of their growing life there must be plans for using the biomass as a source of fuel to offset fossil energies or as very long-lived timber products.* Otherwise, the many years of paying to sequester and protect the carbon in trees will simply be lost as they decay and/or burn uncontrollably.

Solid biofuels could provide a significant part of the energy demand if appropriate technologies were introduced. For this reason, many countries around the world have become involved in modern applications of wood and biomass to energy technologies. These are not only research or pilot projects; there are actual investment projects that exploit wood and other biomass fuels to generate heat and/or electricity for use by industries, utilities, communities and single households through more efficient, convenient and modern technologies. These projects prove that biomass energy can be a technically efficient, economically viable, and environmentally sustainable fuel option in the environment in which it operates.

---

* This does not apply to primary forests where preserving biodiversity is of major importance.

1.1 Current status of bioenergy

Table 1.1 illustrates that the current share of bioenergy in various regions in the world is still very limited. The contribution of biomass in industrialized countries is estimated at only 4 per cent.

In developing countries, around 22 per cent of the energy used originates from biomass, but the majority of it is used non-commercially in traditional applications (such as cooking stoves). These traditional cooking stoves are often characterized by low efficiencies and high release of toxic organic compounds. With 1.3 million deaths globally each year due to pneumonia, chronic respiratory disease and lung cancer, indoor smoke in high-mortality developing countries is responsible for an estimated 3.7 per cent of the overall disease burden, making it the most lethal killer after malnutrition, unsafe sex and lack of safe water and sanitation [2]. In a country like Nepal, traditional biomass fuels cover over 90 per cent of the primary energy input.

Table 1.1 *Primary energy consumption by energy source and region in 2006, PJ/year*

	Modern biomass	Traditional biomass	Other renewables	Conventional energy	Total primary energy	Modern biomass as % of primary energy
World	16,611	33,432	13,776	409,479	473,319	3.5%
OECD	8442	42	6783	222,369	237,636	3.6%
OECD North America	4158	–	3276	112,959	120,393	3.5%
US and Canada	3801	–	2898	106,281	112,980	3.4%
Mexico	357	–	399	6678	7392	4.8%
OECD Pacific	882	42	798	36,561	38,283	2.4%
OECD Asia	504	42	525	31,374	32,445	1.6%
OECD Oceania	378	–	252	5208	5838	6.5%
OECD Europe	3402	–	2688	72,828	78,939	4.3%
OECD Europe – EU	3129	–	1785	69,384	74,298	4.2%
Transition economies	693	–	1176	44,688	46,536	1.5%
Russia	273	–	672	26,901	27,867	1.0%
Developing countries	7434	33,432	5817	136,269	182,994	4.1%
China	315	8988	1323	49,602	60,144	0.5%
East Asia	1092	3633	1197	20,202	26,145	4.2%
Indonesia	126	1680	357	5418	7560	1.7%
South Asia	1302	9828	504	18,627	30,261	4.3%
India	1092	8043	357	15,582	25,074	4.4%
Latin America	2394	1239	2373	15,834	21,840	11.0%
Brazil	1680	357	1176	5502	8736	19.2%
Middle East	21	63	105	19,341	19,551	0.1%
Africa	2310	9702	315	12,726	25,074	9.2%

Note: OECD = Organisation for Economic Co-operation and Development.
Source: Interpolated from data in [3], other renewables includes solar, wind, hydro, geothermal, wave and ocean energy; conventional energy includes coal, oil, gas and nuclear energy.

Many countries around the globe have developed a growing interest in the use of biomass as an energy source, and therefore various technological developments in this field are ongoing. Although major technological developments have already been achieved, most bioenergy technologies are not yet commercially feasible without political support. In order to achieve wider application of modern bioenergy technologies, individual countries have set varying targets and implemented promotional policies. As a result of increased support for bioenergy technologies, major progress has been made. Chapter 10 provides an overview of approaches and progress made in selected countries.

1.2 Combustion as main bioenergy technology

Biomass combustion is the main technology route for bioenergy, responsible for over 90 per cent of the global contribution to bioenergy. The selection and design of any biomass combustion system is mainly determined by the characteristics of the fuel to be used, local environmental legislation, the costs and performance of the equipment necessary or available as well as the energy and capacity needed (heat, electricity). Furthermore, the fuel characteristics can be influenced in order to fulfil the technological and ecological requirements of a given combustion technology. The most suitable technology package therefore can vary from case to case but generally, due to *economy of scale* effects concerning the complexity of the fuel-feeding system, the combustion technology and the flue gas cleaning system, large-scale systems use low-quality fuels (with inhomogeneous fuel characteristics concerning, e.g., moisture content, particle size, and ash-melting behaviour), and high-quality fuels are necessary for small-scale systems.

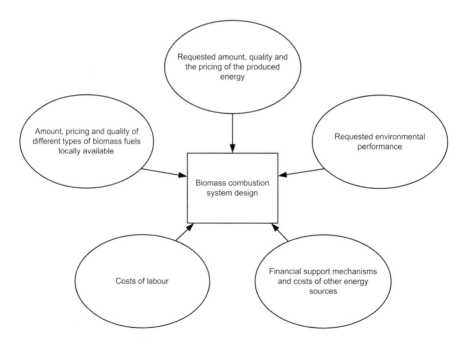

Figure 1.3 *Influencing parameters for the optimal design of biomass combustion systems*

Biomass combustion technologies show, especially for large-scale applications, similarities to waste combustion systems, but especially when chemically untreated (natural) biomass fuels are utilized, the necessary flue gas cleaning technologies are less complex and therefore cheaper. Furthermore, old combustion technologies have proven unable to handle inhomogeneous biomass fuels, and problems concerning emissions and fail-safety have occurred. New fuel preparation, combustion and flue gas cleaning technologies have been developed and introduced that are more efficient, cleaner and more cost-effective than previous systems and can be utilized for multifuel feed. This opens up new opportunities for biomass combustion applications under conditions that were previously too expensive or inadequate, increases the competitiveness of these systems, and raises plant availability. In this respect, knowledge exchange through the IEA, the EU and other international organizations as well as the creation of conducive market mechanisms and legislation are essential for a more widespread introduction of biomass energy systems.

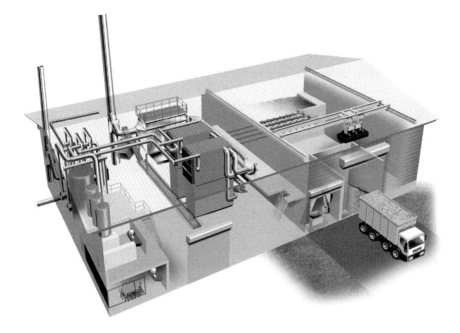

Figure 1.4 *Wood-fired heating plant used for district heating in Wilderswil, Switzerland*

Note: Thermal capacity 6.4MW$_{th}$ on woodchips + 3MW$_{th}$ back-up on fuel oil.
Source: Courtesy of Schmid AG, Switzerland

1.3 This handbook

This handbook describes the current state of biomass combustion technologies for both domestic and industrial use. It is a thorough update of the first edition of the handbook with the same title. The book was carefully compiled through the collaborative work of members of the IEA Bioenergy Agreement, Task 32 'Biomass Combustion and Cofiring', using available literature sources, national information and experiences as well as suggestions and comments from equipment suppliers. As technological developments in the field of biomass combustion occur very rapidly and are often difficult to keep track of, this handbook is not to be regarded as complete. Nevertheless, it represents a comprehensive overview of important issues and topics concerning biomass combustion, and the reader may especially benefit from the large number of international experts in this field who participated as authors.

In Chapter 2, the basic principles of combustion are explained and the various biomass fuels are characterized regarding their physical and chemical parameters and their influence on the combustion process.

Chapter 3 provides information on possible biomass fuel pre-treatment options and fuel-feeding technologies.

Chapter 4 describes currently available biomass combustion technologies for domestic space heating.

Chapter 5 describes the biomass combustion technologies currently applied or under development for industrial utilization of biomass fuels. Moreover, technological possibilities to increase the efficiency of biomass combustion plants are discussed, and technological and economic standards regarding the proper dimensioning of biomass combustion systems are given.

Chapter 6 covers the various technologies for power production based on biomass combustion.

In Chapter 7, various concepts for biomass co-firing technologies and applications are explained and discussed. Typical technical problems are explained and guidelines for co-firing presented.

Inorganic components in biomass have direct influence on the eventual formation of slag deposits, corrosion of boiler components, aerosol formation and utilization options for the ashes formed. Chapter 8 is dedicated to ash characteristics and the behaviour of ash in biomass combustion systems.

Chapter 9 is devoted to environmental aspects of biomass combustion. It provides overviews of average gaseous and solid emissions from combustion installations and describes possible primary and secondary measures for emission reduction. Furthermore, biomass ashes are characterized and possible treatments and utilizations are pointed out. Finally, this chapter offers an overview of the various environmental regulations in 19 European countries regarding emissions limits for biomass combustion facilities.

Chapter 10 provides an overview of trends and policies with regard to the implementation of biomass combustion systems.

Finally, Chapter 11 supplies an overview of needs and ongoing activities concerning research and development in the field of biomass combustion.

1.4 References

1   UN FOOD AND AGRICULTURE ORGANIZATION (FAO) (1997) Bioenergy Terminology and Bioenergy Database, Wood Energy Programme, Wood and Non-Wood Products Utilization Branch (FOPW)/FAO (ed.), Rome, Italy
2   WORLD HEALTH ORGANIZATION, see www.who.int
3   IEA (2004) *World Energy Outlook*, IEA

2
Biomass Fuel Properties and Basic Principles of Biomass Combustion

2.1 Introduction

Biomass can be converted into useful energy (heat or electricity) or energy carriers (charcoal, oil or gas) by both thermochemical and biochemical conversion technologies. Biochemical conversion technologies include fermentation for alcohol production and anaerobic digestion for production of methane-enriched gas. However, in this chapter we will concentrate on thermochemical conversion technologies. These technologies are in varying stages of development, where combustion is most developed and most frequently applied. Gasification and pyrolysis are becoming more and more important. An overview of thermochemical conversion technologies, products and potential end uses is shown in Figure 2.1.

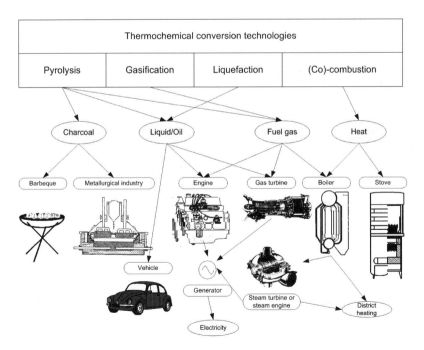

Figure 2.1 *Thermochemical conversion technologies, products and potential end uses*

Source: Adapted from [1].

Four thermochemical biomass conversion technologies for energy purposes exist: pyrolysis, gasification, direct combustion, and liquefaction. The primary products from these conversion technologies may be in the form of energy carriers such as charcoal, oil or gas, or as heat. Various techniques exist for the utilization of these products. Secondary products may be derived through additional processing. In principle, most of the petroleum-derived chemicals currently being used can be produced from biomass, but some will require rather circuitous synthesis routes [2].

2.2 The process of biomass combustion

The process of biomass combustion involves a number of physical/chemical aspects of high complexity. The nature of the combustion process depends both on the fuel properties and the combustion application. The combustion process can be divided into several general processes: drying, pyrolysis, gasification and combustion. The overall combustion process can either be a continuous combustion process or a batch combustion process, and air addition can be carried out either by forced or natural draught. Batch combustion is used in some small-scale combustion units, some of which also use natural draught. This is typical for, e.g., traditional wood-stoves. Medium- to large-scale combustion units are always continuous combustion applications, with forced draught.

Drying and pyrolysis/gasification will always be the first steps in a solid fuel combustion process. The relative importance of these steps will vary, depending on the combustion technology implemented, the fuel properties and the combustion process conditions. A separation of drying/pyrolysis/gasification and gas and char combustion, as in staged-air combustion, may be utilized. In large-scale biomass combustion applications with continuous fuel feeding, such as moving grates, these processes will occur in various sections of the grate. However, in batch combustion applications there will be a distinct separation between a volatile and a char combustion phase, in both position and time. Figure 2.2 shows qualitatively the combustion process for a small biomass particle. For larger particles, there will be a certain degree of overlap between the phases, while in batch combustion processes, as in wood log combustion in wood-stoves and fireplaces, there will be a large degree of overlap between the phases.

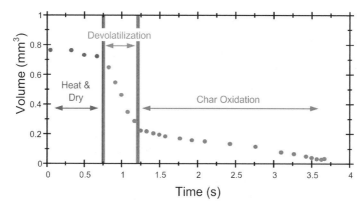

Figure 2.2 *The combustion of a small biomass particle proceeds in distinct stages*
Source: [3].

2.2.1 Drying, pyrolysis, gasification and combustion

Drying: Moisture will evaporate at low temperatures (<100°C). Since vaporization uses energy released from the combustion process, it lowers the temperature in the combustion chamber, which slows down the combustion process. In wood-fired boilers, for example, it has been found that the combustion process cannot be maintained if the wood moisture content exceeds 60 per cent on a wet basis (w.b.). The wet wood requires so much energy to evaporate contained moisture, and subsequently to heat the water vapour, that temperatures are reduced below the minimum temperature required to sustain combustion. Consequently, moisture content is a very important fuel variable.

Pyrolysis can be defined as thermal degradation (devolatilization) in the absence of an externally supplied oxidizing agent. The pyrolysis products are mainly tar and carbonaceous charcoal, and low molecular weight gases. In addition, CO and CO_2 can be formed in considerable quantities, especially from oxygen-rich fuels, such as biomass. Fuel type, temperature, pressure, heating rate and reaction time are all variables that affect the amounts and properties of the products formed.

In Figure 2.3, TGA (thermogravimetric analysis) experiments with four wood species are shown. The four wood species show similar trends for weight and derived weight as a function of temperature. However, the curves differ in certain details. When the temperature is raised, drying of the sample occurs. At 473K, devolatilization starts and the devolatilization rate increases as the temperature is raised. There are two areas of weight loss producing a single peak with a plateau or shoulder located at the lower temperature region. The lower temperature shoulder represents the decomposition of hemicellulose and the higher temperature peak represents the decomposition of cellulose.* Hardwoods (birch, beech white and acacia) contain more hemicellulose than softwoods (spruce), causing the 'hemicellulose shoulder' to be more visible.

At 673K, most of the volatiles are gone and the devolatilization rate decreases rapidly. However, a low devolatilization rate can be observed in the temperature range of 673–773K. This is caused by lignin† decomposition, which occurs throughout the whole temperature range, but the main area of weight loss occurs at higher temperatures. This means that the lignin is mainly responsible for the flat tailing section that can be observed for all the wood species at higher temperatures. In addition, birch, having the highest hemicellulose and lowest lignin content, yields the lowest char residue.

The pyrolysis products can be used in a variety of ways. The char can be upgraded to activated carbon, used in the metallurgical industry, as domestic cooking fuel or for barbecuing. Pyrolysis gas can be used for heat production or power generation, or synthesized to produce methanol or ammonia. The tarry liquid, pyrolysis oil or bio-oil can be upgraded to high-grade hydrocarbon liquid fuel for combustion engines (e.g. transportation), or used directly for electricity production or heating purposes.

Gasification can be defined as thermal degradation (devolatilization) in the presence of an externally supplied oxidizing agent. However, the term gasification is also used for char

---

* Hemicellulose consists of various sugars other than glucose that encase the cellulose fibers and represent 20–35% of the dry weight of wood. Cellulose ($C_6H_{10}O_5$) is a condensed polymer of glucose ($C_6H_{10}O_6$). The fibre walls consist mainly of cellulose and represent 40–45% of the dry weight of wood.

† Lignin (e.g. $C_{40}H_{44}O_{14}$) is a high molecular mass complex non-sugar polymer that gives strength to the wood fibre, accounting for 15–30% of the dry weight.

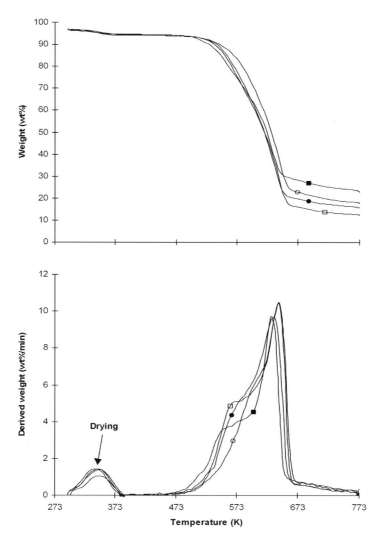

Figure 2.3 *Thermographic analysis of four wood samples*

Note: ○ = spruce, □ = birch, ● = beech white, ■ = acacia. 5mg samples heated at a rate of 10°C/min.
Source: [4].

oxidation reactions with, for example, CO_2 or H_2O. While pyrolysis is usually optimized with respect to a maximum char or tar yield, gasification is optimized with respect to a maximum gas yield. Temperatures of 1073–1373K are used. The gas contains mainly CO, CO_2, H_2O, H_2, CH_4 and other hydrocarbons. Gasification can be carried out with air, oxygen, steam or CO_2 as oxidizing agents.

Air gasification produces a low calorific value gas (gross calorific value (GCV) of 4–7 megajoules per normal cubic metre (MJ/Nm³ dry), while oxygen gasification produces a medium calorific value gas (GCV of 10–18MJ/Nm³ dry). Through synthesis, the fuel gas can be upgraded to methanol, burned externally in a boiler for producing hot water or steam, in a gas turbine for electricity production or in internal combustion engines, such as diesel and spark ignition engines. Before the fuel gas can be used in gas turbines or internal combustion engines, the contaminants (tar, char particles, ash and alkali compounds) have to be removed. The hot gas from the gas turbine can be used to raise steam to be utilized in a steam turbine (IGCC – integrated gasification combined cycle).

Combustion can ideally be defined as a complete oxidation of the fuel. The hot gases from the combustion may be used for direct heating purposes in small combustion units, for water heating in small central heating boilers, to heat water in a boiler for electricity generation in larger units, as a source of process heat, or for water heating in larger central heating systems. Drying and pyrolysis/gasification will always be the first steps in a solid-fuel combustion process.

Liquefaction can be defined as thermochemical conversion in the liquid phase at low temperatures (523–623K) and high pressures (100–200 bar), usually with a high hydrogen partial pressure and a catalyst to enhance the rate of reaction and/or to improve the selectivity of the process. As compared to pyrolysis, liquefaction has a higher liquid yield, and results in a liquid with a higher calorific value and lower oxygen content.

2.2.2 Operational and design variables affecting the combustion process

2.2.2.1 Moisture content

The moisture content of biomass fuels varies considerably, depending on the type of biomass and biomass storage. Typical moisture contents of various biomass fuels are given in Section 2.3. In some cases it will be necessary to dry the biomass fuel prior to combustion to be able to sustain combustion. Increasing moisture content will reduce the maximum possible combustion temperature (the adiabatic combustion temperature) and increase the necessary residence time in the combustion chamber, thereby giving less room for preventing emissions as a result of incomplete combustion.

Figure 2.4 shows the adiabatic combustion temperature as a function of moisture content and excess air ratio for a fuel composition of 50wt% C, 6wt% H and 44wt% O (dry basis (d.b.)).

2.2.2.2 Calorific value

The gross calorific value (GCV) is defined as the heat released during combustion per mass unit fuel under the constraints that the water formed during combustion is in liquid phase and that the water and the flue gas have the same temperatures as the temperature of the fuel prior to combustion (ÖNORM C 1138).

The net calorific value (NCV) is defined as the heat released during combustion per mass unit of fuel under the constraints that the water formed during combustion is in a gaseous phase and that the water and the flue gas have the same temperature as the fuel prior to combustion (ÖNORM C 1138).

The GCV of biomass fuels usually varies between 18 and 22MJ/kg (d.b.) and can be calculated reasonably well by using the following empirical formula [5]:

$$GCV = 0.3491 \cdot X_C + 1.1783 \cdot X_H + 0.1005 \cdot X_S - 0.0151 \cdot X_N - 0.1034 \cdot X_O - 0.0211 \cdot X_{ash} \quad [\text{MJ/kg, d.b.}] \quad (2.1)$$

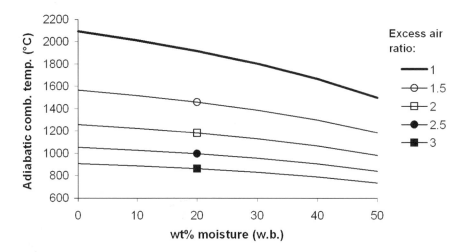

Figure 2.4 *Adiabatic combustion temperature as a function of moisture content (w.b.) and excess air ratio for a continuous combustion process*

Note: Fuel composition: 50wt% C, 6wt% H and 44wt% O, d.b.

where X_i is the content of carbon (C), hydrogen (H), sulphur (S), nitrogen (N), oxygen (O) and ash in wt% (d.b.). For exact determination of GCV, a bomb calorimeter is used according to standardized procedures. As can be seen from the formula, the content of C, H and S contributes positively to the GCV, while the content of N, O and ash contributes negatively to the GCV.

The NCV can be calculated from the GCV taking into account the moisture and hydrogen content of the fuel by applying the equation:

$$NCV = GCV\left(1 - \frac{w}{100}\right) - 2.444 \cdot \frac{w}{100} - 2.444 \cdot \frac{h}{100} \cdot 8.936\left(1 - \frac{w}{100}\right) \quad [MJ/kg, w.b.] \quad (2.2)$$

where:

2.444 = *enthalpy difference between gaseous and liquid water at 25°C*

8.936 = M_{H_2O}/M_{H_2}; *i.e. the molecular mass ratio between H_2O and H_2*

Abbreviations:

- NCV   net calorific value in MJ/kg fuel (w.b.)
- GCV   gross calorific value in MJ/kg fuel (d.b.)
- w   moisture content of the fuel in wt% (w.b.)
- h   concentration of hydrogen in wt% (d.b.)
- guiding values for woody biomass fuels: 6.0wt% (d.b.)
- and for herbaceous biomass fuels: 5.5wt% (d.b.)

In Figure 2.5, GCV and NCV (assuming zero wt% moisture) are shown as a function of wt% C and H (d.b.); O is calculated by difference, neglecting the influence of N, S and ash in the fuel. Charcoal may consist of close to 100 per cent carbon, with a GCV of above 30MJ/kg (d.b.). This is of special importance in batch combustion applications where the fuel composition will change continuously as a function of burnout, from virgin biomass to charcoal. Figure 2.6 shows the NCV as a function of wt% moisture (w.b.) for a fuel composition of 50wt% C, 6wt% H and 44wt% O (d.b.).

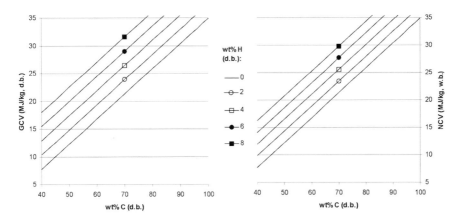

Figure 2.5 *GCV and NCV (assuming zero wt% moisture) as a function of wt% C and H*

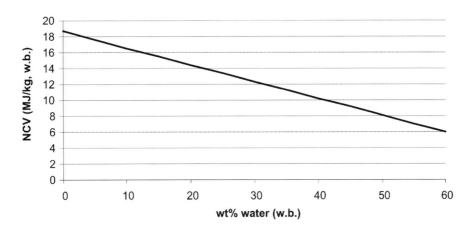

Figure 2.6 *NCV as a function of wt% moisture (w.b.) for a fuel composition of 50wt% C, 6wt% H, and 44wt% O (d.b.)*

The ultimate chemical composition of biomass as compared to other fuels is illustrated in Figure 2.7. Compared to other fuels, biomass contains relatively high levels of hydrogen and oxygen.

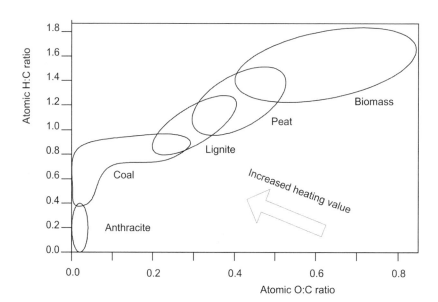

Figure 2.7 *Chemical composition of various solid fuels*

2.2.2.3 Temperature, residence time, stoichiometry and mixing

Emissions caused by incomplete combustion are mainly a result of either:

- inadequate mixing of combustion air and fuel in the combustion chamber, giving room for local fuel-rich combustion zones;
- an overall lack of available oxygen;
- combustion temperatures that are too low;
- residence times that are too short; or
- radical concentrations that are too low, in special cases, for example in the final stage of the combustion process (the char combustion phase) in a batch combustion process.

These variables are all linked together through the reaction rate expressions for the elementary combustion reactions. However, in cases where oxygen is available in sufficient quantities, temperature is the most important variable due to its exponential influence on the reaction rates. An optimization of these variables will in general contribute to reduced emission levels of all emissions from incomplete combustion.

A combustion process can – in a simplified manner – be described by an overall combustion reaction:

$$a \cdot (Y_C Y_H Y_O Y_N Y_S) + n \cdot \left(\frac{1 - Y_{O_2, Air}}{Y_{O_2, Air}} \cdot N_2 + O_2 \right) \quad (2.3)$$

$$\Rightarrow b \cdot CO_2 + c \cdot H_2O + d \cdot O_2 + e \cdot N_2 + f \cdot CO + g \cdot NO$$
$$+ h \cdot NO_2 + i \cdot C_k H_l + j \cdot SO_2 + m \cdot N_2O$$

where Y_i is the volume fraction of carbon (C), hydrogen (H), oxygen (O), nitrogen (N) and sulphur (S) in the fuel. $Y_{O_2,Air}$ is the volume fraction of oxygen in air and the coefficients a, n, b, c, d, e, f, g, h, i, j and m can be found partly by applying element balances (C, H, O, N, S) and partly by making assumptions regarding product composition (e.g. chemical equilibrium). By assuming complete combustion, neglecting dissociation effects at high temperatures and formation of NO, NO_2 and N_2O, the product composition can be found directly from the elemental balances.

However, in reality several hundreds of reversible elementary reactions are needed to describe in detail the combustion process of even light hydrocarbons, such as methane, which is an important intermediate in biomass combustion. An integral reaction flow analysis in a premixed stoichiometrical methane flame is shown in Figure 2.8, illustrating the complexity of chemical kinetics in combustion processes. The reaction rate of the elementary reactions depends on mixing through the availability of reactants in the flame zone, and temperature and residence time through the reaction rate constant, k. This is shown in Eq. 2.4:

$$a \cdot \text{Fuel} + b \cdot \text{Oxidant} \rightarrow c \cdot \text{Product} \tag{2.4}$$

$$\Downarrow$$

$$\frac{d[\text{Product}]}{dt} = k \cdot [\text{Fuel}]^a \cdot [\text{Oxidant}]^b \quad \left[\frac{mole}{m^3 s}\right]$$

$$\text{where} \quad k = A \cdot T^\beta \cdot \exp\left(-\frac{E}{R_u \cdot T}\right)$$

where A, b and E are the pre-exponential factor, the temperature exponent, and the activation energy, respectively. These are constants for a given elementary reaction and are found experimentally. R_u is the universal gas constant and T is the temperature. This handbook is not the place to go into further detail on the elementary combustion chemistry. For further reading the following textbooks are recommended: [6–10]. The International Flame Research Foundation (IFRF) has an extensive list of recommended literature on the internet: www. combustion-centre.ifrf.net/bookstore/.

Radicals, such as OH, H and O, are important intermediates in all combustion processes and are used as oxidants in the elementary reactions, as can be seen in Figure 2.8. The most important radical for CO oxidation is OH. In most cases, radicals are available in sufficient amounts. However, if there is a very dry fuel and low fuel hydrogen content, as in charcoal combustion, the radical concentrations (OH and H) available may be too low to achieve complete gas phase oxidation of CO, formed from heterogeneous charcoal combustion, to CO_2.

Figure 2.9 shows results from chemical kinetics modelling of methane combustion in two ideal flow reactors (PFR = plug flow reactor, PSR = perfectly stirred reactor), utilizing a detailed elementary reaction scheme. In the ideal flow reactors, oxygen is available in sufficient amounts if the excess air ratio (L) is above one. However, the oxygen concentration will still influence the combustion process, with an increasing reaction rate for those elementary reactions involving oxygen with increasing oxygen concentration. As can be seen, the conversion of CH_4, and hence the reaction rates for CH_4 oxidation, increases with increasing temperature. An increasing residence time will increase the time available for oxidation of CH_4, but does not affect the reaction rates directly.

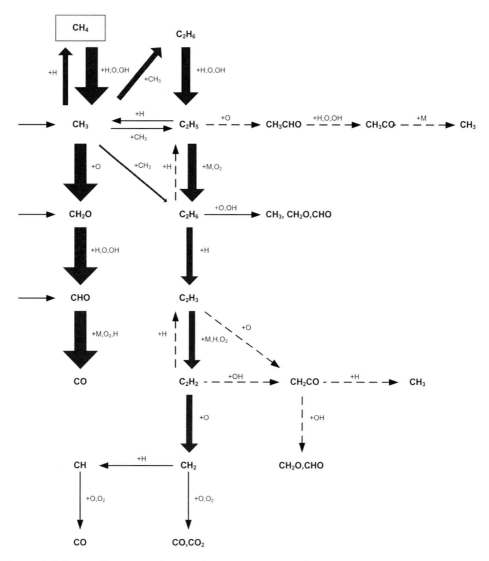

Figure 2.8 *Integral reaction flow analysis in a premixed stoichiometrical methane flame*

Source: [6]. (© Springer-Verlag Berlin Heidelberg, 1999)

Hence, by optimizing any combustion process by adjusting the mixing of fuel and oxidant, temperature and residence time, emissions from incomplete combustion can be minimized. However, in some cases, as for thermal NO_x formation (see Section 9.2.1), an increasing residence time will increase the NO_x emission level.

A detailed understanding of combustion processes requires a detailed knowledge of chemical kinetics, fluid mechanics, turbulence phenomena, etc. This is also the case for biomass combustion. An additional complexity introduced in biomass combustion

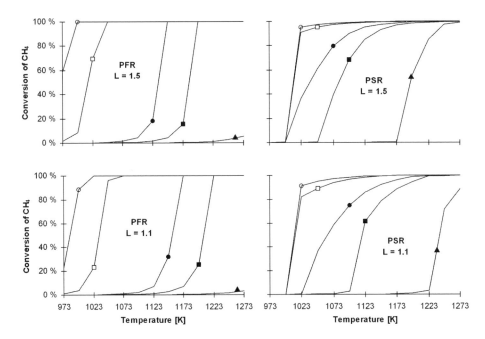

Figure 2.9 *Influence of residence time and temperature on CH_4 conversion at an excess air ratio (λ) of 1.1 and 1.5 for both the PFR and the PSR*

Note: ○ = 1000ms, □ = 500ms, ● = 100ms, ■ = 50ms, ▲ = 10ms)

Source: [4]

compared to gas combustion, is solid fuel combustion. This means that heterogeneous combustion, such as charcoal combustion, is also included. Heterogeneous reactions involving carbon are usually slower than gas phase reactions, as for charcoal combustion. However, in some cases, the heterogeneous reactions may be significantly faster than the equivalent gas phase reactions. This is the case for catalytic converters (see Section 9.2.4.9).

2.2.2.4 Case study – Small-scale combustion applications

To illustrate the complexity of combustion processes involving biomass, the main variables influencing emission levels and energy efficiency in wood-stoves and fireplaces are presented in Figure 2.10. Many of these variables will also be applicable for large-scale biomass combustion applications.

As can be seen, many variables directly or indirectly influence emission levels and energy efficiency. They can be briefly described as follows:

Heat transfer mechanisms: Heat can be transferred by conduction, convection or radiation. To achieve low levels of emissions from incomplete combustion, it is necessary to minimize heat losses from the combustion chamber, which is done by optimizing those variables that directly affect the heat transfer mechanisms. However, to achieve a high thermal efficiency, efficient heat exchange is necessary between the combustion chamber and the chimney inlet. The heat can be transferred directly to the surroundings or can be used to heat water for central heating, as in wood log boilers.

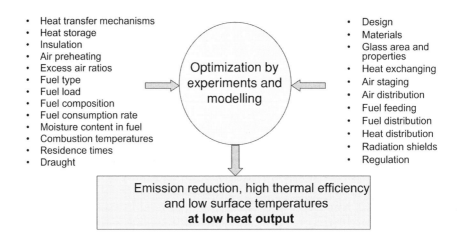

Figure 2.10 *Main variables influencing emission levels and energy efficiency in wood-stoves and fireplaces*

Heat storage: A significant amount of heat will accumulate in the walls of the combustion chamber, stealing heat from the combustion chamber in the start-up phase. This is of special importance in small-scale biomass combustion applications. The stored heat will be transferred to the surroundings with a significant time delay, which is positively utilized in heat-storing stoves (heavy stoves). However, high emission levels of emissions from incomplete combustion are usually found in the start-up phase.

Insulation: Heat is transferred by conduction through the walls of the combustion chamber. Consequently, by improving the insulation of the combustion chamber, a higher combustion chamber temperature can be achieved. The insulation can be improved by either increasing the thickness of the insulation or using a material that insulates better. However, insulation occupies space and is an additional expense, and should be utilized with care. In small-scale biomass combustion applications, some form of insulation is usually necessary to maintain a sufficient combustion chamber temperature.

Air preheating: The combustion chamber temperature can be significantly increased by air preheating. The inlet air is normally preheated through heat exchange with the flue gas, after the flue gas has left the combustion chamber. Stealing heat directly from the combustion chamber for air preheating will have no effect, unless the goal is to reduce the temperature in one part of the combustion chamber, by moving heat to another part. One example is preheating of secondary air at the expense of the fuel bed temperature.

Excess air ratio: In small-scale biomass combustion applications, it is necessary to have an excess air ratio well above one, usually above 1.5, to ensure a sufficient mixing of inlet air and fuel gas. This means that there will be an overall excess of oxygen. The combustion temperature will be significantly reduced, compared to the stoichiometrical combustion temperature, as can be seen in Figure 2.11, mainly due to heating of inert nitrogen in the air.

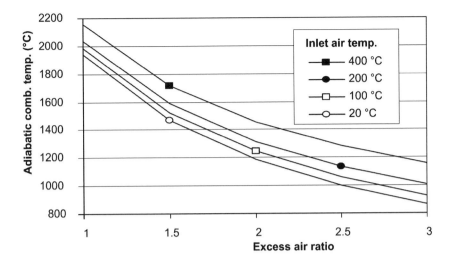

Figure 2.11 *Adiabatic combustion temperature as a function of excess air ratio and inlet air temperature for a continuous combustion process*

Note: Fuel composition: 50wt% C, 6wt% H and 44wt% O, d.b.; Moisture content: 18wt%, w.b.

Hence, an optimal mixing of air and fuel is of the utmost importance, enabling operation at lower overall excess air ratios, with increased combustion temperatures. Optimal design of the air inlets and advanced process control optimization are necessary to ensure sufficient mixing at very low excess air ratios in biomass combustion applications. This can be done cost-effectively in large-scale biomass combustion applications. In small-scale biomass combustion applications, however, advanced process control systems are usually not cost-effective. An additional complication in wood-stoves and fireplaces is their batch-wise operation, and in many cases only natural draught is applied.

Fuel type: The fuel type influences the combustion process through various characteristics of different fuel types, mainly with respect to fuel composition, volatile/char content, thermal behaviour, density, porosity, size and active surface area. The **fuel composition** is important with respect to GCV and emissions, although mainly emissions from complete combustion (see Section 2.3 and Section 9.2.1.1), and ash-related problems. In batch combustion applications, the fuel composition will vary continuously as a function of degree of burnout. Biomass generally contains a high **volatile content** and low **char content** compared to coal, which makes biomass a highly reactive fuel. However, the volatile content varies for different biomass fuels, and influences the **thermal behaviour** of the fuel. The thermal behaviour of a fuel is also influenced by the different chemical structures and bonds present in different biomass fuels. This results in significantly different devolatilization behaviour as a function of temperature. However, for different wood species, as shown in Figure 2.3, similar thermal behaviour can be observed. The **density** of different biomass fuels is highly variable, and a significant difference can also be found between hardwoods and softwoods. Hardwoods, such as birch, have a higher density, which influences the combustion chamber volume to energy input ratio, and also the combustion characteristics of the fuel. The **porosity** of the fuel influences the reactivity (mass loss per time unit) of the fuel, and thereby its devolatilization behaviour. **Fuel size** is an important variable in large-scale biomass combustion applications, especially where entrainment of fuel particles in the flue gas occurs, as in pulverized fuel combustion. Smaller fuel particles will need a shorter residence time in the combustion chamber. The homogeneity of the fuel is also of importance: increasing homogeneity, which improves with decreasing fuel size, enables better process control. Finally, the **active surface area** of the fuel influences the reactivity of the fuel.

Fuel load: The fuel load influences the combustion chamber volume to energy input ratio in batch combustion applications. A decrease of the **fuel consumption rate** below the nominal load the combustion application has been designed for may cause serious problems with respect to emissions from incomplete combustion, thermal efficiency and operational aspects. Decreasing the fuel consumption rate below nominal load increases the emission levels of emissions from incomplete combustion exponentially with decreasing load in small-scale biomass combustion applications [11].

Moisture content: The significance of the moisture content has already been described. However, in batch combustion applications an additional complication is introduced: the moisture content will vary continuously as a function of burnout. The moisture will be released in the devolatilization phase, and the moisture content decreases as a function of burnout. Hence, the moisture content and its negative effects on the combustion process may be substantial in the early stages of the devolatilization phase, resulting in high levels of emissions from incomplete combustion.

Combustion temperature: The significance of sufficiently high combustion temperatures has already been described. However, in batch combustion applications an additional complication is introduced: the moisture content and fuel composition will vary continuously as a function of burnout. This will influence the adiabatic combustion temperature. The adiabatic combustion temperature will increase as a function of degree of burnout at a constant excess air ratio. However, as char is much less reactive than the volatile fraction of biomass fuels, the fuel consumption rate and the oxygen need will be much lower. Since it is usually difficult to effectively control the amount of air supplied in the char combustion phase, especially if natural draught is applied, the excess air ratio will be quite high. This, together with a much lower fuel consumption rate, may decrease the temperature in the combustion chamber below the level needed for complete combustion. However, the higher GCV of char will to some extent compensate for the much lower fuel consumption rate in the char combustion phase. The **residence time** needed for complete combustion is directly influenced by the combustion temperature and to some extent by the mixing time.

Draught: Two possibilities exist, i.e. forced or natural draught. Forced draught is used in all large-scale biomass combustion applications by means of fans or air blowers, which control the amount of air supplied. Forced draught is also widely applied in small-scale biomass combustion applications, such as wood log boilers. However, in wood-stoves and fireplaces, natural draught is most commonly applied, and the possibilities for controlling the combustion process therefore significantly decrease. This usually results in significantly higher levels of emissions from incomplete combustion.

Design: From the variables described above it is clear that the combustion application design significantly influences the combustion process through the construction and operational principle of the combustion chamber, through the choice of materials, and through process control possibilities. The **materials** used, mainly their heat capacity, density, thickness, insulating effect and surface properties, influence the combustion chamber temperatures. The **glass area and properties** are of special importance in small-scale biomass combustion applications, such as wood-stoves and fireplaces, since the radiation loss through the glass will be large per unit surface area, compared to the conductive heat loss through the combustion chamber walls per unit surface area.

Heat exchange: Proper heat exchange, either with the surroundings or with water for central heating, is necessary to achieve high thermal efficiency. This can be arranged in many ways by using different kinds of heat exchangers before the flue gas reaches the chimney. The simplest form is a pipe from the unit to the chimney inlet. However, to be able to control the heat exchange, either passive or active process control systems must be applied. Passive systems control the heat exchange by its principle of construction, such as radiation shields. Active systems use process control variables, for example the amount of water flowing through the boiler. In batch combustion applications, an additional complication is introduced since the combustion chamber temperature is not constant, which usually results in poor thermal efficiency, compared to continuous combustion applications.

Air staging: By applying staged-air combustion, a simultaneous reduction of both emissions from incomplete combustion and NO_x emissions is possible through a separation of devolatilization and gas phase combustion. This results in improved mixing of fuel gas and secondary combustion air, which reduces the amount of air needed, providing a lower local and overall excess air ratio and higher combustion temperatures. Hence, emissions from incomplete combustion are reduced by a temperature increase, which speeds up the elementary reaction rates, and by an improved mixing which reduces the residence time needed for the mixing of fuel and air. See Section 9.2.4.7 for further information.

Air distribution: Efficient air distribution is of the utmost importance to achieve an effective reduction of both emissions from incomplete combustion and NO_x emissions in staged-air systems. The distribution of primary and secondary air, within the combustion chamber and the flame zone, influences the mixing quality of air and fuel, and thus the residence time, and subsequently the combustion temperature needed for complete combustion.

Fuel feeding: Any batch combustion application will benefit from a more continuous combustion process, in which the negative effects of the start-up phase and the char phase are reduced. This is partly achieved manually in wood log boilers, by semi-batch operation.

Fuel distribution: The distribution of fuel inside the combustion chamber, reducing or increasing the active surface area, will influence the combustion process through a decreasing or increasing reactivity, respectively.

Heat distribution: Heat distribution is closely related to heat exchange and fuel distribution, and, in addition to several other variables, influences the temperature in different sections of the combustion chamber, and the heat transfer after the combustion chamber.

Radiation shields: One method of reducing the surface temperatures of small-scale biomass combustion applications is to use radiation shields. Reduced surface temperatures reduce the risk of burns and the need for a firewall. Introducing a radiation shield causes an increasing heat transfer by convection will occur, which is commonly utilized in fireplace inserts, where the main heat transfer to the surroundings is through convection. However, radiation shields can also be utilized for wood-stoves. By optimizing the thermal properties and the displacement of the radiation shields from the combustion chamber walls, a very low outer surface temperature can be achieved with minimized heat loss by radiation.

Regulation: One of the main drawbacks of small-scale biomass combustion applications is the limited possibility for regulating the combustion process. By applying efficient combustion process control, emission levels can be minimized and thermal efficiency can be optimized. Several different methods for combustion process control have been developed (see Sections 5.7 and 9.2.4.6). These can be based on measurements of specific flue gas compounds or temperatures, which will then provide a combustion process controller with the necessary information needed to change the combustion process, for example by changing the amount and distribution of air fed to the combustion chamber. However, combustion process control devices are usually not cost-effective in small-scale biomass combustion applications.

From this case study it can be concluded that the combustion process, and therefore emission levels and energy efficiency, is influenced by a great many variables. These should be kept in mind when designing and operating any biomass combustion application. Small-scale biomass combustion applications are additionally complex due to their batch-wise operation, which limits the possibilities for combustion process control. Large-scale biomass combustion applications are fired continuously, with the potential for advanced combustion process control optimization. However, several additional aspects are of great importance in large-scale biomass combustion applications, such as problems caused by the utilization of low-quality and cheap biomass fuels. This often results in depositions and corrosion on heat exchangers and superheaters, and additional emissions caused by higher nitrogen, sulphur, chlorine, fluorine, potassium and sodium content in the fuel, than there is in wood, which is most commonly used in small-scale biomass combustion applications. See Section 2.3 and Chapter 8 for further information.

2.2.3 Principles of medium- to large-scale combustion applications

Small particles combust relatively rapidly, and can be combusted in entrained flow combustion applications for pulverized fuels. How small the particles need to be depends on a number of factors that influence the overall combustion rate. For pulverized coal combustion applications the particle size should be less than 75μm [7]. Coal can relatively easily be crushed to small particles, while this becomes much more energy intensive for e.g. wood. Hence, pulverized combustion applications for biomass are not very common. Instead larger particle sizes are used in grate combustion applications or fluidized bed combustion (FBC) applications.

In fluidized bed combustion applications there will be an upper particle size dictated by the ability of the particle to fluidize and at the same time achieve complete combustion. By adjusting the air flow, one can influence the fluidization, and at high enough airflows the particle will leave the combustion chamber before it is completely combusted. This is the case in a so-called circulating fluidized bed (CFB), where the unburned particles are collected in a cyclone, and re-injected into the fluidizing bed. In a so-called bubbling fluidized bed (BFB) the air flow is kept low enough to allow for complete combustion without re-injection.

In grate combustion applications the particle size can be rather large. The particle is kept on the grate within the combustion chamber for the length of time necessary to complete combustion. The particles are burned in a bed of particles; hence, the combustion process becomes more like the combustion of a large layer of particles. The layer then goes through the steps of drying, pyrolysis and combustion on two axes, for the layer as a whole and for each single particle.

See Chapter 5 for further information.

2.2.4 Thermodynamics and gas phase kinetics

A combustion process is, as mentioned, influenced or controlled by a number of factors. Any thermal system, with or without chemical reactions, will approach equilibrium as time goes to infinity. The equilibrium composition of a flue gas can easily be calculated for a specific set of conditions, and is only dependent of thermodynamics, i.e. it is independent of chemical kinetics. However, in most combustion processes, and especially for emissions such as NO_x and CO, equilibrium does not prevail. Hence, gas phase kinetics is one very important controlling factor in biomass combustion applications.

However, the thermodynamics is useful for basic engineering calculations, e.g. the calculation of adiabatic flame temperature and thermal efficiency.

2.2.5 Solid fuel kinetics and gas phase interaction

The other main controlling factor is the heterogeneous combustion/gasification of the char remaining after drying and pyrolysis has occurred. Combustion/gasification of the char, which contains mainly carbon, requires that oxygen/oxidant is transported into the char surface by diffusion/convection, where it is adsorbed. The oxygen/oxidant then reacts heterogeneously on the surface with carbon; hereafter the product formed in the reaction desorbs, and is carried away from the particle by diffusion/convection. The product is mainly CO_2 or CO. Final burnout of CO takes place in the gas phase, controlled by gas-phase chemical kinetics. The surface reaction is controlled by heterogeneous chemical kinetics, while the other steps are controlled by physical processes. Compared to gas, phase combustion, heterogeneous combustion is usually slow. This has major implications e.g. for entrained flow combustion applications for pulverized fuels. Here it is essential to allow for a total residence time in the combustion application that satisfies the time requirements for drying, pyrolysis, heterogeneous combustion and gas phase combustion. Hence, the particle size must be small.

Unburned carbon reduces the heat release and therefore the combustion efficiency. Basic engineering calculations can be used to calculate the combustion efficiency.

2.2.6 Basic engineering calculations

Even though solid fuel combustion is extremely complex, there are, as mentioned, a number of basic engineering calculations that can be performed relatively easily based on mass balances and thermodynamics. These calculations focus on the overall fuel conversion reaction taking place, and the thermal performance.

Important parameters are, for example, the flue gas flow and composition, the air flow and the excess air ratio, the calorific value of the fuel, achievable combustion temperatures, and combustion efficiency and thermal efficiency.

We will now introduce the formulae needed to perform these basic engineering calculations.

2.2.6.1 Flows and compositions

The flue gas flow is calculated from the knowledge of the elemental fuel composition and the amount of air introduced. We will here assume complete combustion, i.e. the elements in the fuel and air form only major species in the flue gas, i.e. CO_2, H_2O, SO_2, N_2 and O_2. Biomass contains five main fuel elements: C, H, S, N and O. The O contributes to the oxidation of the other four fuel elements, and does as such reduce the amount of air needed.

The O_2 need:

$$\bar{m}_{O_2,Air}\left[\frac{kg\ O_2}{kg\ fuel\ (waf)}\right] = \left(X_C \frac{M_{O_2}}{M_C} + \frac{X_H}{4}\frac{M_{O_2}}{M_H} + X_S \frac{M_{O_2}}{M_S} - X_O\right)(1-X_{H_2O})\lambda \quad (2.5)$$

where:

$M_i\left[\dfrac{kg}{kmole}\right]$: molecular mass of element i

$M_C = 12.01115$, $M_H = 1.00797$, $M_S = 32.064$, $M_{O_2} = 31.9988$

X_i: mass fraction of element i in dry ash free fuel (daf)

X_{H_2O}: mass fraction of H_2O in wet ash free fuel (waf)

λ: excess air ratio

N_2 from air:

$$\bar{m}_{N_2,Air}\left[\frac{kg\ N_2}{kg\ fuel\ (waf)}\right] = \bar{m}_{O_2,Air}\frac{Y_{N_2,Air}}{Y_{O_2,Air}}\frac{M_{N_2}}{M_{O_2}} \quad (2.6)$$

where:

$Y_{O_2,Air}$: volume fraction of O_2 in air (usually assumed to be 0.21)

$Y_{N_2,Air} = 1 - Y_{O_2,Air}$

$M_{N_2} = 28.0134$

Air:

$$\bar{m}_{Air}\left[\frac{kg\ air}{kg\ fuel\ (waf)}\right] = \bar{m}_{O_2,Air} + \bar{m}_{N_2,Air} \quad (2.7)$$

Flue gas:

$$\bar{m}_{FG}\left[\frac{kg\ flue\ gas}{kg\ fuel\ (waf)}\right] = \bar{m}_{Air} + 1 \qquad (2.8)$$

The flue gas species mass:

$$\bar{m}_{CO_2} = X_C \frac{M_{CO_2}}{M_C}(1-X_{H_2O}), \quad \bar{m}_{H_2O} = X_H \frac{M_{H_2O}}{M_{H_2}}(1-X_{H_2O}) + X_{H_2O} \qquad (2.9)$$

$$\bar{m}_{SO_2} = X_S \frac{M_{SO_2}}{M_S}(1-X_{H_2O}), \quad \bar{m}_{N_2} = \bar{m}_{N_2,Air} + X_N(1-X_{H_2O}),$$

$$\bar{m}_{O_2} = \frac{\bar{m}_{O_2,Air}}{\lambda}(\lambda-1) \quad \left[\frac{kg\ i}{kg\ fuel\ (waf)}\right]$$

$$M_{CO_2} = 44.00995,\ M_{H_2O} = 18.01534,\ M_{H_2} = 2.01594\ M_{SO_2} = 64.0628$$

The flue gas species mass fractions $X_{CO_2}, X_{H_2O}, X_{SO_2}, X_{N_2}$, and X_{O_2}, are found by normalizing the total flue gas mass.

The flue gas volume fractions are found by converting the mass fractions to volume fractions according to:

$$Y_i = \frac{\frac{X_i}{M_i}}{\sum_i \frac{X_i}{M_i}}, \quad where: \sum_i \frac{X_i}{M_i} = \frac{X_{CO_2}}{M_{CO_2}} + \frac{X_{H_2O}}{M_{H_2O}} + \frac{X_{SO_2}}{M_{SO_2}} + \frac{X_{N_2}}{M_{N_2}} + \frac{X_{O_2}}{M_{O_2}} \qquad (2.10)$$

The flue gas molecular mass:

$$M_{FG}\left[\frac{kg}{kmole}\right] = \sum_i Y_i M_i = Y_{CO_2}M_{CO_2} + Y_{H_2O}M_{H_2O} + Y_{SO_2}M_{SO_2} + Y_{N_2}M_{N_2} + Y_{O_2}M_{O_2} \qquad (2.11)$$

The flue gas density:

$$\rho_{FG}\left[\frac{kg}{Nm^3}\right] = \frac{p_0}{\frac{R_u}{M_{FG}}T_0} \qquad (2.12)$$

where:
p_0 : pressure at normal condition = 101325 Pa
T_0 : temperature at normal condition = 273.15 K
R_u : universal gas constant = 8314.32 $J/kmole \cdot K$

The air density:

$$\rho_{Air}\left[\frac{kg}{Nm^3}\right] = \frac{p_0}{\frac{R_u}{M_{Air}}T_0} \quad (2.13)$$

where:

$M_{Air}[kg/kmole]$: air molecular mass $= Y_{O_2,Air}M_{O_2} + Y_{N_2,Air}M_{N_2}$

The flue gas mass flow:

$$\dot{m}_{FG}\left[\frac{kg}{h}\right] = \dot{m}_F \cdot \bar{m}_{FG} \quad (2.14)$$

where:

$\dot{m}_F\left[\frac{kg\ fuel\ (waf)}{h}\right]$: wet ash free fuel consumption rate

The flue gas volume:

$$\bar{V}_{FG}\left[\frac{Nm^3}{kg\ fuel\ (waf)}\right] = \frac{\bar{m}_{FG}}{\rho_{FG}} \quad (2.15)$$

The flue gas volume flow:

$$\dot{V}_{FG}\left[\frac{Nm^3}{h}\right] = \frac{\dot{m}_{FG}}{\rho_{FG}} \quad (2.16)$$

The air mass flow:

$$\dot{m}_{Air}\left[\frac{kg}{h}\right] = \dot{m}_F \cdot \bar{m}_{Air} \quad (2.17)$$

The air volume flow:

$$\dot{V}_{Air}\left[\frac{Nm^3}{h}\right] = \frac{\dot{m}_{Air}}{\rho_{Air}} \quad (2.18)$$

2.2.6.2 Energy calculations

Energy calculations are useful for estimating combustion temperatures, and are also needed for the calculation of efficiencies. Energy stored in the fuel is released in the combustion process and will heat the flue gas to a certain temperature. This temperature will depend on a number of factors. A general energy balance can be set up:

Energy in fuel + Preheat energy = Energy in flue gas + Energy losses (2.19)

The energy in the fuel is the chemically stored energy released as the fuel is combusted and converted to final products in the flue gas. The preheat energy is an energy difference due to a temperature difference between the fuel or air and ambient conditions. The energy in the flue gas is the energy stored in the flue gas at a certain temperature. This energy will be released as the flue gas cools down. Energy losses are losses due to heat loss to the surroundings or due to incomplete combustion (e.g. CO in the flue gas, or carbon in the ash).

The energy in the fuel is given by the NCV (Eq. 2.2). To calculate the preheat energy (for air) and the energy in the flue gas, we need to introduce enthalpy expressions for each of our five gas species.

$$h_i(T)\left[\frac{J}{kg\ i}\right] = T(a_1 + \frac{a_2}{2}\cdot T + \frac{a_3}{3}\cdot T^2 + \frac{a_4}{4}\cdot T^3 + \frac{a_5}{5}\cdot T^4 + \frac{a_6}{T})\frac{R_u}{M_i} \qquad (2.20)$$

where:

T : temperature in gas $[K]$

The coefficients are divided into two sets for each gas species, one for use at temperatures above 1000K and one for use at 1000K and below. The coefficients are:

	CO_2 [12]		H_2O [12]	
	<= 1000K	> 1000K	<= 1000K	> 1000K
a_1	2.35677352	3.85746029	4.19864056	3.03399249
a_2	0.00898459677	0.00441437026	−0.0020364341	0.00217691804
a_3	−0.00000712356269	−0.00000221481404	0.00000652040211	−0.000000164072518
a_4	2.45919022E-09	5.23490188E-10	−5.48797062E-09	−9.7041987E-11
a_5	−1.43699548E-13	−4.72084164E-14	1.77197817E-12	1.68200992E-14
a_6	−48371.9697	−48759.166	−30293.7266999999	−30004.2971

	SO_2 [13]		N_2 [12]	
	<= 1000K	> 1000K	<= 1000K	> 1000K
a_1	2.911438	5.254498	3.298677	2.92664
a_2	0.008103022	0.001978545	0.0014082404	0.0014879768
a_3	−0.00000690671	−0.0000008204226	−0.000003963222	−0.000000568476
a_4	0.000000003329015	1.576383E-10	0.000000005641515	1.0097038E-10
a_5	−8.777121E-13	−1.1204512E-14	−2.444854E-12	−6.753351E-15
a_6	−36878.81	−37568.85	−1020.8999	−922.7977

	O_2 [12]	
	<= 1000K	> 1000K
a_1	3.78245636	3.28253784
a_2	−0.00299673416	0.00148308754
a_3	0.00000984730201	−0.000000757966669
a_4	−9.68129509E-09	2.09470555E-10
a_5	3.24372837E-12	−2.16717794E-14
a_6	−1063.94356	−1088.45772

Calculation of the preheat energy in a solid fuel is more complex, requiring knowledge about the specific heat capacity of the fuel. However, this contribution is usually negligible.

Energy loss due to incomplete combustion can be included, e.g. for CO:

$$\overline{E}_{CO}\left[\frac{kJ}{kg\ fuel\ (waf)}\right] = CV_{CO}\left[\frac{kJ}{kg\ CO}\right]\overline{m}_{CO}\left[\frac{kg\ CO}{kg\ fuel\ (waf)}\right] \quad (2.21)$$

where:
CV_{CO} is the calorific value of CO
$= 10102\ kJ/kg\ CO$ at a reference temperature of $25°C$

Energy loss due to unburned carbon:

$$\overline{E}_C\left[\frac{kJ}{kg\ fuel\ (waf)}\right] = 34910\left[\frac{kJ}{kg\ C}\right] \cdot \overline{m}_C\left[\frac{kg\ C}{kg\ fuel\ (waf)}\right] \quad (2.22)$$

where:
34910 is the calorific value of C given in Eq. 2.1

The general energy balance becomes (J/h):

$$NCV \cdot \dot{m}_F + [h_F(T_F) - h_F(T_{Amb})]\dot{m}_F + [h_{Air}(T_{Air}) - h_{Air}(T_{Amb})]\dot{m}_{Air} = $$
$$[h_{FG}(T_{FG}) - h_{FG}(T_{Amb})]\dot{m}_{FG} + \sum_i Q_i + \sum_j \overline{E}_j \cdot \dot{m}_F \quad (2.23)$$

where:
NCV : Net Calorific Value of the fuel
h_F : fuel enthalpy
h_{Air} : air enthalpy $= X_{N_2}h_{N_2} + X_{O_2}h_{O_2}$
h_{FG} : flue gas enthalpy $= X_{CO_2}h_{CO_2} + X_{H_2O}h_{H_2O} + X_{SO_2}h_{SO_2} + X_{N_2}h_{N_2} + X_{O_2}h_{O_2}$
T_F : fuel temperature; T_{Amb} : ambient temperature; T_{FG} : flue gas temperature

$\sum_i Q_i$: sum of heat losses by radiation, convection and conduction

$\sum_j \overline{E}_j \cdot \dot{m}_F$: sum of heat losses in unburned components

To find the combustion temperature the flue gas temperature is adjusted until the left hand side of the equation equals the right hand side. In the case of no losses and no preheating the general energy balance simplifies to:

$$NCV \cdot \dot{m}_F = [h_{FG}(T_{FG}) - h_{FG}(T_{Amb})]\dot{m}_{FG} \quad (2.24)$$

2.2.6.3 Efficiencies

Thermal efficiency is the ratio between the heat and/or work utilized and the energy supplied. The energy supplied is in principle all energy supplied. However, the common assumption is to use the NCV of the fuel, assuming no condensation of the water vapour in the flue gas. A flue gas condensing plant may then achieve a thermal efficiency above unity. The inclusion of the preheating of air and fuel depends on how the preheating was achieved. External preheating (heat not originating from the fuel conversion) gives a net energy supply, and should then be taken into account. Internal preheating does not give a net energy supply, and should therefore not be included. The heat utilized can be all the heat transferred to a room for a room heating application, or the heat transferred to the boiler in a central heating system. The work may be generated electricity or mechanical work. The heat transferred to a room for a room heating application can be calculated if the chimney inlet temperature is known.

The chimney heat loss:

$$Q_{Ch} = [h_{FG}(T_{Ch}) - h_{FG}(T_{Amb})] \dot{m}_{FG} \tag{2.25}$$

where:

T_{Ch}: chimney inlet temperature

Thermal efficiency:

$$\eta_{th} = 1 - \frac{Q_{Ch} + \sum_i Q_i}{NCV \cdot \dot{m}_F + [h_F(T_F) - h_F(T_{Amb})] \dot{m}_F + [h_{Air}(T_{Air}) - h_{Air}(T_{Amb})] \dot{m}_{Air}} \tag{2.26}$$

Thermal efficiency, if the chimney heat loss is the only heat loss, and for no preheating of fuel or air is given by:

$$\eta_{th} = 1 - \frac{Q_{Ch}}{NCV \cdot \dot{m}_F} \tag{2.27}$$

Combustion efficiency:

$$\eta_{comb} = 1 - \frac{\sum_j \bar{E}_j \cdot \dot{m}_F}{NCV \cdot \dot{m}_F} = 1 - \frac{\sum_j \bar{E}_j}{NCV} \tag{2.28}$$

Total efficiency:

$$\eta_{tot} = 1 - \frac{Q_{Ch} + \sum_i Q_i + \sum_j \bar{E}_j \cdot \dot{m}_F}{NCV \cdot \dot{m}_F + [h_F(T_F) - h_F(T_{Amb})] \dot{m}_F + [h_{Air}(T_{Air}) - h_{Air}(T_{Amb})] \dot{m}_{Air}} \tag{2.29}$$

2.2.7 Basic emission conversion calculations

Various denominators are used for reporting emission levels from biomass combustion applications. This makes it difficult to compare emission levels directly. Typical denominators are mg/Nm³ flue gas at a given vol% O_2 (d.b.), ppm in flue gas at a given vol% O_2 (d.b.), mg/kg fuel (d.b.) and mg/MJ (based on GCV, or NCV). However, it is possible to convert the reported emission levels to the other denominators by applying the following formulae and the procedures given in Sections 2.2.7.1 to 2.2.7.5 for each species.

From: \ To:	A mg/Nm³ at O₂	B mg/Nm³ at O₂r	C ppm at O₂	D mg/kg	E mg/MJ
A mg/Nm³ at O₂	A	$A \cdot \dfrac{V_{FG}}{V_{FGr}}$	$A \cdot \dfrac{V_{Mole}}{M_i}$	$A \cdot V_{FG}$	$A \cdot \dfrac{V_{FG}}{GCV}$
B mg/Nm³ at O₂r	$B \cdot \dfrac{V_{FGr}}{V_{FG}}$	B	$B \cdot \dfrac{V_{Mole}}{M_i} \cdot \dfrac{V_{FGr}}{V_{FG}}$	$B \cdot V_{FGr}$	$B \cdot \dfrac{V_{FGr}}{GCV}$
C ppm at O₂	$C \cdot \dfrac{M_i}{V_{Mole}}$	$C \cdot \dfrac{M_i}{V_{Mole}} \cdot \dfrac{V_{FG}}{V_{FGr}}$	C	$C \cdot V_{FG} \cdot \dfrac{M_i}{V_{Mole}}$	$C \cdot \dfrac{M_i}{V_{Mole}} \cdot \dfrac{V_{FG}}{GCV}$
D mg/kg	$\dfrac{D}{V_{FG}}$	$\dfrac{D}{V_{FGr}}$	$\dfrac{D}{V_{FG}} \cdot \dfrac{V_{Mole}}{M_i}$	D	$\dfrac{D}{GCV}$
E mg/MJ	$E \cdot \dfrac{GCV}{V_{FG}}$	$E \cdot \dfrac{GCV}{V_{FGr}}$	$E \cdot \dfrac{V_{Mole}}{M_i} \cdot \dfrac{GCV}{V_{FG}}$	$E \cdot GCV$	E

A Microsoft Excel spreadsheet for emission level conversion (Fuelsim-Average.xls) can be downloaded from the IEA Biomass Combustion and Co-Firing internet site: www.ieabcc.nl/.

2.2.7.1 Emission conversion: mg/Nm³ ⇔ ppm, mg/kg ⇔ mg/MJ

The conversion from mg/Nm³ to ppm and from mg/kg to mg/MJ is straightforward.

2.2.7.2 Emission conversion: mg/Nm³, ppm ⇒ mg/kg, mg/MJ

For conversion between volumetric and mass units, the dry flue gas volume produced per kg dry fuel is needed. The dry flue gas volume produced per kg dry fuel, assuming that carbon is converted to CO_2 only, is given by:

$$V_{FG}\left[\frac{Nm^3 \text{ dry } FG}{kg \text{ fuel }(daf)}\right] = \frac{\bar{V}_{FG}\left(1 - Y_{H_2O,FG}\right)}{1 - X_{H_2O,Fuel}} \qquad (2.30)$$

where $\bar{V}_{FG}$ is calculated using Eq. 2.15

Alternatively, if the CO_2 concentration in the dry flue gas is known, the calculation becomes rather simple:

$$V_{FG}\left[\frac{Nm^3 \text{ dry } FG}{kg \text{ fuel }(daf)}\right] = \frac{100 \cdot X_C \cdot V_{Mole}}{M_C \cdot CO_2}, \text{ at } O_2 \qquad (2.31)$$

where:
CO_2 : vol% CO_2 in dry flue gas at O_2

A representative dry fuel composition must be assumed if the fuel composition is not stated. The assumption of complete combustion introduces an error in the calculation, which is negligible unless the amount of unburned fuel becomes very high.

2.2.7.3 Emission conversion: mg/Nm³ at $O_2 \Rightarrow$ mg/Nm³ at $O_2 r$

For conversion from mg/Nm³ at a given vol% O_2 to another reference vol% O_2 ($O_2 r$) we need to calculate the dry flue gas volume produced per kg dry fuel at $O_2 r$. This can be done by considering the conversion as a dilution process in which air is added or removed until O_2 equals $O_2 r$. The following can be deduced:

$$V_{FGr} = V_{FG} \frac{Y_{O_2,Air} - Y_{O_2,dry\,FG}}{Y_{O_2,Air} - Y_{O_2 r,dry\,FG}}, \text{ at } O_2 r \qquad (2.32)$$

2.2.7.4 Emission conversion: mg/kg, mg/MJ $\Rightarrow$ mg/Nm³, ppm

For conversion from mg/kg or mg/MJ to mg/Nm³ or ppm at O_2 we need to calculate the dry flue gas volume produced per kg dry fuel at O_2, as shown in Eq. 2.30.

2.2.7.5 Emission conversion: SO_2, particles and trace compounds

Since a significant fraction of the sulphur content in the fuel may be retained in the ashes (see Section 9.2.1.1), the dry flue gas flow should ideally be calculated using a dry fuel composition with a reduced sulphur content, equivalent to the sulphur content emitted as SO_2 (and other gaseous sulphur compounds). However, since the sulphur content in most biomass fuels is low, this is usually not necessary.

Particle emission levels from biomass combustion applications can be quite high, especially from small-scale installations. Particle emissions can be divided into emissions from both complete and incomplete combustion (see Sections 9.2.1.1 and 9.2.1.2). As for SO_2, the amount of C, H, O, N and S emitted as particles should ideally be corrected in the dry fuel composition when calculating the dry flue gas flow. However, this is not necessary if the particle emission level is low.

Emissions of trace compounds can safely be neglected when calculating the dry flue gas flow, since these emission levels do not significantly influence the calculated dry flue gas flow.

2.2.8 Batch versus continuous combustion

The complexity of batch combustion with respect to the main variables influencing emission levels and energy efficiency in wood-stoves and fireplaces was described in the previous section. However, to illustrate the time dependence of combustion processes in batch combustion applications, Figure 2.12 is included, which shows the results of an experiment with an advanced down-draught staged-air wood-stove.

The figure shows that, typically, 80 per cent of the weight loss occurs in the volatile combustion phase and 20 per cent in the char combustion phase. However, GCV in the char combustion phase will be significantly higher than in the volatile combustion phase, and will, to some extent, compensate for the much smaller fuel consumption rate. The excess air ratio can usually be controlled relatively well in the volatile combustion phase. However, if the overall excess air ratio becomes too low, as in this case, the emissions from incomplete combustion, such as CO, will increase considerably due to insufficient mixing of air and fuel. This will be the case even if the combustion chamber temperature is more than sufficient for a complete burnout. When the char combustion phase begins, the combustion chamber temperature decreases, in most cases below a level sufficient for complete oxidation of CO, which is formed from heterogeneous charcoal combustion, to

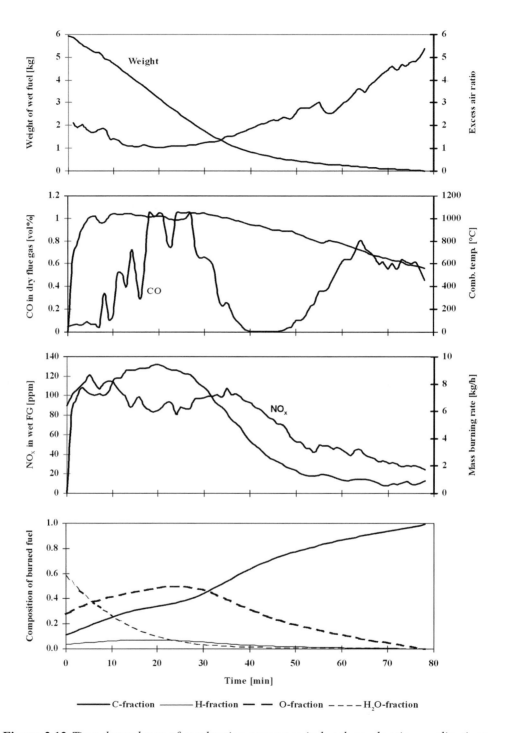

Figure 2.12 *Time dependence of combustion processes in batch combustion applications*
Source: [14].

CO_2, as can be seen. Concerning CO burnout, a temperature level of 800°C is usually needed [15]. The NO_x emission level is only influenced directly by the combustion chamber temperature to a minor degree, and the NO_x emission level in g/kg dry fuel usually increases in the char combustion phase. The fuel composition varies considerably throughout the experiment, with high moisture content at the beginning of the volatile combustion phase, and with close to 100 per cent carbon at the end of the char combustion phase.

These time-dependent phenomena and all the resulting complications will be negligible in continuous combustion applications.

2.2.9 Modelling biomass combustion

Modelling, together with experiments, enables a cost-effective approach for future biomass combustion application design, and can improve the competitiveness of biomass combustion for heat and electricity generation. Various types of modelling tools exist, from simple heat and mass balance models to high user level and time-consuming computational fluid dynamics (CFD) tools, as illustrated in Figure 2.13.

Modelling improves our understanding of the fundamental processes involved in biomass combustion, and may significantly reduce the 'trial and error' development time needed if experiments only are used for design optimization. By combining modelling with experiments, an improved design is possible with respect to the reduction of emissions from both incomplete and complete combustion. Parametric studies can be carried out that reveal the relative influence of different combustion process variables on emission levels and energy efficiency. This enables us to make the correct decisions with respect to the optimal design and operational principles of the biomass combustion application.

The most advanced modelling tools, CFD tools, usually have a high user level, and the amount of computational time needed when combining flow calculations and chemical kinetics is currently very high, even with the most advanced computers. Hence, in reality,

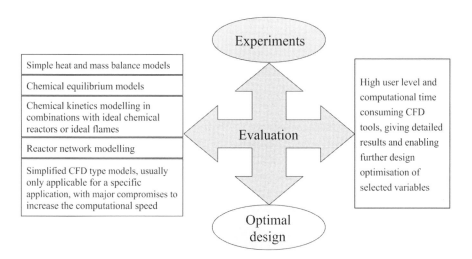

Figure 2.13 *Modelling tools*

simplifications have to be made that reduce the reliability of the modelling results. Also, the fact that biomass is a solid fuel introduces additional complications with regard to devolatilization behaviour and solids combustion, compared to gas combustion. At the moment, research is being carried out in many countries to improve our understanding of the fundamental processes involved in biomass combustion, and in the development of models that can effectively be included in CFD tools.

In the meantime, there are other modelling tools that can be combined with experiments, such as simple heat and mass balance models, chemical equilibrium models, chemical kinetics modelling in combination with ideal chemical reactors or ideal flames, reactor network modelling, and simplified CFD-type models. An overview of the various classes of models available and their various purposes is given in the Table 2.1. Further information and references on modelling of biomass combustion are given in Section 11.8.

Table 2.1 *Overview of the usefulness of various model types for certain purposes*

Application	Type of model						
	Thermo-dynamic	Empirical Regression	Empirical Process identification	Expert	Physical/chemical Stationary	Physical/chemical Dynamic	CFD
System design/optimization	++			+	++	++	
Apparatus design/optimization		+		+	++	++	++
Process control design/optimization			++			++	
Selection of biomass for application in reactor				+	++	++	
Evaluation of operational performance			+			++	

Source: [16]

2.2.10 Ash formation

2.2.10.1 Process and mechanisms of ash formation

Ash-forming elements are present in biomass as salts, bound in the carbon structure (inherent ash) or they are attendant as mineral particles from dirt and clay introduced into the biomass fuel during harvest or transport (entrained ash). The compounds in inherent ash are homogeneously dispersed in the fuel and are much more mobile than the compounds in entrained ash and, therefore, are readily volatile and available for reactions in burning char (see Figure 2.14) [17].

During combustion, a fraction of the ash-forming compounds in the fuel is volatilized and released to the gas phase. The volatilized fraction depends on the fuel characteristics, the gas atmosphere, and the combustion technology in use. For example, high combustion

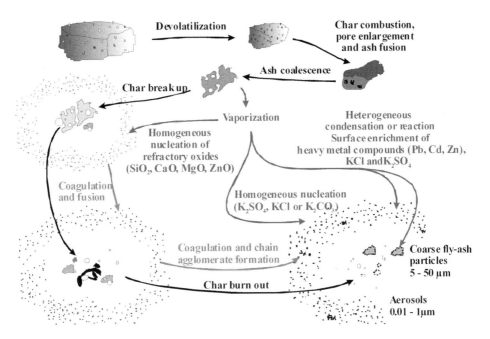

Figure 2.14 *Mechanisms involved in ash formation in biomass combustion*

Source: Adapted from [18].

temperatures and a reducing atmosphere have been reported to enhance the volatilization of the environmentally relevant heavy metals Zn, Pb and Cd [19].

In the reducing conditions and the high temperatures inside and on the surface of burning char, even a small fraction present as refractory oxides, such as SiO_2, CaO and MgO, may convert to more volatile SiO, Ca and Mg and volatilize. When released from the char as vapours, these elements form very small primary particles in the boundary layer of the burning char particles due to re-oxidation and subsequent nucleation [20]. This mechanism is induced when the volatilized compounds are transported from the reducing atmosphere at the surface of the burning char particle through the boundary layer into the surrounding oxidizing gas conditions in the furnace. Chemical equilibrium model calculations suggest that Zn should also follow this mechanism due to its volatile metal but refractory oxide compound. Due to the high volatility of Cd and Pb, even in an oxidizing atmosphere, these metals behave differently than does Zn. Depending on the furnace temperature, Cd and Pb will pass through the boundary layer and enter the surrounding furnace environment in their gaseous state, together with other volatile elements such as Cl, S and K [21].

Primary particles formed by vaporization and subsequent nucleation in the boundary layer are very small in size, about 5–10nm, but on their way in the flue gas they grow by coagulation, agglomeration and condensation [20]. These particles form the basis for the fine mode of the fly ash, characterized by a particle size of < 1µm (see Figure 2.15).

The non-volatile ash compounds remaining in the char may melt and coalesce inside and on the surface of the char, depending on the temperature and chemical composition of

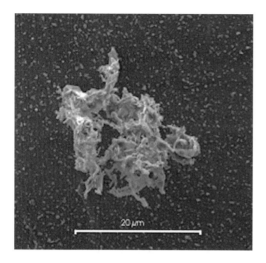 

Figure 2.15 *Images of coarse fly ash (left) and aerosol particles (right) from wood combustion in a grate furnace*

Explanations: Biomass fuel: wood (beech). Equipment used: scanning electron microscopy (SEM) and energy-dispersive X-ray spectroscopy (EDX).

Source: [17]

the particles. This results in residual ash particles with a wide range of compositions, shapes and sizes, related to the characteristic of the parent mineral particles. Depending on the density and size of the residual ash particles, the combustion technology and the flue gas velocity, a fraction of the residual ash will be entrained with the flue gas and form the coarse part of fly ash, while the other fraction will stay on the grate and form bottom ash. In contrast to the fine-mode fly-ash particles from volatilized ash compounds, coarse fly-ash particles are larger, typically exceeding 5μm (see Figure 2.15) [22, 23].

Upon cooling of the flue gas in the convective heat exchanger section, vapours of volatilized compounds condense or react on the surface of pre-existing ash particles in the flue gas. Due to the much larger specific surface of the fine-mode particles compared to the coarse fly-ash particles, the concentrations of condensing or reacting ash-forming elements increase with decreasing particle size. This explains some of the very high heavy-metal concentrations found in aerosol particles from combustion plants [19, 22].

If the concentration of inorganic vapours in the flue gas and the cooling rate in the heat exchanger are both high, supersaturation of these vaporous compounds could occur, causing formation of new particles by nucleation. In biomass combustion the most abundant volatile element is potassium originating from inherent ash (K in entrained ash is typically present as thermally very stable mineral silicate compounds and will not vaporize). According to chemical equilibrium calculations, vaporized K is mainly present as gaseous KCl or KOH at high temperatures in the flue gas. As the gas temperature decreases, the chloride and hydroxide are converted to sulphate by homogeneous gas-phase reactions. Gaseous K_2SO_4 has a very low vapour pressure and becomes highly supersaturated as soon as it is formed, forming high numbers of new primary particles by homogeneous nucleation. However, according to gas phase kinetic considerations, the formation of K_2SO_4 does not always follow equilibrium and only a part of the potassium in the gas phase is converted [18]. The part of the gaseous potassium that does not form sulphates will either nucleate as KCl or K_2CO_3 or condensate on pre-existing particles at significantly lower temperatures than the K_2SO_4. Furthermore, as time proceeds in the flue gas, solid KCl or K_2CO_3 on the particles may undergo heterogeneous reactions with SO_2 (g) and form solid K_2SO_4. This reaction can promote corrosion caused by sub-micron particles, containing KCl, precipitating on the convective surface of heat exchangers. The sulphation of KCl releases corrosive Cl which can catalytically react with the heat exchanger surface by the 'active oxidation' mechanism [24], see also Section 8.6.

2.2.10.2 Ash fractions formed in biomass combustion plants

According to the ash formation process explained, in biomass combustion plants three different ash fractions must normally be distinguished (see Figure 2.16).

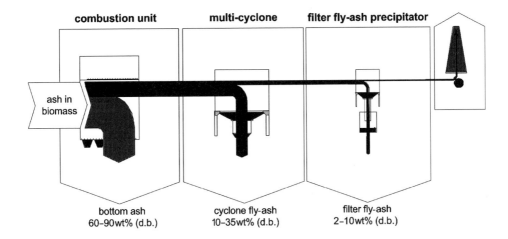

Figure 2.16 *Ash fractions produced in a biomass combustion plant*

Explanations: Amounts of ash for the various fractions are valid for fixed bed biomass combustion units. In fluidized bed combustion plants the fly-ash fractions are quantitatively dominant.

Source: [25]

The bottom ash: Ash fraction produced on the grate and in the primary combustion chamber; often mixed with mineral impurities contained in the biomass fuel like sand, stones and earth or with bed material in fluidized bed combustion plants. These mineral impurities can, especially in bark-fired fixed-bed combustion plants, cause slag formation (due to a lowering of the melting point) and sintered ash particles in the bottom ash.

The cyclone fly ash: Fine, mainly inorganic, ash particles carried with the flue gas and precipitated in the secondary combustion zone, in the boiler and especially in multi-cyclones placed behind the combustion unit. This ash fraction mainly consists of coarse fly-ash particles.

The filter fly ash: Second and finer fly-ash fraction precipitated in electrostatic filters, fibrous filters or as condensation sludge in flue gas condensation units (normally placed behind the multi-cyclone). In small-scale biomass combustion plants without efficient dust precipitation technology, this ash fraction is emitted with the flue gas. A small part of the filter fly ash remains in the flue gas anyway and causes dust emissions (depending on the efficiency of the dust precipitation technology used). This ash fraction mainly consists of aerosols (sub-micron ash particles).

2.3 Physical and chemical characteristics of biomass fuels

2.3.1 Overview

The kind of biomass fuel, its physical characteristics and chemical composition influence the whole process of biomass utilization (fuel supply, combustion system, solid and gaseous emissions). In the last few years comprehensive studies of the physical characteristics and chemical composition of biomass fuels have been carried out and are still ongoing. An overview of the most important characteristics and their effects is shown in Table 2.2.

2.3.2 Fuel characteristics and interaction with the combustion system

The characteristics and quality of biomass as a fuel vary widely, depending mainly on the kind of biomass and the pre-treatment technologies applied. For example, the moisture content of the fuel as fed into the furnace may vary from 25 to 60wt% (w.b.) (bark, sawmill side-products) or drop below 10wt% (w.b.) (pellets, dry wood-processing residues). Also, the ash sintering temperatures of biomass fuels used exhibit a wide range (800–1700°C) and the particle shapes and sizes can be manifold. Fuel quality can be influenced and improved by suitable pre-treatment technologies, but this increases costs.

On the other hand, various combustion technologies are available for different fuel qualities. In this context, it must be noted that less homogeneous and low-quality fuels need more sophisticated combustion systems. Because of this and other 'economy-of-scale' reasons, only medium-scale and large-scale systems are suitable to combust low-quality and cheap biomass fuels. The smaller the combustion plant, the higher the demands concerning fuel quality and fuel homogeneity.

2.3.2.1 Particle dimensions, moisture content, GCV, NCV, bulk density and energy density

Important physical parameters are particle dimensions, bulk and energy density, gross and net calorific value and moisture content. Table 2.3 gives an overview of typical values for physical parameters of various biomass fuels. The GCV and the NCV can be calculated by applying Eq. 2.1 and Eq. 2.2, respectively.

Depending on the fuel preparation process, biomass fuels are available as bulk material (e.g. woodchips, sawdust) or unit material (e.g. straw bales, firewood). For bulk material, the particle dimensions vary from a few millimetres to about 50 cm. Furthermore, the particle size distribution can be homogeneous (e.g. pellets) or inhomogeneous (e.g. untreated bark). Olive residues, an important biomass fuel in Southern European countries, are a homogeneous paste-like material. The particle dimension and particle size distribution determines the appropriate fuel-feeding system and the combustion technology (see Section 3.4.2).

Table 2.2 *Characteristics of solid biomass fuels and their most important effects*

Characteristics	Effects
Physical properties	
Moisture content	Storage durability and dry-matter losses, NCV, self-ignition, plant design
NCV, GCV	Fuel utilization, plant design
Volatiles	Thermal decomposition behaviour
Ash content	Dust emissions, ash manipulation, ash utilization/disposal, combustion technology
Ash-melting behaviour	Operational safety, combustion technology, process control system, hard deposit formation
Fungi	Health risks
Bulk density	Fuel logistics (storage, transport, handling)
Particle density	Thermal conductance, thermal decomposition
Physical dimension, form, size distribution	Hoisting and conveying, combustion technology, bridging, operational safety, drying, dust formation
Fine parts (wood pressings)	Storage volume, transport losses, dust formation
Abrasion resistance (wood pressings)	Quality changes, segregation, fine parts
Chemical properties	
Elements:	
– Carbon C	GCV (see Section 2.2.2.2)
– Hydrogen H	GCV, NCV (see Section 2.2.2.2)
– Oxygen O	GCV (see Section 2.2.2.2)
– Chlorine Cl	HCl, PCDD/PCDF emissions, corrosion, lowering ash-melting temperature
– Nitrogen N	NO_x, N_2O emissions
– Sulphur S	SO_x emissions, corrosion
– Fluorine F	HF emissions, corrosion
– Potassium K	Corrosion (heat exchangers, superheaters), lowering ash-melting temperature, aerosol formation, ash utilization (plant nutrient)
– Sodium Na	Corrosion (heat exchangers, superheaters), lowering ash-melting temperature, aerosol formation
– Magnesium Mg	Increase of ash-melting temperature, ash utilization (plant nutrient)
– Calcium Ca	Increase of ash-melting temperature, ash utilization (plant nutrient)
– Phosphorus P	Ash utilization (plant nutrient)
– Heavy metals	Emissions, ash utilization, aerosol formation

Explanations: NCV = net calorific value, GCV = gross calorific value.
Source: According to [26].

The energy density, resulting from the bulk density and the net calorific value (NCV), influences the fuel logistics (transport, storage) as well as the process control of the furnace's fuel-feeding system.

The moisture content influences the combustion behaviour, the adiabatic temperature of combustion and the volume of flue gas produced per energy unit. Wet biomass fuels need a longer residence time for drying before gasification and charcoal combustion take place, which means bigger combustion chambers are required. Consequently, the knowledge of these parameters is necessary to adjust the temperature control system of the furnace properly (to avoid slagging) and to design the volume and the geometry of the furnace in a

Table 2.3 *Moisture content, gross calorific value, net calorific value, bulk density and energy density of biomass fuels*

	Moisture content [wt% w.b.]	GCV [MJ/kg d.b.]	NCV [MJ/kg w.b.]	Bulk density [kg w.b./m³]	Energy density [MJ/m³]
Wood pellets	10.0	19.8	16.4	600	9840
Woodchips – hardwood – pre-dried	30.0	19.8	12.2	320	3900
Woodchips – hardwood	50.0	19.8	8.0	450	3600
Woodchips – softwood – pre-dried	30.0	19.8	12.2	250	3050
Woodchips – softwood	50.0	19.8	8.0	350	2800
Grass – high-pressure bales	18.0	18.4	13.7	200	2740
Bark	50.0	20.2	8.2	320	2620
Triticale (cereals) – high-pressure bales	15.0	18.7	14.5	175	2540
Sawdust	50.0	19.8	8.0	240	1920
Straw (winter wheat) – high-pressure bales	15.0	18.7	14.5	120	1740
Olive residues (from 2-phase production)	63.0	21.5	6.1	1130	6890
Olive residues (from 3-phase production)	53.0	22.6	8.5	650	5530

Abbreviations: GCV = gross calorific value; NCV = net calorific value; d.b. = dry basis; w.b. = wet basis.

Explanations: Gross calorific values according to [27–29], moisture content of sawmill side-products and fresh woodchips based on Austrian measurements [30], moisture content of pre-dried biomass fuels related to natural air drying over a 9-month period [31], moisture content of straw, grass and cereals according to average values at harvest [32], bulk densities related to Austrian measurements [30, 33], measurements related to olive residues based on Greek samples performed by BIOS BIOENERGIESYSTEME GmbH, Graz, Austria.

way that ensures a sufficient residence time of the flue gas in the hot combustion chambers for a complete combustion. Also, the efficiency of the combustion system (heat output of the boiler/energy input by the fuel (NCV)) decreases as the moisture content of the fuel increases. This effect can be partly compensated for by installing an appropriate heat recovery unit (e.g. flue gas condensation unit).

The GCV of biomass fuels usually varies between 18 and 22MJ/kg (d.b.). The lower values apply to herbaceous fuels, the higher ones to fresh wood fuels and bark. Due to the residual oil content in olive residues, their GCV is even higher.

2.3.2.2 Concentrations and relevance of C, H, O and volatiles

Table 2.4 contains guiding values for the average amounts of C, H, O and the amount of volatiles in biomass fuels. C, H and O are the main components of biomass fuels. C and H are oxidized during combustion by exothermic reactions (formation of CO_2 and H_2O). The organically bound O released through the thermal decomposition of biomass fuels covers a part of the overall O needed for the combustion reactions taking place; the rest is supplied by air injection. C in biomass fuels is present in partly oxidized forms, which explains the low GCV of biomass fuels, in comparison to coal. The C concentrations of wood fuels (including bark) are higher than those of herbaceous biomass fuels, which explain the slightly higher GCV of wood fuels (see Table 2.4). The C concentrations of olive residues are higher than those of wood fuels, which also results in a higher GCV.

Table 2.4 C, H and O concentrations as well as amounts of volatile matter in biomass fuels

Fuel type	C wt% (d.b.)	H wt% (d.b.)	O wt% (d.b.)	volatiles wt% (d.b.)
Woodchips (spruce, beech, poplar, willow)	47.1–51.6	6.1–6.3	38.0–45.2	76.0–86.0
Bark (coniferous trees)	48.8–52.5	4.6–6.1	38.7–42.4	69.6–77.2
Straw (rye, wheat, triticale)	43.2–48.1	5.0–6.0	36.0–48.2	70.0–81.0
Miscanthus	46.7–50.7	4.4–6.2	41.7–43.5	77.6–84.0
Olive residues	51.0–54.9	6.6–7.2	34.1–38.0	n.m.

Abbreviations: d.b. = dry basis; n.m. = not measured.

Source: Data from: [29, 34–38], measurements related to olive residues based on Greek samples performed by BIOS BIOENERGIESYSTEME GmbH, Graz, Austria.

The amount of volatile matter in biomass fuels is higher than in coals, and usually varies between 70 and 86wt% (d.b.). As a result of this high amount of volatiles, the major part of a biomass fuel is vaporized before homogeneous gas phase combustion reactions take place; the remaining char then undergoes heterogeneous combustion reactions. Therefore, the amount of volatile matter strongly influences the thermal decomposition and combustion behaviour of solid fuels.

2.3.2.3 Concentrations and relevance of N, S and Cl

Table 2.5 contains guiding values for the average amounts of N, S and Cl in biomass fuels.

Table 2.5 N, S and Cl concentrations in biomass fuels

	Nitrogen (N) mg/kg (d.b.)	Sulphur (S) mg/kg (d.b.)	Chlorine (Cl) mg/kg (d.b.)
Woodchips (spruce)	900–1700	70–1000	50–60
Woodchips (poplar, willow)	1000–9600	300–1200	100
Bark (spruce)	1000–5000	100–2000	100–370
Straw (winter wheat)	3000–5000	500–1100	1000–7000
Miscanthus	4000–17,000	200–2000	500–4000
Triticale (cereals)	6000–14,000	1000–1200	1000–3000
Hay	10,000–24,000	2000–6000	2500–20,000
Needles (spruce)	11,000–17,000	–	–
Grass	4000–36,000	800–7000	2600–20,000
Waste wood	1000–39,000	300–2000	300–4000
Olive residues	7700–19,400	920–1200	1000–3300

Explanations: d.b. = dry basis.

Source: Data from: [27, 28, 39–43], measurements related to olive residues based on Greek samples performed by BIOS BIOENERGIESYSTEME GmbH, Graz, Austria; waste wood samples represent the quality classes A I and A II according to the German classification guideline.

Research has shown that NO_x formation during biomass combustion processes at temperatures between 800 and 1100°C mainly results from the fuel-bound N [44–46] (see also Section 9.2.1.1). Measurements in Austrian and Swiss furnaces (grate firings and underfeed stokers) show a logarithmic dependence between the concentration of nitrogen oxides emitted and the concentration of N in the biomass fuel used. Swedish research results also indicate a clear dependence of the NO_x formation on the N content in the fuel in CFB furnaces, although the overall NO_x emissions produced are considerably lower than in fixed bed combustion systems – between 30–100mg/Nm^3 (dry flue gas, 11 vol% O_2) for woodchips with 0.15–0.22wt% N (d.b.) as fuel – [47]. Furthermore, the results achieved in various European countries clearly show that the air supply, the geometry of the furnace and the type of furnace are major influencing variables for NO_x formation [48, 49]. In order to reduce NO_x formation, primary measures have been comprehensively investigated for grate firings, BFB and CFB furnaces. The results show that primary and secondary air should be injected in geometrically well separated combustion chambers (grate firings, underfeed stokers) or furnace sections (CFB, BFB). Furthermore, the air fans should be frequency-controlled to be able to ensure a primary air ratio of $\lambda = 0.6$–0.8 in the furnace. Moreover, the excess oxygen amount, determined by the secondary air, should be as low as possible while complete combustion remains guaranteed [47, 50]. Flue gas recirculation can enhance the mixing and also contribute to a well-defined temperature in the primary combustion zone. Both of these points support the efficiency of primary measures.

If the targeted NO_x reduction cannot be achieved using primary measures alone, they can be combined with secondary measures. These are (a) selective catalytic reduction (SCR) by injecting a reducing agent over a catalyst (ammonia between 220–270°C or urea between 400–450°C), and (b) selective non-catalytic reduction (SNCR) by injecting ammonia or urea in a separate reduction chamber at 850–950°C [51, 52]. For the normal operation of chemically untreated biomass fuels, secondary measures are usually not necessary [51, 53, 54] (see also Section 9.2).

Cl vaporizes almost completely during combustion, forming HCl, Cl_2 and alkali chlorides (see also Section 9.2.1.1). With decreasing flue gas temperature, alkali and alkaline earth chlorides will condense in the boiler section on fly-ash particles or on the heat exchanger surfaces. Subsequently, part of the Cl will be bound in the fly ash while the rest will be emitted as HCl in the flue gas. Measurements and calculations of material balances for grate firings showed that 40–85 per cent of the Cl input by the biomass fuel is bound in the ash, depending on the concentration of alkali and alkaline earth metals in the fuel and on the dust precipitation technology used and its efficiency [27, 32].

The importance of Cl results from the HCl emissions and their influence on the formation of polychlorinated dibenzo-p-dioxins and dibenzofurans (PCDD/F) on the one hand and on the corrosive effects of this element and its compounds on the other hand [27, 28, 55].

PCDD/F formation takes place in a heterogeneous reaction on the surface of fly-ash particles in the presence of C, Cl and O at temperatures between 250–500°C (see also Section 9.2.1.2). Consequently, a low amount of fly-ash particles in the flue gas, a combustion that is as complete as possible, low amounts of excess oxygen and low concentrations of Cl in the fuel reduce PCDD/F formation. A secondary means of reducing the PCDD/F emissions with the flue gas is an efficient dust precipitation technology at low temperatures (< 200°C) as the largest part of PCDD/F is normally bound on the surface of the fly-ash particles due to their formation process (about 80 per cent) [27, 55–57].

S forms the gaseous compounds SO_2, SO_3 and alkali sulphates during the combustion process (see also Section 9.2.1.1). As a result, the larger part of S turns to the vapour phase. In the boiler where the flue gas is rapidly cooled, the sulphates condense on the fly-ash particles or at the tube surfaces. Furthermore, SO_2 can be bound to the fly ash by sulphation reactions. Test runs and evaluations of material balances have shown that 40–90 per cent of the total S input by the biomass fuel is bound in the ash; the rest is emitted with the flue gas as SO_2 and, to a minor extent, as SO_3. The efficiency of S fixation in the ash depends on the concentration of alkaline earths (especially Ca) in the ash as well as on the efficiency and technology used for dust precipitation. The importance of S does not primarily result from the SO_2 emissions but from the role S can play in corrosion processes (see also Chapter 8).

2.3.2.4 Concentrations and relevance of the ash content as well as of major ash-forming elements and trace elements

The ash content of different biomass fuels varies in a broad range from around 0.5wt% (d.b.) for some kinds of clean woody biomass up to 12.0wt% (d.b.) for some straw and cereal assortments (see Table 2.6), and even more if the fuel is contaminated with mineral impurities. For woody biomass, the lower values indicated in Table 2.6 are valid for woodchips and sawdust from soft wood, the higher values for hard wood.

Table 2.6 *Fuel-specific ash content of biomass fuels*

Biomass fuel used	Ash content
Bark	5.0–8.0
Woodchips with bark (forest)	1.0–2.5
Woodchips without bark (industrial)	0.8–1.4
Sawdust	0.5–1.1
Waste wood	3.0–12.0
Straw and cereals	4.0–12.0
Miscanthus	2.0–8.0
Olive residues	2.0–4.0

Explanations: Ash content in wt% (d.b.), ash content measurement according to ISO 1171–1981 at 550°C.
Source: Data from [51]

The ash content of the fuel influences the kind of de-ashing and also the combustion technology applied. Transport and storage of the ash produced also depends on the ash content of the fuel. Fuels with low ash content are better suited for thermal utilization than fuels with high ash content, as lower amounts of ash simplify de-ashing, ash transport and storage as well as utilization and ash disposal. Higher ash contents in the fuel usually lead to higher dust emissions and have an influence on the heat exchanger design, the heat exchanger cleaning system and the dust precipitation technology.

Si, Ca, Mg, K, Na and P are the major ash-forming elements occurring in biomass fuels. Their concentration ranges are shown in Table 2.7.

Table 2.7 *Concentration ranges of major ash-forming elements in biomass fuels and ashes*

Element	Woodchips (spruce)	Bark (spruce)	Straw (wheat, rye, barley)	Cereals (triticale)	Olive residues
[mg/kg (fuel d.b.)]					
Si	440–2900	2000–11,000	4200–27,000	6800–12,000	80–1800
Ca	2900–7000	7700–18,000	440–7000	1400–2800	1500–3400
Mg	310–800	960–2400	100–3200	1100–1200	320–710
K	910–1,500	1500–3600	320–21,000	6000–8900	7500–16,000
Na	20–110	71–530	56–4800	19–78	74–440
P	97–340	380–670	110–2900	2200–3400	560–1200
[wt% (ash d.b.)]					
Si	4.0–11.0	7.0–17.0	16.0–30.0	16.0–26.0	0.3–4.9
Ca	26.0–38.0	24.0–36.0	4.5–8.0	3.0–7.0	5.1–13.5
Mg	2.2–3.6	2.4–5.6	1.1–2.7	1.2–2.6	1.3–1.8
K	4.9–6.3	5.0–9.9	10.0–16.0	11.0–18.0	30.5–42.1
Na	0.3–0.5	0.5–0.7	0.2–1.0	0.2–0.5	0.2–1.1
P	0.8–1.9	1.0–1.9	0.2–6.7	4.5–6.8	2.3–3.1

Explanations: d.b. = dry base; values related to laboratory ash determined at 550°C.
Source: Data from: **[28, 39, 41, 43, 56]**, measurements related to olive residues based on Greek samples performed by BIOS BIOENERGIESYSTEME GmbH, Graz, Austria.

K, P and Mg are relevant plant nutrients. Ca is a relevant liming agent. These elements are therefore of importance in the utilization of ashes as fertilizers (see Section 9.3). This is of special relevance if chemically untreated biomass fuels are used.

Ca and Mg normally increase the melting temperature of ashes, K and Na decrease it. Si in combination with K and Na can lead to the formation of low-melting silicates in fly-ash particles **[38]**. These processes are, on the one hand, important to avoid ash sintering and ash melting on the grate or in the bed of BFB and CFB plants and, on the other hand, to avoid fly-ash slagging, causing deposits on the furnace walls or on the heat exchanger surfaces. Table 2.8 gives an overview of the ash-melting behaviour of various biomass fuels, indicating that ashes of straw, cereals and grass, which contain low concentrations of Ca and high concentrations of K, start to sinter and melt at significantly lower temperatures than wood fuels. This fact has to be considered in selecting the necessary temperature control equipment for the furnace (see also Section 8.4).

Furthermore, in combination with Cl and S, K and Na play a major role in corrosion mechanisms. These elements partly evaporate during the combustion process by forming alkali chlorides, which condense on the heat exchanger surfaces and react with the flue gas by forming sulphates and releasing chlorine. It has been shown that chlorine has a catalytic function, which leads to active oxidation of the tube material in the heat exchanger even when the wall temperature of the tubes is low (100–150°C) **[58–61]**. Research results show that biomass fuels, especially those with a molar S:Cl ratio below 2, cause corrosion problems because the formation of chlorides becomes significant **[62]**. Furthermore, the volatilization and subsequent condensation of volatile metals leads to the formation of sub-micron fly-ash particles (aerosols), which are difficult to precipitate in dust filters, form

Table 2.8 *Ash-melting behaviour of biomass fuels*

	Sintering temp. [°C]	Softening temp. [°C]	Hemisphere temp. [°C]	Melting temp. [°C]	Number of samples
Wood (beech, Austria)	1140	1260	1310	1340	1
Wood (spruce, Austria)	1110–1340	1410–1640	1630– >1700	>1700	3
Bark (spruce, Austria)	1250–1390	1320–1680	1340– >1700	1410– >1700	3
Bark + mineral impurities (spruce, Austria)	1020	1100	>1700	>1,700	1
Miscanthus (Austria)	820–980	820–1160	960–1290	1050–1270	27
Miscanthus (Switzerland)	–	980	1210	1320	n.s.
Straw (winter wheat, Austria)	800–860	860–900	1040–1130	1080–1120	3
Straw (winter wheat, Switzerland)	–	910	1150	1290	n.s.
Cereals (winter wheat, Austria)	970–1010	1020	1120–1170	1180–1220	3
Grass (Austria)	890–980	960–1020	1040–1100	1140–1170	3
Grass (Germany)	830–1130	950–1230	1030–1280	1100–1330	9
Grass (Switzerland)	–	960	1040	1120	n.s.

Explanations: n.s. = not stated; measurements taken according to DIN 51730.

Sources: [29, 39, 63].

Table 2.9 *Concentration ranges of minor ash-forming elements in biomass fuels*

Element [mg/kg (d.b.)]	Woodchips (spruce) [mg/kg (d.b.)]	Bark (spruce) [mg/kg (d.b.)]	Straw (wheat, rye, barley) [mg/kg (d.b.)]	Cereals (triticale) [mg/kg (d.b.)]	Olive residues [mg/kg (d.b.)]
Fe	64–340	280–1,200	42–860	37–150	200–4,300
Al	79–580	330–1,500	11–810	42–93	74–770
Mn	63–900	430–1,300	14–44	27–31	9–49
Cu	0.3–4.1	1.5–8.0	1.1–4.2	2.6–3.9	n.m.
Zn	7–90	90–200	11–57	10–25	6–15
Co	0.1–0.7	0.7–2	0.4	0.3–0.5	n.m.
Mo	0.8–1.5	0.1–5	0.9–1.1	0.5–1.5	n.m.
As	0–1.5	0.2–5	1.6	0.6	n.m.
Ni	1.7–11	1.6–13	0.7–2.1	0.7–1.5	n.m.
Cr	1.6–17	1.6–14	1.1–4.1	0.4–2.5	n.m.
Pb	0.3–2.7	0.9–4.4	0.1–3	0.2–0.7	n.m.
Cd	0.06–0.4	0.2–0.9	0.03–0.22	0.04–0.1	n.m.
V	0.6–1.4	0.6–5	0.2–3	0.7	n.m.
Hg	0.01–0.17	0.01–0.17	0.01	0–0.02	n.m.

Explanations: d.b. = dry basis; n.m. = not measured.

Source: [43], measurements related to olive residues based on Greek samples performed by BIOS BIOENERGIESYSTEME GmbH, Graz, Austria.

deposit layers on boiler tubes and cause ecological and health risks. Consequently, the smaller the amounts of K and Na in the fuel, the better.

Minor ash-forming elements occurring in various biomass fuels are Fe, Al, Mn and heavy metals. Typical concentration ranges of Fe, Al and Mn as well as of some heavy metals in various biomass fuels are shown in Table 2.9.

After the combustion of biomass fuels the majority of the heavy metals are bound in the ash. To a smaller extent the heavy metals remain gaseous or as aerosols in the flue gas. For volatile heavy metals such as Hg this amount can be comparatively high.

The heavy-metal concentrations in biomass ashes are of considerable importance for a sustainable ash utilization. The ecologically relevant elements are primarily Cd and, to a smaller extent, Zn if only chemically untreated biomass fuels are considered. By means of primary measures it is possible to considerably reduce the concentrations of these heavy metals in the major amount of ash and to upgrade them in a small ash fraction that has to be separately collected and disposed of or industrially utilized [32, 56, 64]. This aspect is discussed in more detail in Section 9.3.2. Straw, cereals, grass and olive residue ashes contain significantly lower amounts of heavy metals than wood and bark ashes. This can be explained by the long rotation period of wood, which enhances accumulation, the higher deposition rates in forests, and the lower pH value of forest soils that increases the solubility of most heavy metals.

2.3.2.5 Guiding values and guiding ranges for elements in biomass fuels and biomass ashes for unproblematic thermal utilization

Based on the discussions in the previous sections, guiding values and guiding ranges for elements in biomass fuels and biomass ashes for unproblematic thermal utilization can be determined. Table 2.10 sums up guiding values and guiding ranges for concentrations of combustion-relevant elements in biomass fuels and ashes. In the current situation, biomass fuels within the given guiding concentration ranges can be used in modern combustion plants without problems. For fuels with compositions outside the given ranges, additional technological requirements should be considered. Table 2.10 must not be regarded as complete but has to be continuously adjusted to the latest state of research and development. It should help engineering companies and operators of plants to detect possible problems as early as possible.

However, due to complex interactions between K, Cl, S, Si and Ca, each element cannot always be evaluated individually. In addition, the limiting factors are highly dependent on the type of combustion plant, operational conditions, emission limit values, steam parameters, national rules for ash recycling, and so on. For example, a content of Cl below 0.1wt% (d.b.) is not sufficient to avoid superheater corrosion problems at high steam conditions. Influences and interactions of different element concentrations are also discussed in Chapter 8. Consequently, although the guiding values shown in Table 2.10 are derived from long-term experiences with different biomass combustion plants, the values cannot be applied in any case for other biomass combustion plants. They should only be used as a first guideline. In addition, application-specific investigations are recommended and needed.

Table 2.10 *Guiding values and guiding ranges for elements in biomass fuels and ashes for unproblematic thermal utilization*

Element	Guiding concentration in the fuel wt% (d.b.)	Limiting parameter	Fuels affected outside guiding ranges	Technological methods for reducing to guiding ranges
N	< 0.6	NO_x emissions	Straw, cereals, grass, olive residues	Primary measures (air staging, reduction zone)
	< 2.5	NO_x emissions	Waste wood, fibre boards	Secondary measures (SNCR or SCR process)
Cl	< 0.1	Corrosion	Straw, cereals, grass, waste wood, olive residues	– fuel leaching – automatic heat exchanger cleaning – coating of boiler tubes – appropriate material selection
	<0.1	HCl emissions	Straw, cereals, grass, waste wood, olive residues	– dry sorption – scrubbers – fuel leaching
	< 0.3	PCDD/F emissions	Straw, cereals, waste wood	– sorption with activated carbon
S	< 0.1	Corrosion	Straw, cereals, grass, olive residues	See Cl
	< 0.2	SO_x emissions	Grass, hay, waste wood	See HCl emissions
Ca	15–35	Ash-melting point	Straw, cereals, grass, olive residues	Temperature control on the grate and in the furnace
K	< 7.0	Ash-melting point, depositions, corrosion	Straw, cereals, grass, olive residues	Against corrosion: see Cl
	–	Aerosol formation	Straw, cereals, grass, olive residues	Efficient dust precipitation, fuel leaching
Zn	< 0.08	Ash recycling, ash utilization	Bark, woodchips, sawdust, waste wood	Fractioned heavy metal separation, ash treatment
	–	Particulate emissions	Bark, woodchips, sawdust, waste wood	Efficient dust precipitation, treatment of condensates
Cd	< 0.0005	Ash recycling, ash utilization	Bark, woodchips, sawdust, waste wood	See Zn
	–	Particulate emissions	Bark, woodchips, sawdust, waste wood	See Zn

Explanations: Guiding values for ashes related to the biomass fuel ashed according to ISO 1171–1981 at 550°C; analytical method recommended for ash analysis: pressurized acid digestion and inductively coupled plasma mass spectometry (ICP) or flame atomic absorption spectometry (AAS) detection; N and S analysis recommended: combustion/gas chromatographic detection; Cl analysis recommended: bomb combustion/ion chromatographic detection. d.b. = dry basis.

2.3.3 Databases on biomass fuel characteristics

Since biomass fuel resources are extremely heterogeneous and numerous, a sound prediction of the respective fuel characteristics and the results of quality manipulations is essential for planning and control purposes, particularly in the case of highly specialized energy conversion processes. For this purpose, the Task on Biomass Combustion and Co-Firing within the IEA Bioenergy Agreement has prepared an extensive database on compositions of samples from biomass fuels and ashes, collected from actual installations [43]. This database is available on the internet at www.ieabcc.nl and currently contains 1560 different biomass and ash samples.

Two other databases on biomass characteristics, also accessible via the internet, are briefly described below.

The **BIOBIB** database for biomass fuels was developed by the Institute of Chemical Engineering, Fuel and Environmental Technology at the University of Technology, Vienna, Austria [65].

BIOBIB includes data concerning the proximate and ultimate analysis of biomass fuels, also covering several minor and trace elements as well as data about ash content and ash-melting behaviour. BIOBIB covers not only analyses of various types of wood, straw and energy crops, but also waste-wood samples and biomass wastes of various biomass-treating industries (e.g. from the wood-processing industry, the pulp and paper industry, and the food industry). Currently, the database contains 647 different biomass fuel samples of many different species from mainly European plants. The database is accessible through the internet: www.vt.tuwien.ac.at.

The **Phyllis** database is designed and maintained by the Netherlands Energy Research Foundation (ECN), research area Fuels, Conversion & Environment [66]. Phyllis is a database containing information on the composition of biomass and waste fuels. In addition to data from literature and from the BIOBIB database, it contains data on fuels and materials analysed at ECN. Phyllis is also available on the internet at: www.ecn.nl/Phyllis.

In Phyllis, the various materials are divided into groups and subgroups. This division is based on a combination of plant physiological and practical considerations. Each data record has a unique ID number and shows information (if available) on:

- type of material (group, e.g. untreated wood, treated wood, straw, etc.);
- subgroup (e.g. for untreated wood: beech, birch, spruce, etc.);
- ultimate analysis: carbon, hydrogen, oxygen, nitrogen, sulphur, chlorine, fluorine, bromine;
- proximate analysis: ash content, moisture content, volatile matter content, fixed carbon content;
- calorific value;
- (alkali) metal content;
- composition of the ash;
- remarks (specific information).

For each data record, the source (reference) is indicated. Data on fossil fuels, including peat, are added as a reference in a separate group. Phyllis contains about 2275 data records at present and is updated and extended regularly.

2.3.4 Standardization of biomass fuels

Standardization makes material or immaterial objects uniform. This is done by a community of interests for the benefit of the general public (DIN 820).

With regard to the standardization of biomass fuels, the main objectives are to overcome market barriers caused by inhomogeneous fuel properties in order to help create a market in which solid biomass fuels can be traded among producers (farmers, foresters, fuel companies) and users (utilities, district heating companies, industry, end consumers). An overview of existing national standards and guidelines for solid biomass fuels is listed below.

2.3.4.1 Standardization of solid biomass fuels in Europe

The development of standards for solid biomass fuels on a European level started in 1998 with a workshop on solid biomass fuels established by the European Committee for Standardization (CEN) under a programming mandate from the European Commission. Beside the member states of the European Union (EU) (the standardization is being performed via two EU research and development projects), the US and Switzerland were also involved via the former IEA Bioenergy Task 28 'Standardisation of Solid Biofuels'. In the last years the Technical Committee TC 335 'Solid Biofuels' prepared 30 standards in the following five fields [67, 68]:

1. terminology, definitions and description;
2. fuel specifications, classes and quality assurance;
3. sampling and sample reduction;
4. physical and mechanical test methods; and
5. chemical test methods.

The following biomass fuels are covered by these standards [69]:

- products from agriculture and forestry;
- vegetable waste from agriculture and forestry;
- vegetable waste from the food processing industry;
- wood waste, with the exception of wood waste which may contain halogenated organic compounds or heavy metals as a result of treatment with wood preservatives or coating, and which includes in particular such wood waste originated from construction and demolition waste;
- fibrous vegetable waste from virgin pulp production and from production of paper from pulp; and
- cork waste.

As of March 2006, 25 standards were already published, the remaining five standards were then expected to be published during the course of 2006 [70]. A complete list of all standards can be found in Annex 3.

In addition to the standardization activities of the Technical Committee TC 335 related to 'Solid Biofuels', the Technical Committee TC 343 is currently preparing standards for 'Solid Recovered Fuels', where waste-wood fractions, which are excluded in the previously mentioned standards, will also be covered.

2.3.4.2 List of existing European and national standards or guidelines for solid biomass fuels and solid biomass fuel analysis

Annex 3 provides standards for solid biomass fuels and biomass fuel analysis in European countries. This overview does not claim to be complete and should be updated at regular intervals in order to keep in step with national developments.

Research and development concerning appropriate analysis methods for biomass fuels and biomass ashes is still ongoing within the EU. Several international round-robin tests have already been completed [71, 72] and reports, including recommendations for the appropriate selection and application of methods, have been given [73, 74].

2.4 References

1. GRØNLI, M. A. (1996) Theoretical and Experimental Study of the Thermal Degradiation of Biomass, PhD Thesis, Norwegian University of Science and Technology, ITEV report 96:03
2. BRIDGWATER, A.V. (1991) *An Overview of Thermochemical Biomass Conversion Technologies*, Proceedings of Wood, Fuel for Thought Conference, Bristol, United Kingdom, 23–25 October
3. BAXTER, PROF. L. (2000) IEA Bioenergy Task 19 meeting, Goldcoast, Australia
4. SKREIBERG, Ø. (1997) Theoretical and Experimental Studies on Emissions from Wood Combustion, PhD thesis, Norwegian University of Science and Technology, ITEV report 97:03
5. GAUR, S. and REED, T. B., (1995) *An Atlas of Thermal Data for Biomass and Other Fuels*, NREL/TB-433-7965, UC Category:1310, DE95009212
6. WARNATZ, J., MAAS, U. and DIBBLE, R. W. (1999) *Combustion – Physical and Chemical Fundamentals, Modelling and Simulation, Experiments, Pollutant Formation*, 2nd ed, Springer
7. TURNS, S. R. (2000) *An Introduction to Combustion – Concepts and Applications*, 2nd ed, McGraw-Hill, Inc.
8. BORMAN, G. L. and RAGLAND, K. W. (1998) *Combustion Engineering*, McGraw-Hill, Inc.
9. KUO, K. K. (1986) *Principles of Combustion*, John Wiley & Sons, Inc.
10. GLASSMAN, I. (1996) *Combustion*, 3rd ed, Academic Press
11. KARLSVIK, E., HUSTAD, J. E. and SØNJU, O. K. (1993) 'Emissions from wood-stoves and fireplaces', in *Advances in Thermochemical Biomass Conversion*, Blackie Academic & Professional, pp690–707
12. GRI-Mech 3.0 thermodynamic database: www.me.berkeley.edu/gri_mech/version30/files30/thermo30.dat
13. Chemkin 4.0.2 thermodynamic database: www.reactiondesign.com/lobby/open/index.html
14. SKREIBERG, Ø., HUSTAD, J. E. and KARLSVIK, E. (1997) 'Empirical NO_x-modelling and experimental results from wood-stove combustion', in *Developments in Thermochemical Biomass Conversion*, Blackie Academic & Professional, pp1462–1476
15. KARLSVIK, E., HUSTAD, J. E., SKREIBERG, Ø. and SØNJU, O. K. (1998) 'Greenhouse gas and NO_x emissions from wood-stoves', in *Energy, Combustion and the Environment*, vol 2, no A, Gordon and Breach, pp539–550
16. VAN BERKEL, A. I., BREM, G. and VAN LOO, S., (1997) *Dynamische Modellering van een CFG-STEG voor de Vergassing van Biomassa*, fase 1: inventarisatie en programma van eisen, TNO-MEP, October 1997
17. OBERNBERGER, I., DAHL, J. and BRUNNER T. (1999) 'Formation, composition and particle size distribution of fly ashes from biomass combustion plants', in *Proceedings of the 4th Biomass Conference of the Americas*, September 1999, Oakland, CA, Elsevier Science Ltd., Oxford, UK, pp1377–1385
18. CHRISTENSEN, K. A., (1995) The formation of submicron particles from the combustion of straw, PhD thesis, Department of Chemical Engineering, Technical University of Denmark
19. OBERNBERGER, I. and BIEDERMANN, F. (1998) 'Fractionated heavy metal separation in Austrian biomass grate-fired combustion plants approach, experiences, results', in *Ashes and*

Particulate Emissions from Biomass Combustion, volume 3 of Thermal Biomass Utilization series, BIOS, Graz, Austria, dbv-Verlag

20 KAUPPINEN, E., LIND, T., KURKELA, J., LATVA-SOMPPI, J. and JOKINIEMI, J. (1998) 'Ash particle formation mechanisms during pulverised and fluidised bed combustion of solid fuels', in *Ashes and Particulate Emissions from Biomass Combustion*, volume 3 of Thermal Biomass Utilization series, , BIOS, Graz, Austria, dbv-Verlag

21 DAHL, J. (1999) Chemistry and behavior of environmentally relevant heavy metals in biomass combustion, PhD dissertation, Department of Chemical Engineering, Graz University of Technology, Austria

22 LIND, T., VALMARI, T., KAUPPINEN, E., MAENHAUT, W. and HUGGINS, F. (1998) 'Ash formation and heavy metal transformations during fluidised bed combustion of biomass', in *Ashes and Particulate Emissions from Biomass Combustion*, volume 3 of Thermal Biomass Utilization series, , BIOS, Graz, Austria, dbv-Verlag

23 BRUNNER, T., OBERNBERGER, I., BROUWERS J. J. H. and PREVEDEN, Z. (1998) 'Efficient and economic dust separation from flue gas by the rotational particle separator as an innovative technology for biomass combustion and gasification plants', in *Proceedings of the 10th European Bioenergy Conference*, June 1998, Würzburg, Germany, CARMEN, Rimpar, Germany

24 OBERNBERGER, I., BIEDERMANN, F., WIDMANN, W. and RIEDL, R. (1996) 'Concentrations of inorganic elements in biomass fuels and recovery in the different ash fractions', *Biomass and Bioenergy*, vol 12, no 3, pp221–224

25 OBERNBERGER, I., BIEDERMANN, F. and KOHLBACH, W., (1995) FRACTIO – *Fraktionierte Schwermetallabscheidung in Biomasseheizwerken*, annual report, Institute of Chemical Engineering, Graz University of Technology, Austria

26 HARTMANN, H. (1998) 'Influences on the quality of solid ciofuels – Causes for variations and measures for improvement', in *Proceedings of the 10th European Bioenergy Conference*, June 1998, Würzburg, Germany, CARMEN, Rimpar, Germany

27 OBERNBERGER, I., WIDMANN W., WURST, F. and WÖRGETTER M. (1995) *Beurteilung der Umweltverträglichkeit des Einsatzes von Einjahresganzpflanzen und Stroh zur Fernwärmeerzeugung*, annual report; Institute of Chemical Engineering, Graz University of Technology, Austria

28 SANDER, B. (1997) 'Properties of Danish biofuels and the requirements for power production', *Biomass and Bioenergy*, vol 12, no 3, pp177–183

29 SCHMIDT, A., ZSCHETZSCHE, A. and HANTSCH-LINHART, W. (1994) *Analysen von biogenen Brennstoffen, Endbericht zum gleichnamigen Forschungsprojekt, Bundesministerium für Wissenschaft*, Forschung und Kunst (ed.), Vienna, Austria

30 OBERNBERGER, I., NARODOSLAWSKY, M. and MOSER, F. (1994) 'Biomassefernheizwerke in Österreich: Entwicklung, Stoff- und Energieflüsse, Umweltverträglichkeit', *Österreichische Ingenieur- und Architektenzeitschrift*, Heft 3

31 SCHUSTER, K., (1993) 'Überbetriebliche Aufbringung und Vermarktung von Brennhackschnitzeln', *Landtechnische Schriftenreihe* 179, Österr. Kuratorium für Landtechnik, Vienna, Austria

32 OBERNBERGER, I., BIEDERMANN, F. and KOHLBACH, W. (1995) *FRACTIO – Fraktionierte Schwermetallabscheidung in Biomasseheizwerken*, annual report for a research project funded by the ITF and the Bund-Bundesländerkooperation, Institute of Chemical Engineering, Graz University of Technology, Austria

33 LAUCHER, A. (1995) 'Biomasse-Ortszentralheizung – Technische und betriebswirtschaftliche Überlegungen', in *Schriftenreihe 'Nachwachsende Rohstoffe'*, Band 5, Landwirt-schaftsverlag Münster, Germany

34 KALTSCHMITT, M. H. and HARTMANN, H. (eds.) (2001) *Energie aus Biomasse – Grundlagen, Techniken und Verfahren*, Springer Verlag, Berlin-Heidelberg-New York

35 WILEN, C., MOILANEN, A. and KURKELA, E. (1996) *Biomass Feedstock Analysis*, VTT Publications 282, VTT Technical Research Centre of Finland, Espoo

36 NIELSON, L. and SCHMIDT, E. R. (1995) 'Feedstock Characterisation in Denmark', Danish Report for IEA BA Task XII Activity 4.1 'Feedstock Quality and Preparation', in *Minutes of the Activity Workshop in Portland, Oregon*, August 1995, Elsamprojekt A/S, Fredericia, Denmark

37 GAUR, S. and REED, T. B. (1998) *Thermal Data for Natural Sciences and Synthetic Fuels*, Marcel Dekker AG (ed.), New York
38 MILES, T. R. (1996) *Alkali Deposits Found in Biomass Power Plants*, research report NREL/TP-433-8142 SAND96-8225, volumes I and II, National Renewable Energy Laboratory, Oakridge, US
39 NUSSBAUMER, T. (1993) *Verbrennung und Vergasung von Energiegras und Feldholz*, annual report 1992, Bundesamt für Energiewirtschaft, Bern, Switzerland
40 OBERNBERGER, I. (2005) 'Thermochemical Biomass Conversion', lecture, Department for Mechanical Engineering, Section Process Technology, Technical University Eindhoven, the Netherlands
41 LEWANDOWSKY, I. (1996) 'Einflußmöglichkeiten der Pflanzenproduktion auf die Brennstoffeigenschaften am Beispiel von Gräsern', in *Schriftenreihe 'Nachwachsende Rohstoffe'*, Band 6, Landwirtschaftsverlag Münster, Germany
42 BRUNNER, T., OBERNBERGER, I. and WELLACHER, M. (2005) 'Altholzaufbereitung zur Verbesserung der Brennstoffqualität – Möglichkeiten und Auswirkungen', in *Tagungsband zur internat. Konferenz 'Strom und Wärme aus biogenen Festbrennstoffen'*, June 2005, Salzburg, Austria, VDI-Bericht Nr. 1891, VDI-Verlag GmbH Düsseldorf, Germany
43 OBERNBERGER, I., BIEDERMANN, F. and DAHL, J. (2000) *BioBank, Database on the Chemical Composition of Biomass Fuels, Ashes and Condensates from Flue Gas Condensors from Real-Life Installations*, BIOS BIOENERGIESYSTEME GmbH, Graz and Institute of Chemical Engineering Fundamentals and Plant Engineering, Graz University of Technology, Graz, Austria (available at /www.ieabcc.nl)
44 KELLER, R. (1994) *Primärmaßnahmen zur NOx-Minderung bei der Holzverbrennung mit dem Schwerpunkt der Luftstufung*, research report no. 18 (1994), Laboratorium für Energiesysteme (ed.), ETH Zürich, Switzerland
45 NUSSBAUMER, T., (1989) *Schadstoffbildung bei der Verbrennung von Holz*, Forschungsbericht Nr. 6 (1989), Laboratorium für Energiesysteme, ETH Zürich, Switzerland
46 MARUTZKY, R. (1991) *Erkenntnisse zur Schadstoffbildung bei der Verbrennung von Holz und Spanplatten*, WKI-Bericht Nr. 26 (1991), Wilhelm-Klauditz-Institut, Braunschweig, Germany
47 LECKNER, B. and KARLSSON, M., (1993) 'Gaseous emissions from circulating fluidised bed combustion of wood', *Biomass and Bioenergy*, vol 4, no 5, pp379–389
48 KILPINEN, P. (1992) Kinetic modelling of gas-phase nitrogen reactions in advanced combustion processes, Report 92–7, PhD thesis, Abo Akademy University, Abo, Finland
49 DE SOETE, G. (1983) 'Heterogene Stickstoffreduzierung an festen Partikeln', VDI-Bericht 498, pp171–176, VDI Verlag GmbH, Düsseldorf, Germany
50 SALZMANN, R. and NUSSBAUMER, T. (1995) *Zweistufige Verbrennung mit Reduktionskammer und Brennstoffstufung als Primärmaßnahmen zur Stickoxidminderung bei Holzfeuerungen*, research report, Laboratorium für Energiesysteme, ETH Zürich, Switzerland
51 OBERNBERGER, I. (1997) *Nutzung fester Biomasse in Verbrennungsanlagen unter besonderer Berücksichtigung des Verhaltens aschebildender Elemente*, volume 1 of Thermal Biomass Utilization series, BIOS, Graz, Austria, dbv-Verlag der Technischen Universität Graz, Graz, Austria
52 NUSSBAUMER, T. (1997) 'Primär- und Sekundärmaßnahmen zur NOx-Minderung bei Biomassefeuerungen', VDI Bericht 1319, *Thermische Biomassenutzung – Technik und Realisierung*, VDI Verlag GmbH, Düsseldorf, Germany
53 NUSSBAUMER, T. (1996) 'Primary and secondary measures for the reduction of nitric oxide emissions from biomass combustion', volume 2, *Proceedings of the International Conference 'Developments in Thermochemical Biomass Conversion'*, May 1996, Banff, Canada, Blackie Academic and Professional, London, UK
54 BRUNNER, T. and OBERNBERGER, I., (1996) 'New technologies for NOx reduction and ash utilization in biomass combustion plants – JOULE THERMIE 95 demonstration project', in *Proceedings of the 9th European Bioenergy Conference*, vol 2, Elsevier Science Ltd, Oxford
55 WEBER, R., MOXTER, W., PILZ, M., POSPISCHIL, H. and ROLEDER, G. (1995) *Umweltverträglichkeit des Strohheizwerkes Schkölen – Teil Emissionen, Forschungsprojektbericht*, Thüringer Landesanstalt für Umwelttechnik, Jena, Germany
56 RUCKENBAUER, P., OBERNBERGER, I. and HOLZNER, H. (1996) *Erforschung der*

Verwendungsmöglichkeiten von Aschen aus Hackgut- und Rindenfeuerungen, Final Report, part II research project No. StU 48, Bund-Bundesländerkooperation, Institute of Plant Breeding and Plant Growing, University for Agriculture and Forestry, Vienna, Austria

57 NUSSBAUMER, T. and HASLER, P. (1996) 'Formation and reduction of polychlorinated dioxins and furans in biomass combustion', in *Proceedings of the International Conference 'Developments in Thermochemical Biomass Conversion'*, May 1996, Banff, Canada, volume 2, Blackie Academic and Professional, London

58 RIEDL, R. and OBERNBERGER, I. (1996) 'Corrosion and fouling in boilers of biomass combustion plants', in *Proceedings of the 9th European Bioenergy Conference*, volume 2, Elsevier Science, Oxford

59 REICHEL, H. and SCHIRMER, U. (1989) 'Waste incineration plants in the FRG: Construction, materials, investigation on cases of corrosion', *Werkstoffe und Korrosion*, vol 40, pp135–141

60 SPIEGEL, M. and GRABKE, H., (1995) 'Hochtemperaturkorrosion des 2.25Cr-1Mo-Stahls unter simulierten Müllverbrennungsbedingungen', *Materials and Corrosion*, vol 46, pp121–131

61 REESE, E. and GRABKE, H. (1992) 'Einfluß von Chloriden auf die Oxidation des 21/4 Cr-1 Mo-Stahls', *Werkstoffe und Korrosion*, vol 43, pp547–557

62 SALMENOJA, K. and MÄKELÄ, K. (2000) 'Chlorine-induced superheater corrosion in boilers fired with biomass', in *Proceedings to the 5th European Conference on Industrial Furnaces and Boilers*, Espinho, INFUB, Rio Tinto, Portugal, April 2000

63 OBERNBERGER, I. (1996) *Durchführung und verbrennungstechnische Begutachtung von Biomasseanalysen (Heuproben) als Basis für die Vorplanung des Dampferzeugungsprozesses auf Biomassebasis in Neuburg/Donau, Ergebnisbericht*, BIOS Research, Graz, Austria

64 OBERNBERGER, I., WIDMANN, W., ARICH, A. and RIEDL, R. (1995) 'Concentrations of inorganic elements in biomass fuels and recovery in the different ash fractions', *Biomass and Bioenergy*, vol 12, no 3, pp211–224

65 REISINGER, K. et al, (1996) 'BIOBIB – a database for biofuels', in *Proceedings of the THERMIE – Conference: Renewable Energy Databases*, Harwell, UK

66 VAN DER DRIFT, A. and VAN DOORN, J. (1998) *Phyllis ECN. Operationeel maken – fase 2* (in Dutch). ECN-CX—98-138, Netherlands Energy Research Foundation, Petten, Netherlands

67 LIMBRICK, A. J. (2000) *Draft Work Programme for a CEN Technical Committee for Solid Biofuels*, report to the European Commission, Directorate General 'Energy' and CEN/BT/WG108 'Solid Biofuels', THERMIE Contract STR-2066-98-GB, Green Land Reclamation Ltd (ed.), Berkshire, UK

68 HEIN, M. and KALTSCHMITT, M. (Hrsg.) (2004) 'Standardisation of Solid Biofuels – Status of the ongoing standardisation process and results of the supporting research activities (BioNorm)', in *Proceedings of the International Conference Standardisation of Solid Biofuels*, Leipzig, Germany

69 ALAKANGAS, E. (2004) 'The European pellets standardisation', in *Proceedings of the European Pellets Conference 2004*, O.Ö. Energiesparverband, Linz, Austria

70 CEN (2006) Homepage www.cenorm.be/CENORM, European Committee for Standardization, Brussels, Belgium

71 OBERNBERGER, I., DAHL, J. and ARICH, A. (1997) *Round Robin on Biomass Fuel and Ash Analysis*, European Commission

72 WESTBORG, S. and NIELSON, C. (1994) *Analysis of Straw and Straw Ashes*, report prepared for the IEA Biomass Utilisation Task X, Danish Energy Research Project No. 1323/91-0015, dk-TEKNIK (ed.), Soborg, Denmark

73 HEEMSKERK, G. C. A. M. et al, (1998) *Best Practice List for Biomass Fuel and Ash Analysis*, EWAB-9820, KEMA Power Generation (ed.), Arnhem, April

74 CURVERS, A. and GIGLER, J. K. (1996) *Characterisation of Biomass Fuels*, report ECN-C-96-032, ECN, Petten, Netherlands

3
Biomass Fuel Supply and Pre-treatment

The process of producing energy from biomass is a chain of interlinking stages (fuel supply, energy conversion, energy utilization, residue disposal or recycling), which are mutually dependent. Within this chain, the purpose of 'fuel supply' is to produce a biomass fuel from various biomass resources that meets the requirements of the combustion plant with regard to fuel quality (e.g. moisture content, particle size, ash content) and fuel costs (see Figure 3.1). This chapter provides information on the physical options available for fuel supply, pre-treatment and storage. For thermochemical conversion routes of biomass fuel treatment (such as flash pyrolysis, hydrothermal upgrading, gasification, carbonization, etc.), the reader is referred to Chapter 2 and the other tasks in the IEA Bioenergy Agreement (see www.ieabioenergy.com).

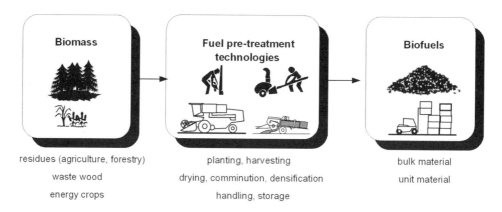

Figure 3.1 *Fuel supply chain for woody biomass*

Source: Diagram from [1], © dbv-Verlag, 1998

Some possible fuel resources relevant to biomass combustion:

- residues from:
  - agriculture (e.g. straw),
  - forestry (e.g. branches, treetops, whole trees from early thinnings, prunings),
  - wood processing industry (e.g. bark, sawdust),
  - food processing industry (e.g. olive stones);
- energy crops (e.g. short rotation coppice, miscanthus, switchgrass); and
- waste wood (e.g. demolition wood, fibreboard residues, railway sleepers).

Figure 3.2 illustrates various paths for wood fuels from the forest to the end-user.

The fuel supply chain consists of a series of sequential steps such as planting, growing, harvesting, comminution, densification, drying, storage, transport and handling. The combination of these steps depends on the kind of biomass as well as on the needs of the combustion plant and determines feedstock form and costs. The aim is to have the whole chain functioning smoothly and cost-effectively with a minimum of losses along the way for a given system. The sequence of steps in the chain can be varied (alternative supply strategies) to meet the needs of a given conversion technology and cost structure [2].

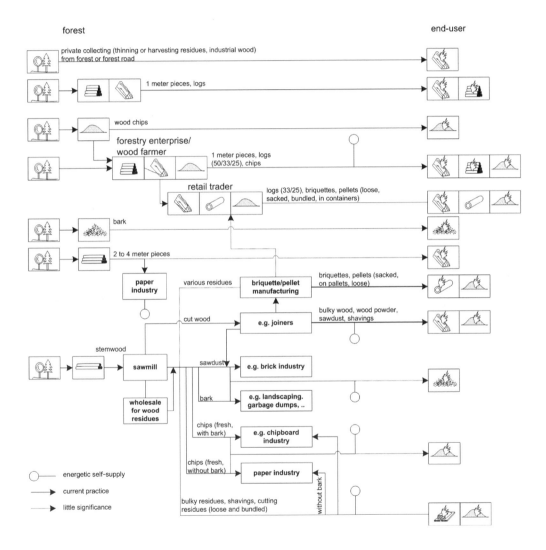

Figure 3.2 *Wood fuel flows from the forest to the end-user*

Ssource: [3], © Landtechnik Weihenstephan, Freising, Germany, 1997.

Within the IEA Bioenergy Agreement, a great amount of information regarding the production of biomass fuels has been made available. In this regard, special attention is given to conventional forestry systems for bioenergy production (Task 31) and short rotation crops for bioenergy production (Task 30). For more information, see www.ieabioenergy.com.

3.1 Quality-related influences along the supply chain of biomass fuels

As shown in Figure 3.3, three phases can be distinguished in the life cycle of biomass fuels. During the growing phase, it is mainly the chemical characteristics of biomass fuels and the desired yield that can be influenced (see Section 3.1.1), while during the supply phase the physical characteristics of biomass fuels are the area of interest (see Section 3.1.2).

The final fuel quality (in terms of its physical and chemical characteristics) is relevant to fuel handling, storage and feeding, as well as to the combustion process itself. Therefore, possible improvements in fuel quality by primary measures (appropriate agricultural engineering) are of increasing interest from a technological as well from an economic point of view.

In order to ensure a fuel quality as required by the respective biomass combustion plant, quality control measures are needed at the point of receipt of the biomass fuel at the plant or quality control has to take place all the way through the production process, following a clearly defined procedure (e.g. in case of pellet production, according to certain standards, see Section 3.1.3).

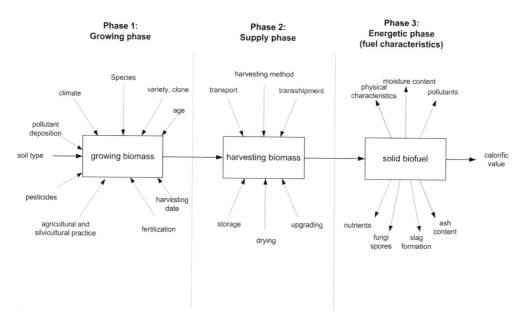

Figure 3.3 *Factors influencing the quality of solid biomass fuels*

Source: [4].

3.1.1 Influences on the fuel quality during the growing phase

The selection of the biomass species is one of the most important determining factors for fuel properties. Due to plant-specific composition and physiological needs (concerning the absorption capacity for nutrients and trace elements), there are essential differences between the chemical compositions of biomass fuels (see Table 2.5). In this context, plant breeding and genetic improvements of energy crops form one possibile way to optimize yield and fuel quality in future (e.g. to grow Salix and Miscanthus species with high yield and low S, Cl and K concentrations).

Fertilizing is another important aspect. For both parameters – quantity and type – of fertilizer, a number of interactions have to be considered. In particular, Cl concentrations in the fuel depend on the kind and amount of nutrient supplied through fertilizers (see Figure 3.4). By the application of chlorine-free fertilizers, the chlorine content in straw can be reduced significantly, without increasing the content of K or S. Also the amount of N supplied with fertilizers will influence the concentration of N in the fuel. Of course, a certain level of key plant nutrients (N, K, P) is essential for proper plant growth and an acceptable yield, but the quantity and type of fertilizer should be carefully considered [5, 6].

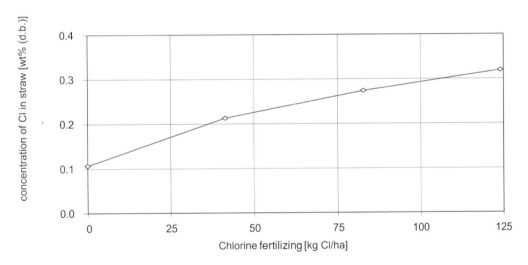

Figure 3.4 *Content of Cl in straw as a function of chlorine supply through fertilizers*

Source: [6].

The climate is especially important for energy crops such as straw or miscanthus as it influences the moisture content as well as the elementary composition. In particular, undesired and easily water-soluble elements such as K and Cl can be leached out by rainfall in the weeks prior to or the days after harvesting (see Figure 3.5). This effect, in combination with the drying of straw during field storage for several days after harvesting, can improve fuel quality significantly.

Danish research on quality improvements of biomass fuels by means of appropriate leaching methods has identified various possible approaches for straw. One approach is leaching of the cut straw after harvesting on the field by rainfall. The disadvantage of this approach is that it is dependent on the weather plus there is the danger of straw degradation. A second approach is to harvest only the corn and leave the straw standing in the field in order to leach it out. The disadvantage of this approach is the need for a second straw harvest. A third approach leaches the straw after conventional harvest by leaching it in warm water right at the combustion plant, by means of a newly developed pre-treatment technology. This approach has the advantage that it is independent of climatic and weather conditions, and that the amount of quality improvement can be determined, but it has the disadvantage of higher investment and operating costs [5, 7–9].

The time of year at which harvesting takes place can also be an important parameter for annual fuel crops, especially in terms of moisture content and yield as shown in Figure 3.6. For short rotation coppice, the age of the crop influences the content of N and the ash content due to the varying proportion of wood/bark in the plant and depending on the amount of small branches. Forest residues show, due to their long rotation time, accumulations of environmentally relevant heavy metals such as Cd and Zn.

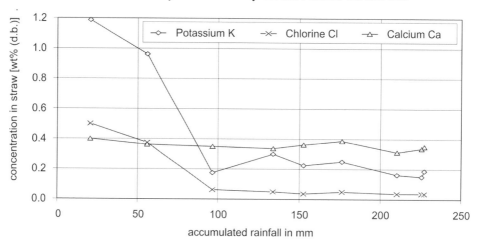

Figure 3.5 *Leaching of barley straw by rainfall during storage on the field*

Source: [5].

3.1.2 Influences on the fuel quality during the supply phase

The fuel supply and fuel pre-treatment influence physical parameters such as particle size, moisture content and energy density. A detailed discussion of technological possibilities and their potential/influence is given in Section 3.2.

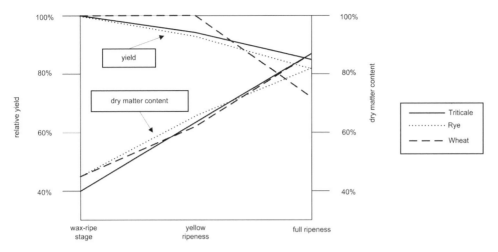

Figure 3.6 *Yield and moisture content at different harvesting dates for various cereals (whole crops)*

Source: [10].

3.1.3 Fuel quality control and price calculations

Large amounts of different biomass sources are available for thermal utilization with a wide variation in fuel quality (see Section 2.3). Therefore, in order to ensure an appropriate fuel quality, the application of appropriate fuel-quality assurance systems at the fuel delivery point at the combustion plant are essential.

A very important step towards the achievement of an acceptable fuel quality is having fuel supply contracts with the respective fuel suppliers, in which fuel specifications including particle size, water content, ash content, type of fuel and other important quality parameters (as required) are clearly defined.

At the fuel delivery point at the combustion plant, a visual inspection of the biomass fuels should be performed. Deliveries that are obviously contaminated and inappropriate, containing mineral contaminants (e.g. sand, earth or stones), waste or other problematic ingredients, containing particle sizes exceeding the specifications, or containing unspecified material (e.g. waste wood instead of woodchips) should be rejected. If the delivery passes the visual inspection, a value assessment process should then follow.

The delivered biomass should be assessed and paid for based on its energy content wherever possible. A calculation of the energy content based on weight and net calorific value (NCV) is recommended. A typical method is to weigh the load at a scale house and to measure the moisture content of samples from the fuel delivered using a moisture meter that provides quick results. The NCV can then be calculated using the gross calorific value (GCV) of the delivered fuel and the moisture content obtained from the measurements (see Equation 2.2 in Section 2.2.2.2). The GCV of each fuel source should be determined by chemical analysis prior to the first delivery. The energy content, which determines the price of the fuel delivered, is calculated by the NCV times the weight of the biomass fuel.

If a scale house is not available, an alternative but less accurate assessment method is based on a measurement of the delivered volume rather than the weight. The volume can be converted to weight by using the average value of the bulk density of the specific biomass fuel. The bulk densities of all delivered biomass fuels should be determined prior to the first delivery in order to improve the accuracy of the assessment process. Guiding values for bulk densities of biomass fuels can be found in Table 2.3. The uncertainty of this method is that (a) the estimation or measurement of the volume delivered is in most cases less precise than a weight measurement and (b) the bulk density of biomass fuels varies, especially with particle size, and also depends on the compression of the material. Moreover, mixtures of certain biomass fuels again result in changed bulk densities.

A third alternative is to calculate the fuel price based on the amount of heat produced from the plant according to heat meter measurements. This calculation must take the efficiency of the combustion plant into account, so that a reasonable market price for the fuel is achieved. The advantages of this method are that no costs for mass, volume or moisture content determination occur and that the determination of the heat produced with calibrated heat meters is very accurate. A disadvantage of this method arises if there is more than one fuel supplier, as the heat produced must be exactly allocated to a specific fuel amount from a specific supplier, which is rather difficult. A further disadvantage is the influence of operating conditions on plant efficiency. For instance, insufficient boiler cleaning can decrease the efficiency, which would reduce the fuel price. Finally, during storage the energy content of the fuel can decrease, which would also result in a reduced fuel price.

The determination method for the price basis should be defined in the fuel delivery contract in as much detail as possible. Depending on the method applied, the determined GCV, the bulk density of the fuel, the calculation methods for the NCV, the moisture content and the fuel price or the plant efficiency must be stated.

Payment based solely on volume or weight is not recommended, since biomass fuels of fluctuating quality and from different sources will most likely be used in the combustion plant. Furthermore, random tests regarding quality parameters defined for delivered fuels should be performed. Biomass deliveries which do not comply with the specified quality should be rejected.

If very homogeneous biomass fuels or biomass fuels produced and certified according to certain standards are utilized, the quality checks can be reduced and the assessment methods became easier (e.g. pellets according to a certain standard have a defined moisture content and bulk density).

3.2 Biomass production

A biomass supply chain consists of several steps. The main steps of biomass production are growing, managing and harvesting the resources. The kind of production steps and the sequence can vary, depending on the type of material and local conditions. For instance, the feedstock can be comminuted before or after storage, depending on the supply strategy chosen.

3.2.1 Forest residues

Strategies have been developed for cost-effective harvesting of biomass from conventional forestry.

Forest residues are recovered following clear felling of the stand. This is especially common in Sweden and Finland. There are a range of options available, all of which result in the production of chips in a size and form similar to those used as feedstock in the pulp and particle-board industries. Forest residues can be collected and stored in the forest prior to comminution or can be comminuted directly at harvest. Another option is to bale and transport the residues to a central location where the material is comminuted using a large-scale chipper (see Section 3.3.3).

When biomass fuels are produced from **thinnings**, whole tree harvesting and comminution systems are normally employed.

With **integrated harvesting systems** the energy component (e.g. tops) is harvested at the same time as the timber. Trees are felled and handled as whole trees, with product separation at the edge of the forest. The energy component is separated from the stem wood and comminuted, usually in the same or with a closely coupled operation. In this way, the supply costs are reduced significantly.

For each of these systems, different scales of operation and equipment can be utilized. The main points for consideration are the type and size of tree, the type of terrain (flat or steep), and the rate of working (production capacity) of the individual items of equipment and the overall system. Hence, relatively wide ranges of costs for harvesting forest residues can be incurred [2].

Integrated harvesting systems are most promising for delivering a reasonably homogeneous product at an acceptable cost. In integrated harvesting systems of both thinning and clear felling, the individual operations can be categorized as:

- felling;
- extraction (removal of timber or residues from the stump to the roadside);
- processing (including de-limbing and bucking); and
- comminution (chipping or hogging of residues).

3.2.2 Herbaceous biomass fuels

The harvesting of biomass fuels such as straw, whole crops, grass or miscanthus consists of cutting the feedstock. An overview of various harvesting processes is given in Figure 3.7.

Straw is a side-product of the corn harvest and, therefore, a further process step for baling is necessary. After the harvest, the bales are stored on the field or are transported to a bale store or the combustion plant. If the whole crop is used, the harvesting step can be combined with a baling or chopping step in order to upgrade the fuel in one process. If the steps are not combined, it is possible to store the crop on the field in order to dry it or to leach out undesired components using rainfall (see Section 3.3.5 and 3.4.1).

Among the herbaceous energy crops, miscanthus is the most well known. Miscanthus can be grown at low fertilizer and pesticide input levels and shows biomass yields of above 15t (dry matter) per annum. The negative aspects of herbaceous crops in comparison to woody energy crops are the high concentrations of elements influencing the combustion process concerning ash melting, emissions and corrosion (see Chapters 2 and 8). Research is still ongoing in plant breeding to make the plant suitable for the various ecological zones in Europe and to optimize the biomass yield and fuel characteristics [11, 12].

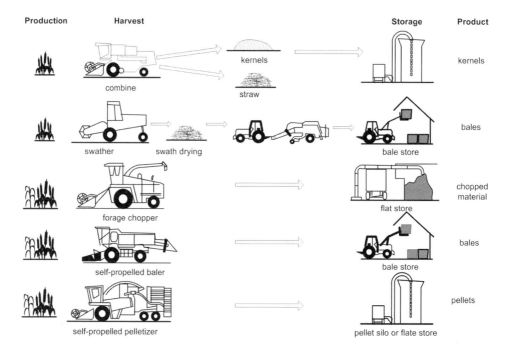

Figure 3.7 *Harvesting processes for herbaceous biomass fuels*

Source: Diagram from [13].

3.2.3 Short rotation woody crops

For short rotation coppice (SRC) plantations, willow (Salix) is the main crop to be considered. In Sweden, Salix is planted on about 17,000ha at present. Usually, 15,000–20,000 cuttings per hectare are needed once in 20 years. The shoots are harvested every 4–5 years with an annual average productivity of about 10 tonnes (d.b.) per hectare [14]. Besides Salix, field tests have been carried out and are still ongoing with poplar, alder and birch within the framework of the IEA Bioenergy Agreement.

In addition, resource management, which includes fertilizing (industrial fertilizers, waste water, sewage sludge, biomass ash) and pest control are necessary. Table 3.1 gives an overview of recommended amounts of fertilizers for Salix plantations in Denmark.

Harvesting SRC crops cost-effectively is still a matter of much concern. Equipment is being developed; some is currently available commercially. However, it does not necessarily work efficiently, it can damage the site in the long run and be expensive to operate. SRC harvesting machines that cut and produce bundles of cut coppice shoots or cut and comminute the crop and produce chips, chunks or billets are available (see Figure 3.8). Modified fodder, maize or sugar cane harvesters for harvesting short rotation coppice have brought about significant reductions in harvesting costs. Crops harvested as shoots can be stored in bundles or after comminution [2], see also Section 3.4.1).

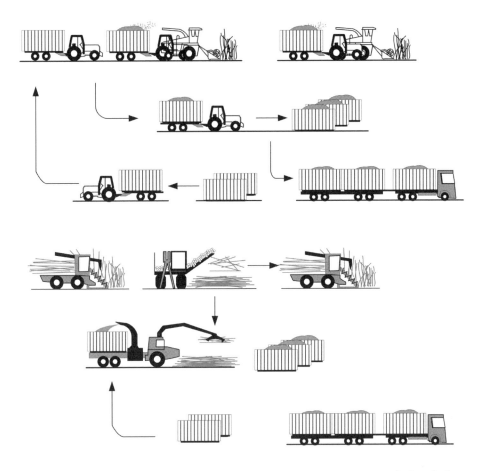

Figure 3.8 *Processes for harvesting short rotation coppice. Top: cut and chip, below: cut and bundle*

Source: Diagrams from [15].

Table 3.1 *Fertilizing plan of Salix plantations in Denmark*

Year of application	Nitrogen (N) kg per hectare	Phosphorus (P) kg per hectare	Potassium (K) kg per hectare
Year of planting	–	0–30	80–130
1st year after planting	45–60	–	
2nd year after planting	100–150	–	
3rd year after planting	90–120	–	
1st year after harvest	60–80	0–30	80–160
2nd year after harvest	90–110	–	
3rd year after harvest	60–80	–	

Source: [16].

3.3 Fuel pre-treatment

Fuel pre-treatment includes all the steps necessary to produce an upgraded biomass fuel (woody or herbaceous biomass) from a harvested biomass resource or from various kinds of waste wood. It is aimed at the following:

- reduction of the plant's investment, maintenance and personnel costs by using a homogeneous fuel that is suitable for an automatic fuel-feeding combustion system;
- reduction of storage, transport and handling costs by increasing the energy density; and
- reduction of impurities contained in the fuel (e.g. stones, earth, sand, metal parts, glass, plastics) in order to improve the fuel quality and the availability of the combustion plant (especially relevant for waste wood and contaminated biomass fuels).

3.3.1 Comminution of woody biomass

Wood fuels may be shrubs, bushes, forest residues (branches, tops), stems and bark up to a diameter of about 50cm, sawmill residues (slabs, edgings, trimmings) and cleaned demolition wood. These raw materials are available in very different shapes and particle sizes. Particle size reduction can take place by:

- chunking 50–250mm pieces,
- chipping 5–50mm pieces,
- grinding 0–80mm pieces.

Common types of chippers include disc chippers and drum chippers (see Figure 3.9). Disc chippers consist of a heavy rotating disc with a diameter of about 600–1000mm and two to four knives. The chip size can be modified by adjusting the knives and the anvil. Disc chippers produce fairly uniform chips, as the cutting angle in relation to the fibre direction of the tree remains unchanged regardless of the thickness of the stem.

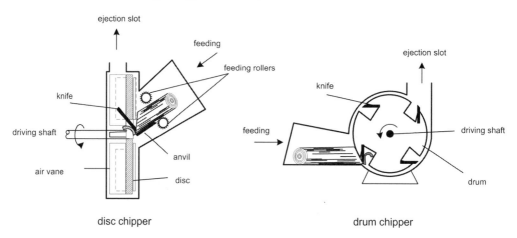

Figure 3.9 *Disc and drum chippers*

Source: [17].

Drum chippers consist of a rotating drum with a diameter of about 450–600mm in which the knives are embedded in two to four longitudinal grooves in the curved surface. As in the disc chipper, the knives pass a fixed anvil and the chip size is modified in the same way. As a result of the circular movement of the drum chipper, the cutting angle in relation to the fibre direction of the tree changes with the diameter of the stem. Consequently, the chips produced are less uniform than those produced by a disc chipper [16].

The energy need for chipping is about 1–3 per cent of the energy content of the wood fuel. It is less for material with a high moisture content, due to a lower friction factor.

Chippers are available in various sizes and can be self-propelled (combustion engine) or powered by a tractor. The fuel feeding can be operated manually (small-scale chippers) or by crane and/or a feeder table. Chippers can be locally fixed within a production system (e.g. chippers for sawmills) or mounted on a forwarder or trailer to operate on site. The characteristics of various classes of chippers are shown in Table 3.2.

Table 3.2 *Characteristics of various types of chippers*

Size	Productivity (m^3 bulk per hour)	Diameter of the material (cm)	Fuel-feeding system	Power consumption (kW)
Small-scale	3–25	8–35	Manual, crane	20–100
Medium-scale	25–40	35–40	Crane	60–200
Large-scale	40–100	40–55	Crane	200–550

Source: [17].

In the production of forest woodchips, there are various approaches to lowering chipping costs. As far as chipping material from early thinnings is concerned (stems with a diameter less than 15–20cm and tops), wood harvesting and chipping in a one-process step is more promising because of lower costs (see Figures 3.10 and 3.11). When utilizing forest residues (branches, tops from clear fellings) the use of a large-scale central chipper at the combustion plant or central storage facility seems advisable.

A spiral chunker consists of a helical cutter mounted on a horizontal shaft (see Figure 3.12). The pitch of the helix is constant over the length, but its diameter ranges from zero at the beginning to a maximum at the end. As the shaft rotates, the spiralling blade engages and slices the material. The device is self-feeding. The advantage of a chunker is its low power consumption, but the particle sizes it produces are significantly more varied than those produced by chippers.

In order to produce particle sizes below 5mm, fine grinding mills or hammer mills have to be used. Fine grinding mills (see Figure 3.13) consist of numerous knives mounted on a grinding drum. The material is pressed through the fixed screening ring by centrifugal forces. The particle size is determined by the perforation of the screen [17].

A hammer mill (see Figure 3.14) consists of a high speed rotor with floating mounted crushing tools (hammers, carbide-tipped). Comminution is caused by shearing forces between the top of the hammers and the cage. Hammer mills are very robust and, in comparison to chippers, are less sensitive to small metal parts (e.g. nails) [17].

Figure 3.10 *Mobile chipper with a crane efficiently collecting roadside thinnings for fuel*
Source: Courtesy of Bruins & Kwast, the Netherlands.

Figure 3.11 *Mobile chipper*
Source: Courtesy of TNO, the Netherlands.

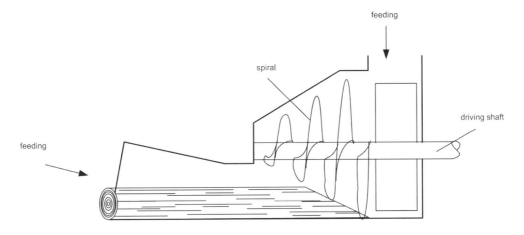

Figure 3.12 *Diagram and operating principle of a spiral chunker*
Source: [17].

Figure 3.13 *Fine grinding mill*
Source: [17].

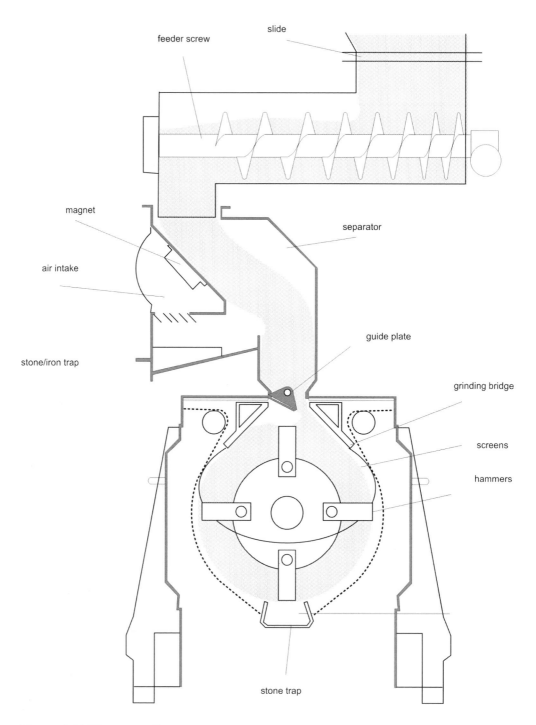

Figure 3.14 *Hammer mill*

Source: [18].

3.3.2 Pre-treatment of waste wood

Due to the fact that waste wood is a low-cost biomass fuel and therefore is very attractive for large-scale biomass combustion plants, a significant trend towards the thermal utilization of waste wood has been recognized in Europe within the last years. Waste wood in this context comprises demolition wood, pallets, fibre boards, residues from the wood processing industry, railway sleepers, pylons, etc. Due to these different sources, waste wood is an extremely inhomogeneous fuel and, therefore, its chemical composition as well as its content of impurities vary significantly depending on local constraints. Waste wood contains elements and compounds that, during combustion, cause the formation of both environmentally harmful emissions and deposits in furnaces and boilers.

European standards concerning the classification of waste wood do not yet exist, but they are under development by CEN TC 343 'Solid Recovered Fuels' (see also Section 2.3.4.1). In Austria woody biomass is categorized into seven different quality classes [19]:

Q1: chemically untreated wood;
Q2: bark;
Q3: binder-containing and halogen-free coated wood;
Q4: surface-treated wood;
Q5: creosoted timber;
Q6: salt-impregnated wood;
Q7: halogen plastics containing wood composite materials.

In Germany waste wood is classified according to the guideline BGBl. I 2002/3302 into classes A1, A2, A3, A4 and PCB-waste wood (PCB: polychlorinated biphenyls).

In comparison with the Austrian regulations, A1 and A2 can be seen to equate to Q1 to Q4 while A3 and A4 are comparable with Q5 to Q7. Waste wood, according to the qualities Q1 to Q4 (A1 and A2 respectively), is often used in medium-scale biomass combustion and combined heat and power (CHP) plants ($< 20MW_{th}$) and is referred to as 'quality sorted waste wood', while waste wood according to classes A3 and A4 is mainly used in large-scale biomass combustion and waste incineration plants.

As the list of different waste wood qualities indicates, waste wood assortments can significantly differ in terms of impurities. Consequently, a broad range of chemical compositions is possible. To reduce the risks regarding the problems mentioned above, concentrations of crucial elements in waste wood should be limited. Typical ranges of parameters determining the combustion behaviour of waste wood as well as recommended limitations for quality sorted waste wood to be utilized as a fuel in biomass combustion plants are shown in Table 3.3. The impact of these parameters on the combustion process and harmful emissions is discussed in Section 2.3.2.

Recent investigations [20] show that the concentrations of impurities and crucial ash-forming elements in most cases increase with decreasing fuel particle size. Consequently, most of these elements can be effectively removed from the fuel by discharging undersized fuel particles. However, this does not apply to Cl and Cd. These elements show no tendency towards increasing concentrations with decreasing fuel particle size. In this case, an appropriate tool to reduce the concentration of these elements is a more precise reception inspection (see Section 3.1.3) and an accurate sorting of the waste wood according to quality classes prior to processing it. For all other parameters the investigations clearly indicate that an appropriate waste-wood processing plant, which contains a facility to

Table 3.3 *Parameters affecting the combustion behaviour of waste wood, their typical ranges and recommended guiding values*

Parameter	Typical range	Recommendation for quality sorted waste wood
Ash [wt% (d.b.)]	1.5–10.0	< 4.0
Cl [mg/kg (d.b.)]	300–4000	< 1000
S [mg/kg (d.b.)]	300–2000	< 1000
Fe [mg/kg (d.b.)]	250–3000	< 500
Al [mg/kg (d.b.)]	400–5000	< 800
Si [mg/kg (d.b.)]	1500–15,000	< 4000
Na [mg/kg (d.b.)]	200–3000	< 500
K [mg/kg (d.b.)]	800–2500	< 1500
Zn [mg/kg (d.b.)]	200–1200	< 200
Pb [mg/kg (d.b.)]	50–400	< 100
Cd [mg/kg (d.b.)]	0.3–3.0	< 1.0
Hg [mg/kg (d.b.)]	up to 0.4	< 0.3
Particle size [mm]	<< 7 – >> 100	7–100

Source: [20].

separate undersized particles, as well as units for metal and non-ferrous metal separation, can significantly reduce the concentrations.

Based on the experiences from the tests and analyses mentioned above [20], a fuel processing plant has been developed that is based on four main working steps. First, the waste wood is shredded with a low speed wood shredder including a 100mm screen basket. Subsequently, iron pieces are removed by an overband magnet and magnetic roller. The waste wood is then screened with a 10mm mesh size and, finally, non-ferrous metals are separated. As a result, wood waste that is almost ferrous and non-ferrous metal free and sized between 10 and 100mm can be produced. An image of the fuel processing unit is shown in Figure 3.15.

Figure 3.15 *Pilot plant for processing waste wood*

Explanations: 1 = fuel input, 2 = shredder, 3 = iron removal, 4 = 10mm sieve, 5 = discharge of undersized particles, 6 = non-ferrous metal removal, 7 = processed fuel.

Source: [20].

Tests with the waste wood processing plant shown in Figure 3.15 showed that waste wood processing significantly increases the fuel quality and, therefore, positively influences the operation of waste wood combustion plants. In addition to the separation of ferrous and non-ferrous metals, the discharge of fuel particles <7mm is strongly recommended. This undersized fuel fraction could then be densified and used in waste incineration plants or in waste combustion plants using A1 to A4 (Q1 to Q7) waste wood.

A further benefit of this waste wood processing is a reduction of costs for ash disposal, as lower amounts of ashes are produced during combustion. Combustion tests performed at a grate-fired combustion unit with a fuel power capacity of 10MW showed that the separation of the undersized fuel fraction (ash-rich and small particle size), causes a significant reduction of the fuel particle and charcoal entrainment from the grate and, consequently, also of deposit formation in the furnace and the boiler. This results in an increased availability and reduced operating costs for the combustion plant.

These investigations have shown that, when using an appropriate fuel processing plant, the properties of waste wood can be significantly improved, which considerably reduces deposit formation problems during combustion.

In order to upgrade large amounts of various kinds of waste wood, centralized waste-wood pre-treatment plants are used. The productivity can be up to 100 tonnes per hour. For small throughput rates or less sophisticated applications, so-called low-speed breakers are widely used (see Figure 3.16). A breaker consists of a filling compartment into which the wood is dumped. At the bottom of this compartment, a hydraulically operated slide pushes the material against a rotor, equipped with tempered steel knives. The product is then sieved to the desired size. The advantages are a relatively low driving power, ease of operation, low acoustic emissions, low maintenance costs and high flexibility. Breakers are available in varying capacities from 50 up to 10,000kg per hour.

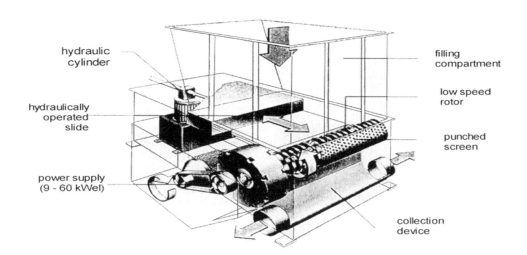

Figure 3.16 *Breaker for waste wood*

Source: [17].

3.3.3 Baling and bundling of biomass fuels

3.3.3.1 Herbaceous biomass fuels

Herbaceous biomass fuels such as straw, hay or cereals are pressed into bales to increase their energy density and to ease handling. Baling of straw is usually done as an extra process step after grain harvesting or feedstock cutting. For energy crops such as miscanthus or cereals, one-step combined harvesting and baling is possible. If the straw is dried on the field before collection, leaching of water-soluble compounds such as Cl and K by rainfall can be achieved, which improves the combustion quality of the fuel (see Figure 3.5).

The size of the bales depends on the machine used. The density of the bales for a given machine depends on the product to be baled. Regarding the products of baling, the following forms and sizes can be distinguished:

- square bales (small bales produced with high density balers, large 'Hesston' bales produced with big balers);
- round bales (produced with roll balers); and
- compact bales (produced with compact balers).

For square and round bales, towed machines are available. Big balers are also available as self-propelling machines. Compact balers are still in the development phase and not yet commercially available. An overview of technical data is given in Table 3.4.

Table 3.4 *Technical data for various baling technologies*

Parameter		Small square	Round	Large square	Compact
Driving power	kW	> 25	> 30	> 60	> 70
Throughput	t/h	8–20	15–20	15–20	14
Density (unit)	kg (d.b.)/m³	120	110	150	300
Shape	–	cuboid	cylindrical	cuboid	cylindrical
Density (storage)	kg (d.b.)/m³	120[1]/90[2]	85	150	270[1]/165[2]
Dimensions	cm	40 x 50 x 50–120	Ø 120–200 x 120–170	120 x 130 x 120–250	Ø 25–40 any length
Weight[3]	kg (w.b.)	8–25	300–500	500–600	any weight

Explanations: [1] stacked, [2] not stacked, [3] moisture content 20wt% (w.b.).
Source: [21, 22].

3.3.3.2 Forest residues

Baling of forest residues before transport, in combination with centralized chipping, is a comparatively new approach to reduce the production costs of forest woodchips. Initial activities in this direction were reported in the mid 1990s [23]. The bales are similar to round straw bales, with a diameter of 1.2m and 1.2m high. Their weight is about 600kg (w.b.) at a moisture content of 45wt% (w.b.). Table 3.5 compares the transport of various pre-treated forest residues in Sweden.

Table 3.5 *Comparison of payload and energy content for truck transportation of forest residues in Sweden*

	Volume/ no. of bales	Payload weight tonnes	Energy content MWh
Woodchips	105m³ (bulk)	33	85
Unchipped logging residues	135m³	18	46
Unchipped logging residues in bales	60 bales	37	95

Explanations: fuel moisture content 45wt% (w.b.), gross vehicle weight 60 tonnes.
Source: [23].

The transportation of baled forest residues can be up to 50 per cent cheaper than that of unchipped material, due to a considerably higher payload. The transport of baled forest residues is also about 10 per cent cheaper than the transport of conventional woodchips.

A new and innovative approach to bundling forest residues has been developed in Finland [24]. Working from logging residues, directly at the stand, the bundling machine (see Figure 3.17) produces bundles that are about 3.3m long and about 0.6–0.8m in diameter (see Figure 3.18). The weight of the bundles is 0.4–0.7t. About every 40cm the bundles are wrapped with strings; 20–30 bundles can be produced per hour, each bundle with an energy content of roughly 1MWh. Based on a one shift operation with one bundler, between 20,000 and 60,000 bundles can be produced annually (equal to about 20–60 GWh/a).

Figure 3.17 *Bundler*

Source: [24]

Figure 3.18 *Bundles stored at the roadside*

Source: [24].

The bundle transportation in the forest to the nearest forest road can be done by a standard forwarder. Ordinary trucks can be used to transport the bundles from the roadside to the combustion plant. Between 60 and 70 bundles can be transported in one load (based on Finnish framework conditions). Due to the increased density of the bundles, the specific transportation costs decrease. Train transportation is an additional and efficient option for long distances. Examples are shown in Figure 3.19.

At the plant the bundles are usually stored in intermediate storage. The bundles have the advantage that during storage their tendency to biological degradation is less accentuated than that of loose slash piles. In addition, their sensitivity to moisture is low. Therefore, covering the intermediate storage is not necessary. A disadvantage of this technology is the higher nutrient removal from the forest (due to the needles on the residues). In addition, the applicability of the technology in mountainous regions is limited. Before combustion the bundles are crushed in high-speed crushers or chippers.

3.3.4 Pellets and briquettes

In order to compress fine wood particles such as sawdust and shavings into larger shapes and to produce a homogeneous biomass fuel with a high energy density, one can apply briquetting and pelleting. Pellets are cylinders with typical of dimensions Ø 6–10mm, while briquettes have typical of dimensions Ø 30–100mm. Briquettes are primarily used instead of firewood for manually charged domestic stoves. Pellets can be used for automatically charged stoves and boilers due to their good flowability, uniform water content, grain size

Figure 3.19 *Truck and train transportation of bundles*
Source [24].

and chemical composition. Consequently, pellet stoves and boilers for domestic applications offer an ease of operation similar to oil-fired systems. In Sweden pellets are primarily used to substitute coal in large-scale power plants. For this purpose, sawdust is only pelletized in order to reduce transport and storage costs and the pellets are milled before combustion.

An overview of the whole pelletizing process is shown in Figure 3.20. Five production steps can be distinguished [1]:

1 **Drying**: depending on the kind of wood used, the moisture content of the raw material before entering the pellet press must be between 8 and 12wt% (w.b.). Due to the fact that the stability of the pressings is influenced by the friction between gooseneck and raw material, a constant moisture content is essential. If the material is too dry, the surface of the particle may carbonize and binders will get burned before the process is finished. If the wood is too wet, moisture contained in the pressings cannot escape and enlarges the product volume, making it mechanically weak.
2 **Milling**: the particle size of the raw material must be reduced and homogenized depending on the diameter of the pellets produced. For this purpose hammer mills are used.
3 **Conditioning**: by the addition of steam, the particles are covered with a thin liquid layer in order to improve adhesion.
4 **Pelletizing**: for pelletizing wood, flat die or ring die pelletizers are used (see Figure 3.21). The productivity of pellet presses range between some 100kg to about 10 tonnes per hour.
5 **Cooling**: the temperature of the pellets increases during the densification process. Therefore, careful cooling of the pellets leaving the press is necessary to guarantee high durability.

The whole energy consumption of the pelletizing process is about 2.5 per cent of the NCV of the fuel (without drying, raw material: dry sawdust). If the raw material must be dried, the energy consumption increases to approximately 20 per cent of the NCV of the fuel [25]. The production capacity of existing pellet plants in Sweden, Denmark and Austria varies in a broad range between some 100 and 100,000 tonnes per year. As of 2006, global pellet production capacity amounted to approximately 3 million tonnes per year.

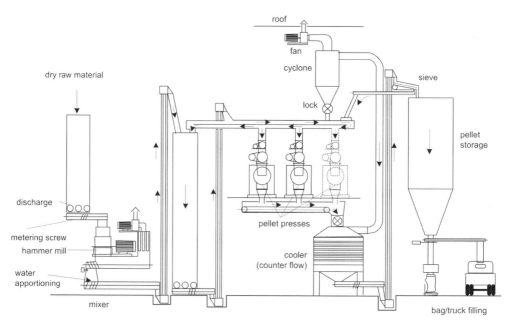

Figure 3.20 *Pelletizing plant*

Source: [17].

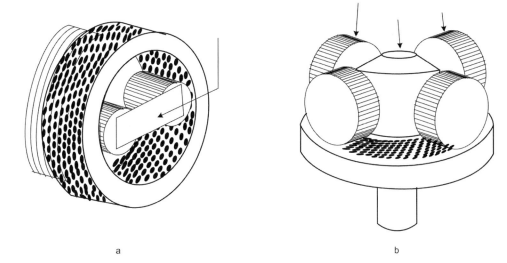

Figure 3.21 *Ring die pelletizer (a) and flat die pelletizer (b)*

Source: [26].

Several European countries have national standards for pellets and briquettes, e.g. Austria, Germany and Sweden. In addition, respective standards on a European level are also available (see also Annex 3). As an example, the chemical and physical properties of pellets and briquettes in accordance with the Austrian standard ÖNORM M 7135 are shown in Table 3.6.

Table 3.6 *Chemical and physical fuel properties in accordance with ÖNORM M7135*

Parameter	Unit	Compressed wood		Compressed bark	
		Wood pellets	Wood briquettes	Bark pellets	Bark briquettes
Diameter D	mm	4–10	40–120	4–10	40–120
Length	mm	< 5 x D[1]	< 400	< 5 x D[1)]	< 400
Particle density	t/m3	> 1.12	> 1.00	> 1.12	> 1.10
Moisture content	wt% (w.b.)	< 10	< 10	< 18	< 18
Ash content	wt% (d.b.)	< 0.5	< 0.5[2)]	< 6	< 6
NCV	MJ/kg (d.b.)	> 18.0	> 18.0	> 18.0	> 18.0
Sulphur content	wt% (d.b.)	< 0.04	< 0.04	< 0.08	< 0.08
Nitrogen content	wt% (d.b.)	< 0.30	< 0.30	< 0.60	< 0.60
Chlorine content	wt% (d.b.)	< 0.02	< 0.02	< 0.04	< 0.04
Fines	wt% (w.b.)	< 2.3	–	< 2.3	–
Bio-additives	%	< 2	< 2	< 2	< 2

Explanations: [1] a diameter of up to 7.5 x D is allowed for a maximum of 20wt% of the compressed wood or bark; [2] an ash content of 0.8% is permitted if the natural untreated wood already has a higher ash content.

Further research is necessary to improve pellet stability and resistance to abrasion. For domestic utilization, reduction of dust emissions during handling and storage is important. Adding binders such as black-liquor can solve these problems, but deteriorate fuel characteristics (e.g. increase sulphur content). Because of this, the use of chemical additives for pellet production is prohibited in several countries (e.g. Austria, Germany, Denmark).

The basic technologies for briquetting are shown in Figure 3.22. A constant moisture content of the raw material (12–14wt% w.b.), a constant back-pressure in the briquette press and sufficient cooling are essential for the stability of the briquettes produced.

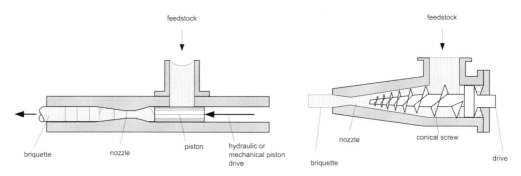

Figure 3.22 *Two briquetting technologies*

Source: [27].

3.3.5 Drying of biomass

The content of water in biomass fuels can be described by the *moisture content w on a wet basis* (Eq. 3.1) or the *moisture content u on a dry basis* (see Eq. 3.2). Eq. 3.3 shows how to convert between these two moisture contents.

$$\text{moisture content } w = \frac{\text{mass water}}{\text{mass water} + \text{mass biofuel (d.b.)}} \times 100\% \ [\text{wt}\% \ (\text{w.b.})] \quad (3.1)$$

$$\text{moisture content } u = \frac{\text{mass water}}{\text{mass biofuel (d.b.)}} \times 100\% \ [\text{wt}\% \ (\text{d.b.})] \quad (3.2)$$

$$w = \frac{100 \times u}{100 + u} \qquad u = \frac{100 \times w}{100 - w} \quad (3.3)$$

The moisture content of biomass fuels varies widely, depending on the kind of material, the time of harvesting, the form of pre-treatment, and the method and duration of storage. For example, the moisture content of fresh wood is over 50wt% (w.b.), whereas the moisture content of waste wood or straw is usually below 15wt% (w.b.). The most important reasons for drying biomass are:

- The energy content of the fuel (based on NCV) depends on the moisture content of the fuel. Therefore, the efficiency of the combustion system increases with decreasing moisture content (see Chapter 2).
- To optimize the combustion process (minimum emissions, maximum efficiency), the moisture content of the fuel should be as constant as possible. Fuel with varying moisture content requires a more complex combustion technology and process control system and therefore pushes up investment costs.
- The long-term storage of wet biomass fuels causes problems with dry-matter loss and hygiene (fungi production due to biological degradation). Therefore, the moisture content of the fuel should be below 30wt% (w.b.) for domestic applications.
- For small-scale furnaces or stoves, the moisture content of the fuel should be about 10–30wt% (w.b.) for technological, economic and ecological reasons.
- For the production of pellets or briquettes, the moisture content of the raw material must be about 10wt% (w.b.) (see Section 3.3.4).

Drying of biomass influences the total fuel costs significantly and should be done as simply as possible. Due to the fact that there are no economically attractive biomass drying methods available for large-scale applications, the combustion systems are mostly suitable for a wide range of fuel moisture content.

An effective way of drying fresh wood is to store the unchipped material (wood logs) in a heap or pile outdoors over the summer. In this way, the moisture content can be lowered from about 50 to 30wt% (w.b.). Drying cut straw for several days on the field before baling will also reduce the moisture content efficiently. The disadvantages of natural drying are mainly unforeseeable weather conditions and logistical problems (e.g. sawmill residues are mainly available as chips, handling of wood logs is more time-consuming, a long storage time is necessary).

When fresh bark or woodchips are stored in a pile, the temperature in the pile will increase due to biological degradation effects. The heat produced by the micro-organisms causes a natural convection process. Air will circulate through the pile and will transport water vapour to the pile surface. Consequently, the fuel in the central parts of the pile dries, while part of the vapour condenses in the upper and cooler regions of the pile. Also, if the pile is stored outdoors, re-moistening of the outer layer by rainfall will usually take place. Figure 3.23 shows the results of indoor and outdoor storage trials with bark in piles 3m high. The tests lasted from August until January and were carried out in southern Austria. While there was hardly any change in the average moisture content of the pile stored outdoors, the pile stored indoors dried from an initial moisture content of 59wt% (w.b.) to an average moisture content of 44wt% (w.b.) over six months.

For evaluating the effects of drying by natural convection on the energy content of the biomass fuel pile, dry-matter losses caused by biological degradation processes also have to be considered. In the storage trials described above, energy losses of 12 per cent were measured for the pile stored outdoors (related to the NCV of the fuel and the total weight loss of the fuel). Indoor storage scarcely had any effect on the energy content of the fuel stored (the drying effect was balanced by the dry-matter losses).

Recent investigations have shown that the length of woodchips stored in a pile plays an important additional role with respect to temperature, drying, mould formation, mass and energy loss in woodchip piles during long-term storage [28]. Mean pile temperature, moisture content, dry mass loss and energy loss decrease with increasing woodchip length.

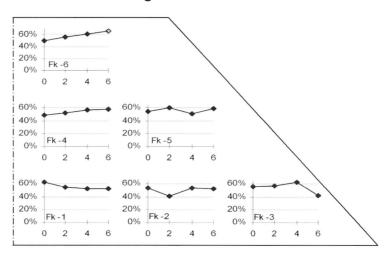
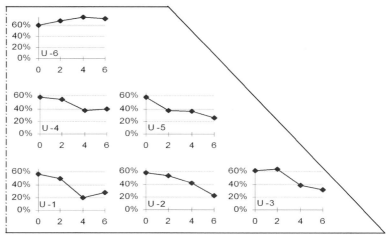

Figure 3.23 *Development of moisture content in bark piles stored outdoors and indoors*

Explanations: The diagrams show half of the cross-section of the bark pile; height of piles: 3m; abscissa: duration of storage in months, ordinate: moisture content in wt% (w.b.); the graphs indicate the position in the respective fuel pile.
Source: [29].

The number of thermophile moulds also shows a decreasing trend with increasing woodchip length. For mesophile moulds this trend has not been determined. The most important parameter with regard to thermal utilization is the energy loss, which amounts to between 20 per cent and 30 per cent per year for fine woodchips (median diameter 16mm). This loss is between –5 per cent and +5 per cent for coarse woodchip fractions with a median diameter of more than 120mm (due to lower dry mass loss and the reduction of the moisture content during storage). The conclusion from this work is that woodchips to be stored in a pile should have a minimum average length of 100mm in order to minimize energy losses and mould formation. If fine or medium-sized woodchip fractions must be stored in a pile, the storage duration should be below 2 weeks in order to keep mould formation low.

Drying of bulk material can be supported by ventilating the pile with ambient or preheated air. In most cases, biomass fuel drying is only economic and efficient if a cheap heat source is available, because not only investment costs but also operating costs have to be considered. Figure 3.24 shows a concept for drying biomass with preheated air using solar energy. Waste heat from the flue gas condensation units of combustion plants can also be used for drying purposes (Figure 3.25). These two drying processes, based on forced convection and a cheap or cost-free heat source, can be economically interesting.

Continuous drying technologies are necessary for conditioning sawdust or woodchips for the production of pellets or briquettes. Several technologies such as belt dryers, drum dryers, tube bundle dryers and superheated steam dryers are available. These can be directly heated with flue gas from a combustion process or indirectly with hot water, steam or thermal oil.

Belt dryers (see Figure 3.26) operate at a low temperature level with gas input temperatures of 90–110°C and gas output temperatures of 60–70°C depending on the technology used. Therefore, problems regarding odour emissions and emissions of volatile organic compounds released from the biomass fuel can be avoided.

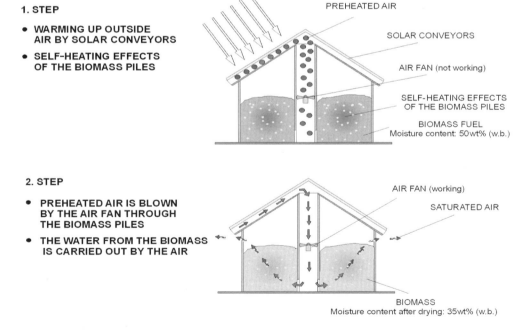

Figure 3.24 *Pulsating solar biomass drying process*

Source: [6]. © dbv-Verlag, Graz, Austria.

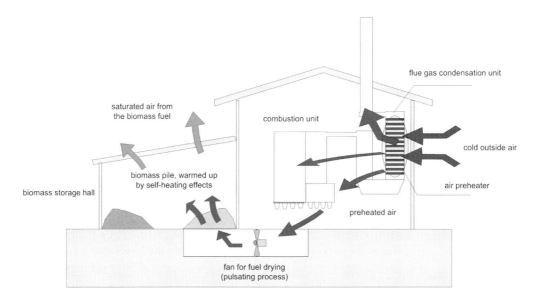

Figure 3.25 *Short-term biomass drying process based on preheated air from a flue gas condensation unit*

Source: [6].

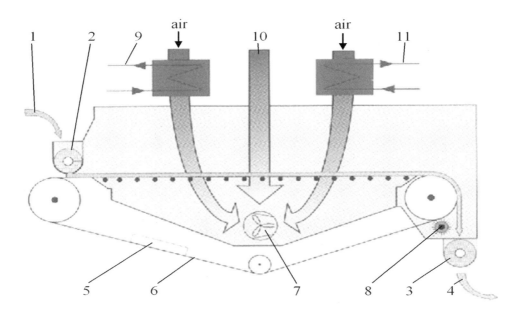

Figure 3.26 *Belt dryer for biomass*

Explanations: 1 = raw material; 2 = feeding screw; 3 = outlet screw; 4 = dried product; 5 = belt washing unit; 6 = dryer belt; 7 = exhaust fan; 8 = rotating cleaning brush; energy supply by steam (9) or waste heat from processes (10) or water (11).

Source: [30].

Usually belt dryers are heated indirectly by using air as a heating medium. In this case, the air is heated by thermal oil, steam, hot water or waste heat from flue gas condensation units. In addition, belt dryers can also be heated directly by waste heat from processes [30].

Drum dryers operate at higher temperatures and are heated directly. The inlet temperatures can be up to 600°C. However, these dryers do not only release wood particles during the drying process, but volatile organic compounds as well. They therefore have to be outfitted with a complex flue gas cleaning system [31]. Furthermore, fly ash contained in the flue gas used in directly heated drum dryers remains in the dried sawdust, which subsequently leads to a higher ash content of the pellets and a certain contamination with heavy metals [32].

Tube bundle dryers are always indirectly heated, which means that there is no direct contact of the heating medium with the material to be dried. This ensures a gentle method of drying at biomass temperatures of about 90°C with no danger of odour or volatile organic compound emissions. Steam, hot water or thermal oil may be used as a heating agent with an inlet temperature of 150–210°C [33].

Superheated steam dryers utilize the drying and heating capacity of superheated steam. The material to be dried is exposed to indirectly heated superheated steam in a closed and pressurized system at temperatures between 115–140°C. The heating steam is usually condensed at a pressure of 10–20 bar in the heat exchanger at a temperature of 180–210°C. Thermal oil as well as flue gas can also be used for heating. The main advantage of superheated steam dryers is the possibility of heat recovery due to the steam generated during the drying process. This excess steam is continuously bled off from the system and is available as saturated steam at a pressure of 2–5 bar, which can be used for other purposes and thus increases the efficiency of this system. Another benefit is that no pollutants such as dust or smell will be emitted to the atmosphere. During the heat recovery, the emissions are collected in the vapour condensate and can be treated separately. Therefore, the condensate from the recovered steam must be discharged in a sewage system [34, 35].

No general rules can be laid out as to whether biomass drying before combustion is feasible. The reason for this is that many influencing variables have to be considered (e.g. fuel prices, size of the plant, combustion and heat recovery technology applied). Usually biomass drying is only feasible if natural heating sources (e.g. solar energy) or waste heat (e.g. from flue gas condensation units) are available. In any case, the total costs of drying have to be critically compared with the economic advantages that can be achieved (e.g. fuel savings, reduction of flue gas volume flow, higher combustion efficiency) in order to evaluate whether an economic surplus can be gained.

3.4 Storage, handling and transport systems

3.4.1 Storage of biomass

Long-term storage of biomass fuels is necessary if there is a time gap between production and utilization. Furthermore, fuel needs to be stored at the combustion plant in order to assure fail-safe operation. Considering the fact that biomass fuels generally have a relatively low energy density, the design of the storage facilities is quite important in order to keep fuel costs low.

Short-term storage with an automatic discharge system is needed for feeding the fuel to the combustion plant. The fuel manipulation between long-term and daily storage facilities is usually performed by cranes or by wheel loaders.

The simplest way of storing biomass is to pile it. Usually, a wheel loader is used for the manipulation of the fuel. When applying this method several aspects have to be taken into account. First, some general points have to be considered when long-term storage of woodchips and bark with a moisture content over 20–30wt % (w.b.) in a pile is performed. Biological and biochemical degradation as well as, in some cases, chemical oxidation processes [36] result in heat development, which can cause self-ignition in certain cases. A schematic diagram of the temperature development due to self-heating in stored biological material is given in Figure 3.27. Second, dry-matter losses, changes in moisture content, and health risks (growth of fungi and bacteria) should be taken into consideration.

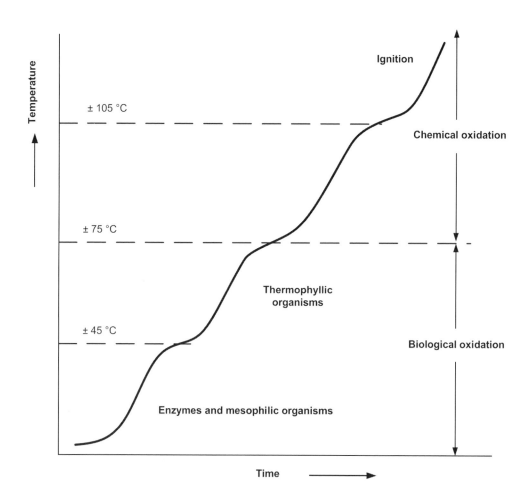

Figure 3.27 *Temperature development and processes responsible for self-heating in stored biological material*

Source: [36].

The effects described above are complex and depend mainly on the particle size of the material (whole shoots, chunk wood, chips, sawdust), the kind of material (bark, wood), the moisture content of the material, the kind of storage (outdoors, outdoors but covered with a sheet, indoors) and the kind of ventilation of the pile (airtight storage, unventilated, active ventilation with ambient air or preheated air). Further information is also given in Section 3.3.5.

Several storage trials have been performed [28, 36–41]. The main conclusions are listed below.

- When storing fresh woodchips or bark, the temperature in the core of the pile normally rises up to 60°C within the first days. No increase of temperature occurs when material with a particle size of more than about 20cm is stored.
- Self-ignition mainly occurs in piles of bark and can be avoided when bark is stored in piles with a maximum height of 8m and a storage time of less than five months. Figure 3.28 shows the temperature development and distribution in a bark pile.
- Biomass piles should not be compacted as this can lead to concentration of moisture on local spots in the pile, leading to an increased tendency for self-heating.
- Biomass piles should contain homogeneous material. In piles of different materials or different batches of the same material (e.g. with different moisture content) the process of self-heating can start in a niche and can spread over the rest of the storage pile.
- Control of biomass piles with regard to the risk of self-heating and self-ignition can be done by a combination of temperature and gas measurements. Temperature measurement points must be widespread over the pile as the self-heating process can start locally. With measurement of CO_2 early stages of the process can be detected. CO measurements can detect the latter stage of the process, where a direct threat of fire already exists and immediate action is necessary.
- When storing freshly chipped wood or bark in a pile, dry-matter losses can amount to 5 per cent per month. Dry-matter losses are highest in the initial storage period and depend mainly on:
  - the moisture content of the fuel (high moisture content increases dry-matter losses);
  - the kind and age of the crops (young crops such as short rotation coppice show higher dry-matter losses than woodchips from forest residues); and
  - the particle size (the smaller the particle size, the bigger the particle surface and the higher the dry-matter losses).
- Ventilation of the pile can lower dry-matter losses, but ventilation is costly and therefore in most cases not used on an industrial level. To avoid dry-matter losses, the storage of uncomminuted material or airtight storage of chips can be recommended.
- When storing fresh biomass in a pile indoors, the moisture content can be reduced if natural convection is possible. Therefore the walls of storage halls should allow air circulation through the piles. Moreover, natural convection of air through the biomass stored is also important in order to avoid self-ignition (this is especially relevant for bark and sawdust). In this respect, biomass in the form of uncomminuted material (e.g. whole logs) can be stored and dried more effectively than biomass in the form of chips.
- Outdoor storage of uncovered dry biomass fuels (moisture content lower than 20–30wt% (w.b.), e.g. straw) should be avoided due to possible re-moistening by rainfall. Indoor storage or rain protection is recommended.

- Outdoor storage of biomass fuels with small particle sizes, like sawdust, will cause dust emissions and is not possible in populated areas. For this type of fuel, storage in closed buildings or silos is necessary.
- Uncovered storage of biomass will cause waste water emissions due to leaching by rainfall. This effect has to be considered, especially for the storage of bark. It can require waste water treatment. Waste water of biomass piles usually shows low pH values (2.5–5.0) due to the dissolution of organic acids [42, 43].
- The surface of the storage area should be paved to avoid contamination of the biomass fuels with mineral impurities (e.g. sand, earth, stones).

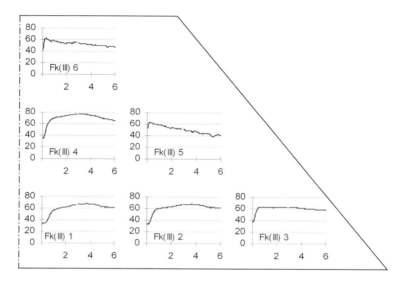

Figure 3.28 *Development of temperature in a bark pile*

Explanations: Diagram shows half of the cross-section of the bark pile; outdoor storage, height of pile 3m; abscissa: duration of storage in months, ordinate: temperature (°C); the graphs indicate the position in the fuel pile.

Short-term biomass storage units are directly connected with the feeding system of the combustion plant. For short-term storage of bulk materials such as bark or woodchips, bunkers with sliding bar conveyors or walking floors can be used. Sliding bar conveyors (see Figure 3.29) show a very robust construction and are especially suitable for fuels such as bark and woodchips. In contrast to sliding bar conveyors, walking floors forward the stored material as a whole. Walking floors are especially suitable for the automatic discharge of long-term storage without additional devices, but have the disadvantage that in order to gain a certain storage height, the feeding of the biomass fuel onto the walking floor has to take place from above (see Figure 3.30).

If no walking floor system is applied, the manipulation of the biomass fuel from the long-term storage hall to the bunker is achieved using wheel loaders or crane systems (see Section 3.4.2). Depending on the material and the discharging technology used, the storage

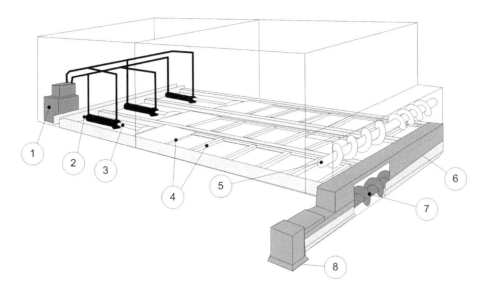

Figure 3.29 *Sliding bar conveyor used for the discharge of bunkers*

Explanations: 1 = hydraulic generator, 2 = bearing for hydraulic cylinder, 3 = hydraulic cylinder, 4 = sliding bars, 5 = control screw, 6 = drop channel, 7 = screw conveyor, 8 = delivery end.

Source: [17].

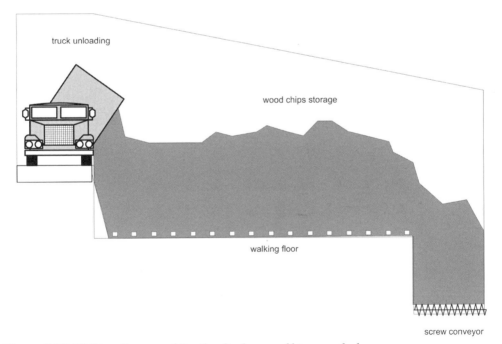

Figure 3.30 *Walking floor used for the discharge of biomass fuels from a long-term storage hall*

Source: [44].

height can be up to 10m and the cross-section of the bunker 10 x 25m, resulting in a storage volume of up to 2500m³ per unit [17].

Sawdust and fine wood waste is preferably stored in closed silos to avoid dust emissions. The diameter of such silos can be up to 15m; the height can be up to 40m. For automatic discharge of these silos, rotating screws or inclined screws with an agitator are used (see Figure 3.31).

When storing biomass fuels in a bunker or silo, the bridging behaviour of the material has to be considered. Bridging can be avoided by a proper design of the bunker and discharging system.

Baled woody biomass fuels can be stored outdoors, as their tendency to biological degradation and their sensitivity to moisture are low (see Section 3.3.3.2). Roofing of the intermediate storage area is thus not necessary. A paved surface is an advantage, as this would avoid mineral contamination with sand or stones.

Baled herbaceous biomass fuels (see Section 3.3.3.1) can be stored in a number of different ways. Their low moisture content is a limiting factor for outdoor storage because they must be prevented from getting wet. The piling of baled herbaceous biomass fuels on the field is only feasible for short-term storage. Long-term storage may result in big losses since the upper and lower layers of the bales get wet, making them unuseable. Covering the bales with a tarpaulin protects the upper layer, while the lower layer is still exposed to moisture. Storage facilities with flying roofs provide better protection, since only the bales on the side are exposed to moisture. Indoor storage is the best – but also most expensive – option for maintaining the quality of baled herbaceous biomass fuels during storage.

Square bales show the lowest storage costs due to their better volume utilization [1].

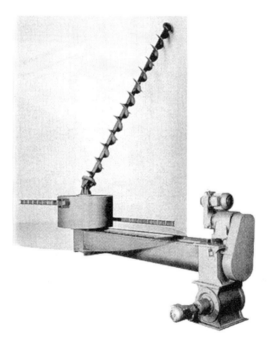

Figure 3.31 *Inclined rotating screw used for the discharge of silos*

Source: [17].

3.4.2 Fuel-feeding and handling systems

Fuel-feeding and handling systems are necessary to transport the fuel from the point of delivery or from the storage to the combustion system. Due to its direct influence on the availability and performance of the combustion system, the fuel feeding needs to be designed carefully and has to be adjusted to the combustion technology used.

The great variety of biomass fuels also requires appropriate transport and handling systems. For their design, the following criteria have to be considered [17]:

- characteristics of the fuel (particle form and size distribution, moisture content);
- distance of transport;
- height difference to be managed;
- noise emissions;
- risk of dust explosions and fire;
- transport capacity;
- investment, operation and maintenance costs; and
- availability of the feeding and handling systems.

Wheel loaders (see Figure 3.32) offer the easiest and most flexible way of fuel handling for nearly all kinds of bulk materials (sawdust, woodchips, bark, waste wood). The volume of the bucket can be up to 5m³. One disadvantage is the fact that personnel are needed for wheel-loader operation, which does not allow for fully automatic plant operation. This also has to be considered in cost calculations.

Figure 3.32 *Wheel loaders are often used for transporting sawdust and bark from the long-term storage to the feeding system of the biomass heating plant*

Source: Courtesy Stadtwärme Lienz, Austria.

Figure 3.33 *Automatic crane for straw feeding*

Source: [45].

Crane systems are used to transport chips or bales from the storage to the combustion feeding unit (see Figure 3.33). Crane systems work well for bales, woodchips and pellets and allow a fully automatic operation (see Figure 3.30). If inhomogeneous particle sizes occur to a high degree, crane feeders can cause problems (e.g. they are not recommended for mixtures of bark, sawdust and woodchips).

Belt conveyors can be used to transport fuels over long distances. An endless belt, wound over two tail pulleys is used for carrying and hauling. Belt conveyors can be used for bulk material or unit loads. The construction is simple, inexpensive and offers the opportunity to install a conveyor belt weigher. However, belt conveyors are not suitable for inclined conveying, and avoiding dust emissions is costly. Moreover, they are sensitive to external influences such as temperature or dirt accumulations on the pulleys.

Figure 3.34 *Cranes are often used for handling woodchips in the intermediate storage bunkers of a biomass heating plant*

Source: Courtesy Schmid AG, Switzerland.

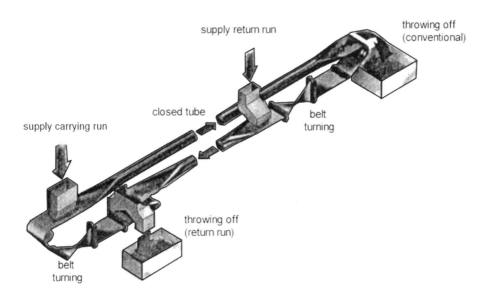

Figure 3.35 *Tube-rubber belt conveyor*

Source: [46].

Tube-rubber belt conveyors enclose the material and therefore dust emissions are avoided (see Figure 3.35). Fuel transport is possible in both directions; the belt is curved and it is possible to overcome height differences. Long and sharp particles may pierce the rubber belt and make a replacement of the whole belt necessary. Tube-rubber belt conveyors are available for transportation distances of up to 2000m.

Chain conveyor systems such as chain trough conveyors or scraping conveyors can be used for sawdust, bark and woodchips. Chain trough conveyors are available for horizontal and also inclined transport (up to 90°) and show a high flexibility in terms of particle size (see Figure 3.36). Charging and discharging of material is possible at any point. In order to avoid dust emissions, the conveyor can be totally encased. The disadvantages are a relatively high power demand, a low conveying capacity (due to low hauling speed) and high wear of the chain and the coating of the trough.

Screw conveyors allow the conveying of bulk materials without dust emissions (see Figure 3.37). They have small dimensions and are relatively cheap. This conveying technology is suitable for short distances and is often used for the material transport of biomass fuels with particle sizes smaller than 50mm (e.g. sawdust, pellets, woodchips). Screw conveyors have a relatively high power demand and are sensitive to metal and mineral impurities in the fuel (pieces of metals and stones can cause malfunctions). Moreover, screw conveyors are not suitable for bark. They can be recommended for relatively fine and clean biomass fuels with well-defined particle sizes.

Hydraulic piston feeders (see e.g. Figure 5.12) are used for the feeding of grate furnaces and are suitable for baled biomass fuels (straw, cereals) and bulk material with inhomogeneous particle sizes (e.g. bark, mixtures of bark and sawdust).

Figure 3.36 *Chain trough conveyor*

Source: [17].

Figure 3.37 *Screw conveyor*

Source: [17].

Figure 3.38 *Bucket elevator*

Source: [47].

Bucket elevators are used for inclined and vertical conveying of small and medium-sized particles (see Figure 3.38). The capacity can be up to 400 tonnes per hour and the maximum transportation height is up to 40m. The particle size of the material is limited by the dimension of the buckets used [47]. Problems due to dust emissions and dirt accumulation may occur if fine particles are elevated at high speeds.

For fuels like sawdust or shredded straw, *pneumatic conveyors* are widely used. They are very suitable for dust burners and complex transport routes. High-pressure systems have a hauling speed of up to 40m/s (Figure 3.39).

A comparison of the indicative process costs and energy consumption levels of a range of fuel pre-treatment options is given in Table 3.7. These costs can vary significantly between different countries.

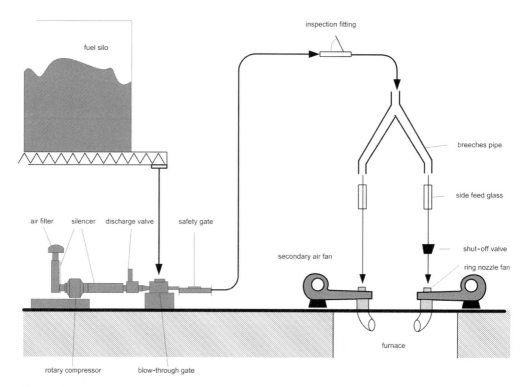

Figure 3.39 *Pneumatic conveyor with low-pressure ventilators*

Source: [48].

Table 3.7 *Typical energy consumption and specific processing costs for biomass fuel pre-treatment under Dutch conditions*

Process	Costs € per tonne input	Energy consumption per tonne input
Breaking, chipping, shredding	6.00	15kWh$_e$
Grinding	12.50	25kWh$_e$
Pulverization	31.0	40kWh$_e$
Mechanical separation by wind sifting, sieving	9.50	10kWh$_e$
Briquetting, pelletizing	7.50–25.00	15–80kWh$_e$
Mechanical dewatering (down to 75wt% (w.b.))	3.00	1.4kWh$_e$
Drying (wood, from 50 to 15wt% (w.b.))	7.0	
Thermal drying (75–15wt% (w.b.) dung, sludge)	31.0 *	3GJ$_{th}$ *
Production of RDF (10wt% (w.b.)) from MSW, yield 50% of input	18.50–34.00 **	15–100kWh$_e$ [3] + 150–700MJ$_{th}$ **

Explanations: w.b. = wet basis, * = per tonne water evaporation, ** = depending on need for drying, pelletizing, MSW = municipal solid waste, RDF = residue derived fuel

Source: [49].

Table 3.8 gives an overview of appropriate fuel delivery and furnace technologies according to the form and particle size of the biomass fuel.

Table 3.8 *Suitable fuel-feeding and combustion technologies according to shape and particle size of the biomass fuel*

Shape	Maximum particle size	Appropriate delivery system	Appropriate combustion technology
Bulk material	< 5mm	Direct injection, pneumatic conveyors	Directly fired furnaces, cyclone burners, CFB
Bulk material	< 50mm	Screw conveyors, belt conveyors	Underfeed stokers, grate furnaces, BFB, CFB
Bulk material	< 100mm	Vibro-conveyors, chain trough conveyors, hydraulic piston feeders	Grate furnace, BFB
Bulk material	< 500mm	Sliding bar conveyors, chain trough conveyors	Grate furnace, BFB
Shredded or cut bales	< 50mm	Cutters/shredders followed by pneumatic conveyors, screw conveyors or belt conveyors	Directly fired furnaces, grate furnaces, BFB, CFB
Bales, sliced bales	whole bales	Cranes, hydraulic piston feeders	Grate furnaces, cigar burners
Pellets	< 30mm	Screw conveyors, belt conveyors	Underfeed stokers, grate furnaces, BFB, CFB
Briquettes	< 120mm	Sliding bar conveyors, chain trough conveyors	Grate furnaces, BFB

Source: [50].

3.4.3 Transport systems for biomass fuels

Biomass fuels have a comparatively low energy density compared to fossil fuels. This means that biomass fuels face higher transportation costs. Transport distances must thus be kept as short as possible to minimize total transport costs, usually leading to the requirement of decentralized utilization of biomass fuels. In addition, the utilization of the available capacity of different means of transport must be optimized, taking legal framework conditions in the respective country into account.

Different means of transport can be used, depending on transport distances and type of biomass fuel:

- tractors with trailers;
- trucks;
- trains; and
- ships.

Tractors with trailers are normally used for short-distance transport of unchipped thinning residues, forest woodchips and different kinds of herbaceous biomass fuels. Average distances for tractor transport in Austria are around 10km.

Trucks are used for medium- and long-distance transport of all kinds of woody biomass fuels (e.g. logs, unchipped thinning residues (bulk or bundled), woodchips, sawdust and bark) and herbaceous biomass fuels (bulk, different bales). Different types of trucks are used depending on the type of biomass fuel transported. Flat bed trucks with side stakes are used for the transportation of logs. Bulk materials like woodchips are usually transported in trucks with side walls and a tilting bed. Pellets can also be transported by such tipper trucks, but special tank trucks are generally used instead. Typical transport distances for the supply of biomass heating and CHP plants in Austria are between 20 and 120km.

Trucks with interchangeable containers are an alternative to tipper trucks for the transportation of forest woodchips. An advantage of such container systems is the higher availability of the trucks, as the containers can be used for intermediate storage directly in the forest. A disadvantage is the decreased transport capacity due to the weight of the container.

Rail transport is used for logs, bundles and industrial by-products in bulk form. Different wagons are available depending on the fuel to be transported. The transport of herbaceous biomass fuels by train is of minor relevance.

Typical transport costs for different biomass fuels, means of transport and transport distances (based on Austrian framework conditions) are shown in Table 3.9. It can be seen that biomass fuel transport by tractor is, in general, comparatively expensive. Trucks are a reasonable means of transport for all kinds of woody biomass fuels. Rail transport is only feasible if the biomass fuels must be transported over longer distances, which is usually not the case in Austria.

Transportation of biomass fuels by ship could be reasonable for long distances and large-scale biomass trade and is especially relevant for the transportation of pellets, as pellets have now become an internationally traded product. In Scandinavia, the transport of pellets by water has become of great relevance, both within Scandinavia and for imports from Canada. Not only pellets, however, but also woodchips and bales or bundles could be transported by ship.

Economic calculations regarding international bioenergy transport focusing on long-distance large-scale transport have been performed [51]. According to this study the transport of woodchips costs between €0.019/(km.t (d.b.)) (ship) and €0.054/(km.t (d.b.)) (train). Bales can be transported over long distances for €0.020/(km.t (d.b.)) by ship and for €0.047/(km.t (d.b.)) by train. Pellets are the most economic woody biomass fuel in terms of transport costs. Long-distance transport of pellets costs between €0.020 and €0.022/(km.t (d.b.)) by train and only between €0.001 and €0.012/(km.t (d.b.)) by ship. A conclusion from this investigation is that international water transport of woody biomass fuels accounts for only a small part of the total fuel costs, whereas international rail transport is considerably more expensive.

3.5 System perspectives on bioenergy

The complexity of environmental problems implies that measures based on a holistic perspective are needed. Focusing on only one problem at a time can at worst make another environmental problem even more serious, or at best prevent advantage being taken of potential synergy effects. There is no strict definition of systems analysis. This section

Table 3.9 *Comparison of transportation costs of different biomass fuel under Austrian framework conditions*

Fuel	Moisture content [wt% (w.b.)]	Bulk density [kg (w.b.)/m³]	Means of transport	Transport capacity [m³]	Energy content [GJ]	Total costs [€/GJ (NCV)]	Specific costs [€/kmt (d.b.)]
Bark	50	320	Truck	70	180	0.40	0.13
	50	320	Train, 20 wagons	1400	3636	0.89	0.29
Bark (dried)	30	230	Truck	100	284	0.26	0.09
Industrial waste woodchips (wet)	50	340	Truck	65	180	0.40	0.13
Industrial waste woodchips (dry)	30	240	Truck	95	281	0.26	0.09
Sawdust	50	240	Truck	95	184	0.38	0.13
Pellets	10	550	Truck, container	35	320	0.46	0.09
	10	550	Train, 20 wagons	1400	12,755	0.38	0.07
Forest woodchips	30	250	Truck, container	35	108	0.28	0.48
	30	250	Tractor, trailer	12	36	1.33	2.33
Forest residues	50	80	Tractor, trailer	12	7	6.30	10.22
Whole trees (thinnings)	50	160	Tractor, trailer	12	14	3.15	5.11
Straw (bulk)	15	60	Tractor, trailer	20	18	3.43	5.84
Straw (cuboid bales)	15	140	Tractor, trailer	28	54	0.89	1.49
	15	140	Truck	110	223	0.32	0.11
Whole crops (cuboid bales)	15	190	Truck	110	302	0.24	0.08
Heating oil (extra light)		840	Tank truck	30	1087	0.34	0.07

Explanations: w.b. = wet basis; d.b. = dry basis; NCV = net calorific value.
Source: [1].

refers to what one may call 'energy systems analysis', which could be (loosely) defined as analysis of energy, environmental and economic questions linked to the energy system by means of methods and models which facilitate an integrated analysis of the adaptation of new technologies and measures to complex energy systems.

Energy systems analysis can be employed with varying scope and aim. It is not axiomatic that a systems analysis approach to defining – and proposing measures to mitigate – environmental problems succeeds in considering all factors of importance, but careful analyses within wisely chosen system boundaries can facilitate a structured analysis and a good insight into problems, with possibilities to draw conclusions that are not obvious at a first glance at the problem.

Biomass production and use for energy purposes is a good example of where a systems perspective must be adopted. One primary aim of a greater use of renewable energy sources such as biomass is to reduce society's influence on the climate by replacing fossil fuels. But several central environmental objectives in contemporary policy, besides climate change mitigation, can be associated with land use. These must be taken into account, therefore, when designing strategies to increase the use of bioenergy, which is regarded as an important element in the reorientation of many countries' energy systems to becoming more environment-friendly.

Here, we will use two illustrative case studies to show how systems analysis can guide the development of strategies for cost-effective and environmentally attractive bioenergy expansion. First, we will outline the concept of multifunctional bioenergy systems and use the example of Salix production in Sweden to show how systems analysis can yield information on how biomass can be produced to provide significant environmental benefits in connection with changing how land is used in forestry and agriculture. Second, we will present a case study on biomass co-firing in Poland to show how systems analysis that considers not only the supply side, but also the state and dynamics of the end-use side, can help to identify cost-effective early options for bioenergy expansion.

3.5.1 Case 1 Multifunctional bioenergy systems (MBS)

If located, designed and managed wisely, energy crop plantations can, besides producing renewable energy, also generate local environmental benefits. Examples of such multifunctional bioenergy systems (MBS) are Salix plantations leading to soil carbon accumulation, increased soil fertility, reduced nutrient leaching and improved hunting potential, representing more general benefits. Another category of plantations are those designed for dedicated environmental services, such as shelter belts for the prevention of soil erosion, plantations for the removal of cadmium from contaminated arable land (phytoextraction), and vegetation filters for the treatment of nutrient-rich, polluted water.

A correct assessment of different bioenergy systems' total environmental impact requires that the entire system be considered through a systems analysis approach. It is from a holistic perspective that potential synergy effects can be identified.

The MBS concept presupposes a well-executed system integration, where the energy-forest plantation is linked with the activities that demand a specific environmental service, and frequently also with the end-user stage for the biomass (such as a local heating plant). The systems analysis approach to identifying and evaluating different multifunctional bioenergy systems also facilitates the implementation phase by elucidating the different functions required for the realization of the full potential of a specific MBS application.

Salix cultivation for cadmium unloading on farmland with high cadmium content in the soil can be taken as an example of the need for integration (Figure 3.40). Due to the unusually high cadmium content in the harvested biomass, special requirements are placed on the end-user. After the cadmium is concentrated in the Salix chips, it must be separated from the fly ash upon incineration. Most suitable, for reasons of cost, is judged to be the use of large incinerator facilities, with combined burnout of the ash's carbon content and, finally, burning in a waste heat boiler where cadmium is separated in a barrier filter after cooling. The cadmium content is judged to decrease by more than 90 per cent by this method, which means that the remaining ash fraction is so pure it can be returned to farmland as fertilizer. The fly-ash fraction, in which the cadmium is concentrated, can then be disposed of, for example in a controlled deposit for permanent storage.

Obviously it is an advantage if MBS applications are established in such a way that the end-user of the biomass is located near the biomass plantations, whose placement is governed by the local demand for specific environmental services. Such local solutions improve the conditions for creating local cycles for plant nutrients. This can also be a strength in the sense that the actors involved share a common local vision, which constitutes a starting point for the collaboration. But it is not an absolute requirement that the end-user be a 'neighbour' to the energy-forest plantation. Biomass can be transported for relatively long distances while maintaining good economy.

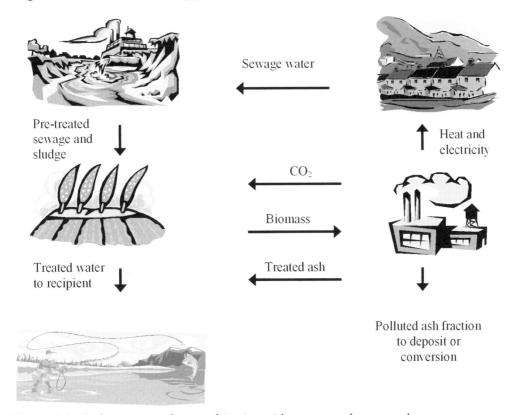

Figure 3.40 *Linking energy-forest cultivation with energy and sewage plants.*

Note: This satisfies the need for both energy raw materials and recirculation of nutrients which would otherwise require different handling, or else go to a deposit or a recipient nearby.

Below, the case of multifunctional Salix production (MSP) in Sweden is briefly outlined. Research projects funded by the Swedish Energy Agency **[52–54]** have been accomplished with the intention to:

- investigate which environmental services could be obtained from multifunctional bioenergy systems;
- estimate how much biomass could be produced in MBS in Sweden based on an inventory of demand for the environmental services that can be provided with such systems;
- estimate the economic value of the environmental services that can be offered with MBS, as well as the production costs for biomass from such systems; and

- identify market- and structure-determining barriers to different MBS, and propose solutions that overcome barriers and strengthen present driving forces for increasing the production of bioenergy.

The analysis has in this case been confined to biomass production on agricultural land, with Salix as the energy crop. This does not mean Salix is the only conceivable energy crop for MBS applications. The focus is explained by the fact that, in Sweden, most research has been done on Salix as an energy crop, and that it is also the most widely cultivated lignocellulosic energy crop. Hence, the best foundation for evaluating MBS in Sweden has been with those based on Salix.

The majority of environmental services are specific to a particular type of soil, localization in the landscape, or geographical region, and therefore cannot be obtained

Figure 3.41 *Salix field irrigated with pre-treated municipal waste water in Enköping*

Note: The inset picture shows measurement equipment used to chart the nitrogen flows in the field. An important question is how much of the nitrogen input is transformed to nitrogen oxide, a powerful greenhouse gas. The investigations until now indicate that the climatic impact of these discharges is small in relation to the climatic benefit of the produced biomass in replacing, for example, fossil fuels in municipal heating plants. Also important are the hygienic aspects of sewage-irrigated Salix production. Experiments show that the risk of spreading infection is low, but that unsuitable locations, such as near to waterways, should be avoided.

Source: Photo by Pär Aronsson, SLU.

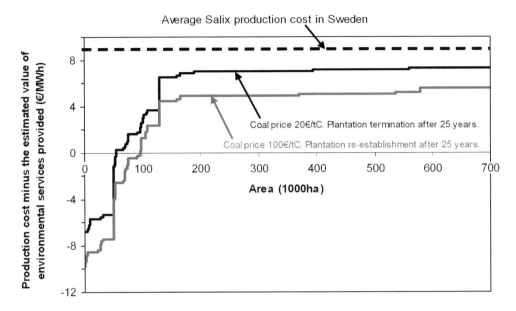

Figure 3.42 *Practical potential and production cost for multifunctional energy plantations on a Sweden-wide level*

Note: The average production cost for the country (excluding land purchase and common business costs for the farmer) is given as a reference. Income that is attributable to soil carbon changes (increased fertility, carbon binding in mineral soils, and decreased emission of carbon dioxide from humus soils) is added to income from other environmental services in the cases where these combinations arise. Not included, however, are those areas where income related to soil carbon changes is the only income.

simultaneously. For instance, Salix cultivation can be established for sewage water purification on both mineral soils and humus soils. But the possibility of combining this purification with cadmium unloading depends on whether mineral soils or humus soils are utilized, because cadmium unloading is assumed to be feasible only on mineral soils.* Moreover, the possibility of combining these two environmental services is dependent on whether the cadmium content of farmland being used is so high that one can consider Salix cultivation for cadmium unloading.

Hence, in order to be able to calculate the total potential of multifunctional energy plantations in Sweden, an analysis is required of which combinations of environmental services can be obtained at the same place of cultivation. Based upon these possible combinations of services and the respective services' potential, the total practical potential is calculable.

Figure 3.42 shows the estimated practical potential (expressed in hectares of Salix cultivation) for MSP in Sweden, together with the estimated production cost that results if the value of the provided environmental services is included as an extra income in the cultivation's calculation. The figure displays two graphs for different cases of the coal price

---

* Cadmium is bound to a higher degree in hard-to-access compounds in humus soils, and the proportion of cadmium accessible to plants is therefore lower than in other soils.

(€20 and €100/t coal) and thus of income for carbon binding in soil and standing biomass. As the figure demonstrates, it is judged possible to grow Salix at a negative production cost in an arear approaching 100,000ha, if diverse potential extra environmental services are fully utilized. For a further 200,000ha or more, Salix cultivation can be done at a production cost half that of conventional cultivation. This may be compared with the approximately, 15000ha of farmland which are used today for growing Salix.

The results of the analysis are very encouraging: the environmental benefits from a large-scale expansion of properly located, designed and managed biomass plantations could be substantial, as the negative environmental impacts from current agricultural practices and also municipal waste treatment could be significantly reduced. Given that additional revenues – corresponding to the economic value of the provided environmental services – can be linked to the MSP systems, the economic performance of such biomass production can improve dramatically. Biomass supply from MSPs could bring substantial improvements in the biomass supply cost and also in other aspects of competitiveness against conventional resources. Establishment and expansion of such plantation systems would also induce development and cost reductions along the whole biomass supply chain. Thus, MSPs could become prime movers and pave the way for an expansion of low-cost perennial crop production for the supply of biomass as industrial feedstock and for the production of fuel and electricity.

3.5.2 Case 2 Co-firing opportunities in Poland

3.5.2.1 Introduction

This section applies a systems analysis in investigating to what extent co-firing of biomass in existing coal-fired power plants can contribute to initiating a market for lignocellulosic biomass, using Poland as an example. More specifically, the example matches regional biomass supply potentials with opportunities for co-firing biomass in existing coal-fired power plants in the near term (2010) and estimates medium-term (2020) potentials in these units. The example given combines available estimates of biomass supply, information on the power plant infrastructure and assessment of co-firing capacity in different types of power plants based on previous experiences. It illustrates a systems approach which combines biomass supply data with data on conversion opportunities. The results of the example call for quick action with respect to implementation of co-firing in order to establish biomass markets.

In the EU, bioenergy and energy crops are seen as key components in strategies for the transformation of the energy system to become more sustainable and less dependent on imported fuels. In December 2005, the European Commission (EC) presented a Biomass Action Plan [55] in order to increase the development of bioenergy, which is hampered by barriers both on the supply side and on the conversion side. The biomass potential from residues and energy crops in most EU countries is significant, but mostly unexploited. On the conversion side, a suitable energy conversion infrastructure is lacking and the introduction of such infrastructure will be costly with regard to direct investments (e.g. for new CHP plants) and indirect investments (e.g. for district heating systems). In addition, the large need for new investments in heat and power generation in several member states may result in a lock-in situation dominated by natural gas and coal-fired power generation, with technologies not appropriate for biomass use.

Thus, both supply-side and demand-side strategies are required to promote growth in bioenergy use. Decreases in biomass supply costs reduce supply-side barriers, and the development of near-term conversion-side options stimulates growth by providing a market for the produced biomass. In more general terms there are both opportunities and limitations imposed by the turnover in capital stock of the existing energy system. Replacement of old stock (e.g. power plants, as this is the scope of this work) opens up the potential for investments in new technologies, but too high a requirement on early replacement opens up a risk of lock-in effects in second-best technologies.

The aim of the example given in this section is to illustrate the use of systems analysis in identifying cost-competitive near-term options for the expanded use of biomass for energy, by combining lignocellulosic supply potentials with near-term conversion options. The methodology is applied to Poland as a model country, and systems for biomass co-firing with coal in the existing power plant infrastructure are considered as conversion-side options. Focusing on the existing system is expected to yield low-cost options and does not require speculation on preferred technologies in future investments. In addition to illustrating the general approach, the analysis gives insights into the possible volume of biomass that can be used in co-firing applications in relation to estimates of medium- to long-term bioenergy potentials in Poland. Figure 3.43 shows the Polish power plant infrastructure with respect to fuel and the age distribution of the plants. The fossil dominance is clear as is the fact that a substantial fraction of these units are old, and large investments in the future are required in order to meet likely post-Kyoto targets on greenhouse gases as well as the Large Combustion Plant Directive (LCPD).

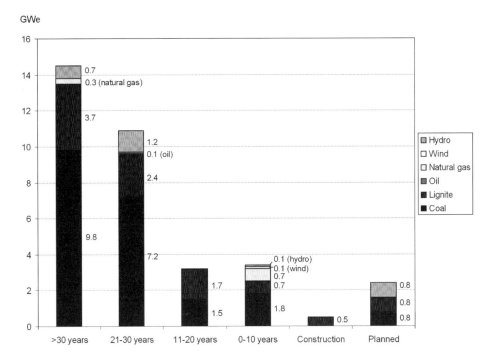

Figure 3.43 *Power generation system of Poland by age and fuel, including hydro*

Source: The data are taken from the Chalmers Energy Infrastructure Database [56, 57].

3.5.2.2 Systems modelling by cost optimization

The analysis is divided into two parts, referring to the years 2010 and 2020, respectively. The year 2010 analysis is based on combining estimates of near-term biomass potential and information on the power plant infrastructure as input for modelling the co-firing potential for each of the 16 regions ('*voivodships*') in Poland. This potential is then compared to Poland's indicative 2010 targets for renewable energy (RES: 7.5 per cent of primary energy mix) and electricity from renewable sources (RES-E: 7.5 per cent of total electricity from a share of about 2 per cent in 2002), adopted as part of the EU accession treaty.

The 2020 analysis compares the anticipated growth in electricity generation **[58]** with the modelled electricity generation in existing power plants, considering turnover in capital stock, i.e. phasing out of the existing power plant infrastructure. This gives insights into the prospective contribution of the presently existing power plant infrastructure to meet the future electricity demand. The year 2020 analysis also constitutes a national-level estimate of the co-firing potential up to the year 2020, which is evaluated in relation to medium-term energy policy targets (RES: 14 per cent of primary energy mix by 2020) and estimates of longer-term bioenergy potentials in Poland. This gives information on the role of biomass co-firing as a contributor to realizing the RES target and longer-term bioenergy potentials.

For the 2010 analysis, energy systems modelling in the form of a cost-optimizing linear programming (LP) model is applied, which is briefly outlined in Figure 3.44 and described in detailed elsewhere **[59]**. The inputs to the LP model are estimates of the biomass supply potential (including straw, forest residues, fuelwood and energy crops) and data on the power plant infrastructure together with an assessment of the capacity for co-firing in the boilers and cost data. Maximum and minimum biomass supply potentials were defined in order to examine the influence of the biomass supply on the co-firing potential. Also, two different scenarios for the transportation of biomass across regional boundaries (between *voivodships*) have been analysed: the first scenario does not allow cross-border transport of biomass, the second scenario allows cross-border transport of biomass. The biomass supply and the power plants are matched by the model, which as outputs gives the potential of electricity in 2010 from co-firing, and the total cost for implementing the co-firing potential together with the CO_2-avoidance cost (Figure 3.44). For the parameters used in the modelling see **[59]**.

The available boiler capacity for co-firing is based on the Chalmers Energy Supply Side database, which contains all power plants in EU25 countries with a capacity generally exceeding 10MW_e **[56, 57]** with the age structure given in Figure 3.43, i.e. with respect to the installed capacity. The 233 boilers less than 30 years old were considered in this work. The co-firing capacity of each boiler is based on the above-mentioned technical assessment and an assumption of load hours for each boiler and fuel type (see **[59]** for details).

3.5.2.3 Modelling results

The result of the year 2010 analysis shows a potential for electricity to be produced from biomass in co-firing at 1.6–4.6 per cent (2.3–6.6TWh_e) of the total projected electricity production in 2010, corresponding to maximum and minimum biomass supply estimates (and depending on biomass transportation scenario). Adding this potential to the existing production of about 2 per cent electricity from renewable energy sources (RES-E) gives an overall contribution of RES-E in the range 3.6–6.6 per cent, i.e., less than the above-mentioned 7.5 per cent RES-E target. The additional cost for the implementation of co-firing is less than €20 per MWh_e, corresponding to a CO_2-avoidance cost somewhat lower than €20 per tonne CO_2.

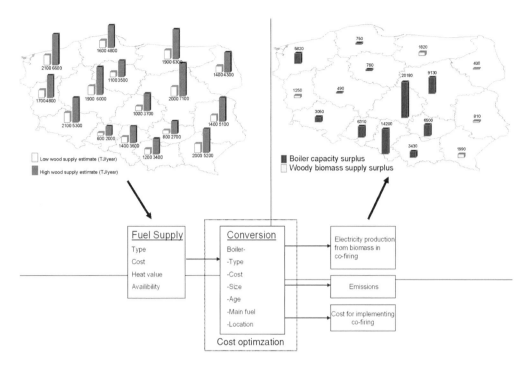

Figure 3.44 *Modelling methodology used in the co-firing example*

Note: The parameters of the LP model (bottom); maximum and minimum estimates of the biomass supply 2010 for the 16 regions in Poland (left); the cost optimization parameters (centre) and model output, i.e. distribution of surplus of either boiler capacity or biomass supply potential (right).

Figure 3.45 compares the modelled electricity generation 2005–2020 in the presently installed power plant infrastructure with the total domestic electricity generation in Poland up to 2020, as given in [58]. From the figure it is clear that – although increases in average load, electricity imports and investments in energy efficiency can reduce the demand for new generation capacity – Poland will experience a large need for investments in new electricity generation. In Figure 3.45, 'Additional electricity' should not be understood as equivalent to additional generation capacity requirement: increased average load, electricity imports as well as investments in energy efficiency measures provide additional opportunities. 'Additional electricity' in 2005 corresponds to the electricity generation from wind and hydro.

Figure 3.46 shows the co-firing potential in currently existing power plants up to the year 2020 with the contribution from individual technologies. The dashed line corresponds to a (hypothetical) case where co-firing is applied in all existing boilers. Figure 3.47 gives an illustrative modelling of the expansion of biomass use for co-firing. Here, it is assumed that biomass co-firing is implemented rapidly in fluidized bed combustion (100 per cent of available boilers by 2010) while full implementation takes a longer time for the other boiler types (100 per cent of available boilers by 2016). Under the assumptions made, biomass

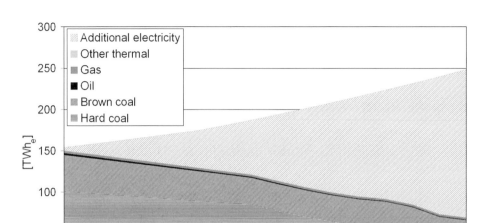

Figure 3.45 *Modelled electricity generation 2005–2020 in boilers less than 30 years old*

Note: Total domestic electricity generation in Poland up to 2020 as given in **[58]** is included for comparison.

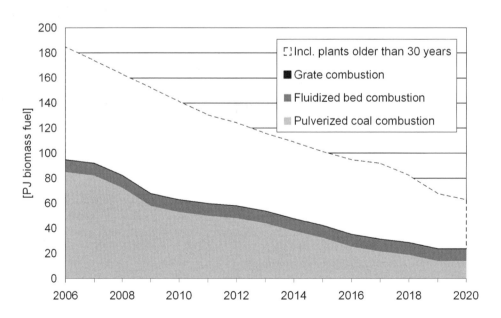

Figure 3.46 *Biomass co-firing potential in Polish power plants up to the year 2020, as estimated in the example*

co-firing in existing plants reaches its maximum in 2014 and then decreases due to the phasing out of the plants. Thus, RES-E generation from biomass co-firing in presently installed power plant infrastructure decreases quite fast due to the relatively large share of old plants, and the contribution to longer-term RES-E targets will be small (unless substantial investments are made in the modernization of existing generation or in new coal-fired power plants suitable for co-firing).

When comparing the modelled biomass used for co-firing in Figure 3.47 with the Polish 2020 RES target (14 per cent of primary energy supply, or some 600–700PJ per year) and with estimates of longer-term bioenergy potentials (some 1–3EJ per year, depending on the amount of land available for energy crop production in the future [60]), it becomes evident that co-firing will have its best role as an early option for expanding biomass use for energy production.

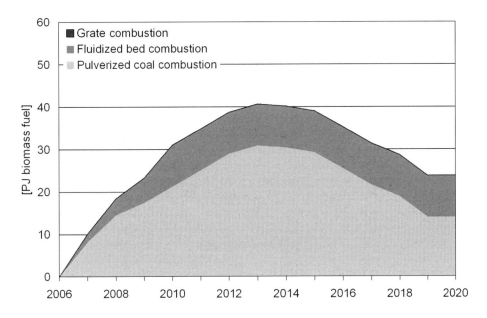

Figure 3.47 *Illustrative modelling of the expansion of biomass use for co-firing in existing units in Poland over the period up to 2020*

The example given in this section illustrates how an energy systems analysis which combines biomass supply data with data on conversion opportunities can be used to estimate to what extent the existing boiler infrastructure can act as a prime mover for expanding the co-firing option. Thus, in this example, the focus is on the system that is already in place as this does not require any speculation on future technology mix.

Obviously, the new investments to be carried out over the period (to fill the gap 'Additional electricity' in Figure 3.45) will also have co-firing as an option, but depend on the choice of future fossil-fuelled technologies. For the example given, it should be mentioned that there have been some apprehensions about the risk of depletion of local biomass markets (biomass supply market in the vicinity of large biomass co-fired boilers).

However, in most regions in Europe the overall challenge is the low use of biomass for energy compared to assessed potentials and there is a great need to establish biomass markets. Biomass co-firing with coal in existing boilers is a cost-efficient way to initiate a biomass market for lignocellulosic biomass. It allows for the diffusion of competitive bioenergy strategies within the existing (or moderately modified) energy capital stock, and a growth in biomass use that is not constrained by the rate at which new bioenergy facilities (conversion technologies) can be put in place. Co-firing strategies must of course be matched with policies for promoting the development of other and more decentralized RES systems, such as biomass-based district heating with CHP.

3.5.3 Conclusions from the two cases

From the case on MSP in Sweden it can be concluded that the production and use of biomass from MBS would not only contribute to the development of more sustainable energy and industrial production, but also to the development of a more sustainable agriculture and to the increased recirculation and efficiency in the societal use of essential resources such as phosphorus and other nutrients. In this way, MBS systems may also become a valuable tool for meeting additional challenges, such as getting the world's water cleaner and preserving the long-term quality of agricultural soils. Systems analysis is essential for correctly identifying and evaluating different MBS concepts.

The Polish co-firing example illustrates how energy systems modelling combined with an assessment of available technologies can be used to investigate biomass co-firing potentials and to estimate the cost of such an option. Thus, for Poland, the implementation of co-firing constitutes a relatively quick way to facilitate conditions for developing a supply system which can grow over time. In addition, it offers opportunities to increase biomass use for energy at a cost corresponding to a CO_2-avoidance cost of less than €20 per tonne CO_2.

3.6 References

1   STOCKINGER, H. and OBERNBERGER, I. (1998) *Systemanalyse der Nahwärmeversorgung mit Biomasse*, volume 2 in Thermische Biomassenutzung series, dbv-Verlag der Technischen Universität Graz, Graz, Austria
2   MITCHELL, P. (1997) 'Harvesting and pre-treatment technologies for different biomass fuels', in *Proceedings of the 2nd Biomass Summer School*, Institute of Chemical Engineering, Graz University of Technology, Austria
3   HARTMANN, H. and MADEKER, U. (1997) *Der Handel mit biogenen Festbrennstoffen – Anbieter, Absatzmengen, Qualitäten, Service, Preise*, Landtechnik Bericht Nr. 28, Landtechnik Weihenstephan, Freising, Germany
4   HARTMANN, H. (1998) ‚Influences on the quality of solid biofuels – causes for variations and measures for improvement', in *Proceedings of the 10th European Conference and Technology Exhibition*, June 1998, Würzburg, Germany, CARMEN, Rimpar, Germany, pp184–187
5   SANDER, B. (1997) 'Properties of Danish biofuels and the requirements for power production', *Biomass and Bioenergy*, vol 12, no 3, pp177–183
6   OBERNBERGER, I. (1997) *Nutzung fester Biomasse in Verbrennungsanlagen unter besonderer Berücksichtigung des Verhaltens aschebildender Elemente*, volume 1 of Thermal Biomass Utilization series, BIOS, Graz, Austria, dbv-Verlag der Technischen Universität Graz, Graz, Austria
7   BAKKER, R. and JENKINS, B. (1996) 'Feasibility of fuel leaching to reduce ash fouling in biomass combustion systems', in *Proceedings of the 9th European Bioenergy Conference*, volume 2, Elsevier Science Ltd, Oxford

8   NIKOLAISEN, L. et al (1998) *Straw for energy production – Technology – Environment – Economy*, 2nd edn, Centre for Biomass Technology, Denmark
9   KNUDSEN, N.O., JENSEN, P. and DAM-JOHANSEN, K. (1998) 'Possibilities and evaluation of straw pretreatment', in *Proceedings of the 10th European Bioenergy Conference*, June 1998, Würzburg, Germany, CARMEN, Rimpar, Germany
10 CLEMENS et al (1996) *Auswertung und Beschreibung des Energiegetreide-Anbauversuchs im Betrieb Franz Schaub*, report of the Lehr- und Versuchsanstalt Emmenhausen-Borler for the year 1994/95, Landwirtschaftskammer Rheinland-Pfalz , Germany
11 LEWANDOWSKI, I. (1998) 'EMI – European Miscanthus Improvement', in *Proceedings of the 10th European Bioenergy Conference*, June 1998, Würzburg, Germany, CARMEN, Rimpar, Germany
12 JORGENSEN, U. (1996) 'Miscanthus yields in Denmark', in *Proceedings of the 9th European Bioenergy Conference*, Volume 2, Elsevier Science Ltd, Oxford, pp1129
13 HARTMANN, H. (1996) 'Characterisation and properties of biomass fuels', in *Proceedings of the 1st Biomass Summer School*, Institute of Chemical Engineering, Graz University of Technology, Austria
14 OLSSON, R. and HADDERS, G. (1996) *European Energy Crops Overview – Country Report for Sweden*, Swedish University of Agricultural Science, Uppsala, Sweden
15 DANFORS, B. (1996) *Salix – biobränsle fran skörd till värmeverk*, Swedish Institute of Agricultural Engineering, Uppsala, Sweden
16 SERUP, H. et al (1999) *Wood for Energy Production, Technology – Environment – Economy*, Center for Biomass Technology, Denmark, available at: www.videncenter.dk
17 MARUTZKY, R. and SEEGER, K. (1999) *Energie aus Holz und anderer Biomasse*, DRW-Verlag Weinbrenner, Leinfelden-Echtlingen, Germany
18 SPROUT-MATADOR (n.d.) company brochure, Sprout-Matador, Esbjerg, Denmark
19 BMUJF (1994) *Branchenkonzept Holz, Bundesministerium für Umwelt*, Jugend und Familie, Vienna, Austria
20 BRUNNER, T., OBERNBERGER, I. and WELLACHER, M. (2004) 'Waste wood processing in order to improve its quality for biomass combustion', in *Proceedings of the 2nd World Conference and Exhibition on Biomass for Energy, Industry and Climate Protection*, May 2004, Rome, Italy, volume I, ETA-Florence, Italy, pp250–253
21 HARTMANN, H. (1995) 'Lagerung, Transport und Umschlag von Halmgütern', in *Proceedings of the Conference 'Logistik bei der Nutzung biogener Festbrennstoffe'*, Stuttgart, May 1995, Landwirtschaftsverlag GmbH, Münster, Germany
22 LUGER, E. (1996) 'Forschungsprojekt Nachwachsende Rohstoffe – Verfahrenstechnik', *Tagungsband zum Workshop: Thermische Ganzpflanzennutzung*, Wieselburg, May 1994, Ministry for Agriculture and Forestry, Vienna
23 ANDERSEN, G. and BRUNBERG, B. (1996) 'Baling of unchipped logging residues', in *Harvesting, Storage and Road Transportation of Logging Residues*, proceedings of a workshop of IEA-BA-Task XII activity 1.2, October 1995 in Glasgow, Scotland, Danish Forest and Landscape Research Institute, Horsholm, Denmark
24 TIMBERJACK (2003) Slash Bundling Technology, http://websrv1.tekes.fi:8080/opencms/opencms/OhjelmaPortaali/Paattyneet/Puuenergia/fi/Dokumentt iarkisto/Viestinta_ja_aktivointi/Julkaisut/PROJEKTIT/PUUY12ESITELMA.pdf [19 May 2006]
25 OBERNBERGER, I. and THEK G. (2005) *Herstellung und Nutzung von Pellets*, volume 5 of the Thermal Biomass Utilization series , Institute for Resource Efficient and Sustainable Systems, Graz University of Technology, Austria
26 FRANK, G. (1989) *Kompakt geht es besser – Pelletieren unter Druck, Maschinenmarkt*, Würzburg 95 Sonderdruck Nr. 32 und 37/1989, Vogel Verlag und Druck KG, Germany
27 NILSSON, B. (1997) 'Biomass pellets – pelletizing technology, logistical aspects. pellet furnace', in *Proceedings of the 2nd Biomass Summer School*, Institute of Chemical Engineering, Graz University of Technology, Austria
28 SCHOLZ, V., IDLER, C., DARIES, W. and EGERT, J. (2005) 'Lagerung von Feldholzhackgut, Verluste und Schimmelpilze', *Agrartechnische Forschung*, vol 11, no 4, pp100–113

29 STOCKINGER, H. and OBERNBERGER, I. (1997) *Langzeitlagerung von Rinde – Meßbericht zu einem Lagerversuch von Rinde bei verschiedenen Randbedingungen*, report for the FWF research project P10669-ÖTE: Life-Cycle Analyse für Bioenergie, Institute of Chemical Engineering, Graz University of Technologyz, Austria
30 SWISS COMBI (2006) Homepage, www.swisscombi.ch, SWISS COMBI W. Kunz dryTec AG, Dintikon, Switzerland
31 BÜTTNER (2001) company brochure, Büttner Gesellschaft für Trocknungs- und Umwelttechnik mbH, Krefeld, Germany
32 ÖHMANN, M., NORDIN, A., HEDMAN, H. and JIRJIS, R. (2002) 'Reasons for slagging during stemwood pellet combustion and some measures for prevention', in *Proceedings of the 1st World Conference on Pellets*, September 2002, Stockholm, Sweden, Swedish Bioenergy Association, Stockholm, Sweden, pp93–97
33 PONNDORF (2001) company brochure, Ponndorf Maschinenfabrik GmbH (Ed.), Kassel, Germany
34 STORK (1995) company brochure, Stork Engineering, Göteborg, Sweden
35 MÜNTER, C. (2004) 'Exergy steam drying for biofuel production', in *Proceedings of the European Pellets Conference 2004*, O.Ö. Energiesparverband, Linz, Austria, pp267–269,
36 MEIJER R. and GAST C. H. (2004) 'Spontaneous combustion of biomass: Experimental study into guidelines to avoid and control this phenomena', in *Proceedings of the 2nd World Conference on Biomass for Energy, Industry and Climate Protection*, 10–14 May 2004, Volume II, Rome, Italy, pp1231–1233
37 KOFMANN, P. D. and SPINELLI, R. (1997) *Storage and Handling of Willow from Short Rotation Coppice*, ELSAMPROJEKT, Denmark
38 KOFMANN, P., THOMSEN, I., OHLSEN, C., LEER, E., SCHMIDT, E. R., SORENSEN, M. and KNUDSEN, P. (1999) *Preservation of Forest Wood Chips*, ELSAMPROJEKT, Denmark
39 JIRJIS, R. (1996) 'Storage and drying of biomass – new concepts', in *Proceedings of the 2nd Biomass Summer School*, Institute of Chemical Engineering, Graz University of Technology, Austria
40 WEINGARTMANN, H. (1991) *Hackguttrocknung*, Österr. Kuratorium für Landtechnik, Vienna
41 BOJER, N. and SCHMIDT, E. R. (1996) 'Experience of procurement of straw and wood fuel for Danish power plants', in *Proceedings of 'Feedstock preparation and quality'; workshop of IEA Bioenergy Task XII*, ELSAMPROJEKT A/S, Fredericia; Dänemark
42 OBERNBERGER, I. (1994) *Zu erwartende Abwasseremissionen des geplanten Rindenfreilagerplatzes des Biomasseheizwerkes Lieboch und daraus ableitbare Empfehlungen hinsichtlich bautechnischer Ausführung und Lagerlogistik*, Umwelttechnischer Bericht, Institute of Chemical Engineering, Graz University of Technology, Austria
43 BJÖRKHEM, U., DEHLEN, R., LUNDIN, L., NILSSON, S., OLSSON, M. and REGNANDER, J. (1977) Storage of pulpwood under water sprinkling – effects on insects and the surrounding area, Research notes 107/177, Royal College of Forestry, Dep. of Operational Efficiency, Garpenberg, Sweden
44 MAWERA (n.d.) company brochure, MAWERA Holzfeuerungsanlagen GmbH&CoKG (ed.), Hard/Bodensee, Austria
45 NIKOLAISON, L. (1998) *Straw for Energy Production – Technology – Environment – Economy*, Center for Biomass Technology, Denmark, available at www.videncenter.dk
46 KONSORTIUM RESTSTOFFHANDBUCH (1996) *Reststoffhandbuch für die Papierindustrie*, Konsortium Reststoffhandbuch, Switzerland
47 PAJER, G. and KUHNT, H. (1988) *Stetigförderer*, Verl. Technik Berlin, Germany
48 LAMBION, (n.d.) company brochure, Lambion Feuerungs- und Anlagenbau GmbH, Arolsen-Wetterburg, Germany
49 ZEEVALKINK, J. A. and VAN REE, R. (2000) *Conversietechnologieën voor de productie van elektriciteit en warmte uit biomassa en afval*, EWAB Marsroutes, Taak 2
50 OBERNBERGER, I. (1996) 'Decentralised Biomass Combustion – State-of-the-Art and Future Development', Keynote lecture at the 9th European Biomass Conference in Copenhagen, *Biomass and Bioenergy*, vol 14, no 1, pp33–56 (1998)
51 HAMELINCK C. and N., SUURS R. A. A., FAAIJ ANDRÉ, P. C. (2005) 'International bioenergy transport costs and energy balance', *Biomass and Bioenergy*, vol 29, pp114–134

52 BERNDES, G., FREDRIKSSON, F. and BÖRJESSON, P. (2004) 'Cadmium accumulation and Salix based phytoextraction on arable land in Sweden', *Agriculture, Ecosystems and Environment*, vol 103, no 1, pp207–223
53 BERNDES, G. and BÖRJESSON, P. (2004) 'Low cost biomass produced in multifunctional plantations – the case of willow production in Sweden', *2nd World Conference and Technology Exhibition on Biomass for Energy, Industry and Climate Protection*, Rome, Italy, 10–14 May
54 BÖRJESSON, P. and BERNDES, G. (2006) 'The prospects for willow plantations for wastewater treatment in Sweden', *Biomass and Bioenergy*, vol 30, pp428–438
55 EC (2005) *Biomass Action Plan*, COM(2005) 628 final, European Commission
56 KJÄRSTAD J., and JOHNSSON, F. (2004) 'Simulating future paths of the European power generation – applying the chalmers power plant database to the british and german power generation system', *7th Greenhouse Gas Control Technologies Conference (GHGT7)*, Vancouver, Canada, 5–9 September
57 KJÄRSTAD, J. and JOHNSSON, F. (2006) 'The European power plant infrastructure: Presentation of the Chalmers energy infrastructure database with applications', *Energy Policy*, vol 35, no 7, pp3643–3664
58 EC (2003) *European Energy and Transport Trends to 2030*, European Commission
59 BERGGREN, M., LJUNGGREN, E. and JOHNSSON, F. (2006) 'Biomass co-firing potentials for electricity generation in Poland – matching supply and co-firing opportunities', *Biomass and Bioenergy*, in press
60 VAN DAM, J., FAAIJ, A. and LEWANDOWSKI, I. (2005) *Biomass Production Potentials in Central and Eastern Europe under Different Scenarios*, Final WP 3 report in the EC-funded project VIEWLS. http://viewls.viadesk.com

4
Domestic Wood-burning Appliances

4.1 Introduction

Wood fire has been used as a heat source for thousands of years. During that time, the methods for burning wood and other biomass have progressed from an open pit to very sophisticated, controlled combustion systems. A number of appliance types have been developed to provide central heating using furnaces or boilers, or more localized heat using stoves and fireplaces. The traditional batch-fired systems burning firewood have been augmented by systems designed to burn pelletized wood wastes, agricultural grains or woodchips. The pressure to develop systems which minimize air pollution and maximize heating efficiency has led to imaginative and innovative new designs.

Provided that firewood is grown in a sustainable manner and used in efficient combustion systems with insignificant hydrocarbon emissions, firewood is a renewable energy source. Already today, it is a significant heating source in most of the world. Due to its potential of being CO_2 neutral, an increased use of small-scale combustion can have a significant impact on reducing greenhouse gas emissions. Developments in Europe and North America are the focus of this chapter, but its usage is much more widespread.

The fuels in use are mainly wood logs, both soft wood and hard wood. In wood stoves and boilers, wood briquettes, wood pellets, peat, peat briquettes and coal briquettes are also used to some extent. Wood pellets are also used in specially designed pellet stoves and pellet boilers.

4.2 Design considerations

In spite of its long history, biomass is the most difficult of the commonly used heating fuels to burn cleanly and efficiently. Residential wood-fired heating systems typically have considerable amounts of particulate, carbon monoxide and other unburned gaseous emissions compared to systems fired by natural gas and oil-fired systems. The more recent development of wood-fired appliances has focused largely on the reduction of these undesirable emissions in order to retain the desirability of wood as a fuel source.

The challenge in designing batch-fired heating systems is to provide a controlled rate of heat output over as long a duration as possible, while capturing as much of the heat released by the fuel as practicable. It is quite common that the heat output is controlled by limiting the supply of primary combustion air to the fuel charge. This results in very high emissions of incomplete combustion products unless an effective secondary burning system is incorporated. Development of efficient and clean-burning systems has preoccupied product designers for decades. An alternative approach has been the use of systems where the heat is accumulated by high thermal mass systems (stoves) or separate heat storage tanks for boilers. These approaches allow the system to operate at a high burn rate in a less air-starved mode, while using the accumulating heat capacity to provide an acceptable rate and duration of heat release.

With the introduction of wood pellets and automatic combustion systems in recent years, it has become much easier to continuously adapt the heat output to variations in heat demand over time.

A major impediment to the design of good appliances has been the lack of commonly accepted test protocols for determining the emissions and efficiency of these appliances. It is only since the 1990s that consensus standards for these evaluations have been developed, and they continue to undergo refinement as more experience with their use is gained.

The basic principles of various domestic batch-fed wood-burning appliances available on the market are described in Section 4.3, and processed-fuel appliances in Sections 4.4 and 4.5. An overview of certification standards for domestic wood-burning appliances is given in Section 4.6.

4.3 Residential batch-fired wood-burning appliances

Domestic batch-fed wood-burning appliances include 'high-efficiency' fireplaces, fireplace inserts, stoves, heat-storing stoves, central heating furnaces and boilers.

Open fireplaces are not considered to be a heating device because of their low thermal efficiency (−10 to + 10 per cent), and are therefore not dealt with in this document. Because of the high amount of combustion air they use, most open fireplaces probably consume more energy than they produce if the outdoor temperature is below 0°C [1] .

4.3.1 Wood-stoves

A wood-stove is a free-standing appliance for heating the space within which it is located, without the use of ducts. Stoves release useful heat energy by radiation and convection to their surroundings. Although all stoves utilize both forms of heat transfer, they are generally classified as radiating or circulating (convection) stoves, depending on the main mode of heat transfer. The firebox walls and the firebox hearth are typically lined with fire-resistant materials. Some stoves are equipped with ash grates, normally with an ash box under the grate. Others have no grate and ashes are taken out directly from the hearth.

The burn rate of wood-stoves is regulated by controlling the supply of primary combustion air. Since reduced primary air would result in incomplete combustion, secondary air is normally supplied outside the primary combustion zone, and much of the effort in wood-stove design is spent on achieving good secondary burning. The combustion chamber can be equipped with horizontal and inclined baffles made of rigid insulating material or steel, and secondary air is normally preheated and introduced at strategic locations. In many stoves, a viewing window is provided in the front door, which not only contributes to the aesthetics of wood heat, but gives the operator a much better opportunity to adjust the stove for optimal combustion.

Wood-stoves can be classified on the basis of air-flow paths through the combustion chamber. Four basic classifications are commonly used, with numerous combinations or modifications available. The air flow of the primary air determines whether the stove is an up-draught, down-draught, cross-draught, 'S'-draught or dual-zone stove. These designs are illustrated in Figure 4.1 [2].

114   *The Handbook of Biomass Combustion and Co-firing*

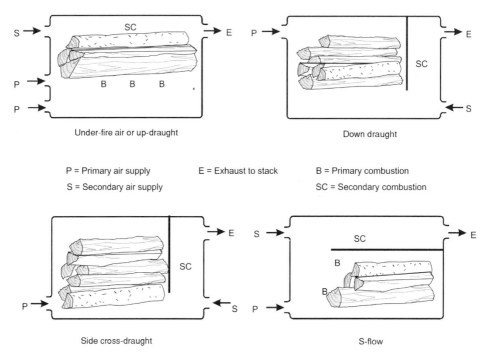

Figure 4.1 *Classification of wood-stoves depending on primary air flow paths*

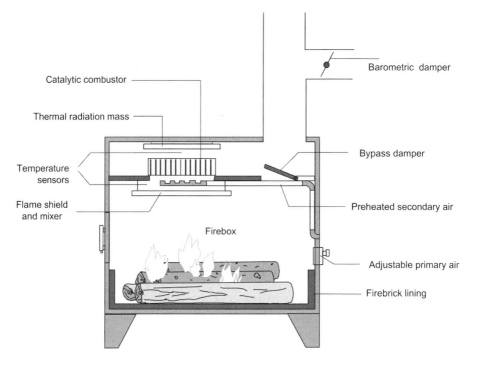

Figure 4.2 *Catalytic stove design*

Normally, smoke will ignite and burn only at temperatures around 540°C or higher. It is difficult to obtain this elevated temperature outside the primary combustion zone. However, a catalytic combustor can reduce this ignition temperature to around 260°C, a temperature more easily obtained. Wood-stoves can be equipped with a catalytic combustor in order to reduce emissions caused by incomplete combustion. The catalytic combustor is usually placed in the flue gas channel beyond the combustion chamber. Figure 4.2 shows a catalytic stove design.

A catalytic combustor has a durable, heat-resistant ceramic composition, which is extruded into a cellular, or honeycomb, configuration. After extrusion, this ceramic monolith is fired and then covered with a very small amount of noble-metal catalyst (usually platinum, rhodium or palladium, or combinations of these) or metal oxides (see Section 9.2.4.9). When wood smoke contacts this catalyst, chemical reactions occur much more quickly on the surface of the catalyst. Catalysts can be damaged by excessive temperatures, and can be blocked, making periodic cleaning necessary. However their durability has improved to the point that some catalyst manufacturers promise a lifetime of about 10,000 hours for a wood-stove catalyst if maintenance is done properly. Catalytic combustors are common in the US.

4.3.2 Fireplace inserts and zero clearance fireplaces

A fireplace insert is basically a wood-stove installed in an existing fireplace. A zero clearance fireplace is in essence a wood-stove designed to be enclosed to give the appearance of a fireplace. Both function in a similar manner to a stove, but require a more elaborate heat transfer system to avoid excessive heat loss. In fireplace inserts, heat is transferred to some

Figure 4.3 *Fireplace insert*

Source: [3].

extent by radiation, mainly from the front door and from the flame and combustion chamber walls through a front door glass. However, the most important heat transfer mechanism is natural or forced convection. Open fireplaces can be changed to more efficient heating appliances using inserts. Figure 4.3 shows an example of a fireplace insert.

4.3.3 Heat-storing stoves

Heavy heat-storing stoves are constructed of pre-fabricated heavy stone plates or purely of stone. For example soapstone ('*speckstein*' in German) is known for its ability to withstand direct flames and to retain heat. During combustion, wood is burned at a high burn rate. After the fire is extinguished, the stove will continue to release the stored heat to the room space for a considerable time (1–2 days). Finnish soapstone stoves typically have a very high mass, from about 700kg to 3000kg or even more. Thus they can store a lot of energy that is then slowly released to the surrounding space. Figure 4.4 shows a design of a Finnish heat-storing fireplace made of soapstone.

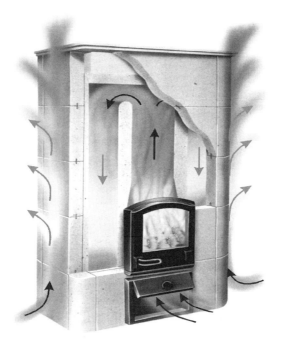

Figure 4.4 *Contraflow system in Finnish heat-storing stoves*

Source: [4].

Another type of heat-accumulating stove is the tiled stove, which is made from ceramic material. This can be manually erected on site or prefabricated in larger blocks. The tiled stove was developed in the 1700s as the first efficient wood-firing device in Sweden and is still quite common.

The contraflow system used in soapstone stoves and tiled stoves (see also Figure 4.4) is used to convey thermal energy from hot flue gases to stone mass. When wood burns in

Domestic Wood-Burning Appliances 117

the firebox, high temperatures are reached quickly, forcing the burning flue gases into the upper combustion chamber below the top lid. The hot gases are then guided down and out into the side channels, where the heat is released into the exterior stones. At the same time, the room air outside the fireplace walls warms and moves up the stone surface in a path opposite to the interior downflow. These two opposing flows are known as contraflow.

Most of the generated heat is transferred evenly into the room in the form of radiant heat. Normally, heat-storing stoves are fired once or twice a day for about two hours during the heating season. The rest of the time heat is released slowly from the stone mass by radiation and convection. Heat-storing stoves are suitable for Nordic cold-climate conditions where the outdoor air temperature does not change very rapidly, because the time response of heat release is quite slow. This type of appliance is not common in North America, where attention has focused more on obtaining clean burning over a range of burning rates.

Another example of a heat storing stove is shown in Figure 4.5. The combustion is based on the down-draught combustion principle. In the first step, the fuel is burned with little excess air. In the second step, combustion is completed with the supply of secondary air. The heating unit is provided with a hearth made of chamotte bricks. In the hearth, two

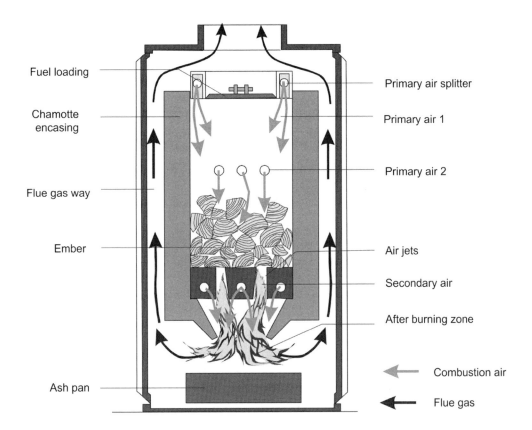

Figure 4.5 *Heat-storing stove using down-draught combustion*

ejectors are constructed, while secondary air holes are placed in the grate. Final combustion takes place in the secondary burning zone, flue gases flow from the bottom to the top of the appliance. There, flue gases transfer their heat to the heat exchangers (not visualized in the figure) and finally flow into the chimney. A large part of the generated heat is accumulated in the heating unit. This type of appliance is not common in North America, where attention has focused more on obtaining clean burning over a range of burn rates.

4.3.4 Wood log boilers

4.3.4.5 Over-fire boilers
Over-fire boilers are the simplest and cheapest domestic boilers for burning wood logs. The principle of a typical over-fire boiler is shown in Figure 4.6. In over-fire boilers, combustion takes place in the whole fuel batch more or less at the same time, as in wood-stoves. The boiler is normally equipped with a primary air inlet under the grate and a secondary air inlet over the fuel batch, into the gas combustion zone. Wood is fed through the upper door and ashes are removed from the lower door. These boilers use natural draught.

The emissions of unburned hydrocarbons from over-fire boilers may be high if they are operated at low burning rates. This is often the case, especially during spring and autumn. An environmentally optimal combustion can only be obtained if the boiler is operated at nominal heat output. In Scandinavian countries, over-fire boilers are usually connected to a water storage tank with a volume of 1–5m^3. Hot tap water is produced by a heat exchanger in the storage tank. In many European countries, over-fire boilers are used with heat storage tanks. This enables firing of the boiler at nominal heat output to minimize emissions.

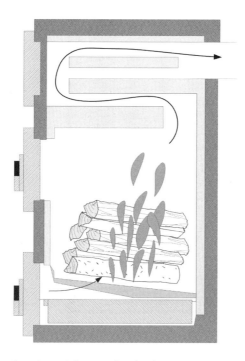

Figure 4.6 *Wood combustion with over-fire boiler*

Electric heaters are quite often placed in storage tanks. In this way, wood combustion and electric heating can be alternated easily, giving more flexibility. In many European countries, the price of electricity is quite low during the summer season and also at night and at the weekend in the winter.

The most common fuels used are wood logs from private forests. Wood briquettes, sod peat and peat briquettes are also used to some extent. A typical over-fire boiler principle is shown in Figure 4.6.

Large versions of these boilers are often installed outdoors in rural parts of North America and used to heat farm buildings, small shops, etc. They have been a major source of complaints of excessive smoke production, and their use is being increasingly restricted or banned outright.

4.3.4.6 Under-fire boilers

In under-fire boilers, gasification and partial combustion take place in only a small amount of the fuel in the bottom of the fuel storage. Final combustion takes place in a separate combustion chamber. Ashes fall through a grate to the ash box.

Fuels commonly used in under-fire boilers are wood logs and woodchips. Wood briquettes, sod peat and peat briquettes are also used to some extent. Wood pellets are not suitable because of their small particle size and high density. Normally, natural draught is used but some models have an air blower or a flue gas fan. The investment costs are about 50 per cent higher than those of over-fire boilers. Combustion in under-fire boilers is more stable than in over-fire boilers, which normally results in lower emissions. The principle of a typical under-fire boiler construction is shown in Figure 4.7.

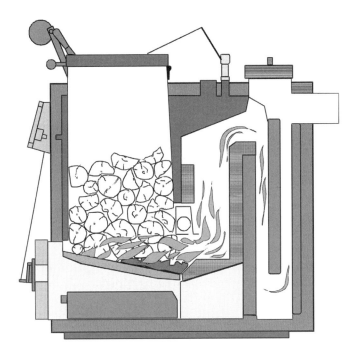

Figure 4.7 *Under-fire boiler*

4.3.4.7 Down-draught boilers

The latest innovations in the wood log combustion sector are down-draught boilers. The basic principle is shown in Figure 4.8. Flue gases are forced to flow down through holes in a ceramic grate. Secondary combustion air is introduced in the grate or in the secondary combustion chamber, which is normally insulated with ceramics. Final combustion takes place at high temperatures. Because the flue gas flow resistance is quite high, a combustion air fan or flue gas fan is needed. This fan also allows the precise introduction and distribution of primary and secondary combustion air in the combustion chamber.

Very strict emission limits in some countries have made it necessary to introduce down-draught boilers. Other new developments are advanced combustion control devices like lambda control probes to measure flue gas oxygen concentration, enabling precise combustion air control and staged-air combustion. Even fuzzy-logic control has been used in an effort to achieve very low emissions. Due to more advanced technology, down-draught boilers are more expensive than simple over-fire boilers and under-fire boilers.

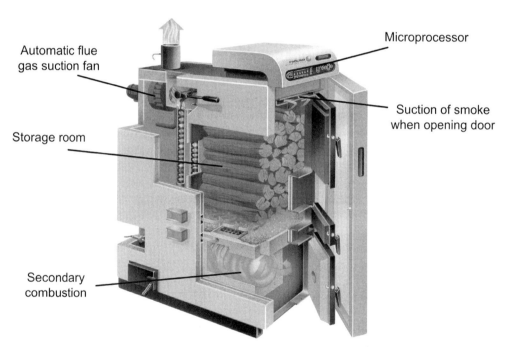

Figure 4.8 *Microprocessor-controlled down-draught boiler for wood logs (Fröling)*
Source: [5].

4.3.4.8. Heat storage tanks

In the Scandinavian countries, wood boilers are usually connected to a heat storage tank with a water volume of 1–5m^3. In many European countries, due to the high emissions from systems without heat storage tanks, wood boilers are required to have a storage tank.

Using a heat storage tank enables optimal combustion at nominal load. The time-dependent demand for space heating and hot water is then supplied from the tank. However, often the heat storage tanks installed are far too small and often poorly insulated too. It is very important that the heat storage tank is large enough to accumulate the total heat released from a wood batch. As a role of thumb, the heat storage tank should be 18 times the firebox volume and a common recommendation is to install heat storage tanks of at least 1.5m^3. A well-insulated tank is a further prerequisite for high overall efficiency of the system.

During the summer season when there is no or limited need for space heating, hot tap water can be produced in other ways. This is often achieved by using electrical heaters placed in the heat storage tanks. In this way, wood combustion and electrical heating can be alternated easily, giving more flexibility. An even more attractive and more sustainable solution is to connect the heat storage tank to solar collectors. In this way, a substantial contribution of solar energy can be obtained during spring and autumn, even in the northern countries, and the wood consumption as well as the adverse environmental effects of combustion can be minimized.

4.3.4.9 Efficiency and emissions

A number of constraints plague the designer of batch-fed appliances. These include the need to maximize duration of heat output, provision for easy fuel loading, the need for low cost, avoidance of fans and electronic controls due to noise and the desirability of being able to operate without electricity, the need for a range of heat outputs, ash build-up and removal provisions, and field operation under a wide range of draught conditions. Maximizing efficiency has traditionally been dealt with by providing a large heat exchange surface area relative to heat output. While this results in adequate efficiency for many applications, it does not address high levels of unburned hydrocarbons and carbon monoxide, which reduce efficiency and result in high pollutant emissions relative to conventional heating systems. With the exception of heat-storage systems, output is regulated by restricting the combustion air supply. However, poor control of the air distribution results in many appliances having high excess air levels, even though the actual combustion process occurs under air-starved conditions. This also contributes to a significant reduction in efficiency.

A major impediment to improving appliance performance was the lack of a generally accepted test protocol, and it was only in the 1980s that consensus test requirements began to emerge. These have allowed developers to better understand and quantify the impact of different performance improvement strategies, and led to significant reductions in appliance emissions. Emission requirements have focused on carbon monoxide and particulates, although emissions of organics are of increasing concern.

Emissions from residential wood burning appliances are in general influenced by combustion technology and process conditions and by fuel properties. Current appliances with low emission levels tend to have a clear delineation of primary and secondary air supply to maintain an air-starved condition in the fuel bed for output control, but still permit adequate burnout of combustible gases. Introducing secondary air under conditions of adequate temperature for ignition and burnout can markedly reduce emission levels, and various strategies have evolved to achieve this objective. These include the use of catalytic materials to reduce the ignition temperature of the combustible gases.

The US regulates particulate emission levels for wood-stoves. National limits are 7.5g/h for non-catalytic and 4.1g/h for catalytic appliances, compared to pre-regulation emissions typically in excess of 20g/h. The best current stoves achieve emission levels of less than 3g/h [6]. Regulation is based on total particulate emissions. Studies have shown that the bulk of emissions are PM2.5, i.e. less than 2.5 microns in diameter. Particulates of less than 10 micron diameter are generally considered to be toxic [7].

Organics emissions from residential wood combustion are of increasing concern to regulators. Polycholorinated dibenzo-p-dioxins and polychlorinated dibenzofurans (PCDDs and PCDFs) have been linked to wood combustion. Studies have shown average emission factors of 250–1900pg toxic equivalent quantity (TEQ)/kg dry fuel [7]. Further research is required to determine the impact of current emission control strategies on the emission of organic compounds.

4.4 Wood pellet appliances and burners

4.4.1 Background

The use of wood pellets for domestic heating is a rather recent phenomenon. Pellet-fired appliances were developed in the 1980s but significant market penetration of pellet stoves has only taken place during the past ten years.

Pellet burners for use in central heating boilers were developed during the 1990s, and now have a significant share of the domestic heating market. In Austria, Germany and Denmark, complete pellet boiler units are the most common appliance type. In other countries, the interest in pellet heating is just awakening. It has increased with rising oil prices. Pellet fuels are of special interest since they have the potential to burn with very low emissions, can offer much of the convenience of conventional fuels, and because pellet burners can replace oil burners in existing boilers, thus reducing payback time.

4.4.2 Technical features

Pellet-fired systems allow continuous automatic combustion of a refined and well-defined fuel, with the burning rate controlled by the rate of fuel supply rather than by restricting the primary air. The fuel is fed automatically into the combustion chamber by means of an auger from a storage hopper. Some burners are equipped with a smaller pellet storage (enough for one or a few days of operation) that can be refilled manually or by an automatic system from a larger storage.

The combustion air is supplied by an electric fan that may also provide a distinct secondary air flow. The fuel feed is typically by means of an auger from a storage hopper. The fuel delivery rate can normally be fine-tuned by on/off timers causing the auger to cycle over a short interval, with the percentage of on-time varied depending on the heat demand. Operation is normally controlled by a thermostat in the heated space, or by an aquastat in a boiler or heat storage tank. Ignition is by means of an electric device or by maintaining a pilot flame.

Minimal maintenance is desired and the best burners can be in operation for more than a week without the manual removal of ash, etc. In fact, the most important development is perhaps to increase the availability and decrease the necessary maintenance. For instance, burners can be installed in boilers with automatic ash removal.

Figure 4.9 *Two wood pellet boilers, used to heat a school in Denmark*

4.4.3 Safety requirements regarding back-burning

Preventing burn-back into the feed auger and storage hopper is a major safety concern for pellet-fired equipment. Burners and stoves can (and should) be equipped with a number of independent safety systems such as:

- a sprinkler system;
- a drop chute into the combustion chamber;
- a thermal sensor in the feed system;
- an airtight fuel storage hopper; and
- a double auger feed system with a firebreak between the two augers.

4.4.4 Emissions

Good pellet burners show very low emissions levels of hydrocarbons and carbon monoxide. However, NO_x emissions are significant despite the rather low nitrogen content in the pellets. In fact, the conversion of fuel nitrogen to NO_x is in many cases close to 100 per cent (Figure 4.10).

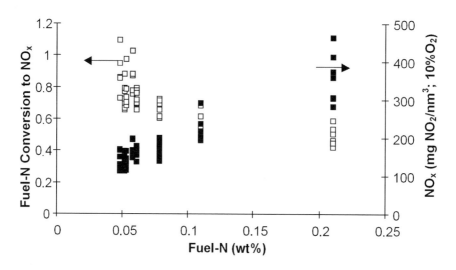

Figure 4.10 *The conversion of fuel nitrogen to NO_x in wood pellet combustion* [8]

Since the emissions of NO_x usually increase as the emissions of unburned hydrocarbons decrease (see for example, Figure 4.11), the challenge is to balance efficient NO_x reduction with low emissions of hydrocarbons. In addition, particulate emissions from small-scale pellet combustion can be significant. The particulates consist of soot from incomplete combustion and ash particles.

4.4.5 Efficiency

Well-designed pellet-fired systems can achieve efficiencies of over 80 per cent. At part load and varying load, the efficiency may decrease somewhat. Many pellet stoves operate with a very high excess air level, which reduces their efficiency to 50–60 per cent.

4.4.6 Certification

In Sweden, the authorities, manufacturers and SP (Swedish National Testing and Research Institute) have, in the process of developing pellet burners, set up a voluntary certification system (the P mark). In order to qualify, a burner must fulfil a number of technical requirements as well as meet tough emission standards (for further details see also Section 4.6) [9].

In North America, a safety certification standard for pellet stoves has been developed by the manufacturers. Certification bodies typically use this standard together with additional in-house requirements. Underwriters' Laboratories of Canada is in the process of developing a consensus standard for pellet stoves for that country. Requirements for residential central heating systems and larger pellet equipment are covered in the CSA (Canadian Standards Association) standards that also cover firewood-fuelled systems, and do not address efficiency or emissions. The US exempts most pellet stoves from emission requirements due to their high excess air level.

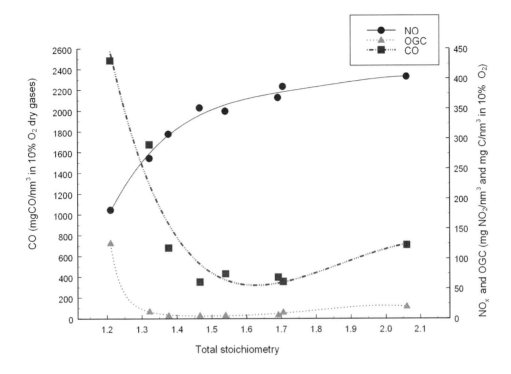

Figure 4.11 *Correlation between unburned compounds (CO and OGC) and NO_x*

Note: OGC are all gaseous hydrocarbons that can be measured using a flame ionization detector (FID) analyser at 200°C

4.4.7 Pellet burners for central heating systems

Pellet burners for domestic use are usually constructed for a nominal thermal output of less than 25kW. Depending on the feeding system, three different types of pellet burners can be identified (see Figure 4.12) with the flame burning either horizontally or upwards. Perhaps the most interesting burners are the horizontally fed ones (no. 2 in Figure 4.12), in which the gases are released within a combustion chamber (tube) during devolatilization and in which a primary combustion zone can be defined, in contrast to the other types of burners. To ensure minimal emissions, it important to obtain stable combustion conditions and minimal stand-by periods, i.e. by avoiding the number of start-ups or operation with a pilot flame. In addition, the overall emissions are generally reduced when using electric ignition rather than a pilot flame.

In some Nordic countries it is quite common to replace an oil burner in an existing heating boiler with a pellet burner. In this case it is important that the pellet burner is properly designed for the boiler. If the flame hits a cold surface, high emissions of soot and hydrocarbons and a low efficiency will result. In addition, the flue gas flow from a pellet burner is higher than that of oil, and the installed thermal output of a pellet burner should therefore be less than that of a corresponding oil burner. This is important since the

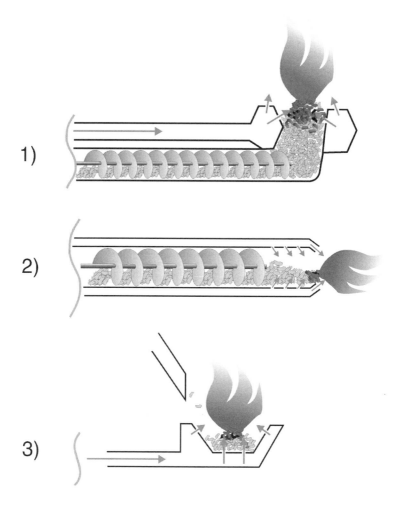

Figure 4.12 *Wood pellet burners: (1) underfed burner, (2) horizontally fed burner, (3) overfed burner*

residence time in the boiler may otherwise be too short, resulting in flue gas temperatures that are too high and emissions of unburned hydrocarbons. This means that a consumer can get slightly less heat output from an oil-fired boiler converted to use a pellet burner.

Normally this does not cause any problems because, for technical reasons, oil boilers are normally over-sized for small houses. Some burners are equipped with an integrated smaller pellet storage (enough for one or a few days of operation) that can be refilled manually. Other burners may have an automatic system, connected to a larger storage. The fuel is usually fed to the burner by means of a screw, but other feeding systems also exist.

In some pellet boilers (pre-manufactured combination burner-boilers), the oxygen concentration measured by means of a lambda sensor is used to control the combustion air. Some burners also operate with a flexible fuel feed controlled by the heat demand.

4.4.8 Wood pellet stoves

Special kinds of stoves have been constructed to combust only pelletized material. Contrary to conventional wood-stoves, pellet stoves are dependent on electricity for their operation. An electric fan controls the combustion process by varying the supply of combustion air. This results in low CO and C_xH_y emissions. Particulate emissions are generally low and consist mainly of inorganic material in contrast to wood-stoves where particle emissions are dominated by soot and tar.

Pellet stoves are typically provided with small fuel storage, fuel feeder, combustion air blower, burner shell and a flue gas discharge system. They are equipped with a rather extensive control system.

Figure 4.13 shows the principle of a pellet stove with a maximum nominal power output of 10kW and a turn-down ratio of a factor of 4 (2.5–10kW). The stove operates as described below.

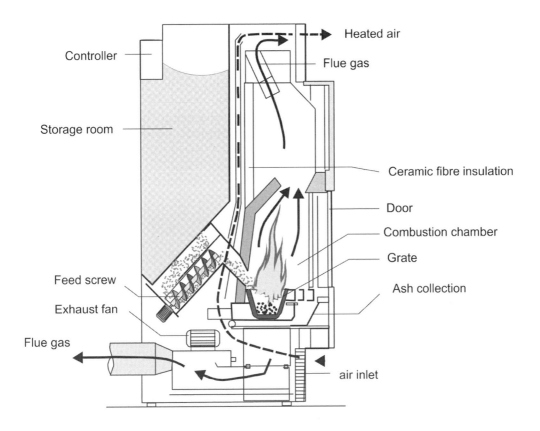

Figure 4.13 *Pellet stove*

The fuel hopper is filled from the top. The burning time with a full hopper at nominal power output is in the order of 6 hours. For transport of the fuel to the burner a fuel-feeding system is used. The speed of the feeder is related to the heat demand. The burner of the stove consists of a shell-like pot in which the pellets are burned. Combustion air is drawn through slots in the inner wall of the shell by means of an exhaust fan, which reduces the pressure in the combustion chamber. The flue gases are discharged at the top of the combustion chamber and in this example flow down towards the chimney constructed in the bottom of the stove, passing through a heat exchanger on the way. Room air enters the stove near the bottom of the unit and leaves the stove at the top.

4.5 Woodchip appliances

Woodchip appliances are also used for domestic heating. They are more common in the countryside, heating larger houses and farms. Advantages of using woodchips instead of firewood are automatic operation and much lower emissions because of the use of feed rate rather than air supply to control heat release rate. The drawback of using woodchips is that making and storing the chips, and in some cases also artificially drying them, requires a greater investment in terms of time, machinery and storage space.

4.5.1 Pre-ovens

Pre-ovens are used when quite moist woodchips are used for combustion. Pre-ovens are well-insulated chambers in which combustion or partial gasification takes place. The basic principle of a woodchip pre-oven is shown in Figure 4.14. Woodchips are fed into the combustion chamber by means of an auger, and combustion air is fed with a fan. Only a small amount of the fuel is burned at a time. The combustion is well controlled and thus emissions are low compared to batch combustion of firewood. Pre-ovens can be connected to an existing boiler or be part of a complete package.

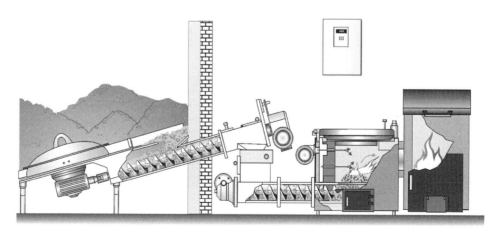

Figure 4.14 *Woodchip pre-oven*

4.5.2 Under-fire boilers

Woodchips can also be burned in under-fire boilers. They are similar to under-fire wood log boilers, but fuel storage is normally made of better material to avoid corrosion. Today, under-fire boilers are not very common for woodchip combustion because automatic combustion units have been replacing them.

4.5.3 Stoker burners

Stoker burners are automatic woodchip combustion devices placed inside the firebox of the boiler. Woodchip stoker burners are quite similar to pellet burners. Under-fed and horizontally fed burners are in fairly common use. In under-fed burners, the flame burns upwards and in horizontally fed burners, it burns horizontally.

Horizontal stoker burners suitable for combustion of biofuels have been on the market for about 20 years. Most of the devices have been designed for a heat output range of 20–40kW and they are used for heating detached houses and farms. Suitable fuels for these devices are woodchips, sod peat and pellets. With a somewhat different design, horizontal stoker burners have also been constructed for higher heat output, up to 1MW.

Using a screw feeder, the fuel handling system feeds the fuel into the burner, where it is combusted. The burner consists of a cast iron, refractory-lined or water-cooled horizontal cylinder. In some burners, water-cooling ensures the durability of the burner materials, and makes insulation of the burner easier. The temperature inside the burner rises to above 1000°C when using dry fuels. The burner is mounted partially inside the furnace, and partially outside it, so the whole firebox of the boiler effectively takes part in radiative heat transfer.

The basic idea in horizontal stoker burners is that fuel is fed precisely according to the heat demand. A blower or several blowers introduce combustion air. This ensures very efficient and clean combustion. The turn-down ratio of this kind of equipment is 0–100 per cent. This means that separate water storage is not needed. A thermostat in the boiler water, using an on/off method in smaller burners and more sophisticated control methods in large burners, controls the burner heat output. The principle of a horizontal stoker burner is shown in Figure 4.15.

4.6 Certification and testing standards

An important tool to facilitate the improvement of appliance performance is the existence of a generally accepted test protocol, and it was only in the late 1990s that consensus test requirements began to emerge. These have allowed developers to better understand and quantify the impact of different performance-improvement strategies, and led to significant reductions in appliance emissions. Emission requirements have focused on carbon monoxide and particulates, though emissions of organics are of increasing concern.

In Sweden, for example, the authorities, manufacturers, and SP (The Swedish National Testing and Research Institute) have developed a voluntary certification system (P-mark) for pellet-burning devices, woodchip stokers and wood-stoves. In order to qualify, the combustion device must fulfil a number of technical requirements as well as meet tough emission standards. The requirements are both wider and more stringent then posted in legislation. In addition, the requirements are revised regularly becoming increasingly

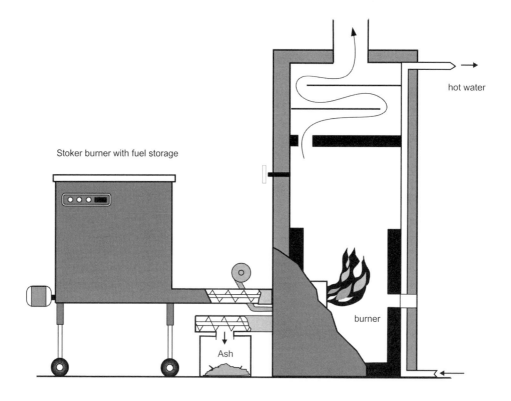

Figure 4.15 *Horizontal stoker burner*

stringent. This means that the product must be improved and the P-marking system has proven to be effective in the process of transferring knowledge from research into industrial applications. An independent, high quality, certification has also been important for the very rapid introduction of new technology such as the pellet burners.

4.6.1 European standards

The European Committee for Standardization (CEN) started a new work item in 1992 through its Technical Committee CEN/TC 295 to develop standards for residential solid fuel-burning appliances.

In 1999, CEN TC 295 approved the first European product standards on most commonly used residential solid fuel-fired appliances in final voting. Since 2002, all CEN member countries have withdrawn their existing national standards to use these standards, which give requirements for construction and safety of the appliances, as well as requirements and instructions to measure heat output, efficiency and emissions. In 2000, TC 295 also started new work to modify these four product standards into harmonized standards (hEN) according to the rules of the EU Directive for Construction Products.

Table 4.1 *Approved EU product standards for residential solid fuel-fired appliances*

EN 12809:2001	Residential independent boilers fired by solid fuel – Nominal heat output up to 50kW – Requirements and test methods
EN 12809:2001/AC:2003	Corrigendum
EN 12809:2001/A1:2004	Amendment, harmonized standard
EN 12809:2001/AC:2006	Corrigendum
EN 12815:2001	Residential cookers fired by solid fuel – Requirements and test methods
EN 12815:2001/AC:2003	Corrigendum
EN 12815:2001/A1:2004	Amendment, harmonized standard
EN 12815:2001/AC:2006	Corrigendum
EN 13229:2001	Inset appliances including open fires fired by solid fuels – Requirements and test methods
EN 13229:2001/A1:2003	Amendment
EN 13229:2001/AC:2003	Corrigendum
EN 13229:2001/A2:2004	Amendment, harmonized standard
EN 13229:2001/AC:2006	Corrigendum
EN 13240:2001	Room heaters fired by solid fuel – Requirements and test methods
EN 13240:2001/AC:2003	Corrigendum
EN 13240:2001/A1:2004	Amendment, harmonized standard
EN 13240:2001/AC:2006	Corrigendum
CEN EN 14785:2006	Residential space heating appliances fired by wood pellets – Requirements and test methods
EN 15250:2007	Slow heat release appliances fired by solid fuel – Requirements and test methods

Later, work has also been undertaken to elaborate European standards for 'Residential space heating appliances fired by wood pellets', i.e. pellet stoves, as well as for 'Slow heat release appliances fired by solid fuel', i.e. soap-stone stoves, tiled stoves etc. Recently, these standards have also been approved as European standards.

The above standards cover quite well the needs of quality assurance of small biomass fuel-fired appliances.

For solid fuel-fired boilers up to 300kW heat output, the standard EN 303-5, 'Heating boilers – Part 5: Heating boilers for solid fuels, hand and automatically stoked, nominal heat output of up to 300kW – Terminology, requirements, testing and marking' has been in use since 1999. It gives constructions and performance requirements and instructions to measure power, efficiency and emissions.

For pellet burners, a draft European standard, prEN 15270, 'Pellet burners for small heating boilers – Definitions, requirements, testing, marking' has been elaborated. This draft will be sent out for formal vote within Europe during 2007.

For biomass systems larger than 300kW, however, there is a need for a simple standard that allows field testing. For large power plants there exists a well-established standard (DIN 1942). As it was originally developed for power plants using fossil fuels, it may need updating so that it is better suited for biomass fuel-fired plants as well.

It should be noted that the requirements for efficiency and emissions given in the above standards are only basic requirements. Any country or part of the country can have tighter demands. Generally approved CEN standards give instructions on how to evaluate these demands. This will help manufacturers who export their products to many countries. In the future, testing is required only in one notified laboratory.

4.6.2 North American standards

In North America, a safety certification standard for pellet stoves has been developed by the manufacturers. Certification bodies typically use this standard together with additional in-house requirements. Underwriters' Laboratories of Canada is in the process of developing a consensus standard for pellet stoves for that country. Requirements for residential central systems and larger pellet equipment are covered in the CSA standards that also cover firewood-fuelled systems. The United States exempts most pellet stoves from emission requirements due to their high excess air level.

The US and Canada are each significant customers for the other's wood-burning appliances. Standards have accordingly tended to be harmonized, in fact if not as policy. Each country has requirements for safety which, while reflecting the experience and concerns of the country, permit testing with a minimum of duplication of core requirements. However, only the US has national emissions requirements, and these apply only to a limited range of appliances, generally those which would be considered to be wood-stoves. In Canada, British Columbia has adopted the US requirements. Canada does have an emissions and efficiency test protocol, but it has yet to be mandated by the regulatory authorities. However, Environment Canada, the federal department responsible for such regulation, has initiated the process necessary for adoption of a national regulation.

Table 4.2 *North American standards for residential solid fuel-fired appliances*

ULC S627 (Can), UL1482 (US)	Wood and pellet stove safety testing. This standard sets minimum requirements for the construction and clearances to combustibles of wood-stoves, including pellet-fired appliances.
ULC S628 (Can), UL737 (US)	Similar to S627 but covering inserts
CSA B365(Can)	Sets out installation requirements of wood-burning appliances, including uncertified equipment
CSA B366.1 (Can), UL 391 (US)	Requirements similar to S627 for central wood-burning systems
CSA B415.1 (Can), EPA Part 40 (US)	Test methods for determination of efficiency and particulate emissions from wood-fired appliances, including conventional and pellet stoves, inserts, central furnaces and boilers, small commercial systems and 'high performance' fireplaces. Maximum particulate emission levels are specified. An ASTM document is also being developed to provide testing requirements for outdoor boilers (WK 5982).
ASTM E1509 (US)	Safety test requirements for pellet-fired heaters
ASTM E1602 (US)	A guide to the construction of masonry heaters

Emissions requirements in North America are based on total particulate emissions as determined by a test protocol modelled on industrial source testing requirements. US emission limits are 7.5g/h for non-catalytic appliances and 4.1g/h for catalyst equipped appliances. These figures compare to typical emissions of approximately 20–30g/h for pre-regulation appliances.

The principal Canadian and American standards dealing with wood heat systems are summarized in Table 4.2.

4.7 References

1   STREHLER, A. (1994) *Emissionsverhalten von Feurungsanlagen fur feste Brennstoffe*, Technische Universität Munchen-Weihenstephan, Bayerische Landesanstalt für Landtechnik, Freising
2   ALLEN, J. M. and COOKE, W. M. (1981) *Control of Emissions from Residential Wood Burning by Combustion Modification*, EPA-600/7-81-091; US Environmental Protection Agency, Research Triangle Park, NC 27711
3   HARVIA OY (n.d.) company brochure, Finland
4   TULIKIVI OY LTD, (n.d.) company brochure, Finland
5   FRÖLING GmbH (n.d.) Heizkessel- und Behälterbau, Austria
6   JARABEK, J. S. and PRETO, F. (2001) *Effect of Appliance Type and Operating Variables on Particulate Matter Emissions from Residential Wood Combustion Appliances*, CANMET NRCan, March
7   JARABEK, J. S. and PRETO, F. (2004) *PCDD and PCDF Emissions from Residential Woodstoves*, CANMET, NRCan, June
8   ESKILSSON, D., RÖNNBÄCK, M., SAMUELSSON, J. and TULLIN, C. (2004) 'Optimisation of efficiency and emissions in pellet burners', *Biomass and Bioenergy*, vol 27, pp541–546
9   'P-marking pellet burners', Certification Rules CR 029, SP Technical Research Institute of Sweden, www.sp.se

5
Combustion Technologies for Industrial and District Heating Systems

5.1 Introduction

This chapter describes combustion systems of a nominal thermal capacity exceeding 100kW. These furnaces are generally equipped with mechanical or pneumatic fuel-feeding systems. Manual fuel-feeding is no longer customary due to high personnel costs and strict emission limits. Moreover, modern industrial combustion plants are equipped with process control systems supporting fully automatic system operation.

In principle, the following combustion technologies can be distinguished:

- fixed bed combustion;
- fluidized bed combustion; and
- pulverized fuel combustion.

The basic principles of these three technologies are shown in Figure 5.1 and described below [1, 2]. The fundamentals of biomass combustion are outlined in Section 2.2.

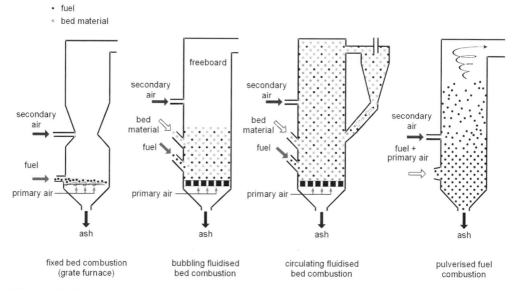

Figure 5.1 *Principal combustion technologies for biomass*

Source: [1].

Fixed bed combustion systems include grate furnaces and underfeed stokers. Primary air passes through a fixed bed, in which drying, gasification and charcoal combustion take place. The combustible gases produced are burned after secondary air addition has taken place, usually in a combustion zone separated from the fuel bed.

Within a fluidized bed furnace, biomass fuel is burned in a self-mixing suspension of gas and solid-bed material into which combustion air enters from below. Depending on the fluidization velocity, bubbling fluidized bed (BFB) and circulating fluidized bed (CFB) combustion can be distinguished.

Pulverized fuel (PF) combustion is suitable for fuels available as small particles (average diameter smaller than 2mm). A mixture of fuel and primary combustion air is injected into the combustion chamber. Combustion takes place while the fuel is in suspension and gas burnout is achieved after secondary air addition.

Variations of these technologies are available. Examples are combustion systems with spreader stokers and cyclone burners.

5.2 Fixed bed combustion

5.2.1 Grate furnaces

There are various grate furnace technologies available: fixed grates, moving grates, travelling grates, rotating grates and vibrating grates. All of these technologies have specific advantages and disadvantages, depending on fuel properties, so careful selection and planning is necessary.

Grate furnaces are appropriate for biomass fuels with a high moisture content, varying particle sizes (with a downward limitation concerning the amount of fine particles in the fuel mixture) and high ash content. Mixtures of wood fuels can be used, but current technology does usually not allow for mixtures of wood fuels and straw, cereals and grass, due to their different combustion behaviour, low moisture content and low ash-melting point. Only special grate constructions are able to cope with woody and herbaceous biomass fuel mixtures, e.g. vibrating grates (see Figure 5.11) or rotating grates (see Figure 5.13). Such grates have to ensure a mixture of the fuels across the grate.

A well-designed and well-controlled grate guarantees a homogeneous distribution of the fuel and the bed of embers over the whole grate surface. This is very important in order to guarantee an equal primary air supply over the various grate areas. Inhomogeneous air supply may cause slagging, higher fly-ash amounts and may increase the excess oxygen needed for a complete combustion, resulting in boiler heat losses. Furthermore, the transport of the fuel over the grate has to be as smooth and homogeneous as possible in order to keep the bed of embers calm and homogeneous, to avoid the formation of 'holes' and to avoid the release of fly ash and unburned particles as much as possible.

The technology needed to achieve these aims includes continuously moving grates, a height control system of the bed of embers (e.g. by infrared beams) and frequency-controlled primary air fans for the various grate sections. The primary air supply must be divided into sections in order to be able to adjust the specific air amounts to the requirements of the zones where drying, gasification and charcoal combustion prevail (see Figure 5.2). This separately controllable primary air supply also allows smooth operation of grate furnaces at partial loads down to about 25 per cent of the nominal furnace load and control of the primary air ratio needed (to secure a reducing atmosphere in the primary combustion chamber necessary for low NO_x operation). Moreover, grate systems can be water-cooled to avoid slagging and to extend the lifetime of the materials.

Another important aspect of grate furnaces is that a staged combustion should be obtained by separating the primary and the secondary combustion chambers in order to avoid back-mixing of the secondary air and to separate gasification and oxidation zones. Due to the fact that the mixing of air and flue gas in the primary combustion chamber is not optimal because of the low turbulence necessary for a calm bed of embers on the grate, the geometry of the secondary combustion chamber and the secondary air injection have to guarantee a mixture of flue gas and air that is as complete as possible. The better the mixing quality between flue gas and secondary combustion air, the lower the amount of excess oxygen that will be necessary for complete combustion and the higher the efficiency. The mixing effect can be improved with relatively small channels where the flue gas reaches high velocities and where the secondary air is injected at high speed via nozzles that are well distributed over the cross-section of this channel. Other means of achieving a good mixture of flue gas and secondary air are combustion chambers with a vortex flow or cyclone flow.

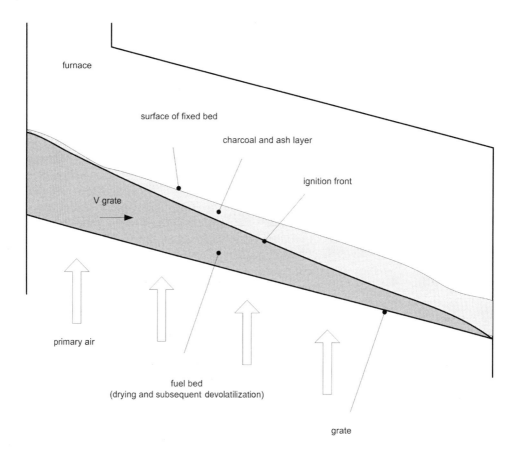

Figure 5.2 *Combustion process in fixed fuel beds*

Source: [3].

Based on the flow directions of fuel and the flue gas, there are three systems of operation for grate combustion plants (Figure 5.3):

1. counter-current flow (flame in the opposite direction to the fuel);
2. co-current flow (flame in the same direction as the fuel); and
3. cross-flow (flue gas removal in the middle of the furnace).

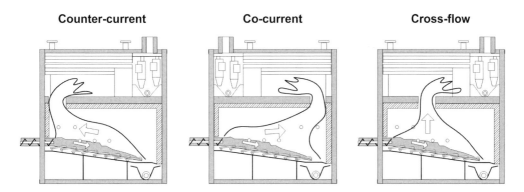

Figure 5.3 *Classification of grate combustion technologies*
Source: [4].

Counter-current combustion is most suitable for fuels with low heating values (wet bark, woodchips or sawdust). Due to the fact that the hot flue gas passes over the fresh and wet biomass fuel entering the furnace, drying and water vapour transport from the fuel bed is increased by convection (in addition to the dominating radiant heat transfer to the fuel surface). This system requires a good mixing of flue gas and secondary air in the combustion chamber in order to avoid the formation of striated flows enriched with unburned gases entering the boiler and increasing emissions.

Co-current combustion is applied for dry fuels like waste wood or straw or in systems where preheated primary air is used. This system increases the residence time of unburned gases released from the fuel bed and can improve NO_x reduction by enhanced contact of the flue gas with the charcoal bed on the backward grate sections. Higher fly-ash entrainment can occur and should be impeded by appropriate flow conditions (furnace design).

Cross-flow systems are a combination of co-current and counter-current units and are also especially applied in combustion plants with vertical secondary combustion chambers.

In order to achieve adequate temperature control in the furnace, flue gas recirculation and water-cooled combustion chambers are used. Combinations of these technologies are also possible. Water-cooling has the advantage of reducing the flue gas volume, impeding ash sintering on the furnace walls and usually extending the lifetime of insulation bricks. If only dry biomass fuels are used, combustion chambers with steel walls can also be applied (without insulation bricks). Wet biomass fuels need combustion chambers with insulation bricks operating as heat accumulators and buffering moisture content and combustion temperature fluctuations in order to ensure a good burnout of the flue gas. Flue gas recirculation can improve the mixing of combustible gases and air and can be regulated more accurately than water-cooled surfaces. However, it has the disadvantage of increasing the flue gas volume in the furnace and boiler section. The flue gas to be recirculated should be extracted from the main stream after fly-ash precipitation in order to avoid dust depositions in the recirculation channels. Moreover, flue gas recirculation should not be operated in stop-and-go mode, to avoid condensation and corrosion in the channels or on the fan blades.

Water cooled grates can be applied in order to increase the lifetime of the grate and in order to decrease ash melting and slag formation, which is especially relevant for herbaceous biomass fuels and waste wood.

5.2.1 Travelling grates

Travelling grate furnaces are comprised of grate bars forming an endless band (like a moving staircase) moving through the combustion chamber (see Figure 5.4). Fuel is supplied at one end of the combustion chamber onto the grate, e.g. by screw conveyors, or is distributed over the grate by spreader stokers injecting the fuel into the combustion chamber (see Figure 5.5). The fuel bed itself does not move, but is transported through the combustion chamber by the

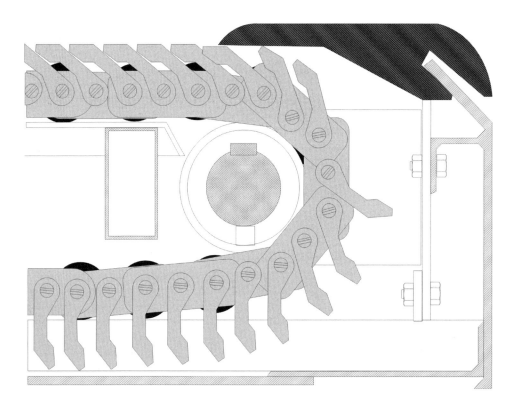

Figure 5.4 *Technological principle of a travelling grate*

Source: [1].

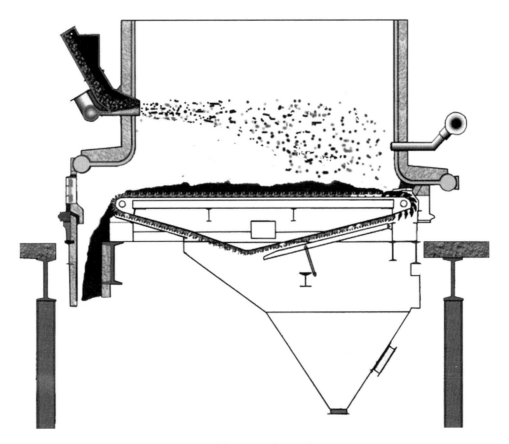

Figure 5.5 *Travelling grate furnace fed by spreader stokers*

Source: [5].

grate, contrary to moving grate furnaces where the fuel bed is moved over the grate. At the end of the combustion chamber the grate is cleaned of ash and dirt while the band turns around (automatic ash removal). On the way back, the grate bars are cooled by primary air in order to avoid overheating and to minimize wear. The speed of the travelling grate is continuously adjustable in order to achieve complete charcoal burnout.

The advantages of travelling grate systems are uniform combustion conditions for woodchips and pellets and low dust emissions, due to the stable and almost unmoving bed of embers. Also, the maintenance or replacement of grate bars is easy to handle.

In comparison to moving grate furnaces, however, the fact that the bed of embers is not stoked results in a longer burnout time. Higher primary air input is needed for complete combustion (which implies a lower NO_x reduction potential by primary measures). Moreover, non-homogeneous biomass fuels imply the danger of bridging and uneven distribution over the grate surface because no mixing occurs. This disadvantage can be avoided by using spreader stokers, because the fuel-feeding mechanism causes a mixing of the fuel bed.

5.2.1.2 Fixed grate systems

Fixed grate systems are only used in small-scale applications. In these systems, fuel transport is managed by fuel feeding and gravity (caused by the inclination of the grate). As fuel transport and fuel distribution over the grate cannot be controlled well, this technology is no longer applied in modern biomass combustion plants.

5.2.1.3 Inclined moving grates and horizontally moving grates

Moving grate furnaces usually have an inclined grate consisting of fixed and movable rows of grate bars (see Figure 5.6). By alternating horizontal forward and backward movements of the movable sections, the fuel is transported along the grate. Thus unburned and burned fuel particles are mixed, the surfaces of the fuel bed are renewed and a more even distribution of the fuel over the grate surface can be achieved (which is important for an equal primary air distribution across the fuel bed). Usually, the whole grate is divided into several grate sections, which can be moved at various speeds according to the different stages of combustion (see Figure 5.7). The movement of the grate bars is achieved by hydraulic cylinders. The grate bars themselves are made of heat-resistant steel alloys. They are equipped with small channels in their side-walls for primary air supply and should be as narrow as possible in order to distribute the primary air across the fuel bed as well as possible.

In moving grate furnaces, a wide variety of biomass fuels can be burned. Air-cooled moving grate furnaces use primary air for cooling the grate and are suitable for wet bark, sawdust and woodchips. For dry biomass fuels or biomass fuels with low ash-sintering temperatures, water-cooled moving grate systems are recommended.

In contrast to travelling grate systems, the correct adjustment of the moving frequency of the grate bars is more complex. If the moving frequencies are too high, high concentrations of unburned carbon in the ash or insufficient coverage of the grate will result. Infrared beams situated over the various grate sections allow adequate control of the moving frequencies by checking the height of the bed.

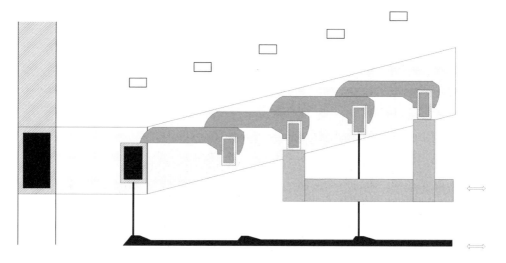

Figure 5.6 *Inclined moving grate*

Source: [6].

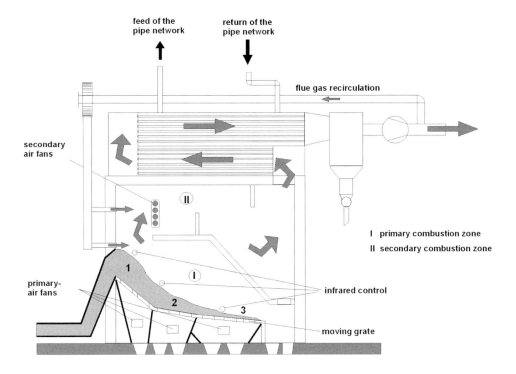

Figure 5.7 *Modern grate furnace with infrared control system, section-separated grate and primary air control*

Explanations: 1 drying; 2 gasification; 3 charcoal combustion.

Source: [1].

Figure 5.8 *Inclined moving grate furnaces*

Explanations: Two furnaces with $3.2MW_{th}$ each; designed for woodchips; used for district heating in Interlaken.

Source: Photograph courtesy of Schmid AG, Switzerland.

Ash removal takes place under the grate in dry or wet form. Fully automatic operation of the whole system is common.

Horizontally moving grates have a completely horizontal fuel bed. This is achieved by the diagonal position of the grate bars (see Figure 5.9). Advantages of this technology are the fact that uncontrolled fuel movements over the grate by gravity are impeded and that the stoking effect by the grate movements is increased, thus leading to a more homogeneous distribution of material on the grate surface and impeding slag formation as a result of hot spots. A further advantage of the horizontally moving grate is that the overall height can be reduced. In order to avoid ash and fuel particles falling through the grate bars, horizontally moving grates should be preloaded so that there is no free space between the bars.

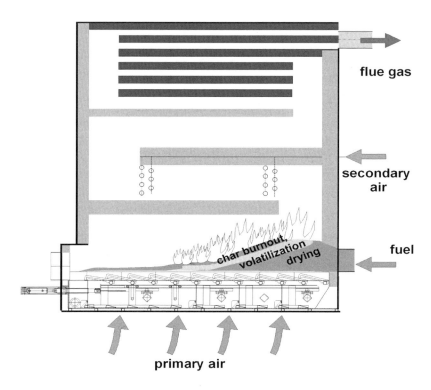

Figure 5.9 *Diagram of a horizontally moving grate furnace*

Source: [7].

5.2.1.4 Vibrating grates

Vibrating grate furnaces consist of an inclined finned tube wall placed on springs (see Figure 5.11). Fuel is fed into the combustion chamber by spreaders, screw conveyors or hydraulic feeders. Depending on the combustion process, two or more vibrators transport fuel and ash towards the ash removal. Primary air is fed through the fuel bed from below through holes located in the ribs of the finned tube walls [1]. Due to the grate vibrating for short periods, the formation of larger slag particles is inhibited, which is the reason why this

Combustion Technologies for Industrial and District Heating Systems 143

Figure 5.10 *Horizontally moving grate furnace*

Explanations: 7.2MW$_{th}$, 1.1MW$_{el}$, designed for waste wood; used for combined heat, cold and power production in Fussach, Austria

Source: Photograph courtesy of MAWERA Holzfeuerungsanlagen GesmbH, Austria.

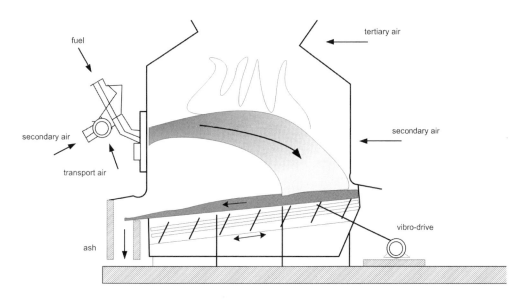

Figure 5.11 *Vibrating grate fed by spreader stokers*

Source: [1].

grate technology is especially applied with fuels showing sintering and slagging tendencies (e.g. straw, waste wood). Disadvantages of vibrating grates are the high fly-ash emissions caused by the vibrations, the higher CO emissions due to the periodic disturbances of the fuel bed and an incomplete burnout of the bottom ash because fuel and ash transport are more difficult to control.

5.2.1.5 Cigar burners

In Denmark, cigar burners have been developed for straw and cereal bale combustion (see Figure 5.12).

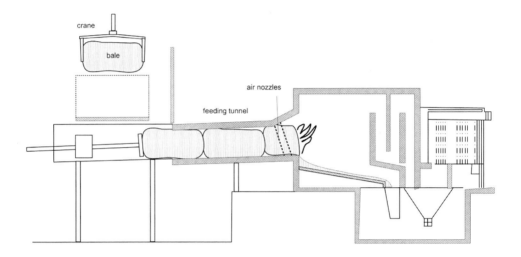

Figure 5.12 *Cigar burner*

Source: [1].

Straw and cereal bales (as a whole or sliced) are delivered in a continuous process by a hydraulic piston through a feeding tunnel on a water-cooled moving grate. Upon entering the combustion chamber, the fuel begins to gasify and combustion of the charcoal follows while the unburned material is moved over the grate. Grate and furnace temperature control are very important for straw and cereal combustion, due to their low ash sintering and melting points and the high adiabatic temperature of combustion caused by their low moisture content. Therefore, the combustion chambers have to be cooled either by water-cooled walls or by flue gas recirculation (combinations of these two techniques are also possible). Furnace temperatures should not exceed 900°C for normal operation. Furthermore, in straw and cereal combustion very fine and light fly-ash particles as well as aerosols are formed from condensed alkali vapours. An automatic heat exchanger cleaning system is required to prevent ash deposit formation and corrosion. Systems for shredded or cut straw also exist and operate in a similar way to the technology described – only the fuel preparation and feeding are different.

Semi-continuous systems such as whole bale combustion furnaces, in which the bales are fed in a batch-wise operation into the furnace, are not recommended due to the temperature and CO peaks caused when a new bale is delivered. Current process control systems are not able to prevent these unsteady combustion conditions.

5.2.1.6 Underfeed rotating grate

Underfeed rotating grate combustion is a new Finnish biomass combustion technology that makes use of conical grate sections that rotate in opposite directions and are supplied with primary air from below (see Figure 5.13). As a result, wet and burning fuels are well mixed, which makes the system suitable for burning very wet fuels such as bark, sawdust and woodchips (with a moisture content up to 65wt% (w.b.)). The combustible gases formed are burned out with secondary air in a separate horizontal or vertical combustion chamber. The horizontal version is suitable for generating hot water or steam in boilers with a nominal capacity between 3 and 17MW$_{th}$ and is used for biomass heating plants. The vertical version is applied for biomass CHP plants with a capacity between 7.7–13.5MW$_{th}$ and 1.0–4.5MW. The fuel is fed to the grate from below by screw conveyors (similar to underfeed stokers), which makes it necessary to keep the average particle size below 50mm. The fuel moves to the periphery of the circular grate. At the edge of the grate ash falls into a water-filled ash basin underneath the grate.

Figure 5.13 *Underfeed rotating grate*

Source: [8].

Underfeed rotating grate combustion plants are also capable of burning mixtures of solid wood fuels and biological sludge. The system is computer-controlled and allows fully automatic operation.

5.2.2 Underfeed stokers

Underfeed stokers (see Figure 5.14) represent a cheap and operationally safe technology for small- and medium-scale systems up to a nominal boiler capacity of 6MW$_{th}$. The fuel is fed into the combustion chamber by screw conveyors from below and is transported upwards on an inner or outer grate. Outer grates are more common in modern combustion plants because they allow for more flexible operation and an automatic ash removing system can be applied more easily. Primary air is supplied through the grate, secondary air usually at the entrance to the secondary combustion chamber. A new Austrian development is an underfeed stoker with a rotational post-combustion, in which a strong vortex flow is achieved by a specially designed secondary air fan equipped with a rotating chain (see Figure 5.15).

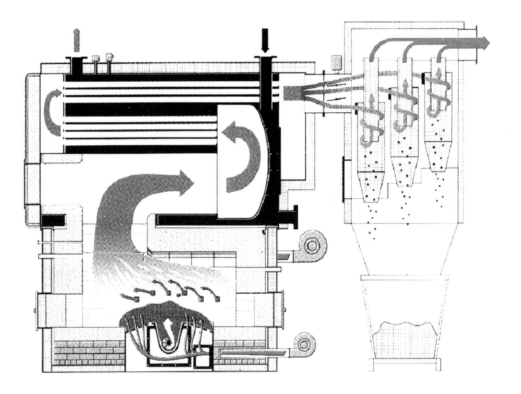

Figure 5.14 *Underfeed stoker furnace*

Source: [7].

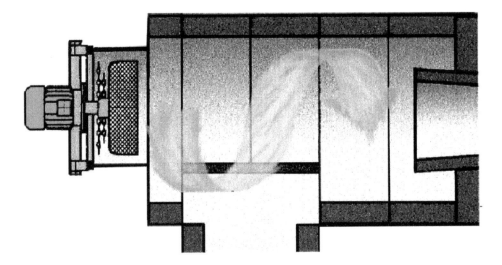

Figure 5.15 *Post-combustion chamber with imposed vortex flow*
Source: [9].

Underfeed stokers are suitable for biomass fuels with low ash content (woodchips, sawdust, pellets) and small particle sizes (up to 50mm). Ash-rich biomass fuels such as bark, straw and cereals need more efficient ash removal systems. Moreover, sintered or melted ash particles covering the upper surface of the fuel bed can cause problems in underfeed stokers, due to unstable combustion conditions when the fuel and the air break through the ash-covered surface. An advantage of underfeed stokers is their good partial-load behaviour and their simple load control. Load changes can be achieved more easily and quickly than in grate combustion plants, because the fuel supply can be controlled more easily and the fuel mass in the furnace is comparatively low.

5.3 Fluidized bed combustion

Fluidized bed combustion (FBC) systems have been applied since 1960 for combustion of municipal and industrial wastes. Since then, over 300 commercial installations have been built worldwide. Regarding technological applications, bubbling fluidized beds (BFB) and circulating fluidized beds (CFB) have to be distinguished (see Sections 5.3.1 and 5.3.2).

A fluidized bed consists of a cylindrical vessel with a perforated bottom plate filled with a suspension bed of hot, inert and granular material. The common bed materials are silica sand and dolomite. The bed material represents 90–98 per cent of the mixture of fuel and bed material. Primary combustion air enters the furnace from below through the air distribution plate and fluidizes the bed so that it becomes a seething mass of particles and bubbles. The intense heat transfer and mixing provides good conditions for complete combustion with low excess air demand (λ between 1.1 and 1.2 for CFB plants and between 1.2 and 1.3 for BFB plants). The combustion temperature has to be kept low (usually 650–900°C) in order to prevent ash sintering in the bed. This can be achieved by internal heat exchanger surfaces, by flue gas recirculation, by water injection or by sub-stoichiometric bed operation (in fixed bed combustion plants combustion temperatures are usually 100–200°C higher than in FBC units).

Due to the good mixing achieved, FBC plants can deal flexibly with various fuel mixtures (e.g. mixtures of different kinds of woody biomass fuels can be burned) but are limited when it comes to fuel particle size and impurities contained in the fuel. Therefore, an appropriate fuel pre-treatment system for particle size reduction and separation of metals is necessary for fail-safe operation. Usually a particle size below 40mm is recommended for CFB units and below 80mm for BFB units. Another critical point is related to the utilization of high alkali biomass fuels (see Section 2.3 and Chapter 8) due to possible ash agglomeration. Conventional fluidized bed furnaces cannot operate with high proportions of straw and similar high alkali fuels (e.g. herbaceous biomass fuels) without special measures. However, modern BFB furnaces with low bed temperatures of 650–850°C can burn fuels with low ash-melting temperature without any sintering problems in the bed [10].

Fluidized bed combustion systems need a relatively long start-up time (approximately 8–15 hours) for which oil or gas burners are used. With regard to emissions, low NO_x emissions can be achieved owing to good air staging, good mixing and a low requirement of excess air. Moreover, the utilization of additives (e.g. limestone addition for S capture) works well due to the good mixing behaviour. The low excess air quantities necessary increase combustion efficiency and reduce the flue gas volume flow. This makes FBC plants especially interesting for large-scale applications (normal boiler capacity above $20MW_{th}$). For smaller combustion plants the investment and operation costs are usually too high in comparison to fixed-bed systems. One disadvantage of FBC plants is posed by the high dust loads entrained with the flue gas, which make efficient dust precipitators and boiler cleaning systems necessary. Bed material is also lost with the ash, making it necessary to periodically add new material to the plant. From the extracted bed material/ash mixture, coarse parts can be separated from the fine particles and sand in an air classifier, and the fine material can be returned into the bed. Thus the bed material consumption of the boiler can be lowered [10].

5.3.1 Bubbling fluidized bed (BFB) combustion

For plants with a nominal boiler capacity of over $20MW_{th}$, BFB furnaces start to be of interest. In BFB furnaces (see Figures 5.16 and 5.17), bed material is located in the bottom part of the furnace. The primary air is supplied over a nozzle distributor plate from below and fluidizes the bed. The bed material is usually silica sand of about 0.5–1.0mm in diameter; the fluidization velocity of the air varies between 1.0 and 2.0m/s. The secondary air is introduced through several inlets in the form of groups of horizontally arranged nozzles at the beginning of the upper part of the furnace (called the freeboard) to ensure a staged-air supply to reduce NO_x emissions. In contrast to coal-fired BFB furnaces, the biomass fuel should not be fed onto, but into, the bed by inclined chutes from fuel hoppers because of the higher reactivity of biomass in comparison to coal. The fuel amounts only to 1–2 per cent of the bed material and the bed has to be heated (internally or externally) before the fuel is introduced.

The advantage of BFB furnaces is their flexibility concerning particle size and moisture content of the biomass fuels. Furthermore, it is also possible to use mixtures of different kinds of biomass or to co-fire them with other fuels. In modern BFB furnaces a sub-

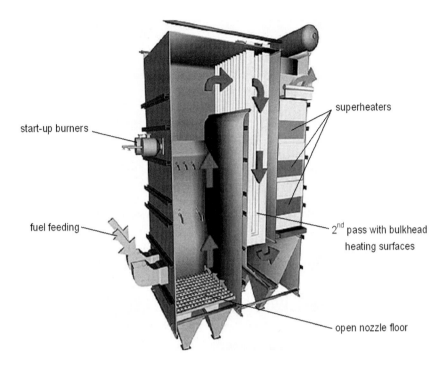

Figure 5.16 *BFB furnace*

Source: [11].

stoichiometric bed operation (λ about 0.35) is possible, which allows the bed temperature to be controlled in the range of 650–850°C. Therefore, fuels with low ash-melting temperature can also be burned [10]. Corrosion problems can be minimized by using refractory-lined superheaters in the first and second pass. In modern BFB furnaces part-load operation in the range of 60–100 per cent of nominal load is possible.

5.3.2 Circulating fluidized bed (CFB) combustion

By increasing the fluidizing velocity to 5–10m/s and using smaller sand particles (0.2–0.4mm in diameter) a CFB system is achieved. The sand particles will be carried with the flue gas, separated in a hot cyclone or a U-beam separator and fed back into the combustion chamber (see Figure 5.18). The bed temperature (750–900°C) is controlled by external heat exchangers cooling the recycled sand, or by water-cooled walls. The higher turbulence in CFB furnaces leads to a better heat transfer and a very homogeneous temperature distribution in the bed. This is of advantage for stable combustion conditions, the control of air staging and the placement of heating surfaces right in the upper part of the furnace. The disadvantages of CFB furnaces are their larger size and therefore higher price; the even greater dust load in the flue gas leaving the sand particle separator than in BFB systems; and the small fuel particle size required (0.1–40mm in diameter), which often causes higher investments in fuel pre-treatment. In view of their high specific heat transfer

150 *The Handbook of Biomass Combustion and Co-firing*

Figure 5.17 *25MW$_e$ woodchip-fired power plant with BFB boiler, in Cuijk, the Netherlands*

capacity, CFB furnaces start to be of interest for plants of more than 30MW$_{th}$, due to their higher combustion efficiency and the lower flue gas flow produced (boiler and flue gas cleaning units can be designed to be smaller).

5.4 Pulverized fuel combustion

In pulverized fuel combustion systems, fuels such as sawdust and fine shavings are pneumatically injected into the furnace. The transportation air is used as primary air. Start-up of the furnace is achieved by an auxiliary burner. When the combustion temperature reaches a certain value, biomass injection starts and the auxiliary burner is shut down. Fuel quality in pulverized fuel combustion systems has to be quite constant. A maximum fuel particle size of 10–20mm has to be maintained and the fuel moisture content should normally not exceed 20wt% (w.b.). Due to the explosion-like gasification of the fine biomass particles, the fuel feeding needs to be controlled very carefully and forms a key technological unit within the overall system. Fuel/air mixtures are usually injected tangentially into the cylindrical furnace muffle to establish a rotational flow (usually a vortex flow). The rotational motion can be supported by flue gas recirculation in the

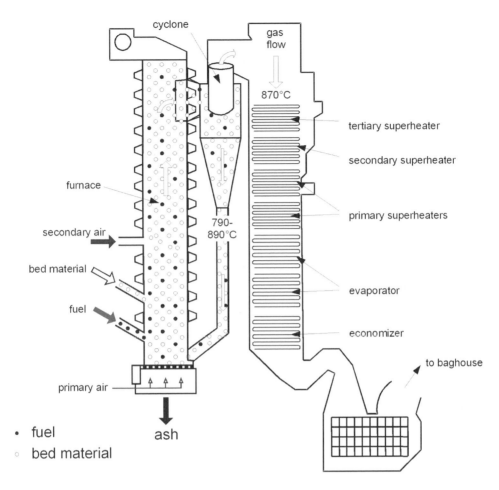

Figure 5.18 *CFB furnace*

Source: Adapted from [12].

combustion chamber. Due to the high energy density at the furnace walls and the high combustion temperature, the muffle should be water-cooled. Fuel gasification and charcoal combustion take place at the same time because of the small particle size. Therefore, quick load changes and an efficient load control can be achieved.

Muffle dust furnaces are being used more and more for fine wood wastes originating from the chipboard industry. Figure 5.19 shows a muffle dust furnace in combination with a water tube steam boiler. This technology is available for thermal capacities of 2–8MW [1]. The outlet of the muffle forms a neck, where secondary air is added in order to achieve a good mixture with the combustible gases. Due to the high flue gas velocities, the ash is carried with the flue gas and is partly precipitated in the post-combustion chamber. Low excess air amounts ($\lambda = 1.3$–1.5) and low NO_x emissions can be achieved by proper air staging. Besides muffle furnaces, cyclone burners for wood pulverized fuel combustion are also in use (see Figure 5.20). Depending on the design of the cyclone and the location of

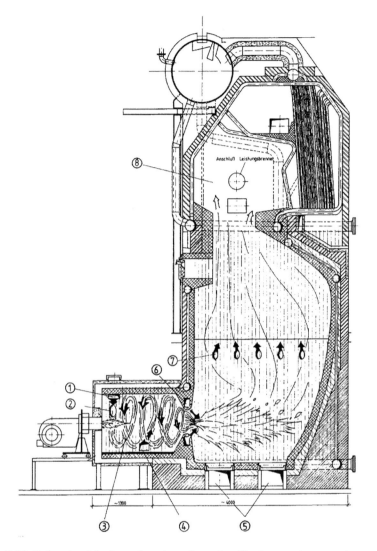

Figure 5.19 *Pulverized fuel combustion plant (muffle furnace) in combination with water-tube steam boiler*

Explanations: 1 primary air supply, 2 fuel feed, 3 gasification and partial combustion, 4 flue gas recirculation, 5 ash disposal, 6 secondary air supply, 7 tertiary air supply, 8 water-tube boiler.
Source: [1]

fuel injection, the residence time of the fuel particles in the furnace (and their burnout) can be controlled well.

A disadvantage of muffle furnaces and cyclone burners is that insulation bricks wear out quickly due to thermal stress and erosion. Therefore, other pulverized fuel combustion systems are being built without rotational flow, where dust injection takes place as in a furnace fired by fuel oil or natural gas (see Figure 5.21).

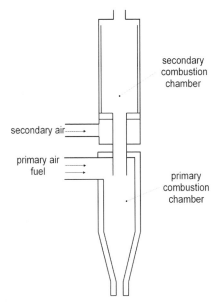

Figure 5.20 *Two-stage cyclone burner*

Source: [13].

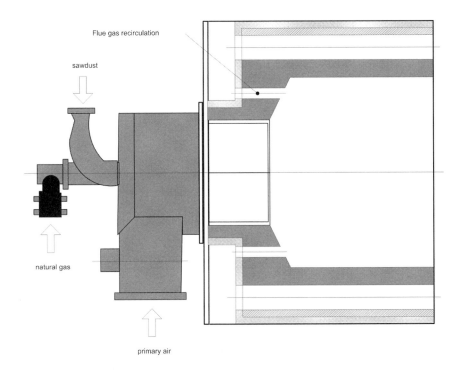

Figure 5.21 *Dust injection furnace*

Source: [1].

5.5 Summary of combustion technologies

Table 5.1 gives an overview of the advantages and disadvantages as well as the fields of application of various combustion technologies, which have been described in Sections 5.2–5.4. Regarding gaseous and solid emissions, BFB and CFB furnaces normally show lower CO and NO_x emissions due to more homogeneous and therefore more controllable combustion conditions. Fixed bed furnaces, in turn, usually emit fewer dust particles and show a better burnout of the fly ash [14–16].

Table 5.1 *Overview of advantages, disadvantages and fields of application of various biomass combustion technologies*

Advantages	Disadvantages
Grate furnaces	
• low investment costs for plants < 20MW_{th} • low operating costs • low dust load in the flue gas • less sensitive to slagging than fluidized bed furnaces	• usually no mixing of wood fuels and herbaceous fuels possible (only special constructions can cope with such fuel mixtures) • efficient NO_x reduction requires special technologies (combination of primary and secondary measures) • high excess oxygen (5–8vol%) decreases efficiency • combustion conditions not as homogeneous as in fluidized bed furnaces • low emission levels at partial load operation require a sophisticated process control
Underfeed stokers	
• low investment costs for plants < 6MW_{th} • simple and good load control due to continuous fuel feeding and low fuel mass in the furnace • low emissions at partial load operation due to good fuel dosing • low flexibility in regard to particle size	• suitable only for biomass fuels with low ash content and high ash-melting point (wood fuels) (< 50mm)
BFB furnaces	
• no moving parts in the hot combustion chamber • NO_x reduction by air staging works well • high flexibility concerning moisture content and kind of biomass fuels used • low excess oxygen (3–4 Vol%) raises efficiency and decreases flue gas flow	• high investment costs, interesting only for plants > 20MW_{th} • high operating costs • reduced flexibility with regard to particle size (< 80mm) • utilization of high alkali biomass fuels (e.g. straw) is critical due to possible bed agglomeration without special measures • high dust load in the flue gas • loss of bed material with the ash without special measures

Table 5.1 *continued*

Advantages	Disadvantages
CFB furnaces	
• no moving parts in the hot combustion chamber • NO_x reduction by air staging works well • high flexibility concerning moisture content and kind of biomass fuels used • homogeneous combustion conditions in the furnace if several fuel injectors are used • high specific heat transfer capacity due to high turbulence • use of additives easy • very low excess oxygen (1–2vol%) raises efficiency and decreases flue gas flow	• high investment costs, interesting only for plants > 30MW_{th} • high operating costs • low flexibility with regard to particle size (< 40mm) • utilization of high alkali biomass fuels (e.g. straw) is critical due to possible bed agglomeration • high dust load in the flue gas • loss of bed material with the ash without special measures • high sensitivity concerning ash slagging
Pulverized fuel combustion	
• low excess oxygen (4–6vol%) increases efficiency • high NO_x reduction by efficient air staging and mixing possible if cyclone or vortex burners are used • very good load control and fast alteration of load possible	• particle size of biomass fuel is limited (< 10–20mm) • high wear rate of the insulation brickwork if cyclone or vortex burners are used • an extra start-up burner is necessary

5.6 Heat recovery systems and possibilities for increasing plant efficiency

Table 5.2 gives an overview of the potential of various options for increasing the efficiency of biomass combustion plants **[17, 18]**. Biomass drying is one interesting method, though the efficiency increase and cost savings are usually moderate. Advantages that can be achieved are the prevention of auto-ignitions in wet bark piles, the reduction of dry-matter losses due to microbiological degradation processes during storage and a reduction of the necessary storage volume at the plant. However, biomass drying processes must be carefully examined for their economic advantage, considering the additional investment costs as well as the operating costs in the form of electricity consumption, and man and machine hours, necessary to run the process. In most cases, biomass drying is only economic if preheated air is available at low or no cost (examples are solar air collectors and the utilization of preheated air from flue gas condensation units – see Section 3.3.5). By drying biomass in piles for several months, a significant amount of the water evaporates by natural convection. Through biological degradation processes, however, biomass piles also loose up to 5wt% of organic matter per month. As a net result, the total energy content usually decreases, which makes outdoor storage unattractive

Table 5.2 *Influence of various measures on the thermal efficiency of biomass combustion plants*

Measures	Potential thermal efficiency improvement, as compared to the NCV of one dry tonne of fuel (%)
• drying from a moisture content of 50wt% to 30wt% (w.b.)	+8.7%
• decreasing the O_2 content in the flue gas by 1.0vol%	about +0.9%
• bark combustion: reducing the organic carbon content in the ash from 10.0 to 5.0wt% (d.b.)	+0.3%
• decreasing the flue gas temperature at boiler outlet by 10°C	+0.8%
• flue gas condensation (compared to conventional combustion units)	average +17% maximum +30%

Explanations: Calculations performed for woodchips and bark as fuel; GCV = 20MJ/kg (d.b.).
Abbreviations: d.b = dry basis; w.b. = wet basis; NCV = net calorific value; GCV = gross calorific value; efficiency = heat output (boiler)/energy input fuel (NCV).

Reducing the excess oxygen content in the flue gas is an effective measure for increasing the efficiency of the combustion plant, as shown in Figure 5.22.

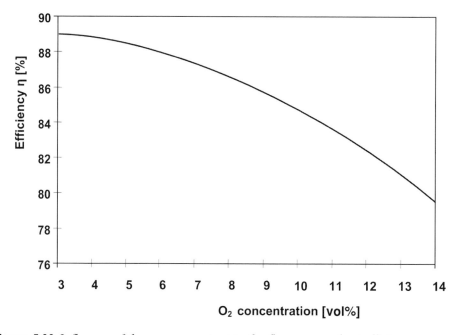

Figure 5.22 *Influence of the oxygen content in the flue gas on plant efficiency*

Explanations: moisture content of the fuel 55.0wt% (w.b.); biomass fuel used: woodchips/bark; H- content 6.0wt% (d.b.); gross calorific value of the fuel 20,300kJ/kg (d.b.); flue gas temperature at boiler outlet 165°C, efficiency related to the NCV of the fuel; O_2 concentration related to dry flue gas; efficiency = heat output from boiler/energy input with the fuel (NCV).

There are two technological possibilities for decreasing the excess air ratio and ensuring a complete combustion at the same time. On the one hand, an oxygen sensor is coupled with a CO sensor in the flue gas at the boiler outlet to optimize the secondary air supply (CO-λ-control); on the other hand, there are improvements that can be made in the mixing quality of flue gas and air in the furnace (as already explained). In addition, a lower excess oxygen concentration in the flue gas can also significantly improve the efficiency of flue gas condensation units, due to the fact that it increases the dew point and therefore raises the amount of recoverable latent heat of the condensing water at a certain temperature (see Figure 5.23).

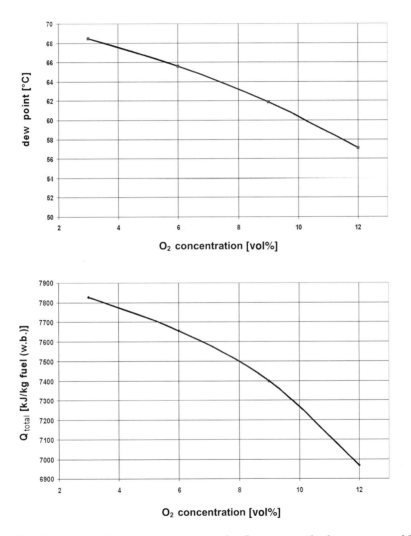

Figure 5.23 *Influence of the oxygen content in the flue gas on the heat recoverable in flue gas condensation plants*

Explanations: dew point calculated for water in the flue gas of a woodchips and bark combustion plant; moisture content of the biomass fuel 55.0wt% (w.b), H content 6.0wt% (d.b.); GCV of the fuel 20MJ/kg (d.b.); Q_{total} = total recoverable heat from 1.0kg biomass fuel (w.b.) burned when the flue gas is cooled down to 55°C.

Furthermore, reducing the excess oxygen concentration in the flue gas also decreases the flue gas volume flow, which limits pressure drops and reduces the sizes of boilers and flue gas cleaning units. Of course, one has to take into consideration that a reduction of the excess oxygen concentration in the flue gas also increases the combustion temperature, which underlines the importance of an appropriate furnace temperature control system.

Low carbon-in-ash values are of minor importance for the efficiency of the plant but of major importance for the possibility of sustainable ash utilization, because the concentration of organic contaminants in biomass ashes normally increases with higher carbon concentrations [15, 19] (see also Chapter 9).

The most effective way to recover energy from the flue gas – and in many cases also an economically interesting technology – is flue gas condensation. In addition to the high energy recovery potential of this process (up to 20 per cent of the energy input by the biomass fuel related to the NCV), dust precipitation efficiencies of 40–75 per cent can be achieved. Furthermore, there is the possibility of preventing condensation of the flue gas in the chimney down to ambient temperatures of about −10°C [20, 21]. In Denmark, the majority of biomass district heating plants are equipped with flue gas condensation units. In Sweden, Finland and Austria, the number of installations is rapidly increasing; Italy, Germany and Switzerland also have several plants already in operation. Figure 5.24 shows the principle of a flue gas condensation unit. The whole plant normally consists of three

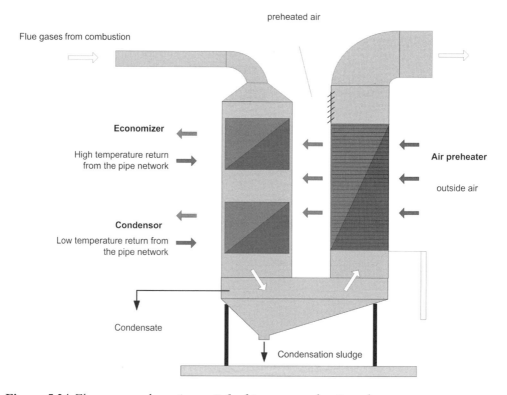

Figure 5.24 *Flue gas condensation unit for biomass combustion plants*

Source: [22].

parts: the economizer (recovery of sensible heat from the flue gas), the condenser (recovery of sensible and latent heat from the flue gas) and the air preheater (preheating the combustion air and the air used to dilute the saturated flue gas before entering the chimney).

The amount of energy that can be recovered from the flue gas depends on the moisture content of the biomass fuel, the amount of excess oxygen in the flue gas (as already explained) and the temperature of the return of the network of pipes. The lower the temperature of the return water, the higher the amount of latent heat that can be recovered when the flue gas is cooled below the dew point (see Figure 5.25). Consequently, the energy recovery potential strongly depends on the quality of the heat exchangers, the hydraulic installations and the process control systems installed by the clients, because they determine the return temperature.

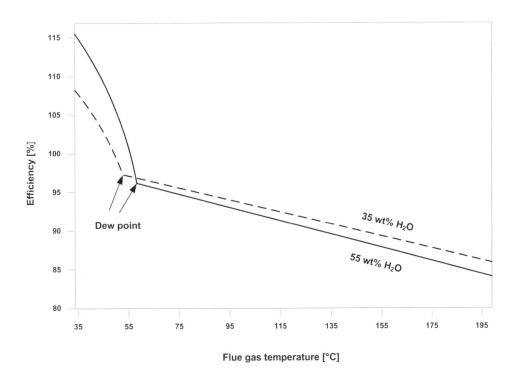

Figure 5.25 *Efficiency of biomass combustion plants with flue gas condensation units as a function of flue gas temperature*

Explanations: oxygen concentration of the flue gas 9.5vol% (d.b.); biomass fuel used: woodchips/bark; H content 6.0wt% (d.b.); GCV of the fuel 20MJ/kg (d.b.); dew point calculated for water; efficiency = heat output (boiler + flue gas condensation unit)/energy input fuel (NCV).

The dust precipitation efficiency of 40–75 per cent mentioned can be significantly increased by placing a simple aerosol electrostatic filter behind the condensation unit. Early test runs showed a dust precipitation efficiency of 99.0 per cent at temperatures below 40°C [23]. Due to the low flue gas temperature, the electrostatic precipitator (ESP) unit can also be kept small and therefore economically acceptable. Furthermore, not only aerosols but also water droplets entrained with the flue gas achieve efficient precipitation, lowering the amount of dilution air that has to be added to the saturated flue gas leaving the condensation unit. The condensation sludge has to be separated from the condensate (by sedimentation units) because it contains significant amounts of heavy metals upgraded in this fine fly-ash fraction. It has to be disposed of or industrially utilized (see Section 9.4). Moreover, research has shown that the separation of sludge and condensate should be done at pH values > 7.5 in order to prevent dissolution of heavy metals and to meet the limiting values for a direct discharge of the condensate into rivers [24].

Flue gas condensation systems can also be operated as quench processes. The disadvantage of such systems is that quenching slightly lowers the amount of heat that can be recovered and requires even lower temperatures of the return in order to be energetically efficient. An interesting option for increasing the heat recovery potential of flue gas condensation units is a Swedish approach with an air humidifier integrated into the system (see Figure 5.26).

This technology foresees the pre-treating and moistening of combustion air by injecting condensate water into the air. By this means, the moisture content of the flue gas rises and the heat recovery potential increases accordingly. When applying this air humidification unit, one has to ensure that the spray nozzles produce very fine water droplets (in order to give them enough time for evaporation) and to design the nozzles in a way that prevents malfunctions due to impurities in the condensate [25].

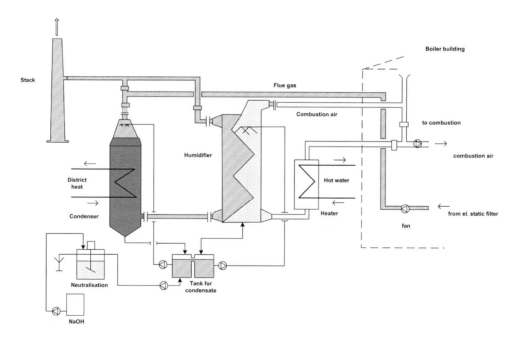

Figure 5.26 *Flue gas condensation unit with an integrated air humidifier*

Source: [25].

With regard to economic aspects, flue gas condensation is generally recommendable for biomass combustion plants. Flue gas condensation units are of interest if wet biomass fuels are utilized (average moisture content 40–55wt% (w.b.)), if the return of the network of pipes is below 60°C and if the nominal boiler capacity is above 2MW$_{th}$.

5.7 Process control systems for biomass combustion installations

5.7.1 Control objectives

The main control objectives for a biomass combustion installation are:

- rejection of the disturbances caused by variations in the fuel; and
- forcing towards the process at its optimal operating points and maintaining it there.

The timescales of these objectives are different. Disturbance rejection requires a quickly reacting control system to effectively cope with fast fluctuations of fuel composition. This control strategy operates in a relatively high frequency domain and tries to minimize the reaction of the process to disturbances and keeps operating points close to their set points. This objective also minimizes the deviation of the most relevant process variables from their set points. When these set points match the designed operating points of an installation, such an installation is often assumed to have optimal (minimal) emissions.

The second objective mentioned involves a slower timescale. Traditional process control strategies operate with fixed setpoints, but more advanced process control strategies like model predictive control (see Section 5.7.4.2) may use an online optimization that minimizes a certain cost function. Such a cost function may contain a weighted mix of several control objectives, such as:

- hard upper or lower limits on certain variables due to environmental or safety restrictions (e.g. maximum steam pressure in the steam drum or minimum oxygen concentration/temperature in the flue gases to prevent emission of CO or dioxins);
- maximizing heat production;
- minimizing flue gas flow (could be demanded by legislation); and
- minimizing superheater temperature (due to corrosion aspects, see also Chapter 8).

5.7.2 Process dynamics

Increased insight into process dynamics can lead to a better understanding of why a process is behaving the way it is. This knowledge can help to improve control strategies or process configurations.

In this section typical moving grate installation process dynamics are explained. The dynamics will vary from case to case, depending on the size, configuration and fuel properties of the installation, but the principles apply in general.

Figure 5.27 shows the dynamic responses of a large incinerator installation to changes in fuel dosage, grate speed, primary air and secondary air. In these experiments, the other control parameters were kept constant, the experiments can therefore said to represent the **open-loop** response.

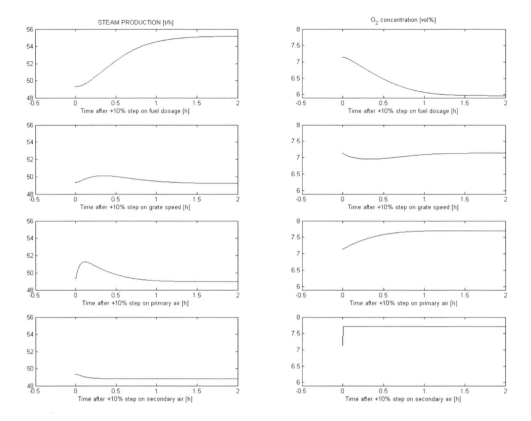

Figure 5.27 *Dynamic responses of a large incinerator installation to fuel dosage, grate speed, primary air and secondary air changes*

Explanations: dynamical responses to 10% (stepwise) increases in fuel dosage (first row), grate speed (second row), primary air (third row) and secondary air (last row) on steam production (left column) and oxygen concentration (right column).

The first row in Figure 5.27 depicts the characteristic dynamic responses to a 10 per cent stepwise increased fuel dosage. The steam production will increase gradually following a mainly first order response. As it takes some time for the newly added fuel flow to catch fire, only after approx. 45 minutes is 66 per cent of the steady state gain reached.* Smaller installations may react more quickly, but the general shape of the response will exhibit great similarities. Oxygen concentration shows an opposite behaviour. Because more fuel will be combusted with the same amount of air, the resulting oxygen concentration in the flue gas will drop.

The second row in Figure 5.27 depicts the responses of steam production and oxygen concentration to a 10 per cent stepwise increased grate speed. In contrast to the responses to increasing fuel flow, both parameters show no static gain as an increased grate speed will not lead to more fuel being combusted (the fuel flow towards the grate will remain

---

* This particular response is therefore said to have a characteristic time constant of about 45 minutes.

constant). In the response of steam production to grate speed, the 'stirring effect' is illustrated. Once the grate starts moving faster, the periods of fuel layer mixing occur more frequently and the fire will intensify. After a certain time, all fuel is on fire again and the feed flow determines the steam production. Again, oxygen shows an opposite trend. Compared to the influence of the fuel dosage, the dynamics of the grate speed are significantly faster and could therefore more effectively be used to react to fast disturbances.

The third row shows the effect on steam production and oxygen concentration of a stepwise increase in primary air. As the amount of fuel burning remains constant, there is no net effect on heat production. When successive increased amounts of air are used, there is a possibility of cold-blowing the fire, thus decreasing the steam production and having to cope with unburned fuel at the end of the grate. Primary air has a positive static gain on oxygen, because more oxygen is provided for the same amount of fuel to be combusted.

The fourth and final row shows the effect of increasing secondary air input. As for increasing primary air, this has hardly any effect on steam production, because secondary air is not directly related to the amount of heat released by the combustion process. Secondary air, however, will dilute the flue gases, leading to an immediate increase in oxygen concentration and cooling of the flue gas. As the steam system will subsequently cool down, steam production might also slightly decrease when secondary air is increased.

5.7.3 State-of-the-art process control

A process control system of a modern biomass combustion plant usually consists of the following control loops:

- load control;
- combustion control;
- furnace temperature control; and
- furnace pressure control.

The load control in biomass furnaces is usually guided by the feed water (or steam or thermal oil) temperature and determines fuel and primary air feed.

For each furnace and fuel moisture content a specific excess air ratio (λ) exists, where the CO emissions are minimal. Below and above this specific excess air ratio the CO emissions increase (see Figure 9.3). In addition, the CO/λ characteristic depends not only on the moisture content of the fuel but also on the actual load conditions of the furnace. Higher fuel moisture contents and decreasing load of the furnace usually increase the optimum excess air ratio and vice versa [26].

Both the moisture content of the fuel as well as the thermal output needed can vary across a broad range in biomass furnaces. Therefore, a fixed set point for the excess air ratio (as applied in very simple applications) can lead to dramatic increases of the CO emissions, if the moisture content of the fuel and/or the load conditions of the furnace change. Consequently, combustion control only on the basis of CO or excess air control would not lead to a desired result.

Therefore, a combined CO/λ control provides best results with regard to combustion control and CO emission reduction. A block diagram of a combined combustion (CO/λ) and load control is shown in Figure 5.28. By this approach the excess air ratio can be varied until the minimum of the CO emissions is found. If the CO emissions change (due to a

164 The Handbook of Biomass Combustion and Co-firing

change in fuel moisture content or load conditions) this procedure is repeated. Consequently, the furnace can be adjusted to any fuel moisture content and load condition, which ensures optimized burnout and CO emissions as well as optimized efficiency of the plant.

Furnace temperature control should be achieved by flue gas recirculation and/or water-cooled furnace walls.

Finally, the negative pressure in the furnace is usually measured by appropriate pressure sensors and is controlled by an induced draught fan.

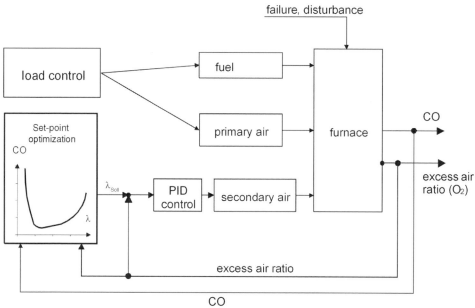

Figure 5.28 *Block diagram of combustion (CO/λ) and load control*
Source: [26].

5.7.4 Advanced process control

Up until now, the most common way to improve control of the combustion process has been by performing improvements in practice based upon practical experiences (trial and error). As combustion processes are multivariable processes (multiple-input, multiple-output), which are often subject to large disturbances in fuel composition, it is difficult to understand the different input–output relations. Due to this complexity, it is therefore difficult to design or improve control systems by practical experiences. This trial and error method will consequently lead to sub-optimal solutions.

Model-based and model predictive process control technologies are technologies that can contribute to these growing needs. These control technologies make extensive use of available knowledge on the dynamic behaviour of processes. Model predictive control enables revision of the control strategy on a sample-by-sample basis using the latest information on the status of the plant and its environment. The knowledge of process behaviour is represented in the form of a mathematical model of the process dynamics. This model is explicitly used in the controller for predicting future process responses to past input manipulations and measured disturbances and to calculate best future input manipulations that satisfy the control objectives. The model reflects the significant dynamic properties of process input–output behaviour and enables simulation of the future process outputs on the basis of known past process inputs within a pre-specified operating envelope of the process with limited inaccuracy and uncertainty [27]. Model-based control uses the simulator to develop off line the most promising control concept. Mostly this will mean optimization of the classical PID (proportional-integral-derivative) controller loops.

5.7.4.1 Model-based process control
Control optimization uses a process model (preferably validated) and a computer model of the controller to be tested/optimized. Process knowledge is used at the design stage of the controller. In this way, different strategies and tuning parameters can be tested and knowledge about their properties can be used to adapt a controller to the real process.

5.7.4.2 Model predictive control
A model predictive control (MPC) system is the ideal tool to control multivariable processes. Multivariable processes are processes whose inputs influence more than just one process output simultaneously. A characteristic of MPC is that the control strategy can be adjusted for each calculation of a next control action. As a result MPC is very flexible for changing conditions, such as, for example, changing requirements, switching-off and/or failure of sensors and actuators. Moreover, MPC can deal with constraint requirements, i.e. it can keep both manipulated variables as well as controlled variables within certain predefined ranges. MPC has been developed within the industry, emerging from the need to operate processes more tightly within the operational and physical constraints of the process and the applied equipment as well as closer to the operating constraints that maximize margins. From its initial development [27], MPC has grown to be a widely proven technology, especially in the oil refining industry. The dominant use of MPC in oil refining applications implies robustly pushing the controlled process to operating conditions that maximize margins and minimize process variability. For most refinery applications this results in maximization of the throughput of a certain product mix. In, for example, glass manufacturing, the benefits mostly stem from tight control of product quality, increase of average furnace load, increase of efficiency, tight control of emissions and minimization of energy consumption.

The success of MPC within industry is in large part due to the fact that MPC meets industrial requirements. These requirements can be roughly categorized into three groups:

- operational requirements: processes have to be operated within a predefined region (safety, emissions, wear, etc.);
- product quality requirements: products have to be produced at specifications (Cpk values, 6-sigma ranges, etc.); and
- economic requirements: products must be produced in such a way that margins are maximized, without violating operating constraints.

Figure 5.29 shows a block diagram of a MPC control system. Initially, MPC did not explicitly take constraints into consideration. Refinements of the technology developed at the end of the 1980s to allow constraints on both input and output variables to be considered in the formulation of the control strategy.

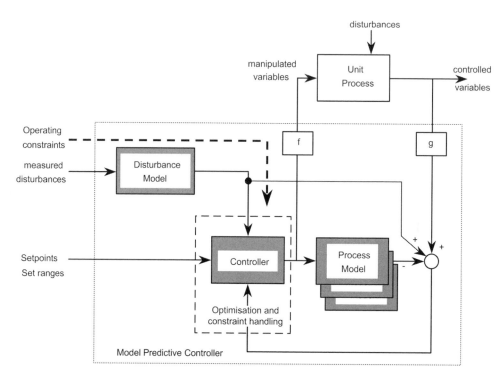

Figure 5.29 *Schematic representation of a model predictive process control system*

From Figure 5.29 it can be seen that there are three main blocks in MPC systems. The controller is the general block, while the other two, process model and disturbance model, depend on the specific application. In these blocks the process knowledge plays an important role.

5.7.4.3 (Neuro-)fuzzy control
Operator knowledge and other knowledge of the process are described in fuzzy sets rules. The type of knowledge is often vague and qualitative by nature. For processes which are hard to identify or have dynamic properties which change over time, fuzzy control can be a better way to cope with these effects than traditional (PI(D)) control. Fuzzy controllers may use fuzzy process models or contain fuzzy control rules which may evolve from human behaviour and can be extracted by interviews.

5.7.4.4 Advanced sensoring
Soft sensoring techniques are software based techniques that are used to estimate an (un)measurable process quantity based on existing process measurements. The term 'soft

sensoring' is used because computer models are used to estimate the sensor output. These techniques could yield valuable additional information about a process quantity which would normally not be available due to:

- a method for a direct measurement may technically not exist;
- a measurement can only be available after a time-consuming laboratory analysis – for controlling applications, this information may be available too late; or
- a direct measurement may be too costly.

An example of a soft sensor which can be used for biomass combustion installations is the calorific value sensor, which has been developed and demonstrated by TNO (the Netherlands Organisation for Applied Scientific Research) [28]. This sensor estimates the properties of the biomass which is being combusted in real time, based upon measurements of the composition of the flue gases and some additional process measurements. An automatic controller (or an operator who is controlling the process manually) may then use the following sensor outputs, in order to manipulate the process:

- calorific value of the biomass;
- density of the biomass; and
- moisture content of the biomass.

Observers show some similarity to the soft-sensoring techniques, because observers are also software-based techniques to estimate process variables online.

The difference with soft sensors is that observers are mostly used to estimate state variables of the process (a state variable is generally not a process output, but an internal process state property such as the total amount of mass on a grate in a biomass combustion installation). These properties can normally not be measured, but an observer is able to estimate these quantities based upon existing process measurements.

The application of observers also differs from soft sensors, as observers are used to adapt process models online to changes in the process dynamics. In MPCs, observers are used to adapt the process model online to changes in the process (for example fouling in a heat exchanger in the steam system will lead to a decreased heat transfer, which will lead to a change in process dynamics). Such an observer keeps the process model up to date by tuning parameters in the process model to the real world.

5.8 Techno-economic aspects of biomass combustion plants design

Biomass combustion plants are complex systems with numerous components. In order to ensure a sustainable and economic operation of such plants, professional dimensioning and engineering are essential.

The engineering process consists of several steps as described briefly below [1, 29].

1 identification of the bases of the biomass combustion plant;
2 feasibility study;
3 planning of the design;
4 approval procedure;

5 planning of the execution;
6 initializing and placing orders;
7 supervision of the construction work; and
8 commissioning test and documentation.

The main components of a biomass combustion plant are (alternative components in brackets):

- fuel storage (long-term storage, daily storage);
- fuel-feeding and handling system;
- biomass furnace;
- boiler (hot water, steam, thermal oil);
- back-up or peak-load boiler (e.g. oil-fired boiler);
- heat recovery system (economizer or flue gas condensation unit);
- ash manipulation and pre-treatment;
- flue gas cleaning system;
- stack;
- control and visualization equipment;
- electric and hydraulic installations;
- (heat accumulator);
- (CHP unit); and
- (network of pipes for district heating plants).

5.8.1 Technical and economic standards for biomass combustion and district heating plants

In Austria, technical and economic standards have been defined for biomass district heating plants in order to secure an economically reasonable investment. Keeping to these standards is a requirement for new biomass district heating or CHP projects in Austria; otherwise no investment subsidies are granted. Relevant technical and economic parameters of these standards are defined and explained in [30, 31].

$$\text{Simultaneity factor [\%]} = \frac{\text{effective peak heat load - district heat network}}{\sum \text{consumer nominal connection capacities}}$$

$$\text{Boiler full-load operating hours [h/a]} = \frac{\text{boiler heat produced per year}}{\text{boiler nominal capacity}}$$

$$\text{Annual utilization rate of biomass combustion plant [\%]} = \frac{\text{boiler heat produced per year}}{\text{fuel heat input (NCV) per year}} \times 100\%$$

$$\text{Network heat utilization rate [kWh/m]} = \frac{\text{heat sold per year [kWh]}}{\text{length of pipe network [m]}}$$

$$\text{Annual utilization rate of district heating network [\%]} = \frac{\text{heat sold to final consumers per year}}{\text{heat output from heating plant per year}} \times 100\%$$

$$\text{Specific investment (boiler) [Euro/kW]} = \frac{\text{Investment cost of total system [Euro]}}{\text{nominal capacity of biomass boiler [kW]}}$$

$$\text{Heat generation costs [Euro/MWh]} = \frac{\text{(annualized capital costs + other payments) per year [Euro]}}{\text{heat sold per year [MWh]}}$$

5.8.2 Plant dimensioning/boiler size

The nominal thermal capacity of a biomass district heating or heat-controlled CHP plant is determined by the energy demand (heat, electricity) and has to allow for future developments. Therefore, as a first step, a detailed and precise survey of capacity and heat requirements in the supply area is necessary. Moreover, the simultaneity of heat demand of the district heating clients, described by the *simultaneity factor*, has to be taken into consideration. This factor depends on the number and type of consumers and fluctuates between 0.5 (large district heating networks) and 1 (micro-networks) [31].

In most cases, the energy demand is not constant over the whole year. The heat load of district heating networks especially varies during the year, reaching a maximum during the winter and a minimum during summer. Therefore, on the basis of the results of the survey of capacity and heat requirements, the annual heat output line has to be calculated (see Figure 5.30). In boiler planning, a distinction must be made between base load and peak load for economic reasons. Base load is covered by one or more biomass boilers, peak load boilers are usually run on fossil energy or liquid biofuels for economic reasons. The installation of heat accumulators can also contribute to peak load coverage. This distinction between base load and peak load is necessary to achieve a high number of *full-load operating hours* of the biomass boiler and to decrease total *heat generation costs*. The correct determination of the boiler sizes depends on the capital costs of the combustion unit as well as on the operating costs (mainly fuel costs – see Table 5.3).

Table 5.3 *Comparison of specific investment and fuel costs for biomass and fuel oil-fired combustion systems*

combustion system	Specific investment costs	Fuel costs
Biomass	high (about €160/kW[1])	low (about €15–25/MWh$_{NCV}$ [3])
Fuel oil	low (about €20/kW[2])	high (about €55–65/MWh$_{NCV}$)

[1] 5MW$_{th}$ biomass combustion unit (fuel feeding, furnace, boiler, multicyclone, ESP, stack); updated to June 2006 level;
[2] 5MW$_{th}$ fuel oil boiler with oil tank, burner and stack; specific investment costs related to nominal boiler capacity; updated to June 2006 level;
[3] untreated woodchips.

Source: [30, 32].

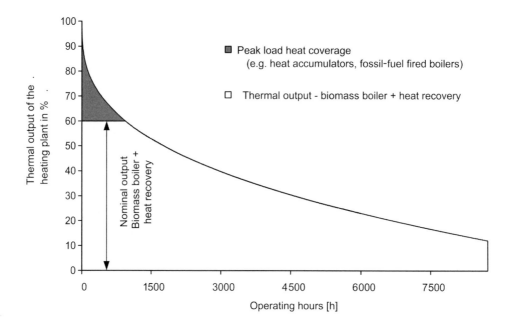

Figure 5.30 *Example of distribution between base load and peak load on the basis of the annual heat output line*

Source: [31].

5.8.3 Annual utilization rate of the biomass system

The *annual utilization rate of the biomass system* (biomass boiler + heat recovery) in the overall plant should be at least 85 per cent. Therefore, the installation of a heat recovery system (e.g. economizer or flue gas condensation unit) is recommended.

For biomass CHP plants, mainly heat-controlled operation is recommended. The annual utilization rate should be at least 75 per cent. Therefore, biomass CHP applications should only be realized on sites where the heat produced can be utilized in a reasonable way. This guideline is also valid for co-firing plants.

5.8.4 Size of the fuel storage unit

The fuel storage unit should be small and should be designed for just-in-time operation (capacity of the biomass storage unit less than 10 per cent of annual fuel consumption). Care should be taken to arrange for appropriate fuel supply contracts, organized fuel purchase and regional coordination. If appropriate long-term fuel supply contracts cannot be established, the fuel storage should be designed for a higher storage capacity, depending on the regional framework conditions.

5.8.5 Construction and civil engineering costs

The costs of the buildings should be less than €150 per m³ converted space; the costs of the storage unit should be less than €80 per m³ of converted space (price status June 2006).

5.8.6 Heat distribution network

The costs of the heat distribution network account for 35–55 per cent of the total investment costs of complete district heating plants. Thus, it is important to calculate the network correctly in order to achieve high rates of utilization and to concentrate on a small and efficient network of pipes. For biomass district heating networks, the *network heat utilization rate* should exceed 800kWh/m; the targeted value is 1200kWh/m. Moreover, a maximum temperature spread between feed and return should be achieved. The targeted value for biomass district heating plants is 40°C or higher. The *annual utilization rate of district heating networks* should exceed 75 per cent.

5.8.7 Heat generation costs and economic optimization

The calculation of heat generation costs is preferably based on the VDI Guideline 2067 **[33]**. This cost calculation scheme distinguishes four types of costs:

1 capital costs (depreciation, interest costs);
2 consumption-based costs (fuel, materials like lubricants);
3 operation-based costs (personnel costs, costs for maintenance); and
4 other costs (administration, insurance).

In comparison to energy systems run on fossil fuels, investment costs for biomass boilers including fuel supply systems and flue gas cleaning are high (see Table 5.3). Typical values for total investment costs for biomass combustion plants in Austria and Denmark are shown in Figure 5.31. Therefore, optimal plant utilization is necessary to decrease heat generation costs. Figure 5.32 illustrates the influence of the boiler full-load operating hours on the capital costs of biomass combustion units. In order to take advantage of the decline of marginal unit costs, the *boiler full-load operating hours* of the biomass combustion unit should exceed 4000 hours per year. For biomass CHP plants in heat-controlled operation, the target is 5000 boiler full-load operating hours or more.

It should be noted that the principles stated in this section are not only valid for small- and medium-scale biomass heating and CHP plants but also for large biomass CHP and co-firing units. However, with increasing plant size the importance of the investment costs decreases (the economy-of-scale effect already indicated in Figure 5.31 continues for larger plant sizes) and the relevance of the fuel costs on the energy generation costs increases.

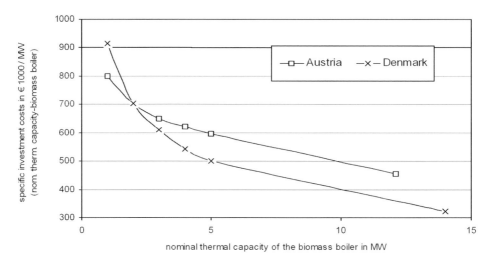

Figure 5.31 *Specific investment costs for biomass combustion plants in Austria and Denmark as a function of biomass boiler size*

Explanations: Data related to biomass grate furnaces, investment costs include: biomass grate furnace for woodchips, hot water fire-tube boiler, back-up boiler (fuel oil), fuel storage, fuel-feeding system, flue gas cleaning (ESP), stack, buildings, hydraulic and electric installations, engineering and construction costs (network of pipes is not included); price level June 2006

Source: [30, 34].

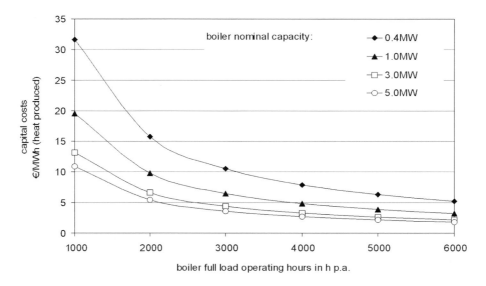

Figure 5.32 *Specific capital costs for biomass combustion systems as a function of boiler capacity and boiler utilization*

Explanations: Biomass moving grate furnace (including hot water fire-tube boiler, fuel feeding and stack), interest rate 7% p.a., lifetime 20 years, calculations according to VDI Guideline 2067; price level June 2006.

Source: [30].

5.9 References

1. MARUTZKY, R. and SEEGER, K. (1999) *Energie aus Holz und anderer Biomasse*, DRW-Verlag Weinbrenner, Leinfelden-Echtlingen, Germany
2. OBERNBERGER, I. (1996) Decentralised Biomass Combustion State-of-the-Art and Future Development, keynote lecture at the 9th European Biomass Conference in Copenhagen, *Biomass and Bioenergy*, vol 14, no 1, pp33–56 (1998)
3. HARTNER, P. (1996) Entwicklung eines Computerprogramms zur eindimensionalen Simulation von heterogenen Festbett- und Vorschubreaktoren, PhD thesis, Institute of Chemical Engineering, Graz University of technology, Austria
4. NUSSBAUMER, T. and GOOD, J. (1995) *Projektieren automatischer Holzfeuerungen*, Bundesamt für Konjunkturfragen, Bern, Switzerland
5. DETROIT STOKER COMPANY (1998) company brochure, Bulletin No. 14-05, Detroit Stoker Comp., Monroe, MI, US
6. EUROTHERM (1995) company brochure, Eurotherm, Viby, Denmark
7. MAWERA (n.d.) company brochure, MAWERA Holzfeuerungsanlagen GmbH&CoKG, Hard/Bodensee, Austria
8. WÄRTSILÄ (2006) Homepage www.wartsila.com, accessed 17 May 2006 (Finland)
9. WEHINGER, F. and KÖB, S. (1998) '"Pyrot" – Automatische Holzfeuerung mit Rotationsgebläse', in *Proceedings of the 5th. Holzenergie-Symposium*, 16 October, ETH Zürich, Bundesamt für Energie, ENET, Bern, Switzerland
10. BOLHAR-NORDENKAMPF, M., TSCHANUN, I. and KAISER, S. (2006) 'Operating experience from two new biomass fired FBC-plants', in *Proceedings of the World Bioenergy 2006 Conference & Exhibition on Biomass for Energy*, May/June 2006, Jönköping, Sweden, Swedish Bioenergy Association, Stockholm, Sweden, pp174–181
11. AE&E (2006) Homepage, www.aee.co.at/ (accessed 19 July 2006), Austrian Energy & Environment AG, Graz, Austria
12. NATIONAL RENEWABLE ENERGY LABORATORY, SANDIA NATIONAL LABORATORY, UNIVERSITY OF CALIFORNIA, FOSTER WHEELER DEV. CORP., US BUREAU OF MINES, T. R. MILES (1996) *Alkali Deposits found in Biomass Power Plants*, research report NREL/TP-433-8142 SAND96-8225 vols I and II, National Renewable Energy Laboratory. Oakridge, US
13. SENGSCHMIED, F. (1995) Ein Beitrag zur Entwicklung einer druckbeaufschlagten Brennkammer für die zweistufige Verbrennung von Holzstaub. PhD thesis at the Vienna University of Technology, Vienna, Austria.
14. LECKNER, B. and KARLSSON, M. (1993) 'Gaseous emissions from circulating fluidised bed combustion of wood', in *Biomass and Bioenergy*, vol 4, no 5, pp379–389
15. RUCKENBAUER, P., OBERNBERGER, I. and HOLZNER, H. (1996) *Erforschung der Verwendungsmöglichkeiten von Aschen aus Hackgut- und Rindenfeuerungen*, final report, part II of the research project No. StU 48 – Bund-Bundesländerkooperation, Institute of Plant Breeding and Plant Growing, University for Agriculture and Forestry, Vienna, Austria
16. WESTERMARK, M. (1994) *Termisk kadmiumrening av träbränsleaskor, preliminär version*, report, Vattenvall Utveckling AB, Vällingby, Sweden
17. OBERNBERGER, I. (1996) 'Combustion – State-of-the-Art and Future Developments', in *Proceedings of the Conference on Applications of Bioenergy Technologies in New Zealand*, 12–14 March, Roturoa, New Zealand, Energy Efficiency and Conservation Authority, Auckland, New Zealand
18. BIEDERMANN, F. (1994) EDV-gestützte Bilanzierung und Bewertung von Biomasseheizwerken, MS thesis, Institute of Chemical Engineering, Graz University of Technology, Austria
19. OBERNBERGER, I. WIDMANN, W., WURST, F. and WÖRGETTER, M. (1995) *Beurteilung der Umweltverträglichkeit des Einsatzes von Einjahresganzpflanzen und Stroh zur Fernwärmeerzeugung*, annual report, Institute of Chemical Engineering, Graz University of Technology, Austria
20. OBERNBERGER, I., BIEDERMANN, F. and KOHLBACH, W. (1995) *FRACTIO – Fraktionierte Schwermetallabscheidung in Biomasseheizwerken*, Jahresbericht zum gleichnamigen ITF-Projekt

mit Unterstützung der Bund-Bundesländerkooperation, Institute of Chemical Engineering, Graz University of Technology, Austria
21 ORAVAINEN, H. (1993) 'Condensing systems', in *Minutes of Meeting of the IEA Bioenergy Agreement, Task X, Activity 1*, May, Vienna, Sintef, Division of Thermal Energy , Trondheim, Norway
22 BRUNNER, T. and OBERNBERGER, I. (1997) 'Trocknung von Biomasse – Grundlagen und innovative Technik', in *Proceedings of the 3rd International Conference 'Energetische Nutzung nachwachsender Rohstoffe'*, September, Freiberg, Technische Universität Bergakademie Freiberg, Freiberg, Germany
23 OBERNBERGER, I., BIEDERMANN, F. and KOHLBACH, W. (1996) *FRACTIO – Fraktionierte Schwermetallabscheidung in Biomasseheizwerken*, Jahresbericht des 2. Projektjahres zum gleichnamigen ITF-Projekt mit Unterstützung der Bund-Bundesländerkoope-ration, Institute of Chemical Engineering, University of Technology Graz, Austria
24 OBERNBERGER, I., PANHOLZER, F. and ARICH, A. (1996) *System- und PH-Wert-abhängige Schwermetalllöslichkeit im Kondensatwasser von Biomasseheizwerken*, final report, research project of the State Government of Salzburg and the Ministry of Science, Research and the Arts, Institute of Chemical Engineering, Graz University of Technology, Austria
25 BOLIN, P. and LENNERAS, G. (2000) 'Abgaskondensation aus Holzverbrennungsanlagen in Fernwärmenetzen im Temperaturbereich 60°C–90°C mit Fallbeispielen', in *Proceedings of the VDI Seminar, Stand der Feuerungstechnik für Holz, Gebrauchtholz und Biomasse*, January, Salzburg, VDI Bildungswerk, Düsseldorf, Germany
26 OBERNBERGER I. (1997) *Nutzung fester Biomasse in Verbrennungsanlagen unter besonderer Berücksichtigung des Verhaltens aschebildender Elemente*, volume 1 of Thermische Biomassenutzung series, dbv-Verlag der Technischen Universität Graz, Graz, Austria
27 CUTLER, C. R. and RAMAKER, B. L. (1980) 'Dynamic matrix control – a computer algorithm', Proc. ACC, paper WP5-B. pp1–6
28 VAN KESSEL, L. B. M., ARENDSEN, A. R. J., BREM, G. (2004) 'Online determination of the calorific value of solid fuels', *Fuel*, vol 83, no 1, pp59–71
29 N. N. (1995) *Verordnung über die Honorare für Leistungen der Architekten und der Ingenieure (Honorarordnung für Architekten und Ingenieure)*, Fassung vom 21. September 1995 (BGBl. I S. 1174), Bauverlag, Germany
30 STOCKINGER, H. and OBERNBERGER, I. (1998) *Systemanalyse der Nahwärmeversorgung mit Biomasse*, volume 2 of Thermische Biomassenutzung series, dbv-Verlag der Technischen Universität Graz, Graz, Austria
31 JÜNGLING, G., OBERNBERGER, I., RAKOS, CH. and STOCKINGER, H. (1999) *Techno-Economic Standards for Biomass District Heating Plants*, 1st edn, ÖKL – Federal-Provincial Working Group Eco-Energy Fund, No. 67, Vienna, Austria
32 EUROPEAN COMMISSION (1998) *Biomass Conversion Technologies – Achievements and Prospects for Heat and Power Generation*, European Communities, Belgium
33 RICHTLINIE VDI 2067, (1983) *Betriebstechnische und wirtschaftliche Grundlagen – Berechnung der Kosten von Wärmeversorgungsanlagen*, VDI-Verlag GmbH, Düsseldorf, Germany
34 OBERNBERGER, I. and THEK, G. (2004) *Techno-Economic Evaluation of Selected Decentralised CHP Applications based on Biomass Combustion in IEA Partner Countries*, final report of the related IEA Task 32 project, BIOS BIOENERGIESYSTEME GmbH, Graz, Austria

6
Power Generation and Co-generation

6.1 Overview of power generation processes

Power generation by combustion can be divided into closed thermal cycles and open processes. In **closed thermal cycles**, among which the steam turbine is the most important application, the combustion process and the power generation cycle are physically separated by a heat transfer from the hot combustion gas to a process medium used in a secondary cycle. Thanks to the separation between fuel and engine, the engine is solely in contact with a clean process medium and thus undesired elements in the fuel and flue gas such as fly-ash particles cannot cause damage to the engine. Hence closed cycles are well suited for solid fuels and widely applied for power production from coal, biomass and municipal solid waste.

Open cycles are commonly applied for gaseous and liquid fuels used in internal combustion engines and gas turbines. The fuel is burned either directly inside an internal combustion engine, which is operated cyclically as a four-stroke or two-stroke engine, or it is burned continuously in an external combustion chamber and then led through an open gas turbine for expansion. The use of solid fuels in internal combustion engines is technically not feasible and their application in open gas turbines is regarded as complex. Nevertheless, two technologies for the direct use of biomass are being considered in open gas turbines:

1. directly fired gas turbines by pressurized combustion of pulverized biomass with consecutive expansion of the purified flue gas to atmosphere in a gas turbine; and
2. directly fired gas turbines by atmospheric combustion of pulverized biomass with expansion of the purified flue gas to vacuum, followed by gas cooling and a compression of the cold gas to enable gas exhaust to the atmosphere.

However, the necessary separation of particles and metals from the hot flue gases is regarded as a relevant disadvantage of these processes. Therefore, the directly fired gas turbine cycles for biomass are in an early stage of development without commercial applications in the near future, and thus are not further discussed in the present handbook.

As an alternative to the direct use of biomass in open gas turbine cycles, the application of producer gas from biomass gasification is regarded as a promising technology for both types of open processes, i.e. internal combustion engines and gas turbines. The development and demonstration of the technology of power production from biomass by gasification has been widely investigated and is well described in the literature [1–6].

6.2 Closed thermal cycles for power production

Since biomass fuels and the resulting flue gases contain elements that may damage engines, such as fly-ash particles, metals and chlorine components, the technologies for power production through biomass combustion used nowadays are based on closed thermal cycles. The processes and engine types are [7]:

- **Steam turbines** and **steam engines** used as expansion engines in the Rankine cycle, where water is evaporated under pressure to high-pressure steam that is then expanded to low pressure in the expansion engine;
- **Steam turbines** used in an **organic Rankine cycle** (ORC) with use of an organic medium instead of water, used in a tertiary cycle separated from the heat production (the combustion heat is transferred to a thermal oil in the boiler which is fed to an external evaporator for the organic medium with a lower boiling temperature than water);
- **Stirling engines** (indirectly fired gas engines using the Stirling cycle) which are driven by a periodic heat exchange from the flue gas to a gaseous medium such as air, helium or hydrogen;
- **Closed gas turbines** using a closed cycle with air, helium or hydrogen which is compressed, heated, and then expanded to drive a turbine as an expansion engine (similar to a Stirling engine);
- **Closed gas turbines** using a heat transfer to compressed air, which is expanded in a gas turbine as an expansion engine and then fed to the boiler as combustion air (hence the thermodynamic cycle corresponds to a closed gas turbine, although the mass flow through the gas turbine is not physically closed).

Table 6.1 gives an overview of the working cycles for power generation from biomass. In the steam cycles the high enthalpy difference between liquid phase and gas phase results in a high energy density of the process medium. In contrast to steam processes, the process medium does not undergo a phase change in the closed gas cycles using Stirling engines and closed gas turbines. Hence heat exchangers and engines exhibit significantly larger dimensions for closed gas cycles than for steam cycles. The different technologies cover a wide capacity range from a few kW_e (Stirling engines) to several hundred MW_e (steam turbines used nowadays and closed gas turbines developed earlier). The current state of the different processes varies from concept to proven technology.

Table 6.1 *Closed processes for power production by biomass combustion*

Working medium	Engine type	Typical size	Status
Liquid and vapour (with phase change)	Steam turbine	$500kW_e$ – $500MW_e$	Proven technology
	Steam piston engine	$25kW_e$ – $1.5MW_e$	Proven technology
	Steam screw engine	Not established, estimated range from $500kW_e$ – $2MW_e$	One demonstration plant with $730kW_e$ and turbine from commercial screw compressor
	Steam turbine with organic medium (ORC)	$400kW_e$ – $1.5MW_e$	Some commercial plants with biomass
Gas (without phase change)	Closed gas turbine (hot air turbine)	Not established, similar size as steam turbine, probably large due to cost and efficiency	Concept and development
	Stirling engine	$1kW_e$ – $100kW_e$	Development and pilot

6.3 Steam turbines

6.3.1 Working principle

Power generation using steam turbines is a highly developed technology for applications in thermal power stations and for combined heat and power (CHP) production [8–10]. Heat generated in a combustion process is used to produce high-pressure steam in a boiler (typically 20–250 bar). The steam is expanded to a lower pressure through the expansion engine and delivers mechanical power to drive an electricity generator.

For small turbines, axial and radial flow type machines exist, while large turbines are built as axial flow type only. Due to cost and size limitations, small units are built as single-stage turbines or as turbines with only a few stages. The pressure and hence the enthalpy difference for a single-stage expansion is limited. Consequently, small turbines usually exhibit a live steam pressure of 20–30 bar and a back-pressure of more than 1 bar, thus resulting in moderate efficiencies, typically of less than 15 per cent.

Figure 6.1 *Single-stage radial flow steam turbine with gear shaft and generator used in a biomass-fired CHP plant of approximately $5MW_{th}$ and $0.7MW_e$*

Source: [11]. Manufacturer: KKK.

Large turbines in power stations are built as multistage expansion machines with up to more than 20 stages and live pressures of up to about 223 bar (for plants operated with supercritical water), and with back-pressures of less than 0.1 bar (vacuum). In such multistage turbines, high-pressure ratios between inlet and outlet steam and hence high efficiencies of up to about 40 per cent are achieved. For industrial applications and for small power stations based on biomass, typical live steam pressures of 70–80 bar and typical live steam temperatures of 400–500°C enable efficiencies of 20–30 per cent.

At the turbine entrance, the steam velocity is typically around 60m/s. Due to stationary blades, this axial speed is redirected and accelerated into a radial speed of about 300m/s, while the steam pressure decreases. The kinetic energy of this steam is converted into rotational energy of the rotor. Expanded steam leaves the turbine in an axial direction.

Figure 6.2 *Axial flow steam turbine with five stages, typical for industrial applications of a few MW$_e$*

Source: [12]. Manufacturer: Tuthill-Nadrowski.

Steam plants are based on the Rankine cycle described in Figure 6.3 with water as the working medium. For practical applications, three design types and operational modes of steam plants can be distinguished:

1   **Back-pressure** plants with utilization of the total waste heat from the condensation of the steam; in such plants, the back-pressure of the turbine corresponds to the temperature needed for the heat utilization, i.e. typically close to or greater than 1 bar and 100°C.

2   **Condensing** plants for power production without heat utilization; the back-pressure and the corresponding temperature are as low as possible to achieve maximum electrical efficiency, hence the temperature is close to ambient temperature while the back-pressure is significantly below 0.1 bar corresponding to 46°C.
3   **Extraction** plants (used as extraction condensing plants or extraction back-pressure plants), as shown in Figure 6.5, for power production with variable heat output. Such a plant enables a variable extraction of steam at an intermediate pressure and temperature level for heat utilization, with the remaining steam being utilized to drive an additional low-pressure section of the turbine in condensing mode. It combines the advantages of back-pressure and condensing plants but exhibits higher complexity.

CHP production, or **co-generation**, is performed by back-pressure plants and by extraction plants. Condensing plants exclude heat utilization and are thus for dedicated power production. Reasonable efficiencies of condensing plants demand medium- or, better, large-scale power plants ranging from at least 25MW$_e$ to more than 500MW$_e$, since smaller applications achieve only poor electrical efficiencies. Back-pressure steam turbines with utilization of the total waste heat from the condensation of the steam are typically in the range 0.5–5MW$_e$, while extraction turbines are typical for applications of greater than 5MW$_e$.

6.3.2 Rankine cycle

In back-pressure plants based on the Rankine cycle (Figure 6.3), the total condensation enthalpy of the steam is used for process heat or for heating purposes. To start the process, water at low pressure (e.g. 2 bar) is led through the feed pump, which increases the pressure to the high pressure level (e.g. 85 bar) (point 1 to point 2 in Figure 6.3). The pressurized water is heated up to the evaporation temperature in the boiler (point 3), then evaporated to saturated steam (point 4), and finally superheated in the superheater section to dry steam (point 5). The live steam is expanded to the back-pressure at point 6 in the turbine, which provides the kinetic energy to the generator. The back-pressure steam is led to the condenser, where the remaining condensation enthalpy needs to be transferred to the heating medium. At point 1, water at low pressure leaves the condenser and is led to the feed pump to start the cycle again.

The process shown in Figure 6.3 with expansion into the two-phase region (point 6) is typical for large steam turbines, which allow a certain concentration of droplets in the turbine (typically 10–15 per cent of wetness). Small turbines have to be operated with dry steam (point 6 is not allowed to be in the two-phase region), which limits their efficiency.

Back-pressure plants are mainly used for applications with an almost constant heat demand for several 1000 hours per year and are of interest, for example, in wood industries with a continuous heat demand. Within a limited range of operation, the plant load can be controlled as a function of the heat demand. In addition, an emergency cooler is needed for safety reasons; this enables a heat transfer to the ambient air to avoid overheating of the plant in cases of interrupted heat demand.

The electrical efficiency of the Rankine cycle depends on the enthalpy difference before and after the turbine and therefore on the difference between inlet and outlet pressure and temperature. Hence high live steam pressures and temperatures are needed to achieve high efficiencies. On the other hand, high pressure and temperature increase the investment costs.

Further, the live steam temperature in biomass plants is limited because severe corrosion and fouling can occur at high temperature due to alkali metals, sulphur and chlorine in the fuel. Consequently, plants of up to 10MW$_e$ based on biomass combustion are typically designed for live steam temperatures between 400°C and 500°C. However, the most critical component in the system is the superheater section, being in contact with the flue gas. Hence one possible measure to increase the electrical efficiency of steam plants based on biomass or waste is the application of topping up with natural gas by a separate gas-fired superheater.

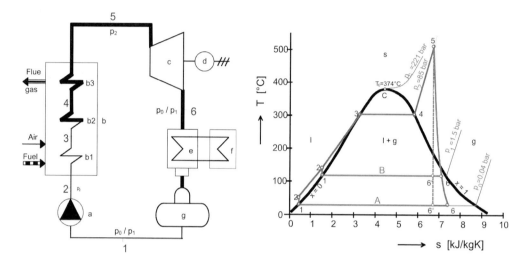

Figure 6.3 *Flow and T/S diagrams of back-pressure plants based on the Rankine cycle*

Left: Principle of a power plant based on a **Rankine cycle with steam superheating** [13].
Right: In the T/s-diagram (temperature versus entropy) case **A** describes a **condensing plant** with component f in the flow diagram being an air cooler or a cooling tower for heat transfer to the ambient. **Case B** describes a **back-pressure plant** with utilization, of the heat at a higher temperature level, hence component f corresponds to a heat exchanger used for heating purposes or process heat. In a back-pressure plant, the condensation temperature $T_6 = T_1$ is increased to enable heat utilization, thus reducing the electrical efficiency.

For small turbines, the back-pressure steam needs to be saturated, as droplets are not allowed. Hence point 6 in the T/s diagram needs to be on the condensation line, thus resulting in a reduced efficiency.

Components in the flow diagram: a = feed water pump, b = boiler with b1 = water preheater, b2 = evaporator, and b3 = superheater, c = steam turbine, d = generator, e = condenser, f = heat exchanger to transfer condensation enthalpy in a secondary circuit to the ambient (case A corresponding to a condensing plant) or to a heat consumer (case B corresponding to a back-pressure plant), g = feed water tank.

Abbreviations in the T/s diagram: s = entropy, t = temperature, l = liquid, g = gas (vapour), s = supercritical, c = critical point, p = pressure, p_2 = high pressure of the live steam, p_0 = back-pressure for a condensing plant A, p_1 = back-pressure for back-pressure plant B.

Process (numbers in both diagrams):

1–2 Adiabatic pressure increase of the water in the feed water pump
2–3 Heating of the water to evaporation temperature in the water preheater
3–4 Evaporation of the water in the evaporator in the boiler
4–5 Superheating of the steam in the superheater
5–6' Isentropic expansion of the steam (ideal process, not achievable in reality)
5–6 Polytropic expansion of the steam in the steam turbine (real process with turbine losses)
6–1 Condensation of the steam, with either heat transfer in a secondary circuit either to ambient (condensing plant) or heat utilization (back-pressure plant).

In condensing plants, the condensation temperature needs to be as low as possible for a high electrical yield (Figures 6.3 and 6.4). If no heat recovery is applied and the condenser is operated with ambient air, the condensation temperature varies with outside temperature and is typically in the order of approximately 30°C, which corresponds to approximately 0.04 Bar. This enables pressure ratios greater than 5000 and electrical efficiencies of more than 40 per cent in large power plants with high steam pressures (up to pressure exceeding 200 bar and, in plants operated with supercritical water, even more than 223 bar). However, electrical efficiencies in this order of magnitude are only reached in large plants (>100MW$_e$) where multistage turbines and additional measures to increase the efficiency, such as feed water preheating and intermediate tapping, are applied.

In comparison to a condensing plant, co-generation demands a higher temperature in the condenser to enable heat utilization, usually 90–140°C, with a back-pressure of approximately 0.7–4 bar. Hence heat utilization in a steam turbine plant leads to a reduction of the electrical efficiency of typically around 10 per cent, as shown in Figure 6.4, since the enthalpy difference is only partly used for power production. However, the overall efficiency indicated as the sum of electrical efficiency and heating efficiency can be increased by co-generation, reaching up to 80 per cent.

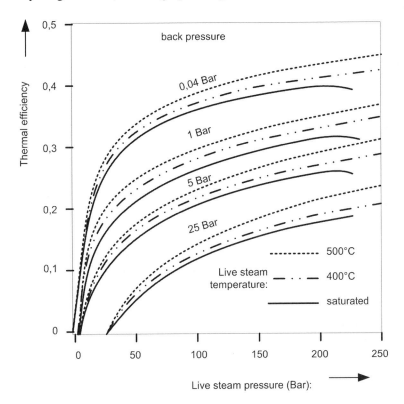

Figure 6.4 *Thermal efficiency of the Rankine cycle as a function of live steam parameters and back-pressure*

Note: The efficiency of the power plant is the product of thermal efficiency, boiler efficiency, turbine efficiency, and generator efficiency and hence lower than the illustrated thermal efficiency of the Rankine cycle.

Source: [9].

To enable heat production allowing for a varying heat demand, steam can be used at an intermediate pressure level for heat production by application of an extraction turbine, as shown in Figure 6.5. This enables operation of the plant at maximum overall efficiency in periods with high heat demand and at maximum electrical efficiency in periods with low heat demand.

6.3.3 Economic aspects

Power plants based on steam cycles are highly sensitive to scale economies. In smaller plants of up to 1MW$_e$, fire tube boilers, which allow steam pressures of only 20–30 bar, are applied instead of water tube boilers for economic reasons. Furthermore, the complexity of these plants is limited (turbines with one or a few stages, dry steam needed at the turbine outlet, no intermediate tapping, etc.). They are often operated as back-pressure turbines for combined heat and power production, which avoids vacuum operation and then leads to electrical efficiencies of approximately 8–12 per cent [13–16]. Efficiencies around 20–25 per cent are reached at condensing plants of 5–10MW$_e$ without co-generation.

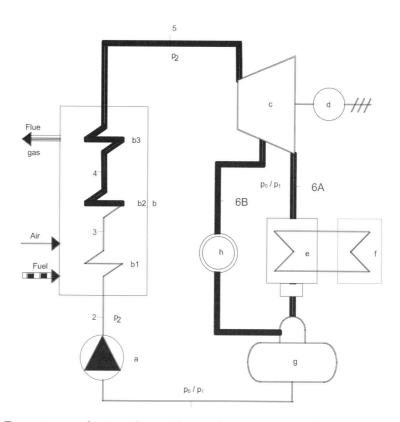

Figure 6.5 *Extraction condensing plant with use of steam at intermediate pressure (6B) for heat production, and condensing operation for the non-utilized part of steam at lower pressure (6A) to drive the low-pressure section of the turbine*

Note: The T/s diagram is given in Figure 6.3. Here, h = heat consumer, and f = heat exchange to ambient.

During the vaporization of the water in the boiler, salts contained in the water remain in the boiler. In order to avoid high salinity and deposits in the boiler, the water must be continuously desalinated and purified. Furthermore, a continuous addition of fresh water to the process is needed to replace steam losses. Since steam turbines are sensitive to contaminants, a relatively costly water treatment is needed in steam turbine plants.

As a result of the high complexity, mainly of the steam boiler and peripheral equipment needed for its operation, small and medium-sized steam plants exhibit high specific investment costs. Furthermore, the operation of steam plants is expensive due to high requirements of safety measures and intensive plant supervision by qualified staff. Since the need for staff is only partially dependent of the plant size, operation cost can be prohibitive for small-scale applications of steam plants.

Since steam turbines have been applied in many situations, research has been done on improved designs and materials. Higher conditions of inlet steam raise the isentropic efficiency, so new materials have been developed that can stand higher inlet temperatures. When wet steam is applied, corrosion can easily occur; therefore materials need to be selected carefully. Research is also being performed on relatively small installations (0.25–10MW$_e$). In this power range, especially, the efficiencies at partial loads are still low. The efficiency in practical operation is lowered further, since the entrance angle of steam on the blades can only be optimized for one single operating situation (steam consumption, pressure, etc.).

Table 6.2 summarizes the main advantages and disadvantages of steam turbines for use in biomass combustion.

Table 6.2 *Advantages and disadvantages of steam turbines for use in biomass combustion*

Advantages	Disadvantages
• Mature, proven technology • Broad power range available • Separation between fuel and thermal cycle, enabling the use of fuel containing ash and contaminants • High pressures and temperatures can be applied enabling high efficiencies for large plants • Co-firing of fossil fuels and biomass is possible to enable high efficiency	• Only limited efficiencies are reached in small, decentralized plants due to investment and technology limitations • High specific investment for low power ranges • High operation costs for small and medium plants • Low part-load efficiencies • Variations in fuel quality lead to variation of steam and power production • Superheater temperature (and therefore efficiency) can be limited due to high temperature corrosion and fouling, especially due to alkali metals, chlorine, and sulphur • High-quality steam is necessary

6.4 Steam piston engines

Steam engines are available with capacities ranging from approximately 25kW$_e$ to 1.5MW$_e$ per unit and therefore can be used in small plants where steam turbines are not available or in medium plants as an alternative to steam turbines.

Steam piston engines show a modular design with one to six cylinders per engine in different configurations. Depending on the steam parameters, single-stage or multistage expansion is applied. The pressure ratio between inlet and outlet is typically around 3, at maximum 6, for one expansion stage. The efficiencies depend on the steam parameters. Typical engine efficiencies are 6–10 per cent for single-stage piston engines and 12–20 per cent for multistage engines, corresponding typically to 4–7 per cent and 8 per cent to a maximum of 14 per cent electric plant efficiency. The intake pressure is typically between 6 and 60 bar, while the back-pressure can range from 0 to 25 bar. For similar steam parameters, the maximum efficiencies are comparable or slightly higher than for steam turbines **[2, 14, 16, 17]**.

For steam engines operated with superheated steam in co-generation mode, the process cycle is the same as shown in Figure 6.3 for steam turbines. If operated with saturated steam, the process cycle is as shown in Figure 6.7 with expansion from point 4 to point 5 into the two-phase region to a wetness of up to 12 per cent.

Figure 6.6 *Steam engine (two cylinders) from Spillingwerk, Germany*

Source: [17].

Steam engines have several advantages in comparison to steam turbines. Steam engines are less sensitive to water droplets in the outlet and a wetness of 12 per cent is acceptable, even for small plants. They can even be operated with low-pressure, saturated steam. Although this reduces the efficiency, investment savings on the steam boiler of up to 30 per cent are obtainable.

Furthermore, steam piston engines have a higher part-load efficiency than turbines. Since they reach up to 90 per cent of the maximum efficiency between 50 and 100 per cent of the nominal power, steam engines are also suitable for varying heat and electricity load. In addition, steam engines are less sensitive to contaminants in the steam than turbines and therefore require less sophisticated boiler water management.

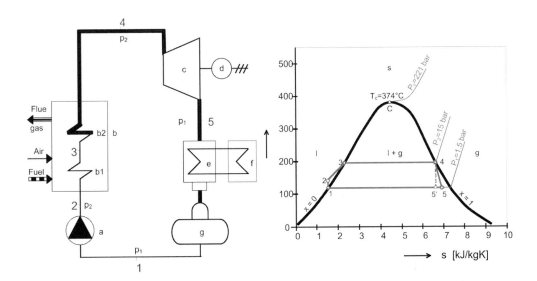

Figure 6.7 *Flow and T/s diagrams for a steam cycle using saturated steam in a steam piston engine or a steam screw-type engine*

Note: The expansion from point 4 to point 5 immediately leads into the two-phase region water/vapour.

Source: [2].

A disadvantage, mainly of older steam piston engines, is the need for the injection of oil into the steam for lubrication before it enters the engine. This oil has to be removed from the condensate before it enters the feed water tank by use of a two-stage process with oil separator and active carbon filter. The oil consumption is around 0.2g/kWh. Since traces of oil can still be found after the oil filter, the steam often cannot be used directly for food processing equipment. Concentrations of oil exceeding 1mg/L can cause problems in the feed water tank and the boiler [18]. Periodic control of the oil concentration and the need to change the oil filters increase maintenance work. To avoid this disadvantage, a new technology introduced in 1999 makes oil-free operation of piston engines possible for new applications and even allows retrofitting of existing engines [17]. Another disadvantage of steam engines, if operated at speeds of 750–1500rpm, is the production of high levels of noise and vibration.

Although saturated and dry steam can be used to operate steam engines, steam consumption and condensation will be less for dry steam (and hence efficiency will be higher), due to the higher enthalpy of dry steam. Table 6.3 indicates the power output for both dry and saturated steam, assuming a constant steam flow of 10t/h.

A new steam engine particularly suitable for small-scale electricity generation is described in Table 6.4. Depending on the desired electrical output, the engine can be supplied with three, four or five cylinders. Table 6.5 summarizes the advantages and disadvantages of steam engines.

Table 6.3 *Output power of a steam engine when using 10t/h of dry and saturated steam*

Entrance pressure [bar]	Exhaust pressure [bar]	Engine power [kW]	
		saturated steam	dry steam
6	0.5	480	740
	2.0	320	500
16	0.5	740	1100
	3.0	460	710
	6.5	310	470
26	0.5	840	1200
	3.0	510	790
	6.5	410	670
	10.5	320	510

Source: [17].

Table 6.4 *Specifications of steam piston engines from Spilling, Germany*

		Three cylinders	Four cylinders	Five cylinders
Boiler pressure	[bar]	11	11	11
Engine head pressure	[bar]	10	10	10
Engine exhaust pressure	[bar]	0.5	0.5	0.5
Steam flow	[t/h]	2.3	3.0	3.7
Electrical capacity	[kW$_e$]	95	125	155
Costs[1]	[Euro]	140,000	160,000	180,000

[1] Including steam engine, foundation, control, oil separation and condensate removal.
Source: [17].

Table 6.5 *Advantages and disadvantages of steam piston engines*

Advantages	Disadvantages
• Suitable for low power ranges starting from 25kW$_e$ • Saturated steam can be used • Efficiency almost independent of partial load • Steam extraction at different pressures possible due to modularity	• Traces of oil in expanded steam for older engines • Maximum power output per steam engine 1.2MW$_e$ • High levels of vibration and noise

6.5 Steam screw engines

An alternative for small-scale power generation is the application of screw-type steam engines [19]. The principle of screw-type compressors, using a male and a female rotor, is used in hundreds of applications worldwide (Figure 6.8). The screw-type engine follows the same principle; however, it is used as an expansion machine instead of as a compression machine. This type of application has been implemented in a 730kW$_e$ biomass-fired demonstration plant. Since screw engines are operated with a closed oil cycle, the outlet steam does not contain oil traces. The engine can also operate under low steam conditions at efficiencies of 10–13 per cent.

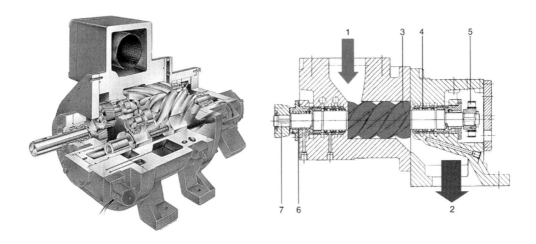

Figure 6.8 *Screw-type compressor and working principle of a steam screw engine*

Notes: 1 live steam inlet, 2 steam outlet, 3 main rotor, 4 sealing, 5 gear box, 6 bearing, 7 shaft.
Source: [20].

Screw-type steam engines are capable of being operated under several different steam conditions. Besides the expansion of superheated steam and saturated steam leading to the cycles described in Figure 6.3 and Figure 6.7, even wet steam or compressed hot water at boiling temperature can be expanded in the two-phase screw-type engine (see Figure 6.10). Although the use of steam with low enthalpy leads to limited efficiencies, the screw-type engine offers specific applications that are not suitable for piston engines or steam turbines. The use of hot water screw engines especially offers a potential for small CHP plants, because no steam boiler is necessary [2, 19]. Figure 6.11 shows the flow diagram with the control device for part-load operation by use of a pressure-reducing valve and a throttle valve.

Figure 6.9 *730 kW$_e$ screw-type engine with high-pressure and low-pressure stage and generator in a demonstration plant in Hartberg, Austria*

Note: Live steam parameters: 255°C/25 bar. Back-pressure steam variable from 0.5 to 1.5 bar/82–112°C.
Source: Engine manufacturer: MAN.

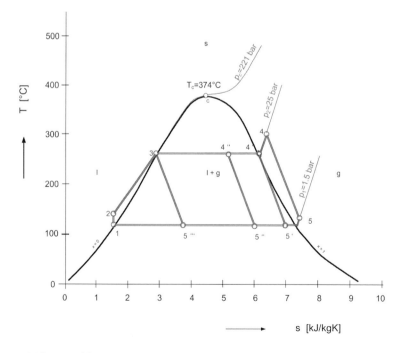

Figure 6.10 *Possible processes with screw engines in the T/s diagram*

Notes: 4–5: Process with superheated steam, 4'–5': process with saturated steam and expansion into the two-phase region, 4'''–5''' process with wet steam and 3–5''' process with hot water at boiling temperature.

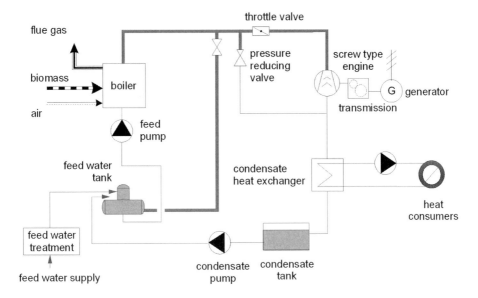

Figure 6.11 *Principle of co-generation using a steam engine controlled by pressure-reducing valve and throttle valve*

Source: [18].

6.6 Organic Rankine cycle (ORC)

The process of the organic Rankine cycle (ORC) is similar to that of the conventional Rankine cycle. However, instead of water, organic oil with a lower boiling temperature is used as the process medium [13, 18, 21, 22]. This enables operation at relatively low temperatures (70–300°C). Therefore, many ORC plants have been installed in geothermal power stations and a few ORC generators are in operation with industrial waste heat. The T/s diagram is shown in Figure 6.12.

Due to the physical properties of the organic oil, the expansion of saturated steam does not lead to wet but to dry steam (in the T/s diagram, the condensation line shifts to the left and not to the right, as for water). For slightly superheated steam, as shown in the diagram, the expansion from D to E leads to dry steam and even the expansion of saturated steam can lead to dry steam.

Due to the low temperature level, the evaporator for the organic oil used in the ORC can be heated in a secondary circuit by thermal oil that is heated in an atmospheric liquid tube boiler (Figure 6.12). The ORC generator is then operated in a completely closed circuit using silicon oil. For low-temperature applications, the organic oil can also be heated directly in the boiler – however, biomass-fuelled plants are operated with a secondary circuit. Hence no steam boiler is needed and therefore investment costs and maintenance of the boiler are considerably lower than for a comparable steam plant. Another advantage in comparison to conventional steam turbine plants is the possibility of part-load operation in the range between 30 and 100 per cent of full load.

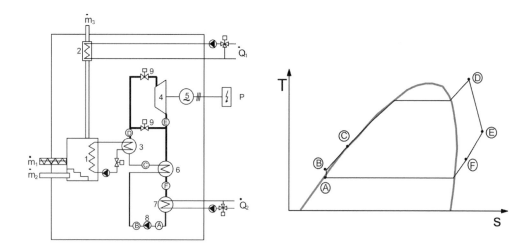

Figure 6.12 *Flow and T/s diagrams for co-generation with an ORC process*

Note: To increase efficiency, a regenerator can be introduced between turbine and condenser to pre-heat the organic oil (6). Furthermore, an economizer can be used for heat extraction from the flue gases after the thermal oil boiler (2).

Explanation: m_1 = fuel input, m_2 = air inlet, m_3 = flue gas exit, Q_1 and Q_2: heat for district heating (depending on heat demand and temperature levels, Q_2 can be used for pre-heating) P = electricity output.

1: Wood furnace with thermal oil boiler, 2 = economizer for heat generation, 3 = evaporator for the organic fluid in the organic Rankine cycle, 4 = expansion turbine, 5 = generator, 6 = organic fluid pre-heater, 7 = condenser for heat generation, 8 = feed pump, 9 = control valves.

A–B Pressure increase of the organic fluid in the feed pump
B–C Preheating of the organic fluid
C–D Pre-heating, evaporation and superheating in the evaporator
D–E Expansion in the turbine
E–F Cooling in the recuperator
F–A Condensation in the condenser

In a plant operated in Switzerland, the following results are obtained: $P_{e,gross}$ = 335kW$_e$, $P_{e,net}$ = 300kW$_e$, Q_2 = 1440kW$_{th}$, Q_1 = 460 kW$_{th}$. Assuming an efficiency of 80% for thermal oil boiler and economizer, the efficiencies according to fuel input are estimated as 11% electrical, 67% thermal and 78% total.

Source of results: [22].

Although the ORC is well known from geothermal applications, there are only a few biomass combustion applications. Currently available generators range from 300kW$_e$ to 1.5MW$_e$. A plant with 300kW$_e$ has been in operation in Switzerland since 1998. Additional plants between 400kW$_e$ and 1.5MW$_e$ per module have been erected since 1999 with a plant with three 1.5MWe generators erected in Austria in 2004 being the largest one. Investment costs are similar or slightly higher than for a steam plant nowadays, due to single unit production. However, wider application might lead to reduced production costs.

Due to the specific properties of the organic oil, the efficiencies of the several process steps, the thermal oil boiler parameters, and the back-pressure in the condenser, gross efficiencies of 10–14 per cent are reached in ORC generators of approximately 300kW$_e$ to

1.5MW$_e$ at thermal oil feed temperatures of 300°C when operated as CHP plants. For optimized processes, up to 17 per cent efficiency is expected. This efficiency, which is slightly higher than that of a steam turbine of a similar size, is due to the fact that well-developed two-stage turbines are available for this application. However, the net efficiency of the ORC plant can be considerably lower than the gross efficiency, due to a relatively high power consumption of the ORC plant.

Figure 6.13 *ORC equipment*

Note: Top: 400kW$_e$ ORC module fired by a biomass grate furnace in Admont, Austria. Bottom: Thermal oil boiler to drive one 1.5MW$_e$ ORC generator out of three in a plant in Leoben, Austria, currently being the largest biomass fired ORC plant

Source: Top: Courtesy of Turboden Srl, Italy.

6.7 Closed gas turbines

The structure of closed gas turbines is similar to that of open gas turbines. In contrast to open gas turbines, in an indirectly fired gas turbine the heat is not transferred to the compressed gas by internal combustion (as with a directly fired gas turbine) but by using a high temperature heat exchanger. Similar to open gas turbines, the mechanical power is produced in a turbine used as an expansion engine. The expanded gas is cooled in a heat exchanger before being compressed again to the gas turbine inlet pressure. Two different types of processes are possible [2, 18, 23–29]:

1. For a completely closed secondary cycle, air, helium or hydrogen is used as the working medium. Since a combination of compressor and turbine is needed for closed gas turbine applications, there is a certain potential for using existing turbocharger units available from large diesel engines to be operated as closed gas turbines in small biomass combustion plants (< 1MW$_e$). However, the existing turbocharger units are not optimized for this application and hence only poor efficiencies of less than 5 per cent can be expected for simple applications with existing machines. To reach significantly higher efficiencies, turbines that can be operated at higher temperatures are needed and much more complex processes have to be applied. Figure 6.14 shows one proposal for a closed gas turbine cycle using three stages of expansion and two stages of compression, combined with recuperation. With the use of optimized machines for this process, more than 20 per cent efficiency is theoretically possible. However, to realize a power plant based on this technology would require far greater capacities than 1MW$_e$ plus specific turbines and compressors for this application.
2. If air is used as the working medium for the closed gas turbine, the exhaust air from the turbine can be used as combustion air for the biomass plant instead of being compressed and used in a closed cycle. In this case, the process medium is replaced continuously and fresh ambient air is being compressed. In this case, the thermodynamic process corresponds to a closed gas turbine, even though the air is being replaced in the process. These plants are frequently called hot air turbines.

Besides the lack of optimized turbines and compressors for such applications, the high temperature gas–gas heat exchange is a complex and expensive part of the system, which is not available as standard technology at present. This component is heavily burdened by high temperatures of up to 1000°C, in combination with possible particles and corrosive components in the flue gas. If a hot particle removal technology is applied to render the heat exchanger unnecessary, the complexity of the plant will again increase as high temperature particle removal technologies are not yet commercially available.

Furthermore, the heat exchanger becomes very large compared to a gas–water heat exchanger to support the large volume flow of the hot gases. Hence the closed gas turbine process for the production of electric power from biomass is at an early stage of research. A 500kW$_e$ test plant is in operation in Belgium [29]. Due to the complexity and the many unresolved problems, commercial application is currently uncertain.

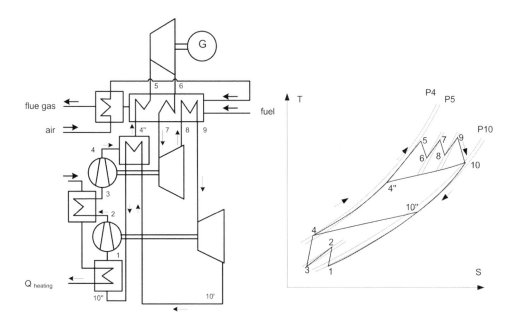

Figure 6.14 *Flow and T/s diagrams of a closed gas turbine cycle with recuperation*

1–2	Polytropic compression of air, compressor 1
2–3	Intermediate cooling, cooler
3–4	Polytropic compression of air, compressor 2
4–4"	Waste heat recuperation, recuperator
4"–5	Heating of air, furnace air heater
5–6	Polytropic expansion of air, power turbine
6–7	Heating of air, furnace/air heater
7–8	Polytropic expansion of air, turbine 2
8–9	Heating of air, furnace/air heater
9–10	Polytropic expansion of air, turbine 1
10–10"	Waste heat recuperation, recuperator
10"–1	Carrying off heat for heating, heat exchanger

Source: [13].

6.8 Stirling engines

Stirling engines are indirectly fired gas engines with air, helium or hydrogen as the process medium to be used in a closed cycle. External heat is transferred to the gas, followed by forced cooling of the gas. Expansion and compression of the medium inside the chamber are allowed over a piston. Figure 6.15 demonstrates the mechanical principles of a Stirling engine and Figure 6.16 is the T/s diagram of an ideal Stirling engine. The Stirling cycle is a thermodynamically ideal process for transforming heat to mechanical energy. It has the same theoretical efficiency as the Carnot process and, in contrast to the Carnot process, it can actually be realized. However, in practice its high efficiency is reduced by friction, limited heat transfer and heat recuperation, pressure losses and other influences [30–32]. Therefore, the actual efficiency for power production is in the range of 15–30 per cent.

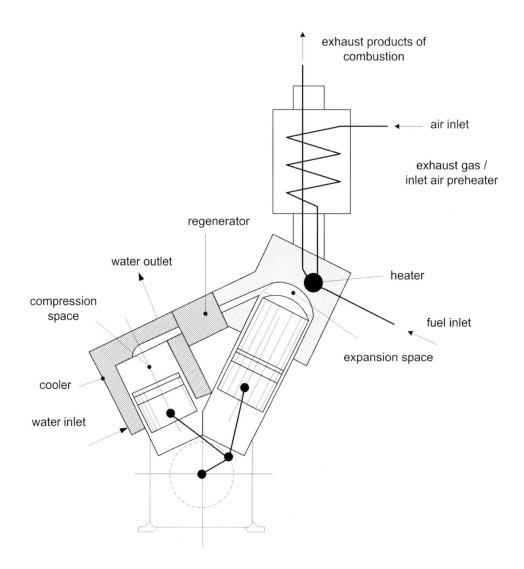

Figure 6.15 *V-shaped Stirling engine*
Source: [30].

An important advantage of the Stirling engine in comparison to the internal combustion (IC) engine, is that any kind of high temperature source can be used to heat the working medium over a heat exchanger. Hot flue gases from the combustion of solid, liquid or gaseous fuels, as well as solar energy can be used. The gases used for heating must be as clean as possible to avoid fast corrosion or fouling of the heat exchanger surface in the Stirling engine. If wood is used as the fuel, the exhaust gases can be corrosive, which requires a special material for the heat exchanger surface. Furthermore, the heat exchanger has to be built in such a way that it can easily be cleaned.

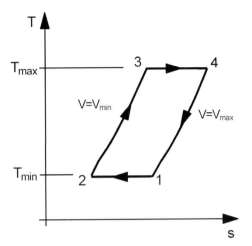

Figure 6.16 *T/s diagram of the Stirling cycle*

2–3 *Heating at constant volume*: The gas is heated while the volume remains constant. The pressure increases.
3–4 *Expansion*: With constant temperature, the gas is allowed to expand over a piston. The piston drives a crankshaft.
4–1 *Cooling*: Gas is allowed to cool, while the volume remains constant. The pressure decreases.
1–2 *Compression*: Gas is compressed while the temperature remains constant. The piston drives the crankshaft further.

Stirling engines are suited small-scale applications and typically designed for 1kW$_e$ up to slightly more than 100kW$_e$. Hence Stirling engines cover a power range which is not feasible for steam cycles or ORC generators. So far, Stirling engines are not available on the open market. However, different Stirling engines are under development or have reached the stage of demonstration for power production from biomass, e.g. in Denmark [31], the UK, Germany, and Austria [27]. In the case of a biomass boiler, heating and cooling of the medium can be done by forcing the gas to flow through a regenerator, where

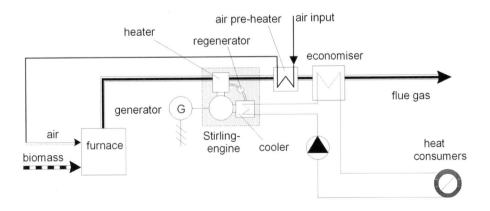

Figure 6.17 *CHP biomass combustion plant with Stirling engine*

Source: [18].

Figure 6.18 *Stirling engines*

Note: Top: 30kW$_e$ Stirling engine developed in Denmark and in use as a demonstration plant in combination with an automatic wood furnace. Hermetical 4-cylinder type engine operated at 4 bar, 620°C/40°C, right side shown with hot gas heat exchanger (upside down). Bottom: 70kW$_e$ Stirling engine with two 4-cylinder engines in series in operation in a biomass-fired combustion plant in Austria since 2005 with electrical efficiency of app. 12%

Source: Top: [30]; bottom: Photo courtesy of BIOS BIOENERGIESYSTEME GmbH, Graz, Austria and Stirling.dk ApS, Denmark

gas is heated on one side with hot flue gases and cooled on the other side with cold water. A 30kW$_e$ Stirling engine has been in operation at an automatic wood furnace in Denmark for several years. The hot gas 4-cylinder type engine operates at 4 bar, 620°C/40°C. At the demonstration plant an electrical efficiency of approximately 18 per cent is reported in CHP mode with a total efficiency of up to 87 per cent [31]. However, a recent application of a 70kW$_e$ Stirling engine in Austria revealed a lower electrical efficiency of approximately 12 per cent.

6.9 Comparison of heat, power and CHP production

Decentralized CHP plants < 1MW$_e$ with a steam turbine or a steam engine deliver about 10 per cent of the fuel energy as electric power. With increasing plant size (and thus increasing steam parameters and plant complexity), electrical efficiencies of approximately 25 per cent are possible in plants with 20MW$_e$. If operated in CHP mode, an overall efficiency η_{tot} of approximately 80 per cent at most can be achieved, where:

$$\eta_{tot} = \eta_e + \eta_h$$

and e stands for electricity and h for heat. The relation between heat and power can also be described with the power-to-heat ratio α which is correspondingly increasing with plant size.

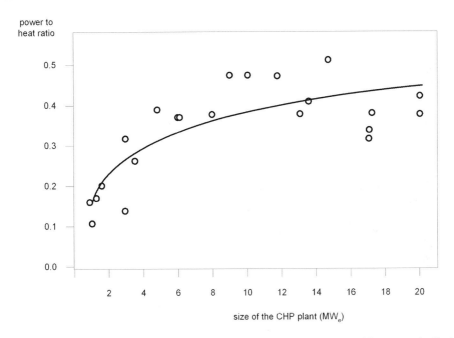

Figure 6.19 *Power-to-heat ratio α as a function of the plant size of biomass-fuelled CHP plants in Finland and Sweden with 1–20MW$_e$*

Source: [33].

To determine the most economic solution for specific boundary conditions for power production, heat production or CHP production, the ratios between fuel price, heat price and electricity price can be used [34]. For CHP production, two operational modes can be distinguished. Small CHP plants with low electrical efficiency should be operated in a heat-controlled mode, while large CHP plants are usually operated in an electricity-controlled mode. Figure 6.20 shows the qualitative efficiencies for heat and electricity that can be achieved, in comparison with a heat plant and an electric power plant.

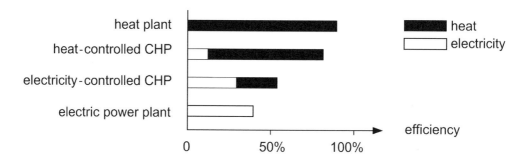

Figure 6.20 *Percentage of heat and electric power production in heating plants, CHP plants and power plants (qualitative figures)*

However, electricity and heat cannot be compared directly, due to their differing exergetic values. To compare the values of different technologies, an exergetically weighted efficiency can be calculated. As a reference scenario, the utilization of heat pumps operated by electricity can be assumed. Depending on the temperature levels of the heat source and the heat consumer, heat pumps can generate more than three units of heat from one unit of electricity. Therefore, it is appropriate to judge the efficiency of the whole energy system, taking an overall view by assuming an application of decentralized heat pumps for the production of space-heat and warm water.

If the electric power is used to run a heat pump with a coefficient of performance ε or COP, an exergetically weighted overall efficiency η_{ex} can be defined as:

$$\eta_{ex} = \varepsilon\, \eta_e + \eta_h$$

For an averaged value, the seasonal performance factor (SPF) has to be considered instead of ε.

Table 6.6 shows typical data achieved at plants installed at present and expected values for future technologies.

If an electric efficiency of 25 per cent is assumed for biomass power plants without heat utilization at current values, the exergetically weighted overall efficiency with utilization of heat pumps with $\varepsilon = 2.5$ reaches 63 per cent. In comparison, an optimized heating plant and well as a heat controlled CHP plant can now achieve higher exergetic efficiencies of up to 85 per cent and up to 98 per cent, respectively.

For future applications, higher electrical efficiencies are sought for both small CHP plants as well as dedicated power plants. If electricity is rated at $\varepsilon = 2.5$, future CHP plants and dedicated power production plants may achieve exergetic efficiencies > 1, and are thus most interesting. With a higher valuation of electricity (corresponding to a higher coefficient of performance of the heat pumps), power plants with high electrical efficiencies such as integrated gasification combined cycles (IGCC) look to be the most promising. Targets of up to 42 per cent have been set for IGCC plants. With improved heat pump technology of $\varepsilon = 4$, an exergetically weighted overall efficiency exceeding 1.5 can be reached without heat utilization in the plant. This is better than a CHP plant with heat utilization but with an exergetically weighted efficiency close to 1.

Table 6.6 *Typical efficiencies for heating plants, CHP plants and power plants today and target values for the future*

		Today		Target for future	
	Heating plant	CHP < 1MW$_e$	Power plant > 10MW$_e$	CHP < 1MW$_e$	Power plant > 10MW$_e$
Technology	Wood boiler	Steam engine, steam turbine, ORC	Steam turbine	Gasifier and IC engine, Stirling engine	IGCC
η_h	0.85	0.68	0	0.6	0
η_e	0	0.12	0.25	0.2	0.42
ϵ	2.5	2.5	2.5	2.5	2.5
$\eta_{tot} = \eta_e + \eta_h$	0.85	0.8	0.25	0.8	0.42
$\eta_{ex} = \epsilon\, \eta_e + \eta_h$	0.85	0.98	0.63	1.1	1.05

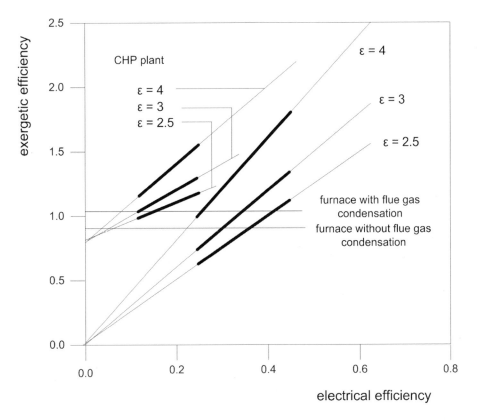

Figure 6.21 *Comparison of heat, CHP and power plant efficiencies by an exergetically weighted efficiency ($\eta_{ex} = \eta_e + \eta_h$)*

Source: [2, 13].

6.10 Summary

For power production through biomass combustion, steam turbines and steam piston engines are available as proven technology. While steam engines are available with capacities ranging from approximately $25kW_e$ to $1.5MW_e$, steam turbines cover a range from $0.5MW_e$ to more than $500MW_e$. The largest biomass-fired plant is around $50MW_e$.

Small-scale steam plants are built with fire tube boilers and hence operated at quite low steam parameters. Furthermore, the turbines are built with single or few stages of expansion and the application of additional measures for efficiency improvement is limited. Due to moderate electrical efficiencies and to avoid vacuum operation, plants smaller than $1MW_e$ are usually operated as back-pressure CHP plants and achieve net electrical efficiencies of typically 10–12 per cent and heating efficiencies of up to 70 per cent. Steam plants operated in CHP mode are mainly feasible for the production of process heat, which enables long operation periods. Steam piston engines can be operated in single- or multistage mode, reaching similar or slightly higher efficiencies than small turbines, i.e. 4–7 per cent for single stage expansion and 8–14 per cent for multistage expansion. Steam engines can also be operated with saturated steam, thus reducing investment costs through reduced electrical efficiency. Large steam turbine plants are operated with water tube boilers and superheaters, thus making high steam parameters and the use of multistage turbines possible. Furthermore, process measures such as feed water preheating and intermediate tapping are implemented for efficiency improvement. This enables electrical efficiencies of around 25 per cent at the size of $5–10MW_e$. In plants of around $50MW_e$ (the largest pure biomass plant) and larger, the electric efficiency can rise to more than 30 per cent in co-generation mode and to more than 40 per cent if operated as condensing plants.

Since the potential of large biomass-fired power stations is limited due to transportation distances, co-firing of biomass in fossil-fired power plants is an interesting option for using biomass, with a high electricity yield. While co-firing of biomass leads to reduced NO_x and SO_x emissions, the negative effects on boiler capacity, efficiency, corrosion and fouling must be considered. Furthermore, the residues from the gas cleaning as well as the ash composition can be negatively influenced due to alkali metals and chlorine in the biomass. Therefore, 5–10 per cent of the heat input is usually provided by biomass, which leads to acceptable effects on the ash and residues. The main application of co-firing is the co-combustion of dry pulverized biofuels in pulverized coal boilers, which usually makes fuel treatment necessary. Fluidized bed boilers, understoker boilers and grate furnaces are also used for co-firing, thus broadening the possibilities in terms of moisture content and fuel size.

As an alternative to conventional steam plants in the range of $0.4–1.5MW_e$, organic Rankine cycle (ORC) plants are also available. These can operate at lower temperatures, so that a combustion plant with a thermal oil boiler can be applied instead of a costly steam boiler. Furthermore, the ORC generator can be operated without a superheater due to the fact that the expansion of the saturated steam of the organic medium leads to dry steam. Therefore, the ORC can have advantages in the areas of process design and operation. Furthermore, similar or slightly higher efficiencies are achieved thanks to the availability of well-designed two-stage turbines for this specific application. ORC plants are a well-proven technology for geothermal applications. A few plants are in operation with biomass combustion and therefore further process improvement and cost reduction for this application is expected. However, ORC plants can exhibit relatively high power consumption, which needs to be taken into consideration in the comparison with water steam cycles, as it can decrease the net efficiency significantly.

An interesting option for small-scale power production is the externally fired Stirling engine with air or helium coupled to a biomass furnace. A critical component in such plants can be the gas–gas heat exchanger operated with hot flue gas from biomass combustion. Native wood is the most suitable fuel, since biofuels with high ash levels are a challenge for the heat exchange. Operational experience exists with 30–70kW$_e$ demonstration plants, reaching more than 12 per cent electrical efficiency. For applications smaller than 100kW$_e$, Stirling engines are the only appropriate technology for use with biomass combustion, thus achieving similar electrical efficiencies as turbines ten times larger in size.

For medium-scale applications, closed gas turbine cycles or hot air turbines are also being considered in research projects. However, the use of existing components for a simple process design will result in moderate efficiency and hence has no relevant market potential. On the other hand, process design to enable high efficiencies is characterized by high complexity for multistage expansion, recuperation, and multistage compression. Furthermore, turbines and compressors for such applications do not exist and hence need to be developed, which is not to be expected in the near future due to a limited potential. Since the high temperature gas–gas heat exchange or hot gas particle separation also remains an unresolved problem, a practical application of closed gas turbines for biomass combustion is uncertain.

6.11 References

1. BRIDGWATER, A. and BOOCOCK, D. (eds.) (1997) *Developments in Thermochemical Biomass Conversion*, Blackie Academic, London
2. SIPILÄ, K. and KORHONEN, M. (eds.) (1999) *Power Production from Biomass III*, Espoo 14–15 September 1998, VTT, Espoo (Finland)
3. KYRITSIS, S., BEENACKERS, A., HELM, P., GRASSI, A. and CHIARAMONTI, D. (Eds.) (2000) *1st World Conference on Biomass for Energy and Industry*, 5–9 June, ETA-Florence and WIP Munich
4. BRIDGWATER, A. (ed.) (2001) *Progress in Thermochemical Biomass Conversion*, Blackwell Science, Oxford
5. PALZ, W., SPITZER, J., MANIATIS, K., KWANT, K., HELM, P. and GRASSI, A. (eds.) (2002) *12th European Conference and Technology Exhibition on Biomass for Energy, Industry and Climate Protection*, 17–21 June, Amsterdam, ETA-Florence and WIP Munich
6. VAN SWAAIJ, W., FJALLSTRÖM, T., HELM, P. and GRASS, A. (Eds.) (2004) *Second World Biomass Conference*, 10–14 May, Rome, James & James London
7. NUSSBAUMER, TH., NEUENSCHWANDER, P., HASLER, PH., JENNI, A. and BÜHLER, R. (1998) 'Technical and Economic Assessment of the Technologies for the Conversion of wood to heat, electricity and synthetic fuels. biomass for energy and industry', *10th European Conference and Technology Exhibition*, 8–11 June, Würzburg (Germany), pp1142–1145
8. STRAUSS, K. (1992) *Kraftwerkstechnik*, Springer
9. KUGELER, K. and PHLIPPEN, P. (1992) *Energietechnik*, Springer
10. SCOTT, P. (1991) *Review of Small Steam Turbines (0.25–10 MWe)*, ETSU, Energy Efficiency Office, Department of the Environment, UK
11. KKK (2000) Product information, Kühnle, Kopp & Kausch (KKK) Ltd., Frankenthal, Germany
12. TUTHILL NADROWSKI (2000) product information, Tuthill Nadrowski Ltd., Bielefeld, Germany
13. NUSSBAUMER, TH., NEUENSCHWANDER, P., HASLER, PH., JENNI, A. and BÜHLER, R. (1997) *Energie aus Holz – Vergleich der Verfahren zur Produktion von Wärme, Strom und Treibstoff aus Holz*, Bundesamt für Energie, Bern, Switzerland

14 DIETLER, R. (1994) 'Wärme-Kraft-Kopplung mittels Dampfprozess bei Holzfeuerungen, in Proceedings of the *3rd Holzenergie-Symposium*, ETH Zürich, 21 October, Bundesamt für Energie, Bern, Switzerland, pp251–274
15 BAUMGARTNER, B. and FINGER, M. (1996) 'Fernwärme-Heizwerk mit Wärme-Kraft-Kopplung in Meiringen, in Proceedings of the *4th Holzenergie-Symposium*, ETH Zürich, 18 October, Bundesamt für Energie, Bern, Switzerland, pp129–154
16 BIOLLAZ, S., RENZ, P. and NUSSBAUMER, Th. (1996) 'Schraubenmotor zur Wärmekraftkopplung mit Holz, in Proceedings of the *4th Holzenergie-Symposium*, ETH Zürich, 18 October, Bundesamt für Energie, Bern, Switzerland, pp239–266
17 SPILLINGWERK (2000) Product information and personel comunication, Spillingwerk GmbH, Hamburg, Germany
18 OBERNBERGER, I. and HAMMERSCHMID, A. (1999) *Dezentrale Biomasse-Kraft-Wärme-Kopplungstechnologien*, Graz University of Technology, Austria
19 PIATKOWSKI, R. and KAUDER, K. (1996) 'Schraubenmotor zur Wärmekraftkopplung, *4th Holzenergie-Symposium*, ETH Zürich, 18 October, Bundesamt für Energie, Bern, Switzerland, pp223–238
20 OCHSNER, E. (1996) 'Kältemittelverdichter von Gram', *Heizung Klima*, vol 9, pp48–51
21 SCHEIDEGGER, K. (1998) 'Wärme-Kraft-Kopplung mit Biomasse', *Heizung Klima*, vol 3, pp78–84
22 ANONYMOUS (1998) 'Ein Holzkraftwerk der neuen Generation', *Heizung Klima*, vol 10, pp76–78
23 TAYGUN, F. and SCHMITT, D. (1972) *Erfahrungen mit konventionellen geschlossenen Gasturbinen und ihre Zukunft in der Nukleartechnik*, Escher Wyss Mitteilungen, pp48–56
24 BAMMERT, K. (1975) *A General Review of Closed-Cycle Gas Turbines using Fossil, Nuclear and Solar Energy*, Thiemig, München, Germany
25 SCHMIDT, L. (1993) Untersuchung zu Kombiprozessen mit geschlossenem Gasturbinenkreislauf zur Bewertung des Wirkungsgradpotentials, PhD thesis, TH-Darmstadt
26 SCHALLER, W. (1995) 'Biomasseprojekte der EFG in Österreich', *Österreichische Zeitschrift für Elektrizitätswirtschaft*, No. 4
27 PODESSER, E. (1996) 'Stand der Technik von zwei Konzepten zur Wärmekraftkopplung mit Holz: Heissluftturbine und Stirlingmotor, in Proceedings of the *4th Holzenergie-Symposium*, ETH Zürich, 18 October, Bundesamt für Energie, Bern, Switzerland, pp177–204
28 DE RUYCK, J., ALLARD, G. and MANIATIS, K. (1996) 'An externally fired evaporative gas turbine cycle for small-scale biomass gasification', *9th European Bioenergy Conference, Biomass for Energy and the Environment*, Copenhagen, Denmark, 24.–27 June, pp260–265
29 BRAM, S., DE RUYCK, J. and NOVAK-ZDRAVKOVIC, A. (2005) 'Status of external firing of biomass in gas turbines', *Proceedings of the Institute of Mechanical Engineers*, vol 219, part A: J. Power and Energy, pp137–145
30 WALKER, G., FAUVEL,O., READER, G. and BINGHAM, E. (1994) *The Stirling Alternative, Power Systems, Refrigerants and Heat Pumps*, Gordon and Breach, Yverdon
31 EDER, F. (1997) 'Einsatz und Marktchancen von Stirling- und Heissgasmotoren', *Brennstoff Wärme Kraft*, vol 49 , no 1/2, pp42–45
32 CARLSEN, H. (1999) 'Status and prospects of small-scale power production based on Stirling engines – Danish experiences', *Power Production from Biomass III*, Espoo, 14–15 September, VTT, Espoo, Finland, pp249–264
33 SIPILÄ, K., PURSIHEIMO, E., SAVOLA, T., FOGELHOLM, C., KEPPO, I. and AHTILA, P. (2005) *Small Scale Biomass CHP Plant and District Heating*, VTT Research Notes 2301, Espoo, Finland
34 NUSSBAUMER, TH. and NEUENSCHWANDER, P. (2000) 'A new method for an economic assessment of heat and power plants using dimensionless numbers', *Biomass and Bioenergy*, vol 18, pp181–188

7

Co-combustion

7.1 Introduction

Increasing concerns about the environmental impacts of power generation from fossil fuels have prompted the development of more sustainable means of generating power. These have included increasing the fraction of renewable and sustainable energy in the national energy supply. Historically, renewable energy sources have struggled to compete with fossil energy, due to their relatively high costs, and high technical risk.

The co-firing of biomass with coal in conventional coal-fired boilers can provide a reasonably attractive option for the utilization of biomass for the generation of power, and in some cases heat. Co-firing makes use of the extensive infrastructure associated with the existing fossil fuel-based power systems, and requires only relatively modest additional capital investment. In most countries, the co-firing of biomass is one of the most economic technologies available for providing significant CO_2 reductions.

Overall, therefore, the principal driver for the increasing demand for the capability to co-fire biomass materials in new and existing coal boiler plants is that co-firing is regarded as representing a very attractive option for biomass utilization, and for the delivery of renewable energy, in terms of the capital investment requirement, security of supply, power generation efficiency and generation cost. This is recognized in the EC Biomass Action Plan (2005), and by a number of member state and other governments, who have introduced specific financial instruments to encourage biomass utilization and co-firing activities at existing and future coal-fired power plants.

7.2 Operational experience

There has been remarkably rapid progress over the past 5–10 years in the development of the co-utilization of biomass materials in coal-fired boiler plants. Co-firing technology is presently being routinely practiced at commercial scales in the USA, Finland, Denmark, Germany, Belgium, The Netherlands, Poland, Austria, Spain, Australia, Japan, Britain and a number of other countries.

An inventory of the application of co-firing worldwide in 2004 [1] indicated that more than 150 coal-fired power plants had experience with co-firing biomass or waste, at least on a trial basis. The power plants involved are in the range 50–700MW$_e$, although a number of very small plants have also been involved. The majority are pulverized coal boilers, including tangentially fired, wall fired, and cyclone fired units. Bubbling and circulating fluidized bed boilers, and stoker boilers, have also been used. The co-firing activities have involved all of the commercially significant solid fossil fuels, including lignites, sub-bituminous coals, bituminous coals, anthracites, and petroleum coke. These fuels have been co-fired with a very wide range of biomass materials, including herbaceous and woody materials, wet and dry agricultural residues and energy crops.

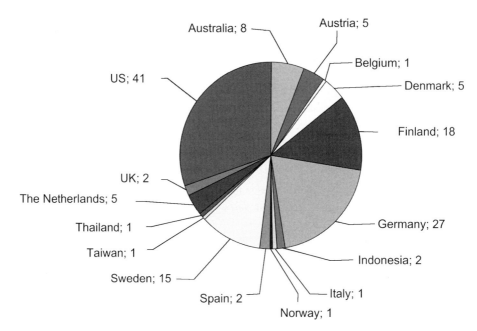

Figure 7.1 *Geographic distribution of power plants that have experience with co-firing biomass with coal, as of 2004*

Source: [1].

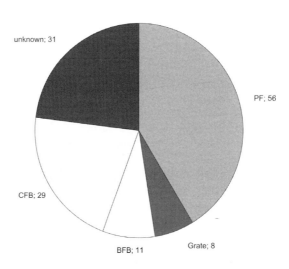

Figure 7.2 *Distribution of firing systems with coal-fired power plants that have experience with co-firing biomass*

Source: [1].

The plant experience to date has indicated that successful biomass co-firing technologies have to be suitable for retrofit to existing coal-fired power plants or for new build plants, simple to operate and control, and involve minimum risk to the normal operation and performance, both technical and environmental, of the plant. The potential costs of significant interference with the coal plant operation are likely to be high compared to any additional revenues from co-firing.

These risks are of two types:

1 reductions in plant availability and flexibility in operation; and
2 increases in maintenance and replacement costs associated with the biomass handling and firing equipment, and with the boiler plant.

The key technical risk areas are:

- fuel preparation, processing and handling issues;
- combustion related issues (e.g. flame stability, burnout, etc) affecting plant operation and control;
- ash related issues (slagging, fouling, corrosion, etc), and;
- emissions and other environmental impacts.

A schematic overview of these issues is presented in Figure 7.3.

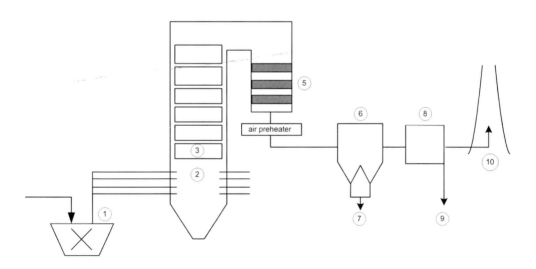

Figure 7.3 *Effects of biomass co-firing at a coal-fired power station*

Notes: 1: grinding equipment: reduced capacity and lifetime, 2: combustion chamber: slagging, 3: superheater: high temperature corrosion, 4: heat exchanger: depositions and erosion, 5: de-NO_x installation: capacity, poisoning, 6: electric precipitator: capacity, 7: ash: utilization, 8 de-SO_x installation: capacity, 9: utilization of residues from desulphurization, 10: flue gas: emissions.

Source: [2].

Co-firing bio-fuels involves risks of increased plant outages, possible interference with the operation of the burners, the furnace, the boiler convective section, and the environmental control equipment. The potential risks of increased plant maintenance costs would be associated principally with increased ash deposition and accumulation within the boiler, the potential for increased corrosion rates of high temperature boiler components, the interference with the operation of NO_x, SO_x and particulate emission reduction equipment, and the impact on ash sales or disposal routes.

The operation of power plants is a highly regulated business, and is subject to authorization from the environmental control and other institutions in most countries. These authorizations represent a deliberately imprecise operating standard for industrial plants, and are subject to regular review and renegotiation on a plant-by-plant basis. Alteration of the fuel supplies to a power station, to include biomass materials, is likely under most circumstances to be regarded as a significant change in plant operations and, as such, may result in a requirement to revise the authorizations to operate. This will generally require plant testing to demonstrate that there are no significant additional environmental impacts associated with the co-firing activities. There are also concerns about the acceptability of mixed coal/biomass ash materials for the current ash utilization and disposal routes, and there may be significant cost and environmental implications.

In this context, there are also concerns about the lack of a secure fuel supply chain for biomass materials to the power plants at stable delivered prices in most countries at the present time. The associated risks are such that the power generator may be obliged to invest directly in the fuel supply infrastructure and to adopt a more pro-active role in creating new supply structures on a site-by-site basis in order to protect the investment in new equipment, to meet contractual and other obligations, and to provide a degree of control over the delivered fuel price and quality. This is not an attractive prospect for most power generators.

This chapter provides an overview of basic co-firing concepts and a summary of major findings from past plant demonstrations and commercial co-firing operations. The technical issues related to fuel preparation, processing and handling as well as combustion related problems are also discussed. More detailed descriptions of the ash-related issues and of the impact on the environmental performance of the power plants are presented in Chapters 8 and 9, respectively.

7.3 Co-firing concepts

The great majority of biomass co-firing worldwide is carried out in large pulverized coal power boilers, and the focus in this section is very much on this type of plant. The basic co-firing options relevant to pulverized coal-fired power plants can be categorized as follows:

- direct co-firing, which involves the direct feeding of the biomass to the coal firing system or the furnace;
- indirect co-firing, which involves the gasification of the biomass and the combustion of the product fuel gas in the furnace; and
- parallel combustion, which involves the combustion of the biomass in a separate combustor and boiler and the utilization of the steam produced within the coal plant steam and power generation systems.

7.3.1 Direct cofiring

The direct co-firing approach can be implemented in a number of ways. The first option involves the mixing of the bio-fuel with the coal upstream of the coal feeders, and generally within the coal conveying system. The mixed fuel is then processed through the installed coal milling and firing system. This is the simplest option and involves the lowest capital cost. This approach has been applied widely for co-firing biomass materials in granular, pelletized and dust forms, generally at relatively low co-firing ratios.

The second option involves separate handling, metering, and comminution of the bio-fuel and injection into the pulverized fuel pipework upstream of the burners or at the burners. This option can permit co-firing at elevated levels.

The third option involves the separate handling and comminution of the bio-fuel with combustion through a number of dedicated burners. This approach involves significant modification of the combustion equipment and the furnace, and represents the highest capital cost direct co-firing option. It is, in principle, possible to inject the pre-milled biomass into the upper furnace as a reburn fuel for NO_x emission control; however this option needs significant further development prior to full scale implementation. Some testwork has been carried out in small scale test facilities.

7.3.2 Indirect co-firing

The indirect co-firing approach is based on the gasification of biomass, with the product fuel gas being combusted directly in the coal-fired furnace. The main product of the gasification process is a low calorific value fuel gas, with the calorific value depending principally on the moisture content of the fuel. The other major products are:

- all of the biomass ash materials, including the alkali metals and trace metals;
- the tars and other condensable organic species, and;
- the Cl, N and S species.

In terms of the nature and cost of the installed equipment, the indirect co-firing is equivalent to the replacement of the comminution equipment by a gasifier, i.e. the gasifier can be regarded as being a form of bio-fuel pre-processing. On the scale of operation relevant to most utility boiler co-firing projects, the preferred systems for biomass gasification are air-blown, atmospheric pressure, circulating fluidized beds. There are a number of gasification technologies of this type, from a number of suppliers, in demonstration or commercial operation.

One of the key issues with indirect co-firing approach is the degree of the fuel gas cleaning prior to co-combustion in the coal-fired furnace. This will be discussed in more detail in Section 7.4.

7.3.3 Parallel co-firing

Parallel firing involves the installation of a separate combustor and boiler for the biomass to produce steam which, in turn, is used in the coal-fired power plant steam circuit. Although parallel firing installations involve significantly higher capital investment than direct co-combustion systems, they may have advantages such as the possibility to use relatively difficult fuels with high alkali metal and chlorine contents and the production of separate coal and biomass ash streams. Again, this approach will be described in more detail in Section 7.4.

7.4 Examples of biomass co-firing in pulverized coal-fired boilers

7.4.1 Direct co-firing of demolition wood waste with coal, Gelderland Power Station, Nijmegen, The Netherlands

One of the earliest and most important demonstrations of the direct co-firing approach was the waste wood co-firing project at Gelderland Power Station in the Netherlands [3]. The justification for the project was based on the fact that in 1992, 240,000 tonnes of waste and demolition wood was sent for landfill disposal in the Netherlands. In the landfill, the waste wood decomposes and releases methane, CO_2 and other greenhouse gases into the environment. If this material could be employed as a boiler fuel, it would replace significant quantities of fossil fuels and the environmental benefits of this are clear and obvious.

It was proposed, therefore, to convert one of the coal-fired power plant boilers to burn 60,000 tonnes per annum of processed waste wood. The boiler unit selected was a 635MW$_e$ pulverized coal-fired boiler at Gelderland Power Station, which had been commissioned in 1981. In 1985–1988, the unit had been fitted with a wet limestone FGD system and in 1994, an SCR system had been fitted for NO$_x$ emission control.

A number of pre-set conditions for the wood co-firing project were established, including:

- there were to be no significant risks to the availability and performance of the boiler,
- all emissions should remain within the limits set by Dutch environmental legislation, and
- the ability to continue commercial utilization of 100 per cent of the boiler fly ashes as a component of construction materials should be retained.

The waste wood material was collected and processed into raw wood chips at three sites in the Netherlands. At these sites, large pieces of tramp material were removed manually, and smaller items of high density were removed by air classification and screening. The wood was then chipped to meet the following specifications:

Table 7.1 *Specifications for chipping of wood for use at Gelderland Power Station*

Bulk density	165–185kg/m^3
Particle size	0–3cm
Moisture content	< 20% (dry basis)
GCV	> 16MJ/kg
Lead content	< 1500mg/kg
Zinc content	< 1400mg/kg
Chlorine content	< 400mg/kg

The chipped material was delivered to the power station via a fuel handling system which is shown schematically in Figure 7.4. The wood chips were unloaded in the reception area and conveyed to the grinding area. Magnetic separation and air classification equipment were employed to provide additional cleaning of the delivered chips. They were then fed to a hammer mill, which reduced the material to a top size of 4mm. After screening out the fines, the oversize material was sent to the milling plant. The mill product and the fines from the hammer mill were combined and powdered in a wood-handling system. There were two mills, each with a capability of producing around 1.8 tonnes per hour of final mill product. The specifications for the powdered wood product were as follows:

Table 7.2 *Specifications for powdered wood product for use at Gelderland Power Station*

Particle size distribution	90% < 0.80mm
	99% < 1.00mm
	100% < 1.50mm
Moisture content	< 8% (dry basis)

The powdered wood was conveyed pneumatically to a storage silo, of 1000m^3 capacity, located adjacent to the boiler. The boiler furnace is opposed-wall-fired with three rows of six burners in both the front and rear walls. A powdered wood metering system then delivered the fuel to the existing pulverized coal burners, where it is mixed just before the burners.

Initially the wood was burned in four separate burner injection systems in the side-walls of the furnace, each rated at 20MWth and capable of delivering 1.1–3.5 tonnes of fuel per hour. This was done to enable completely independent operation of the coal-firing system and the wood firing system, avoiding interference with the boilers' capability to burn coal at full load. However, in practical operation, it turned out that the dedicated burners experienced excessive erosive wear and it was found to be more practical to mix the fuels in the fuel transport lines just before the existing burners.

The wood-firing capacity was around 10 tonnes per hour, which is equivalent to 3–4 per cent of the heat input to the furnace. At this co-firing ratio and with a relatively high-quality fuel, the impacts on the operation, the environmental performance, and the availability of the boiler were small. The system was commissioned in 1995 and, despite considerable initial problems with the wood-handling and milling system, the system was in full commercial operation for a number of years. On average, around 60,000 dry tonnes of wood were fired per annum, replacing around 45,000 tonnes of coal per annum. There was also a reduction of around 4000 tonnes per annum of the fly ash produced, due to the very low ash content of the wood fuel.

The Gelderland project was a very important project in that was the first direct biomass co-combustion demonstration in a large utility boiler in Europe, with significant experience of commercial operation. At an electrical power output from wood firing of 20MW$_e$, it had an operational scale that was of direct value to future co-firing projects.

7.4.2 Direct co-firing of sawdust and woodchips with coal at Wallerawang Power Station, NSW, Australia

Coal fired power stations account for 84 per cent of Australia's electricity generation according to the latest statistics. The growing concern over CO_2 emissions and their impact on global warming have prompted the Federal Government, coal producers, and power

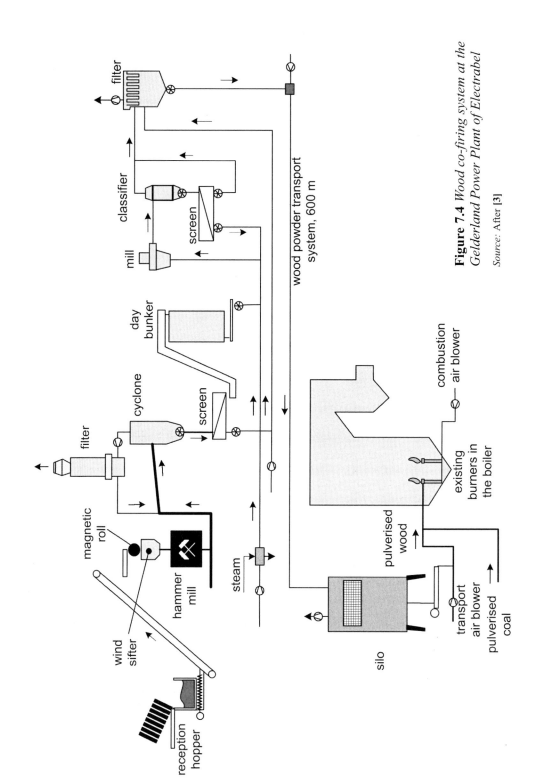

Figure 7.4 *Wood co-firing system at the Gelderland Power Plant of Electrabel*

Source: After [3]

generators to search for new ways of reducing the dependency on fossil fuels. The co-firing of coal and biomass was recognized as an attractive short-term means of diversification of the energy source portfolio of power generators. The Federal Government introduced the Mandatory Renewable Energy Target (MRET) in 1 April 2001 to set the scene for future use of renewable energy sources. The MRET legislation requires all electricity retailers and direct large consumers to source a proportion of their electricity from new renewable energy sources. The total renewable energy target was set at 9500GWh per annum by 2010, with a sliding scale from an initial 300GWh per annum in 2001.

In 2000, Delta Electricity carried out a comprehensive program of co-firing trials at Wallerawang power station [4]. Wallerawang plant has two 500MW$_e$ pulverized coal-fired boilers (Unit 7 and Unit 8). Commissioned in the 1970s, the boilers are dual-furnace tangentially-fired, forced circulation with reheat. Boiler steam pressure is 15,860kPa, steam temperature is 538°C, and steam flow rate is 441kg/s. The main coal supply, around 2.2 million tonnes p.a. is provided by the nearby Angus Place colliery, with some fuel supplied from local privately-owned mines.

Preliminary co-firing tests were carried out by researchers from the University of Newcastle in early 2000, and these were followed by full-scale trials in Unit 7 of the Wallerawang power station in late August, 2000. The objectives of the trials were:

- To gain an understanding of the impacts of co-firing on unit performance and to identify the optimum blending ratio,
- To investigate any changes in emissions due to co-firing, and
- To provide data to the NSW Environmental Protection Agency (NSW-EPA), on which to base a variation of the station operating licence.

Initial trials were performed in Unit 7 by pre-mixing of the biomass and coal at 3, 5, 7wt%, on the conveyor belt prior going into the mills. All six mills of Unit 7 mills were in operation during the trials. Mill performance trends were recorded for each blend and mill product samples were taken for particle size distribution measurement. The mill feeder speed was set at 80 per cent with the cold air damper position fixed to ensure there would be no compensation in the mill temperature control loop. Trials with both sawdust and wood chips were carried out.

The impact of wet sawdust and woodchips on the overall operation of the plant, and particularly the coal mills, were found to be similar. At 3 per cent co-firing on a mass basis, little or no perceptible change was noted in mill performance. At 5 per cent, the mill outlet temperature decayed and the mill power and differential pressure increased. The mill operation was stable and no reduction in the feeder speed was required. At 7 per cent there was a sustained drop in mill temperature, and the mill differential pressure and power consumption increased to unacceptable levels. The feeder speed was reduced to 75 per cent and the mill parameters returned to more normal levels.

The total particulate emissions increased by 48 per cent to 102mg/m^3 during the co-firing trials, but these were still below the NSW EPA limit for Unit 7 of 400mg/m^3. The effect of co-firing on dust burdens was found to be exacerbated during hot weather or continuous high load running.

The measured carbon in ash levels were found to increase during the co-firing tests, and this may have had an impact on the dust emissions. A maximum blending ratio of 5 per cent by mass was recommended as the optimum value for wet sawdust and wood chip co-firing operations at Delta Electricity power plants.

Following the full-scale trials, Delta Electricity decided to go ahead with commercial co-firing operations in two of its power plants, namely Wallerawang and Vales Point power stations. Commercial co-firing at these plants commenced in 2002 on an on-going basis, and continues to date.

7.4.3 Direct co-firing of straw with coal, Studstrup, Denmark

The development of the utilization of cereal straws and other baled biomass materials for energy recovery is more advanced in Denmark than in any other country in the world. This development was initiated by a Danish government decision in 1993 to instruct the power utility companies to burn 1.2 million tonnes of surplus straw per annum by the year 2000. This prompted a significant programme of research and demonstration activity in Denmark, including:

- the construction of a number of small, CHP projects based on straw combustion or co-combustion in fluidized bed boilers, on grates and in purpose-designed combustion equipment, and;
- a programme of research and demonstration projects aimed at the conversion of large pulverized coal-fired utility boilers to co-fire straw.

This programme culminated in a demonstration of the co-firing of straw at Unit 1, Studstrup Power Plant, near Aarhus in Jutland, operated by Midtkraft (currently DONG Energy), starting in 1995. From 1996–1998, straw was co-fired in Unit 1 an older unit, which has since been taken out of service. The long-term demonstration programme was considered to be successful; and in-depth knowledge was gained on the influence of straw co-firing upon boiler plant performance, combustion chemistry, heat exchanger surface ash deposits and corrosion, ash residue quality, and emissions. The experience gained during the Unit 1 trials was instrumental in the conversion of Unit 4 ($824MW_{th}/350MW_e$) in 2002, for the commercial co-firing of straw at up to 10 per cent heat input.

Studstrup Unit 1 was a $150MW_e$ pulverized coal-fired boiler, which was originally commissioned in 1968. The conversion to co-firing consisted of establishing a fully commercial straw reception, storage and pre-processing facility with a capacity of 20 tonnes per hour, corresponding to 20 per cent of the total energy input at full load, and the modification of the coal firing system to permit biomass co-firing.

The system was installed and commissioned during 1994–5, and a demonstration programme started in 1996, focussing on the performance of the straw-handling and firing system, the boiler performance, the process chemistry, i.e. slagging, fouling and corrosion, residue characteristics, and the pilot testing of $deSO_x$ and $deNO_x$ equipment. This involved long-term operation at 10 and 20 per cent straw on a heat input basis.

There were some initial problems with the handling and mechanical conveying of wet straw, and it was found that the system was sensitive to relatively small quantities of straw with moisture content in excess of 25 per cent. The key issue in the solution to this problem was co-operation with the straw suppliers to improve the quality and consistency of the delivered fuel.

The combustion system performed reasonably well, provided that the straw injection velocity through the burner was below 15m/s. Good burnout, even of the more dense components of the straw, was achieved. No particular problems associated with ash deposition, high temperature corrosion, or the environmental performance of the plant were identified at straw co-firing rates up to 20 per cent on a heat input basis [5].

From 1998–2001, the cofiring activities at Studstrup were set on hold due to concerns about the industrial utilization of the fly ash residues. After extensive research, the national standard was revised in 2001 to allow use of fly-ash from biomass co-firing in cement.

Upon completion of the successful demonstration programme in Unit 1, it was decided to convert Unit 4 at Studstrup Power Station, a 350MW$_e$ pulverized coal boiler, to straw co-firing. The boiler is a once-through single reheat type generating steam at 540°C and 250 bar. Twenty four Doosan Babcock Mark III low NO$_x$ burners are arranged in two levels in an opposed wall firing arrangement. Four Deutsche Babcock MPS coal mills, each supplying six burners, are used for pulverization of the fuel. Unit 4 is equipped with a semi-dry desulphurization plant. Four of the burners in the upper burner level on the rear furnace wall were converted for co-firing by relocating the oil lance and the flame scanner in order to clear the burner core air tube for injection of the straw.

The straw storage facility at Studstrup is split into two sections, each with a capacity of 560 Hesston bales of 1.2 x 1.3 x 2.4m, each weighing 450–600kg. The straw delivery trucks are unloaded by an overhead crane. The crane unloads twelve bales in one batch. During unloading, the bales are weighed, the moisture content is measured, using microwave techniques and the data are stored on a central logistics computer.

Figure 7.5 *Straw shredder at Studstrup power plant*

Source: Courtesy of Elsam Engineering.

The processing plant, housed in a separate building, comprises 4 parallel lines, each with a processing capacity of 5 tonnes straw per hour, see Figure 7.6. After the Unit 1 trials, significant modifications were made to the straw shredders and hammer mills. In total the maximum straw co-firing capacity is 20 tonnes per hour, corresponding to a straw share of 10 per cent on energy basis at full boiler load. The straw flow to the boiler is controlled by tier conveyors situated before the shredder. After the bale cords are cut, the straw bales are broken up using a heavy-duty garbage grinding machine. The straw is then sucked trough a stone trap into the hammer mill. In the hammer mill, the straw is cut into lengths of no more than 50–100mm. From the hammer mill the straw is passed through an airlock and pneumatically transported in four parallel lines over a distance of 300m and through the core air tubes of the converted coal burners.

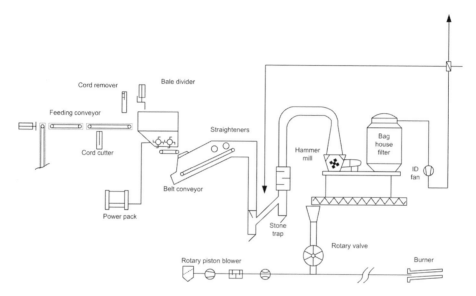

Figure 7.6 *Straw pre-processing equipment at Studstrup Unit 4*

Note: ID = induced draught.
Source: Redrawn from [6].

The results with co-firing straw at 10 per cent of heat input have been very positive. When co-firing straw, the loss on ignition (LoI) in the fly-ash is generally lower than when firing coal alone. The water-soluble alkaline and chlorine contents of the fly-ashes increased significantly, however this is not a significant problem for utilization of fly ash in cement.

The experience at Studstrup indicates that there has been no noticeable negative effect on ash deposition or corrosion. Tests with deactivation of high-dust SCR catalyst show a decrease in performance, however the measured deactivation rate is not larger when cofiring straw as compared to firing coal alone. Ongoing work is focused on long term corrosion tests, the study of the alkali metal-chlorine-sulphur chemistry and the production of a CFD model for optimization of the co-firing plant configuration. Long term, high dust, SCR catalyst deactivation tests are also planned.

The co-firing of straws and other baled materials at Studstrup Unit 4 is still in successful commercial operation, processing around 160,000 tonnes of straw per annum.

7.4.4 Direct co-firing of wood fuels on a grate located directly under the furnace of a PC boiler, St Andra, Austria

This approach to the direct co-firing of biomass in a pulverized coal-fired boiler has been demonstrated at the St Andra power plant in Carinthia, Austria. The plant is no longer in operation for commercial reasons not associated with the biomass co-firing activities. It is included here since it represents an interesting co-firing concept, although this approach has not been widely replicated.

The boiler is a 124MW$_e$ unit with a pulverized coal-fired radiant furnace. The bottom ash hopper was modified to include two travelling grates for the firing of chipped wood materials, and for the handling of the furnace bottom ashes. The arrangement is illustrated schematically in Figure 7.7. The nominal thermal capacity of the biomass grates is 10MW$_{th}$, which is equivalent to around 3 per cent of the total heat input to the furnace.

The co-firing system was commissioned in 1995, and operated for several years without major problems. Minor problems were experienced with the inclusion of tramp materials in the biomass fuel, which caused some blockages of the fuel-feeding system. The travelling grate and the wet ash handling system operated well with the mixed biomass ash/coal bottom ash mixture. The burnout of the fuel was good, with carbon in ash levels in the discards from the grate being less than 5 per cent. Biomass co-firing at this level had no significant impact on the performance and availability of the boiler, nor on the environmental performance of the unit.

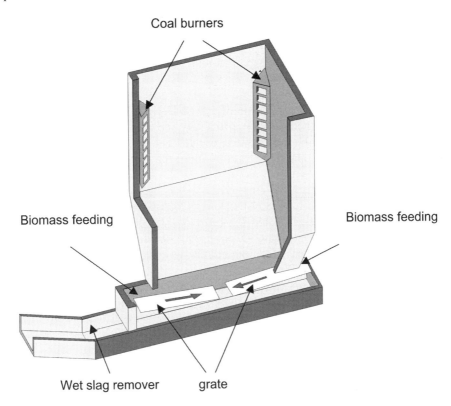

Figure 7.7 *Biomass co-firing system at St Andrea, Austria*

The principal advantage of this approach to biomass co-firing is that the wood fuel needs minimal pre-processing. The fuel, in this case, was wood, chipped to a top-size of 50mm, and at 10–55 per cent moisture, on a wet basis. In this case, the fuel mix includes bark, sawdust, and wood chips, and in principle, this approach can provide a deal of fuel flexibility.

This approach also has the advantage of involving relatively low capital cost, but is only suitable for biomass co-firing at very low co-firing ratios, and can be applied readily only to furnaces which have sufficient clearance under the ash hopper to allow installation of a fuel-feeding and grate system.

7.4.5 Indirect co-firing of biomass fuel gas with coal, Zeltweg Power Plant, Austria

A number of demonstration projects in Europe and elsewhere have been based on the gasification of chipped wood fuels, and the combustion of the product fuel gas in the coal-fired boiler furnace. One important demonstration was carried out in Zeltweg Power Plant in Styria in Austria as part of the BIOCOCOMB project [7].

The Zeltweg Power Plant is a 137MW$_e$ pulverized coal-fired boiler, which was originally commissioned in 1962, and was operated by Verbund. It was built originally to burn lignite, but in the early 1980s, it was converted to the firing of bituminous coals. The environmental control systems were upgraded to include improved particulate collection, an SNCR system with ammonia injection for NO_x emission control, and a CFB reactor for SO_x emission control. The boiler had a tangentially fired furnace and generated steam at 535°C and 185/44 bar. .

This boiler was the host site for a demonstration project involving the gasification of wood materials in an air-blown CFB gasifier unit, and the combustion of the fuel gas in the boiler furnace. A schematic diagram of the system is presented in Figure 7.8.

The wood fuel was principally bark, but the system had a significant degree of fuel flexibility, covering wood wastes, demolition wood, sawdust, and wood chips. The wood fuel was fed through a magnetic separator to a silo, and then through a screw conveyor and weighing belt conveyor to the feeder and to the gasifier. The system could handle material to a top-size of 30 x 30 x 100mm. No pre-drying of the fuel was done.

The CFB gasifier unit, which was rated at 10MW$_{th}$, was of steel construction with a refractory-lined reactor. The CFB employed sand as the bed material and operated at a nominal temperature of 850°C. The system had an internal hot cyclone system. The product gas specification is presented in Table 7.3.

Since the fuel was gasified without pre-drying, the fuel gas had a very low calorific value. The fuel gas exited the gasification unit via the cyclone, and was fed directly through hot gas ductwork to the boiler furnace.

The fuel gas was fired through purpose-designed burner nozzles in the upper furnace and was used as a reburn fuel to help reduce the primary NO_x emission levels. The gas represented around 3 per cent of the heat input to the furnace. The key effect of the use of the biogas as a reburn fuel was a reduction of around 10–15 per cent of the ammonium hydroxide consumption of the SNCR system.

The biomass gasification and gas combustion systems were commissioned in 1997, and underwent extensive performance trials during 1998. The process was in operation through 1999 and the early part of 2000. More than 5000 tonnes of biomass were gasified and combusted in this way, and some trials with other waste fuels were performed. The

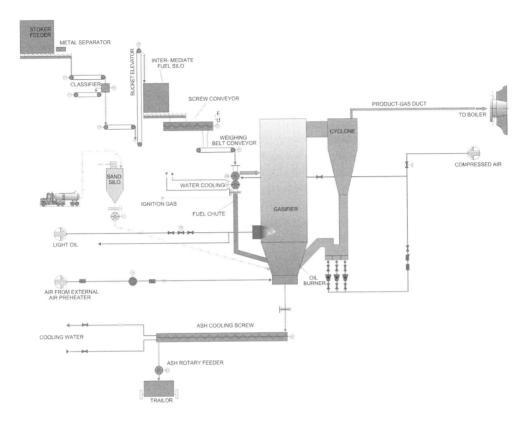

Figure 7.8 *BIOCOCOMB system*

operating experience was good. There were some initial problems with the fuel conveying systems, which were overcome. The operation of the gasifier and the fuel gas combustion systems was reasonably good, and there were no significant impact on the performance and availability of the boiler. The plant is not currently in operation, but this is for commercial reasons, not associated with the biomass co-utilization system.

Table 7.3 *Fuel gas composition for the BIOCOCOMB process*

Gas composition	Units	Calculated	Measured
O_2	Molar%	0.00	0.00
N_2	Molar%	38.12	43 62
CO	Molar%	2.76	2.73
CO_2	Molar%	12.45	13.20
CH_4	Molar%	0.00	1.11
H_2	Molar%	9.03	3.32
H_2O	Molar%	37.64	35.00
Total	Molar%	100.00	100.00
GCV	MJ/kg	1.965	–

Source: [7].

7.4.6 Indirect co-firing of biomass fuel gas with coal, Amer Power Plant, in The Netherlands

Unit 9 of the Amer power plant in Geertruidenberg, the Netherlands, is a 600MW$_e$ pulverized coal fired unit. In October 1998, construction of a 83MW$_{th}$ circulating fluidized bed gasification plant for biomass started, this work was completed in 2000 [8]. In total, the heat input from the gasification system, i.e. both fuel gas and steam from the fuel gas cooler, is equivalent to around 5 per cent of the total heat input to the boiler. The original plant layout, completed in the first half of 2000 and commissioned in 2001 is shown in Figure 7.9.

The system was designed to process 150,000 tonnes of waste wood per annum. Chipped demolition wood is delivered to a wood storage silo. The chips are then transported to the service silo and then to the screw feeder to the gasifier. This is a circulating fluidized bed, with sand bed material, operating at a temperature between 850 and 950°C.

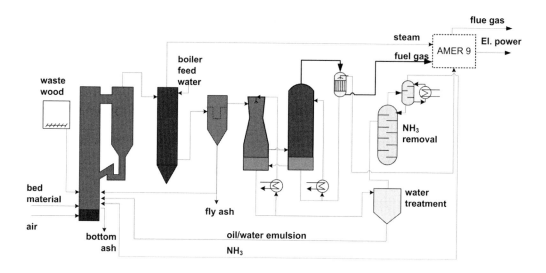

Figure 7.9 *Original layout of the AMERGAS biomass gasification plant at the Amer Power Plant in Geertruidenberg, the Netherlands*

Source: [8].

In the original plant configuration as shown in Figure 7.9, the idea was to cool the fuel gas in a steam generating boiler to an exit temperature of 220–240°C, after which a fabric filter would remove particulate matter and a conventional wet scrubbing unit would remove ammonia and condensable tar materials. The clean gas would then be reheated to 100°C, and fed to the burners in the coal-fired boiler furnace. The fly ashes collected in the fabric filter could then be partially recycled to the gasifier, and the scrubber water stripped of ammonia and then injected into the boiler furnace.

Very rapid and serious fouling of the water tube syngas cooling unit was experienced during commissioning of the system and, in response to these problems, major modifications were made to the syngas cooling and cleaning system. The fouling problems in the syngas cooling unit were associated principally with the deposition of the tars and chars, although the carryover of ashes and elutriated bed material does contribute to the deposition process.

The modified system involves cooling the syngas to around 500°C, i.e. at a temperature above the tar dewpoint, followed by particulate collection in a hot cyclone. The extent of cleaning of the syngas, therefore, is significantly reduced, and this has implications on the fuel flexibility of the system. The modified system has been in operation since December 2002, with reasonably satisfactory performance.

It should also be noted that Amer Unit 9 also has the capability to co-fire directly up to 300,000 tonnes per annum of pelletized biomass materials.

The revised arrangement at Amer Unit 9 is similar in many respects to the co-firing systems at Lahti in Finland and Ruien in Belgium, where hot fuel gases from CFB gasifier units are fired directly into the main coal-fired furnaces, after hot cyclones for coarse particulate matter removal.

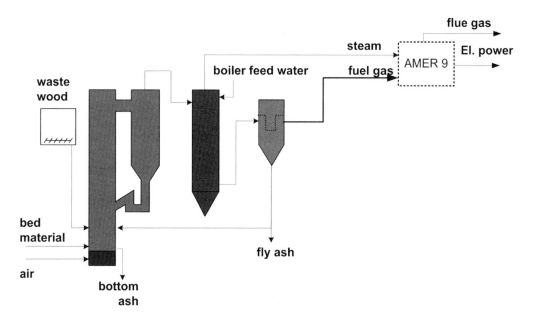

Figure 7.10 *AMERGAS biomass gasification plant after modification*

Source: [10].

7.4.7 Parallel cofiring of biomass and fossil fuels, Avedøre Power Plant, Denmark

One of the more advanced biomass co-utilization projects is at Avedøre Unit 2, near Copenhagen in Denmark [10]. A diagram of the overall multifuel concept is reproduced in Figure 7.11. The main component is a large ultra-supercritical power plant, which fires natural gas, but which has been designed for coal, gas, and oil firing. The plant has a supercritical boiler, rated at 430MW$_e$ with a flue gas clean-up plant, a steam turbine, and an electrical generator. There are also two aero-derivative gas turbines, rated at 51MW$_e$ each, which provide peak load electricity generation, and are used to pre-heat the feedwater to the USC boiler, via exhaust heat recovery units. A biomass boiler, designed originally to burn straw, and rated at 105MW$_{th}$, provides additional steam to the system. The biomass boiler consumes 150,000 tonnes of straw per annum, and generates up to 40kg/s of steam at 583°C and at pressures up to 310 Bar.

The Avedøre Unit 2 power plant has unique operational capabilities, since all of the key components, i.e. the USC boiler, the biomass boiler, the steam turbine and the two gas turbines with their heat recovery units, are integrated into one process. The plant was built by I/s Avedøre 2, a joint venture between Energy E2 (the principal power company in Eastern Denmark) and Vattenfall. It entered into commercial operation by the end of 2001 and is working according to expectations.

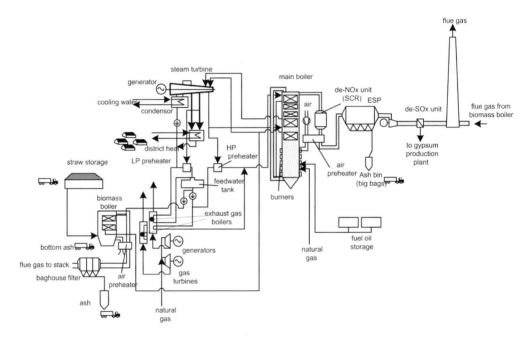

Figure 7.11 *Avedøre 2 multifuel system*

Source: [9].

7.4.8 Summary of experience of biomass cofiring in pulverized coal-fired boilers

It is clear that there has been rapid progress over the past 10–15 years in the development of the co-utilization of biomass materials in pulverized coal-fired boiler plants. Much of this progress has been in Europe, North America and Australia, where research, development and demonstration activities in this subject area have been supported through various funding schemes. The European activities, in particular, have been encouraged by the funding available through the EC Thermie Programme in this area, and by specific subsidy schemes for renewable energy generation introduced by member state governments.

The majority of the key technical options have been demonstrated, and a number are in commercial operation, including:

1. The pre-mixing of the biomass with the coal and the processing of the mixed fuel though the installed coal handling and firing systems,
2. The pre-processing of the biomass materials, and the direct co-combustion of the prepared fuels in the boiler furnace;
3. The gasification of the biomass and the co-firing of the product fuel gas in the boiler furnace;
4. The combustion of the biomass materials in a separate furnace and the utilization of the steam in the main power plant cycle.

A number of other approaches to biomass co-utilization, including the co-firing of biomass materials with coal in circulating fluidized bed boilers and in coal gasification plant, are also now in commercial operation.

In most cases, the projects have involved the retrofit of existing coal-fired boiler plants for biomass co-utilization. In some cases, older power plants, with limited operating lives, have been converted, purely for demonstration purposes, with the result that a number of the projects have provided valuable data and design information, but have not proceeded to full commercial operation. A number of new plants based on very advanced concepts, and involving biomass co-utilization with fossil fuels have also been included. The developments, in turn, have led to commercial-scale operations in many parts of the world, and principally in Europe, North America and Australia.

There are, however, a number of barriers to the wider implementation of biomass co-utilization technologies worldwide, the majority of which are non-technical, viz:

- The electricity supply industries in many countries have been undergoing a process of commercial liberalization, and competition between generators is intensifying. This process can militate against investment in new technologies.
- The legal and political background against which the decisions to invest in renewable energy technologies is uncertain in many countries. The lack of clear and reliable future scenarios for the development of renewable energies has provided an incentive for some electrical utility companies to defer investment decisions. It is instructive that the most progress has been made in the Scandinavian and other Northern European countries where political and social thinking on these matters has tended to be more advanced.
- The changes in the legislative position as regards the environmental performance of power plants and the utilization/disposal of solid wastes can also add to uncertainty.

Technically, there has been major progress in the development of equipment for the handling and pre-processing of biomass materials for power plant applications over the past 15 years. Reliable, commercially demonstrated, systems for the bulk storage, handling, comminution and pneumatic conveying of baled straw, and for the bulk handling and comminution of wood chips and other similar materials are now available.

Although there are still further development requirements, equipment for the co-combustion of both chopped straw and milled biomass materials with pulverized coal is available. The appropriate control philosophies and the related control and monitoring equipment for co-firing have been developed. The performance of CFB gasification equipment for wood materials, and the integration of the gasifiers with coal-fired boilers, have also been demonstrated.

To date, it is clear that the impacts of the co-firing of biomass materials with coal in large pulverized coal boilers at low co-firing ratios have been modest. Relatively few major problems with increased ash deposition or with accelerated corrosion rates of boiler components have been reported. It should be noted, however, that the great majority of the projects to date have operated at co-firing ratios less than 10 per cent on a heat input basis. It is clear from the results of research in this subject area that the impacts are likely to be limited at this co-firing level.

The problems have been more apparent in the demonstration and commercial projects involving the firing or the co-firing of biomass materials in smaller boilers and at higher co-firing ratios. The impacts are also likely to be greater with fast-growing biomass materials and residues from fertilized crops, i.e. the cereal straws and other baleable materials, with short-rotation coppice wood materials, and with agricultural residue materials, since these materials tend to have higher levels of ash, alkali metals, and chloride than do roundwood and high grade pellets.

The environmental performance of the co-firing projects has presented a few more problems. The nature of biomass ash materials is such that the collection efficiencies of the installed particulate collection equipment have been reduced in some cases. There is also evidence that the presence of higher concentrations of alkali metal and phosphate species can have a negative impact on the performance of SCR systems for NO_x emission control. These issues can lead to restrictions on the allowable co-firing ratios for some biomass materials.

The utilization and disposal of the mixed ashes from biomass co-firing with coal are also issues, which required further development. In Denmark, for instance, the utilization of ashes from the co-firing of straw with coal in the cement and concrete industries initially represented a significant barrier to the further development of these technologies. These problems have largely been overcome in Europe by changes in the relevant standards and legislation.

Overall, therefore, it is clear that progress in this area has been very encouraging. A number of the more important technical options for the co-utilization of biomass materials have been successfully demonstrated, and are being replicated elsewhere, and several millions of tonnes of biomass are being utilized per annum for co-firing with coal in large boilers, resulting in significant reductions in fossil fuel use.

The barriers to the further replication of biomass co-utilization techniques are essentially non-technical. There are commercial barriers due to the relatively high delivered prices for many types of biomass, and in particular the clean biomass materials. This will inevitably lead to a demand for equipment and co-firing technologies that provide a greater degree of fuel flexibility, and to an increasing interest, in some countries, in the co-firing of wastes and the more contaminated biomass materials.

There has also been a demand for simplification, and reduction of the costs, of co-utilization technologies. By their nature, demonstration and 'first in class' commercial projects are concerned with the introduction of new technologies, involving a degree of uncertainty and risk, and as such they tend to be over-engineered. There is, therefore, some scope for process simplification and cost reduction.

The costs and risks associated with the poorly developed supply infrastructure for biomass materials in most countries are obviously key barriers to the further development of co-utilization technologies. This has meant, for instance, that there has been significant import of biomass into Northern Europe, from other parts of Europe and elsewhere, as well as the development of indigenous biomass supplies.

7.5 Fuel preparation, processing and handling issues

All industrial liquid and solid fuels require significant bulk storage and handling, and pre-processing, prior to firing in utility boiler plants. In the case of solid fuels, the more important processes are:

- preliminary size reduction;
- bulk handling, storage and transportation
- washing or cleaning to remove tramp material and, in some cases for the reduction of the ash content;
- drying or partial drying, in some cases, and;
- secondary size reduction, i.e. comminution or milling, prior to firing.

In the case of solid biomass materials, similar requirements apply. Fuel processing requirements are dictated by the fuel sources, with potential biomass feedstocks varying widely both in moisture content and physical form.

Figure 7.12 *Biomass fuel handling facility, which directly meters biomass onto the coal conveyor belts, Wallerawang Power Station, Australia*

Source: Courtesy of Delta Electricity, Australia.

Figure 7.13 *Dumping woodchips on a coal conveyor before the mills*

Source: Courtesy of Delta Electricity, Australia.

In this section the general aspects of the material preparation, processing, and handling of biomass and biomass/coal mixtures in pulverized fuel boilers are described. Reference should also be made to Chapter 3 of this Handbook, where there is a more general description of this subject area.

7.5.1 Preliminary size reduction

7.5.1.1 Granular and pelletized biomass materials

The requirement for size reduction depends on the mode of co-firing. For direct co-firing projects involving pre-mixing of the biomass with the coal, and co-milling, biomass materials in granular, dust and pelletized forms may be suitable.

For direct injection co-firing projects, the majority of biomass feedstocks will require size reduction to reduce the particle top size to less than about 4–5mm. Conventional agricultural hammer mills, have been used to reduce a number of biomass materials to this size range, however they have proved to have relatively high maintenance requirements and high power consumption for utility boiler applications [4].

7.5.1.2. Baled herbaceous materials

The pre-processing of baled herbaceous materials at pulverized fuel power plants, e.g. the breaking of the straw bales, and the shredding of the loose straw, proved to be relatively difficult to achieve. A good deal of progress in resolving this problem was made in Denmark throughout the 1980s and 1990s, in association with a number of straw combustion and co-firing trials, and the commercial co-firing projects at Grenaa, in Jutland, which involves the co-firing of straw with coal in a fluidized bed boiler, and at Studstrup Power Station.

This experience has indicated that a modern straw handling and pre-processing system at a pulverized coal-fired boiler plant should include:

- A straw reception and storage hall with an automatic overhead crane. The crane is used to lift the bales from the lorries and deliver them to the store or to the entry conveyors on the straw handling plant. The cranes can also weigh the individual bales and can be fitted with microwave pads for the measurement and recording of the moisture content of individual bales. Special arrangements are made for fire protection of the storage hall, and for vermin control.
- Special equipment is available for the cutting and removal of the bale strings and for the breaking or scarification of the bales. There are a number of designs of slow-rotation, scarifying equipment, which tear the bales apart but do not cut the straw significantly.
- The size of the loose straw is normally reduced using a hammer mill with an outlet screen at 50mm or so.
- The chopped straw can be then conveyed pneumatically to the burners.

7.5.2 Bulk handling

The biomass feedstocks for co-firing can be delivered in a variety of forms. The relatively low bulk density of biomass fuels, the poor flow properties, in some cases, and the variable delivered moisture content can affect the performance of the feeding and handling units, and may lead to the blockage or hang up at transfer points and at points where there is restriction of the flow.

The single most important issue when handling and storing biomass materials has been the generation and accumulation of dust. Dust extraction systems have been used successfully, however the fine materials collected in these systems can be very cohesive in nature, and there can be problems with the disposal of dusts collected, particularly when using wet removal systems. Special arrangements, including explosion vents and fire suppression systems may be required in biomass storage and handling areas.

If the accumulated dust in the fuel store or handling system gets wet, it can swell and there can be mould growth. The experience with water misting systems for dust suppression has been mixed. These systems tend to increase the relative humidity in the store, and this can further encourage mould growth.

7.5.2.1 Wood-based materials
The underlying mechanisms that can cause flow problems was studied under an important recent Australian development project funded by the CCSD [11]. In this study, the flow properties of typical Australian coals and woody biomass materials, and blends of these materials, were studied.

Two types of experiments were carried out, viz:

- direct shear experiments using a Jenike type shear cell; and
- wall shear experiments using a wall yield loci test apparatus.

Four major parameters were obtained from these experiments for evaluating the flow properties of biomass and coal blends, viz:

- the major consolidating stress;
- the unconfined yield stress;
- the effective angle of internal friction; and
- the kinematic angle of internal friction.

From the value of the unconfined yield stress and the major consolidating stress, the flow function graphs for coal and blend samples were then constructed, as illustrated in Figure 7.14. The flow function diagram represents the strength of the material at the free surface. The value of the unconfined yield stress, provides an indication of the ability of a bulk solid to form a cohesive arch.

Generally, for free flowing materials, such as dry sand, the value of the unconfined yield stress is close to zero. Most wet particulate materials, including coals, sawdusts etc., have an appreciable yield stress which increases with increasing consolidating stress. This is illustrated in Figure 7.14 for a coal and for coal-biomass blends. The results of the work indicated that the flow behaviour depends principally on the fines content, the nature of the fines and the moisture content of the material The results indicated that there appear to be no special properties of coal-sawdust blends that would render them significantly more difficult to handle than coals in this respect. Coal-woodchip blends may present more difficulties.

Analysis of the wall friction angle versus normal stress showed that the addition of sawdust does not significantly impact on the frictional properties of the blend as they remain more or less similar to those of the parent coal. However, the addition of woodchips to coal was found to have a significant impact on the frictional characteristics of the bulk solid.

Analysis of experimental data also revealed that the flow properties of sawdust/coal blends are strong functions of the blending ratio at least for ratios below 10 per cent, on a mass basis, whereas for woodchip/coal blends the relationship seems to be weaker.

The main conclusion that can be drawn from the CCSD study is that it is unlikely that coal/sawdust blends would cause flow stoppage in fuel handling units of pulverized fuel power plants. However, blends of woodchip and coal would significantly increase the likelihood of blockage and flow stoppage.

It is clear, therefore, from the results of this work and of other studies that the flow properties of coal/biomass blend are strongly dependent on the physical form of the coal and the biomass materials, and particularly of the fines, and on the moisture content.

7.5.2.2 Herbaceous materials

Herbaceous materials in baled form typically do not pose major bulk handling problems. Equipment for the bulk handling, transportation and storage of large Hesston bales is readily available and is relatively well-developed.

7.5.3 Long-term storage

7.5.3.1 Granular biomass materials

Granular biomass materials are treated in a similar fashion to coal in that they are stored and handled in lump/chip or pelletized form and milled on-site prior to firing. The long-term

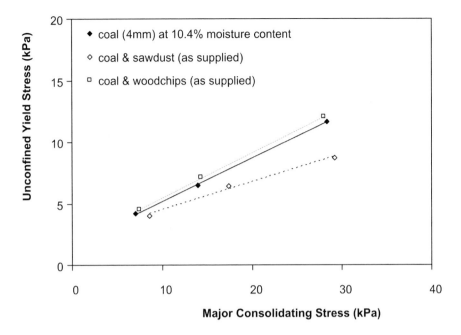

Figure 7.14 *Comparison of flow function for coal and coal/biomass blends*
Source: [11].

storage of wet biomass is problematic because that at moisture contents in excess of around 20 per cent, wet basis, relatively rapid biological activity can lead to heating of the storage pile, loss of dry matter, and a significant deterioration in the physical quality of the fuel. There is also a possibility that high dust and spore concentrations in the stored fuel can give rise to Health and Safety problems during subsequent fuel handling operations. This can be particularly troublesome during and after fuel drying operations, when dust and spores can be released into the working environment. To minimize biological activity during long-term storage of wood fuels, four courses of action are available:

- storage of the biomass in the form of larger pieces to reduce the area of cut surface available for biological activity;
- the use of fungicides and other chemical agents to suppress biological activity;
- pre-drying the fuel to a moisture content at which biological activity is reduced; and
- cooling the stored fuel to temperatures at which biological activity is reduced, by forced-air ventilation.

7.5.3.2 Herbaceous materials

Cereal straws and a number of other baleable biomass materials, which have moisture contents below 20 per cent, wet basis, are not subject to microbial respiration to the same extent as are wetter fuels. Cereal straws are normally stored and handled in the form of large Hesston bales, 1.2 x 1.3 x 2.4 m^3 in size, and weighing 450–500 kg. As long as the straw is baled dry and the bales are stored carefully, they can be kept for relatively long periods of time, including over winter, without serious deterioration in quality or significant dry-matter losses due to microbial respiration.

7.5.4 Drying

The moisture in solid fuels can exist in two forms, viz:

- as free water on the outer surfaces and within the pores of the fuel; and
- as bound water which is absorbed to the interior surface structure of the fuel.

The bound water is bonded to the hydroxyl groups of the major constituents of the fuel, for instance cellulose, hemicellulose and lignin in the case of wood. Biomass materials, are porous in nature' and have both free and bound water. Green wood, for instance, contains about 45–60 per cent water, on an as-received basis, of which a significant portion is bound water contained within the particles.

The relatively high level of moisture content in many biomass fuels produces some technical difficulties with regards to processing and handling of the fuel, and biomass materials are commonly dried prior to long term storage or thermal processing. There are two principal approaches to the drying of biomass fuels:

- Thermal drying; and
- natural drying.

The thermal drying of biomass is widely practised worldwide, particularly for the production of pelletized fuels and for the preparation of biomass materials for long-term storage and transportation as animal feeds, and other high value products. In the main, rotary, moving bed and fluidized bed dryers are employed for chipped materials, and, for fine dusts, superheated steam, ring dryers are proving relatively popular. The direct contact dryers are reasonably efficient, however the costs and the energy requirements for thermal drying are significant.

In some countries, the moisture content of biomass materials can be reduced to levels below 20 per cent by natural drying. Significant research work has been conducted on natural drying of biomass fuels under bench and pilot scale conditions (see also chapter 3.3.5). There are, however, a limited number of large-scale studies. The work carried out by Bauer [12] is one of the most comprehensive pieces of work to date. Bauer conducted two large-scale trials to assess the natural drying of wood chips in outdoor conditions in both summer and winter. The initial moisture content for the summer trial was 44 per cent. The initial level was brought down to about 25 per cent in a time period of 14 weeks. To achieve this, compacted piles of 5m high and 12 wide were setup in order to produce a steady-state temperature difference of 30K between the ambient air and inside the pile.

7.5.5 Secondary size reduction

The milling systems in pulverized coal fired boilers are normally capable of producing coal particles with a topsize of around 300μm with the average particle size being less than 100μm. Generally, most biomass materials are significantly more difficult to grind than coal because of their fibrous nature. However, most of the currently installed types of coal milling equipment, i.e. ball-race mills, roller mills, bowl mills, hammer mills, and tube ball mills, are capable of pulverising biomass fuel particles to a suitable size for pulverized fuel combustion provided that the co-firing percentage is low, generally less than around 10 per cent on a mass basis, without a major impact on the mill performance. Wet biomass materials can be co-milled, however this will have an impact on the mill heat balance, and this can be a limiting factor.

Boylan [13] reported on a programme of co-firing trials conducted at Hammond power plant in USA. The unit was equipped with vertical spindle, ball and race mills, in which initially 5–20 per cent (mass basis) of bark was fed. This resulted in the blockage of the mill feeder and coal feeder outlet hopper. Tree trimmings and yard waste were then tried. The biomass fuel was ground first and then mixed with coal at 3:1 ratio. The mixture was then processed through crushers and into coal bunkers. On average 11.5 per cent by mass of biomass was fed to the mills. It was found that the mill product for the blend was acceptable, but that the mill power consumption increased by 10–15 per cent.

Prinzing and his co-workers [14, 15, 16] studied the impact of co-firing at Shawville power plant in USA. Two units (unit 2 and unit 3) of the power plant were used to carryout the testwork. Unit 2 was a wall fired pulverized coal boiler with a rated capacity of 138MW. It was equipped with four ball-race mills and four rows of four low-NO_x burners. Unit 3 was a 190MW tangentially fired PF boiler equipped with four bowl mills and four rows of eight low-NO_x burners.

A variety of biomass fuels were tested, including sawdust, right-of-way trimmings and a short rotation coppice wood, at a co-milling ratio of 3 per cent, on a mass basis.. The woody crop material had long, stringy fibres and proved to be more problematic. A small but consistent increase in mill power was observed in both units when co-milling the biomass.

As reported by Boylan, the mill product fineness was not seriously affected by the biomass. In the case of the Unit 2 mills, the co-firing of the biomass resulted in a load degradation of around 8–10MW, associated with coal feeder limitations. With the Unit 3 mills, the average mill outlet temperatures decreased significantly and a load reduction of 15MW was required to maintain the mill temperatures.

An important Australian study [17, 18] funded by the CCSD was concerned with an investigation of the comminution of biomass/coal blends at a range of blending ratios on a mass basis. The aim was to gain fundamental insight into the milling behaviour of coal and biomass blends under conditions relevant to coal/biomass co-firing operations. The biomass blending ratio, biomass moisture content, biomass initial size, the types of biomass and of coal used, were investigated at both bench- and pilot-scales using a ball mill and a vertical spindle mill (Figure 7.15), respectively.

Three types of biomass were used, i.e. pine chips, hardwood chips and urban green waste, and these were blended with four Australian domestic coals, i.e. Acland, Collie, Lithgow, and Tarong. The impacts on pulverizer performance were analysed in terms of changes in: general operation of the mill, the product size distribution, the mill power requirement, the inlet and outlet temperatures, etc.

The results of the bench-scale ball mill testwork provided valuable fundamental data on the mechanisms responsible for the size reduction of the biomass particles. This was predominantly by abrasion fracture, rather than the brittle fracture mechanism responsible for the size reduction of the coal and associated mineral matter particles. The biomass fibres were subject to plastic deformation, i.e. they were squashed in the mill, and size reduction occured by the ripping apart of the squashed fibres by the abrasive action of the coal/mineral matter particles.

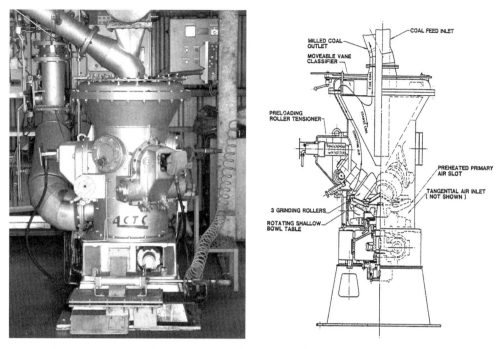

Figure 7.15 *The pilot-scale vertical spindle mill used in the CCSD study. Left: picture of the mill, right: cross-sectional view.*

Source: [17, 18].

Pilot-scale milling experiments, involved an investigation of the impacts of the biomass type, biomass blending ratio, biomass moisture content, biomass feed size and coal type. The impact of co-firing on pulverizer performance was analysed in terms of changes in the general operation of the mill, i.e.

- the feed and product size distributions,
- the mill power requirement,
- the abrasion wear of the grinding elements; and
- the mill inlet and outlet temperatures.

The results indicated that the Hardgrove Index (HGI) did not accurately predict the mill power requirement for high moisture coals and their blends (total moisture > 20 wt% a.r.). The Abrasive Index (AI) was found to be a reasonably good indicator for wear rates associated with milling the coals, but this was not the case for the coal-biomass blends. This is illustrated in Figure 7.16.

The blending of coal and biomass at 5 and 10wt% did not have a significant impact on the product size distribution when compared with pure coals (Figure 7.17). Blending reduced the amount below 75 μm (by 3.7wt% at 5wt% blending and 4.7wt% at 10wt% blending) and increased the amount above 300μm, by 3wt% at 5wt% blending and 4wt% at 10wt% blending.

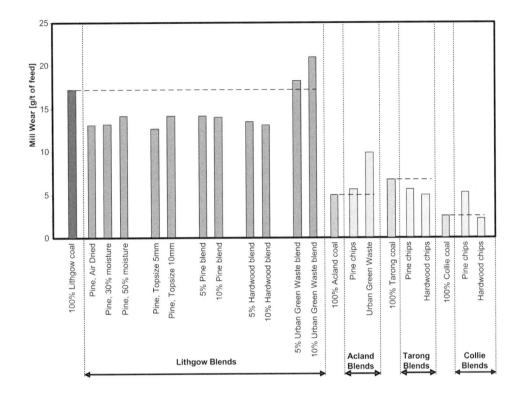

Figure 7.16 *Mill wear rate of pulverized samples*

Note: The wear rates for coals are consistent with their AI while those of blends are inconsistent with the AI.

Source: [17–18].

It was found that the introduction of biomass increased the mill power by as much as 20 per cent, even at a blending ratio around 5 per cent, as illustrated in Figure 7.20. The mill power profiles corresponding to 10wt% blending ratios of woody biomass indicated that steady-state operation of the mill was not possible. These results indicate that blending ratios as high as 10wt% may give rise to significant operational problems in vertical spindle mills. .

It was also found that the grinding pressure also had an influence on the measured mill power consumption, as would be expected. At 5wt% blending ratio the increase in power consumption was about 7 per cent to 20 per cent for low grinding pressure operation, and 20 per cent to 30 per cent for high grinding pressure.

The softwood (radiata pine) and hardwood (spotted gum and ironbark) woody biomass species studied exhibited similar characteristics in terms of their impact on mill power and product size distribution. The hardwood chips were slightly better in terms of the life of the rollers. There were mixed results for the softwood materials. The pine chips samples, for instance, were found to increase the wear rate of only the low wearing coals (e.g. Acland and Collie) while they did not have any notable impact on other coals. The Urban Green Waste generally causes smaller changes in the mill power compared with other biomass

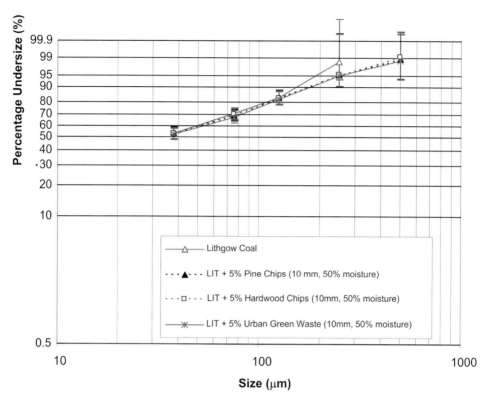

Figure 7.17 *Product size distributions for 5wt% blends of biomass and Lithgow coal*
Source: [17, 18].

fuels. There are concerns, however, about the tendency of the Urban Green Waste to increase the wear rates due to the dirt/soil and other tramp materials in this type of fuel.

It should also be noted that there may be mill safety issues when co-milling biomass with coal, associated specifically with the high reactivity of the biomass materials to combustion. It is commonly necessary to modify the power plant mill operating procedures when co-milling biomass materials, even at relatively low co-milling ratios.

7.6 Operational and environmental Issues

7.6.1 General fuel characteristics of coals and biomass

Biomass fuels used for co-firing may include wood, grassy and straw materials, agricultural residues and energy crops. The biomass fuel properties differ significantly from those of coal in a number of important respects, and show significant variability, viz:

- biomass moisture and ash contents can vary significantly;
- biomass materials have much higher volatile matter contents than coals;

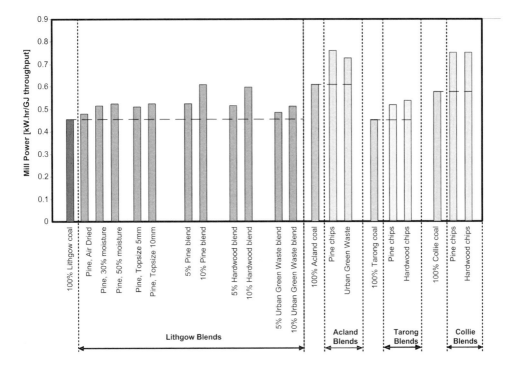

Figure 7.18 *Mill power consumption during size reduction*

Source: [17, 18].

- fuel nitrogen, sulphur and chlorine contents are generally lower than for coals, but can vary significantly;
- biomass materials have high oxygen contents compared to most coals, and hence have relatively low heating values;
- biomass materials have low bulk densities, and hence have very low energy densities compared to coals; and
- the ash chemistries of biomass materials are very different from those of most coals.

A number of these parameters are illustrated in Figures 7.19 and 7.20. Each of these properties will have an impact on the design, operation, and performance of biomass co-firing systems, and on the downstream impacts on the performance of the coal boiler and the associated equipment.

7.6.2 Particle size and residence time

All biomass materials have high high volatile matter contents and are significantly more reactive to combustion than coals. It is not usually necessary, therefore, to reduce the particle sizes to the levels required for pulverized coals to ensure an acceptable level of combustion efficiency. The particle size requirements for a particular biomass fuel have to

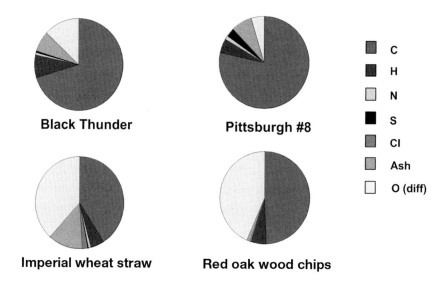

Figure 7.19 *Typical ultimate analyses of biomass and coal*

Note: All of the components are presented clockwise, starting from the top with C.
Source: Courtesy Larry Baxter, US.

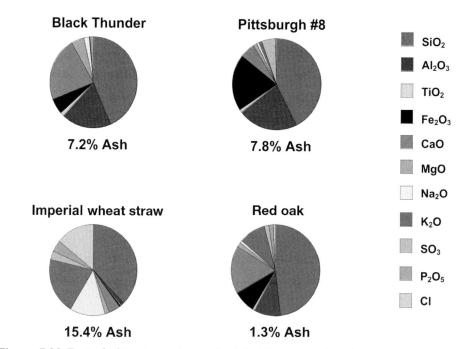

Figure 7.20 *Typical ultimate analyses of ash in biomass and coal*

Note: All of the components are presented clockwise, starting from the top with SiO_2.
Source: Courtesy Larry Baxter, US.

be assessed on a case-by-case basis, and depend principally on the physical characteristics of the biomass particles, i.e. on the particle density and shape, on the moisture content, and on the details of the combustion system.

The moisture content is important because the requirement to dry the particles in the flame before they will heat up and ignite, can delay the ignition process. This can have a negative impact on the ignition and combustion processes.

The impact of the particle size and density is essentially aerodynamic, i.e. very large, denser biomass particles will have a tendency to fall out of the flame under the influence of gravitational forces and will report to the furnace ash hopper as unburned material. Smaller denser particles may remain in the flame and the flue gas stream, as normal but may not burn out completely if the residence time in the flame is insufficient. In this case, the biomass char particles may report as unburned material in the boiler fly ashes.

The general experience with relatively dry biomass materials, less than 10 per cent moisture or so, has been that, provided the mill product is acceptable, i.e. that there are no very large biomass particles, passing to the burners, then the combustion behaviour of the blended fuel has been acceptable. Biomass materials are more reactive in combustion systems than are most coals and in general the unburned carbon levels in bottom and fly ashes are similar to, or less than, those that apply when firing coal alone.

7.6.3 Boiler efficiency

The impact of co-firing on the thermal efficiency of the boiler depends principally on the biomass moisture content and the co-firing ratio. Values reported in the literature [4, 19, 20] show that at a co-firing ratio of 3–5 per cent on a mass basis, there is only a very modest loss of boiler efficiency.

The thermal efficiency of a pulverized coal boiler is normally considered in terms of the efficiency losses, and these are associated principally with the hot gas exhausted to the atmosphere from the chimney, and the unburned fuel in the ash discards. Equation 7.1, derived from boiler heat balance data, has been proposed by Battista et al [20] for evaluation of the efficiency impact of co-firing:

$$\begin{aligned}\text{Actual boiler efficiency} = &\text{ Theoretical thermal efficiency} \\ &- A \times (\% \ O_2 \text{ at economizer exit}) \\ &- B \times (\% \text{ unburned carbon in fly ash}) \\ &- C \times (\% \text{ biomass co-fired on mass basis})\end{aligned} \qquad (7.1)$$

where A, B and C are constants.

The co-firing, by pre-mixing and co-milling, of biomass materials, and particularly of wet biomass, can have an impact on the maximum achievable boiler load, depending on the mill constraints, and on the boiler efficiency. At low biomass co-firing ratios and with dry (<10 per cent moisture content) biomass materials these constraints and the impacts on boiler efficiency are modest.

7.6.4 Rate of combustion and char burnout

The rate at which fuel particles burn in a suspension flamedepend on a number of key parameters which have been discussed in some details by Kanury [21]. The most important step in determining the rate of the solid fuel combustion process is the char combustion. When devolatilization of the raw fuel particles is completed, the char and ash particles remain. Biomass materials produce relatively low quantities of char and the chars from biomass materials are highly porous and hence are very reactive in combustion processes. In general, therefore, the char burnout process is relatively rapid compared to that of coal particles of similar size.

A series of coal/biomass co-firing combustion experiments was carried out under pilot-scale conditions in a 150kW Boiler Simulation Furnace in the ACIRL facility, Australia [18]. Four types of Australian coal and three types of biomass fuels were used in the form of chips (pine, eucalyptus and urban green waste). The objective of the study was to investigate the impacts of: coal and biomass types, biomass blending ratios, biomass moisture content and biomass feed size, on the combustion process, and hence on boiler operation.

The char burnout was quantified by taking fly ash samples at the furnace exit during each test run and measuring the quantity of unburned carbon in the collected ash. Figure 7.21 shows the comparison of the measured burnout efficiency against the biomass blending ratio.

Overall, the results indicated that the impact of a 5–10wt% biomass blending ratio on coal combustion can be considered as being negligible. At higher blending ratios, a greater impact was observed.

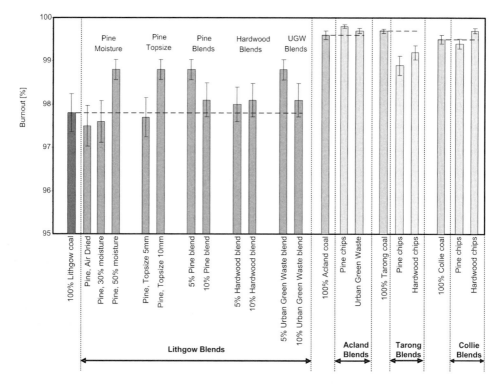

Figure 7.21 *Char burnout from CCSD pilot-scale co-firing studies*

Note: Coal/biomass blend properties are 5wt% biomass with 10mm biomass top size and 50% biomass moisture level (unless otherwise specified).

7.6.5 Flame Stability

In pulverized coal flames, stability is achieved largely by aerodynamic means. A recirculation zone is created in the near-burner region in order to increase the gas temperatures, promote mixing of the fuel and air, prolong particle residence times and engender stable ignition. The major part of volatile matter release occurs within this zone. The energy release required to ensure stabilization is affected by:

- the influence of the burner geometry and operating conditions on the near field aerodynamics; and
- the quality of the fuel.

The importance of burner aerodynamics on pulverized coal flame stability has been treated extensively in the literature.

When biomass is co-fired with coal additional factors affect the flame stabilization processes. These are:

- the injection mode of the biomass;
- the fuel particle characteristics; and
- the blending ratio.

Provided that the biomass particles are not too wet and the biomass particle size is acceptable, the stability limits are generally broadened with increasing biomass blending ratio. This is due to much higher volatile matter content of biomass.

7.6.6 System integration and control issues

For biomass co-firing by pre-mixing and co-milling, the mill and boiler control issues are relatively unaffected, provided that the co-firing ratio is low enough not to have a major impact on the mill operation and performance.

In direct injection co-firing retrofit projects, one of the key issues is that there should be minimum interference with the normal operation of the coal-fired boiler, and this is particularly the case as regards the boiler and combustion control systems. As far as is possible, the emphasis is on simplicity of control of the biomass firing system, consistent with safe and operationally robust approach. The unit and the mill groups involved in co-firing are normally started up and shut down on coal-firing alone. The biomass burner 'permit to fire' system will interface with the coal mill control logic and the boiler tripping system, and all the existing coal interlocks must be satisfied to permit biomass firing. The system will require that the biomass injection can only occur into stable coal flames in a hot furnace, with the appropriate level of coal combustion.

In the case of a new, purpose-built boiler designed for biomass co-firing, this will not be such an important issue, since both the mechanical and the control requirements of the co-firing system can be built into the boiler and combustion systems from the start. Optimization of the system may be possible in a way that is not available for retrofit projects.

7.6.7 Ash deposition

The ash-related issues associated with biomass co-firing with coal are described in some detail in Chapter 8. Ash deposition rates from biomass fuels could be quite different from those of coals (see Figure 7.22). When normalized for ash content, the ash deposition propensities of some herbaceous materials can exceed those of coal by large margins, whereas those for low ash wood materials tend to be lower.

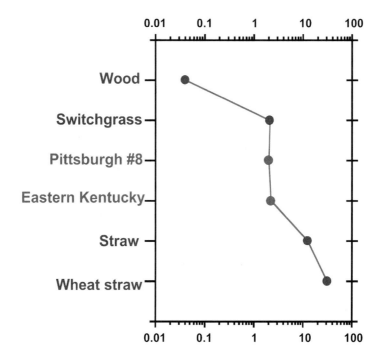

Figure 7.22 *Ash deposition rate for various fuels in g deposit per kg fuel*
Source: Courtesy of Larry Baxter, US.

The following general trends have been observed:

- Ash deposition rates should not be affected significantly when co-firing wood or similar low-ash, low-alkali, low-chlorine fuels; even at elevated co-firing ratios.
- Ash deposition rates should increase when co-firing high-chlorine, high-alkali, high-ash fuels such as many herbaceous and agricultural residue materials.
- Deposition rates depend strongly on both individual fuel properties and interactions between the co-fired fuels.

Deposition rates from blends of coal and biomass can vary strongly with coal and biomass ash properties. More information on the fundamental mechanisms behind ash deposition and related processes can be found in Chapter 8.

7.6.8 Gas-side corrosion of boiler components

The co-firing of biomass materials with coal in large coal-fired boilers can introduce significant changes to the chemistry of the ash materials deposited on boiler surfaces, and this can have a significant impact on the chemical reactions that occur at the metal/oxide/ash deposit interface. This issue is discussed in more detail in Chapter 8.

Two important changes to the deposit chemistry are relevant in this context, viz [27, 28]:

- the introduction of increased levels of alkali metal species into the vapour phase and their subsequent condensation on boiler surfaces can lead to enrichment of particularly potassium compounds at the metal/oxide/ash deposit interface; and
- some biomass materials, in particular the herbaceous or intensely cultivated crops, have very low sulphur contents and can have relatively high chlorine contents. The release of significant HCl concentrations into the boiler flue gases can lead to enrichment of chloride at the metal/oxide/ash deposit interface.

Experimental investigations with a wide range of fuels at pilot-scale indicate that there are a number of fuel-based parameters that indicate when chlorine concentrations exceed the capacity of sulphation mechanisms to control surface chlorine, and hence the corrosion rates, see Figure 7.23). These data are consistent with the limited commercial-scale data available. Generally, the molar ratio of sulphur to available alkali and chlorine should exceed 5, with greater values providing decreased corrosion potential. Available alkali is the fraction of total alkali that is either water-soluble or ion-exchangeable. This fraction is generally close to the total alkali for biomass fuels that do not contain soil or other extraneous sources of the alkali metals.

The sulphur to alkali/chloride ratios should generally be based on fuel compositions through individual burners, not total fuel composition. Most pulverized coal boilers do a relatively poor job of mixing flue gases, and the downstream gas flows are distinctly striated in nature. In cases in which biomass is introduced non-uniformly across the burners, local flue gas compositions in the convection pass can be determined by individual burner composition, not the overall fuel composition.

7.6.9 Impact on emissions

The gaseous and gas-borne emissions of primary concern from the co-combustion of biomass and coal in pulverized fuel firing systems are: SO_x, NO_x; CO, particulate (fly ash); acid halides (HBr, HCl, HF, etc); the organic compounds, such as the VOCs, PAHs, chlorinated dioxins (PCDD) and furans (PCDF); and trace metals.

Because of the low sulphur contents of most biomass fuels (typically less than 0.5 per cent on a dry basis), SO_x emissions generally decrease in line with the co-firing ratio (see Figure 7.24). An additional incremental reduction beyond the amount anticipated on the basis of fuel sulfur content is sometimes observed. This is due to retention of sulfur in coal by alkali based compounds in biomass ashes.

The nitrogen contents of most biomass materials are less than 1 per cent, with many less than 0.5 per cent, on a dry basis. Therefore, combustion of most biomass materials can be expected to result in lower NO_x emissions. However, experimental characterization of NO_x emissions during combustion of coal, biomass, and various blends of the two fuels, particularly at low co-firing ratios, has resulted in NO_x emissions from biomass-coal blends both greater and less than those from coal alone.

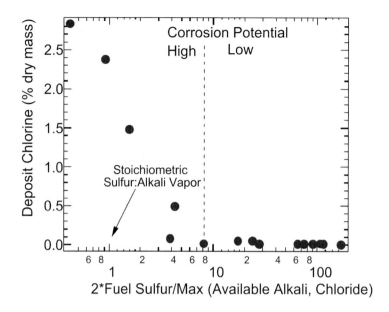

Figure 7.23 *The molar ratio of sulphur to available alkali and chlorine is an indicator of the chlorine corrosion potential*

Source: Courtesy of Larry Baxter, US.

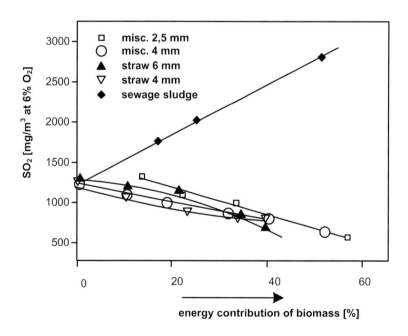

Figure 7.24 *Influence of co-firing various biofuels on SO_2 emissions*

Source: [29].

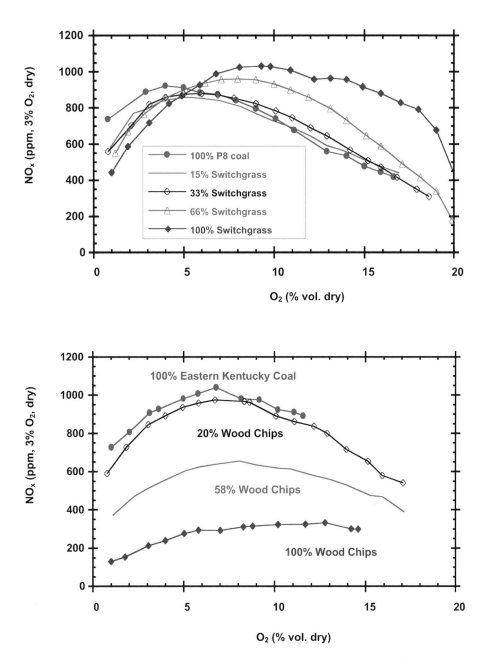

Figure 7.25 *Effect on NO$_x$ emissions when co-firing switchgrass (top) and wood (bottom) with coal*

Note: NO$_x$ emissions can increase or decrease when co-firing biomass. Fuel nitrogen content: wood = 77g/GJ, switchgrass = 33lg/GJ; coal = 430–516g/GJ.

Source: Courtesy of Larry Baxter, US.

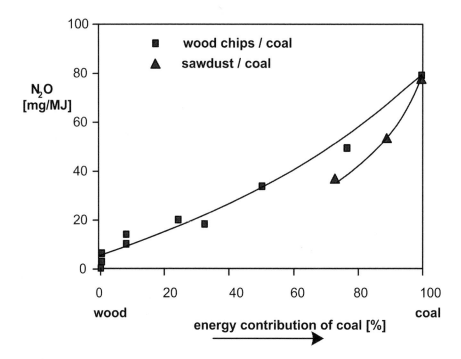

Figure 7.26 *Influence of mixed combustion of wood and coal on N_2O emissions in a circulating fluidized bed boiler*

Source: [29].

Biomass fuels produce greater volatile yields than coals and, hence, they can create larger fuel rich regions than coal in the near-burner region. Biomass fuels are, therefore, expected to enhance the performance of low NO_x burners. Biomass may also have some potential as reburn fuels for NO_x reduction from the coal combustion, which gives a further potential for significant decrease of the NO_x emissions.

In addition to the NO and NO_2, N_2O emissions can also be significantly reduced by co-firing of biomass in coal-fired fluidized bed boilers (Figure 7.26).

The available data on particulate emissions are contradictory. Some plant data indicate a sharp increase in particulate emission when co-firing biomass, whereas some show no significant change. This will be a function of the performance of the installed particulate collection equipment, i.e. will be highly site-specific,

Some biomass materials have relatively high chlorine contents, and the effect of co-firing can be an increase in the total chlorine input to the boiler, and a net increase in the uncontrolled HCl emissions levels.

The emission levels of CO and the organic pollutant species are dependent largely on the quality of the combustion process. Provided that the co-firing of the biomass does not have a significant impact on this, i.e. the particle size and moisture content of the biomass are acceptable, the method of introduction of the biomass is appropriate and designed properly, and the system is operated properly, then the emission levels of these species should be no higher than for coal firing.

It should also be noted that the measurement of the CO emission levels is a good indicator of the quality of the combustion, and provides a reasonably good surrogate for the organic species. CO emission measurement is cheaper and simpler than measurement of the more complex organic species.

7.6.10 The performance of NO_x and SO_x emissions abatement equipment

The great majority of the large coal-fired power boilers have installed primary and/or secondary NO_x emission abatement equipment. These can be classified under four categories, viz:

- low NO_x burners;
- advanced primary NO_x reduction techniques, such as Two-Stage Combustion (TSC) and gas or coal reburn technologies;
- selective Non-Catalytic NO_x Reduction (SNCR) techniques; and
- selective Catalytic NO_x (SCR) techniques.

The co-firing of biomass materials is likely to have little or no impact on the performance of low NO_x coal burners, provided that the combustion equipment has been purpose-designed for co-firing or has been modified appropriately. The only significant risk area is concerned with the possibility of increased ash deposition on the burner quarls due to co-firing, which may interfere with burner operation. This will be dependent on the co-firing ratio, the fusion behaviour of the mixed coal and biomass ashes, and on a number of site-specific factors. In general terms, operation at co-firing ratios less than 10 per cent on a heat input basis is not likely to cause significant problems in this regard.

The advanced primary NO_x emission control techniques, and in particular the deeply staged TSC techniques, involve operation with a significant portion of the furnace volume under reducing conditions. Plant experience with the application of these techniques in both retrofit and new projects has indicated that there are significant risks of increased slag deposition and of accelerated corrosion of furnace wall tubes. The co-firing of biomass materials may act to increase these problems due to the relatively low fusion temperatures of biomass ashes. The risks of furnace wall corrosion may be increased by the co-firing of those biomass materials which have relatively high chlorine contents. The risks of operational problems due to increased ash deposition and furnace wall corrosion will be highly site-specific, and will depend on the co-firing ratio, the fusion behaviour of the mixed coal-biomass ashes, and the chlorine content of the biomass. This is a significant risk area, which will need to be examined carefully on a case-by-case basis.

The co-firing of biomass materials is not likely to have any significant impact on the design and operation of SNCR systems, but may interfere significantly with the performance of SCR systems. This issue will become increasingly important as increased numbers of boilers install SCR system to comply with lower NO_x emission limits. The performance of SCR systems is strongly dependent on the catalyst activity, and on the replacement rate of catalyst material necessary to maintain acceptable NO_x emission levels. SCR catalysts are susceptible to 'poisoning' due to the condensation of volatile inorganic species on the catalyst surfaces. The co-firing of biomass materials will generally lead to an increase in the concentration of these species, and particularly of the alkali metals, and in some cases phosphates, in the flue gases.

There is a significant risk, therefore, that the rate of SCR catalyst poisoning will increase, and that it will be necessary to increase the rate of replacement of catalyst material in order to maintain the required NO_x emission levels, or to limit the co-firing ratios for particular biomass materials. This will add significantly to the costs of operation of the SCR system, and/or limit the fuel flexibility of the system. This is a subject of current interest within the industry, and one of the future challenges to the catalyst suppliers will be to develop catalyst systems, that are more tolerant to poisoning by alkali metal and phosphate condensation. This is a significant area of uncertainty, which will require further development. See Section 8.5.3 for more information.

The conventional approach to the control of SO_x emissions has been the installation of wet limestone-gypsum flue gas desulphurization (FGD) equipment. As has been stated above, most biomass materials have relatively low sulphur contents, and biomass co-firing will, in general, result in a reduction of the SO_x concentration at the inlet to the FGD plant, and reductions in both the limestone requirement and the gypsum production rate, depending on the co-firing ratio and the sulphur content of the biomass.

Some biomass materials have relatively high chlorine contents and there may be an impact on the design and the performance of the FGD plant, depending on the co-firing ratio and the chlorine contents of the coal and the biomass. HCl is very reactive to the limestone slurry in an FGD plant and will be absorbed very efficiently. It may be necessary to modify the design of the FGD system and the wastewater treatment plant in situations where the total chlorine burden in the system is high.

7.6.11 The efficiency of particulate emissions abatement equipment

The characteristics of the solid products of the combustion of biomass materials are very different from those resulting from the combustion of coal. The chemistries of the ashes are very different, as are the physical characteristics. The nature of the inorganic material in most biomass materials is such that a significant amount of submicron inorganic fume and vapour material can be generated in the flame.

The co-firing of biomass materials with coal will generally lead to a reduction in the total fly-ash dust burden, due to the lower ash content of the biomass. However, the mixed ash may contain significant levels of very fine aerosol material, which may present problems to conventional particulate emissions abatement equipment. When electrostatic precipitators are employed, the result may be lower collection efficiencies and an increase in the particulate emissions level from the chimney, compared to that measured when firing coal alone. This effect is likely to be highly site-specific and will be of particular interest in retrofit projects, where the existing electrostatic precipitators have been designed for coal-firing alone. When fabric filters are employed for particulate collection, there is a tendency for the very fine aerosol/fume material to blind the fabric, resulting in cleaning difficulties and an increased pressure drop across the system.

The experience to date in the great majority of the European demonstration and commercial projects, and the straw/coal co-firing projects in Denmark, has been that compliance with the relevant particulate emissions consent limits, normally 20–50 mg/Nm^3, dry at 6 per cent O_2, has not presented any significant difficulties. This issue, however, is likely to be highly site-specific. There are reports from North America and Australia, for instance, of incidences of increased particulate emissions in biomass co-firing retrofit projects.

7.6.12 Fly-ash utilization

More than one third of the fly ashes generated in US coal power stations, and up to 100 per cent of the fly ash in many other countries, is utilized as feed materials in secondary markets. One of the large, high-value markets is the market for concrete additives. Some current standards for the use of fly ash as a concrete additive preclude using fly ash from any source other than coal. This may pose a significant financial barrier to co-firing for some plants. This issue has both technical and regulatory aspects.

The technical case for precluding fly ash from co-firing wood with coal as a concrete additive appears to be unjustified. The results of laboratory tests of concrete made from cement containing commercially produced fly ash and laboratory fly ash indicate that there are no short- or long-term deleterious effects on the properties of the concrete when using wood-based co-fired ash. Less information is available for herbaceous biomass fuels. The existing data and theoretical considerations suggest that alkali, chorine, and other properties may compromise several important concrete properties. Investigation of this issue is ongoing in a number of countries.

The regulatory issues are associated with the specific wording of national and international standards that, while not actual regulations, are the basis for regulations and policy for many institutions and markets. Strict interpretation of all national and international standards that address this issue could preclude all fly ash from use in concrete if it contains any amount of non-coal-derived material. This would include co-fired fly ash as well as fly ashes from dedicated coal systems that involve other clean-up processes (scrubbers, sulphur-injected precipitators, etc.). These standards are under active revision, but such revisions often take many years to complete. There are also many markets/ customers that do not explicitly require adherence to these standards.

In Europe, for instance, there have been major changes to the specification of pfa/fly ash for utilization as a component of concrete, included in the recent revisions to EN450 (2005). New rules for the ashes from the co-combustion of materials other than coal have been added. These new rules include restriction of the allowable co-firing ratio, and the standard requires that suppliers apply additional test methods to ensure that the ash from co-combustion behaves similarly to coal ash. These tests include:

- reactive silica;
- total oxides;
- alkali content;
- magnesium content;
- soluble phosphate; and
- initial setting time.

Looking in more detail at some of the technical issues, Figure 7.27 illustrates the impact of fly ash addition on the required amount of aerating agent to establish ASTM-compliant air entrainment levels in concrete. Air entrainment in concrete is essential to prevent failure during freeze-thaw cycles. As is apparent from the data, the quantity of aerating agent required increases with increasing herbaceous biomass content. This dependence arises from the effect of water-soluble components (higher in herbaceous biomass than in coal fly ash) in tying up the aerating agent (generally surfactants), and preventing them from forming films that support bubble growth. The data indicate that if fly ashes from co-fired units were treated the same way as fly ashes from coal, the resulting concrete would likely

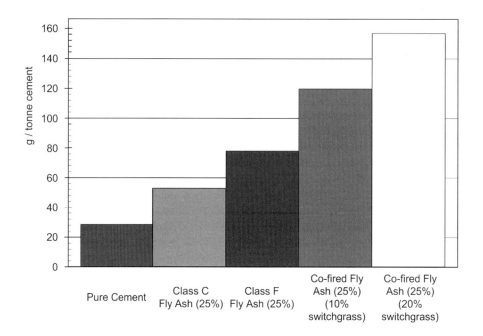

Figure 7.27 *Amount of aerating agent required to generate air entrainment within ASTM specifications for a variety of fly-ash compositions*

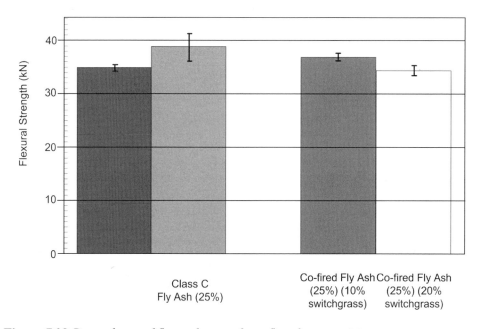

Figure 7.28 *Dependence of flexural strength on fly ash composition*

fail under freeze thaw cycles. Increasing the surfactant to an acceptable level is not expensive, but the failure to recognize the need to adjust it could have a significant impact on the performance of the product.

Figure 7.28 illustrates the impact of biomass-coal commingled fly ash on flexural strength. In these test little significant difference is seen among the various samples. Additional data on set time and compressive strength indicate that all fly ashes delayed set time by 2–4 hours compared to concrete made from cement only. The biomass-containing fly ash did not delay set times significantly more than the non-biomass containing fly ash. Early compressive strength (in the first month or so) was compromised by all fly ashes, again with the biomass-containing fly ash similar to coal fly ashes. However, late strength (longer than 2 months or so) was enhanced by the presence of all fly ashes.

In conclusion, there appear to be only modest impacts of biomass-containing fly ash on concrete properties based on these preliminary data, with the increased aerating agent requirement being an example of one issue that requires further monitoring. Otherwise, biomass-containing fly ash behaves qualitatively similar to coal fly ash with no biomass in terms of structural and performance properties when incorporated into concrete.

More information on fly-ash utilization can be found in Section 8.7.

7.7 References

1   BAXTER, L. and KOPPEJAN, J. (2004) 'Co-combustion of Biomass and Coal', *Euroheat and Power*, vol 1, pp34–39
2   SPLIETHOFF, H., SIEGLE, V. and HEIN, K. (1996) 'Zufeuerung von Biomasse in Kohlekraftwerken: Auswirkungen auf Betrieb, Emissionen und Rückstände, Technik und Kosten', in Th. Nussbaumer (ed.), *Feuerungstechnik, Ascheverwertung und Wärme-Kraft-Kopplung*, ENET, Bundesamt für Energie
3   PENNINKS, F. W. M. (2000) 'Coal and wood fuel for electricity production', *Proceedings of the EU Seminar on the Use of Coal in mixture with Wastes and Residues II*, Cottbus
4   MOGHTADERI, B. and DOROODCHI, E. (1999) *Co-firing Biomass with Coal at Mt Piper and Wallerawang Power Stations*, report prepared for Delta Electricity
5   RASMUSSEN, I. and OVERGAARD, P. (1996) 'General overview over recent results and plans concerning co-combustion of biomass and coal', in Chartier, P., Ferrero, G. L., Henius, U. M., Hultberg, S., Sachau, J. and Wijnblad, M. (eds.), *Biomass for Energy and the Environment: Proceedings of the 9th European Bioenergy Conference,* 24–27 June, Copenhagen, vol 1, Pergamon, Oxford, pp158–163
6   OVERGAARD, P., SANDER, B., JUNKER, H., FRIBORG, K. and LARSEN, O. H. (2004) *Two Years' Operational Experience and Further Development of Full-scale Co-firing of Straw*, Elsam Engineering A/S
7   MORY, A. and TAUSCHITZ, J. (2000) 'BIOCOCOMB – gasification of biomass and co-combustion of the gas in a PF boiler in Zeltweg power plant', *Proceedings of the EU Seminar on the Use of Coal in Mixture with Wastes and Residues II*, Cottbus
8   WILLEBOER, W. (2000) 'Amergas biomass gasifier starting operation', *Proceedings of the EU Seminar on the Use of Coal in Mixture with Wastes and Residues II*, Cottbus
9   OTTOSEN, P. (2000) 'Coal and biomass (multifuel concept for Avedore Unit 2)', *Proceedings of the EU Seminar on the Use of Coal in Mixture with Wastes and Residues II*, Cottbus
10   IEA (2006) IEA Bioenergy Task 32 database of biomass cofiring initiatives, available at www.ieabcc.nl/database/cofiring.html
11   ZULFIQUAR, M. H., MOGHTADERI, B. and WALL, T. F. (2006) 'Flow properties of biomass and coal blends', *Journal of Fuel Processing Technology*, vol 87, pp281–288
12   BAUER, F. (1995) 'Combined combustion of biomass/sewage sludge and coals of high and low

rank in different systems of semi-industrial and industrial scale', in Bemtgen, Hein and Minchener (eds.), *APAS Clean Coal Technology Program 1992–94, Vol. II, Combined Combustion of Biomass/Sewage Sludge and Coals, Final Report*, University of Stuttgart, Institute for Process Engineering, Germany, Paper B6

13 BOYLAN, D. M. (1996) 'Southern company tests of wood/coal co-firing in pulverised coal units', *Biomass and Bioenergy*, vol 10, no 2/3, pp139–147
14 HUNT, F. E., PRINZING, D. E., BATTISTA, J. J. and HUGHES, E. (1996) 'The Shawville coal/biomass cofiring test: A coal/power industry cooperative test of direct fossil fuel mitigation', *Energy Conversion and Management*, vol 38 (Suppl), ppS551–S556
15 PRINZING, D. E., HUNT, F. E. and BATTISTA, J. J. (1996) 'Co-firing biomass with coal at Shawville', *Proceedings of Bioenergy '96*, pp121–128
16 PRINZING, D. E. and HUNT, F. E. (1998) 'Impacts of wood cofiring on coal pulverization at Shawville Generating Station', *Fuel Processing Technology*, vol 54, no 1/3, pp143–157
17 ZULFIQUAR, M. H., MOGHTADERI, B. and WALL, T. F. (2006) 'Co-milling of coal and biomass in a pilot-scale vertical spindle mill', *Technology Assessment Report 49, CRC for Coal in Sustainable Development*, March, QCAT Technology Transfer Centre, Pullenvale, Queensland, Australia
18 ZULFIQUAR, M. H. (2006) A fundamental study on pilot-scale characteristics of coal and biomass blends for co-firing applications, PhD thesis, University of Newcastle, Australia
19 ZULFIQUAR, M. H., MOGHTADERI, B., and WALL, T. F. 'Co-firing of coal and biomass in 150kW pilot-scale boiler simulation furnace', *Technology Assessment Report, CRC for Coal in Sustainable Development*, QCAT Technology Transfer Centre, Pullenvale, Queensland, Australia
20 BATTISTA, J., TILLMAN, D. A. and HUGHES, E. E. (1998) 'Cofiring wood waste with coal in a wall-fired boiler: Initiating a 3-year demonstration program', *Proceedings of BioEnergy '98*, pp243–251
21 KANURY, A. M. (1994) 'Combustion characteristics of biomass fuels', *Combustion Science and Technology*, vol 97, pp469–491
22 MOGHTADERI, B. (1999) 'Pyrolysis of coal/biomass blends', *Proceedings of the 1999 Australian Symposium on Combustion and the Sixth Australian Flame Days*, Newcastle, Australia, pp91–96.
23 AIMAN, S. and STUBINGTON, J. F. (1993) 'The pyrolysis kinetics of bagasse at low heating rates', *Biomass and Bioenergy*, vol 5, no 2, pp113–120
24 STUBINGTON, J. F. and FENTON, H. (1984) 'Combustion characteristics of dried and pelletized bagasse', *Combustion Science and Technology*, vol. 37, pp285–299
25 STUBINGTON, J. F. and AIMAN, S., (1994) 'Pyrolysis kinetics of bagasse at high heating rates', *Energy and Fuels*, vol 8, pp194–203
26 ABBAS, T., COSTEN, P., GLASER, K., KANDAMBY, N. H., LOCKWOOD, F. C. and OU, J. J. (1994) 'The influence of burner injection mode on pulverized coal and biomass co-fired flames', *Combustion and Flames*, vol 99, pp617–625
27 HENRIKSEN, N. and LARSEN, O. H. (1997) 'Corrosion in ultrasupercritical boilers for straw combustion', *Materials at High Temperature*, vol 14, pp227–236
28 HANSEN, P. F. B., ANDERSEN, K. H., WIECK-HANSEN, K., OVERGAARD, P., RASMUSSEN, I., FRANDSEN, F. J., HANSEN, L. A. and DAM-JOHANSEN, K. (1998) 'Co-firing straw in a 150MWe utility boiler: In-situ measurements', *Fuel Processing Technology* vol 54, pp207–225
29 LECKNER, B. and KARLSSON, M. (1992) *Emissions from Combustion of Wood in a Circulating Fluidised Bed Boiler*, report A 92-200, Department of Energy Conversion, Chalmers University of Technology, Göteborg

8

Biomass Ash Characteristics and Behaviour in Combustion Systems

8.1 Introduction

As discussed in Chapter 2, in common with most solid and liquid fuels, biomass materials have significant levels of inorganic matter as impurities, and many of the practical problems encountered with the combustion of biomass materials, or the co-combustion of biomass materials with coal and other fossil fuels, are associated with the nature and behaviour of the biomass ash and the other inorganic constituents.

In practical terms, the ash-related problems in biomass combustors and boilers, and in plants co-firing biomass with more conventional fossil fuels, have commonly been associated with:

- the formation of fused or partly fused ash agglomerates and slag deposits at high temperatures within furnaces;
- the formation of bonded ash deposits at lower gas temperatures on the heat exchange surfaces in the boiler convective sections, and elsewhere;
- the accelerated metal wastage of boiler components due to gas-side corrosion and erosion;
- the formation and emission of sub-micron aerosols and fumes; and
- the handling and the utilization/disposal of ash residues from biomass combustion plants, and of the mixed ash residues from the co-firing of biomass in coal-fired boilers.

In very general terms, the nature of the problems and the impact on plant performance depend both on the characteristics of the biomass fuel, i.e. principally on the ash content and the ash chemistry, and on the design and operation of the combustion equipment and the boiler.

There has been a good deal of technical research and development work on the nature and behaviour of biomass ashes and biomass–coal ash mixtures over the past 10–20 years, both at laboratory scale and in combustion test facilities. There is also increasing operating experience of the combustion and co-combustion of biomass materials, and of the behaviour of biomass ashes and biomass–coal ash mixtures. This experience is relevant to all scales of operation, i.e. in small domestic appliances, in industrial and commercial combustion equipment and boiler plants and in large utility boiler plants.

An attempt is made in this chapter to present a short summary of the key findings of the most relevant biomass ash-related development work and of the practical experience of the impact of ash-related problems on the operation of biomass combustion systems and boiler plants.

8.2 Biomass ash characteristics

8.2.1 Introduction

There is extensive technical literature on the nature and behaviour of the inorganic constituents of solid fuels going back more than a century. This work has been concerned principally with the characteristics and behaviour of the ashes produced by the combustion of coals, lignites and peats, which have been most widely utilized as both domestic and industrial solid fuels. The more recent research and development activities on biomass ash materials have benefited from this body of work, principally from some of the ash characterization techniques, from the use of the combustion and other test facilities, developed over the years for the study of fossil fuels, and from the detailed understanding of the reactions undergone by the inorganic species in solid fuels in combustion and other high temperature processes.

In very general terms, the inorganic materials in most solid fuels, including biomass, can be divided into two broad categories:

1. The **inherent inorganic material**, which exists as part of the organic structure of the fuel, and is most commonly associated with oxygen-, sulphur- and nitrogen-containing functional groups within the organic structure. These functional groups can provide suitable sites for the inorganic species to be associated chemically in the form of cations or chelates. Biomass materials tend to be relatively rich in oxygen-containing sites, and a significant fraction of the inorganic material in most biomass fuels is in this form. It is also possible for certain inorganic species to be present in very fine particulate form within the organic structure of some of the fuels, and to behave essentially as an inherent component of the fuel.
2. The **extraneous inorganic material**, which has been added to the fuel through geological processes, or during harvesting, handling and processing of the fuel. Biomass fuels, for instance, are commonly contaminated with soil and other materials, which have become mixed with the fuel during collection, handling or storage.

A listing of the major inorganic species found in the higher plants, and hence in the inherent inorganic material in most biomass materials, is presented in Table 8.1. The data in this table provide a rough quantitative overview of the major element speciation in biomass, in three useful categories:

1. water soluble, i.e. in free ionic form;
2. organically associated; and
3. precipitated as relatively pure compounds, in crystalline or amorphous forms.

It is clear from the data presented in Table 8.1 that much of the inherent inorganic material in biomass is in the form of simple inorganic salts, and principally as the oxides and hydroxides of silicon, and the nitrates, sulphates, chlorides, phosphates and oxalates of the alkali and alkaline earth metals. It is also clear from these data that a significant fraction of the inorganic material, and particularly of the alkali metals, is in water-soluble form.

The extraneous inorganic material can be in many forms. In most cases, however, it takes the form of contamination with sand, soil or other mineral materials, tramp metal components, etc.

Table 8.1 *Speciation of inorganic materials in higher plants, according to the classes 'water soluble', 'organically associated' and 'precipitated'*

Element	Compound	Formula	Share of total element
Class 1 – water soluble (free ionic form)			
Na	Sodium nitrate, chloride	$NaNO_3$, $NaCl$	>90%
K	Potassium nitrate, chloride	KNO_3, KCl	>90%
Ca	Calcium nitrate, chloride, phosphate	$Ca(NO_3)_2$, $CaCl_2$, $Ca_3(PO_4)_2$	20–60%
Mg	Magnesium nitrate, chloride, phosphate	$Mg(NO_3)_2$, $MgCl_2$, $Mg_3(PO_4)_2$	60–90%
Si	Silicon hydroxide	$Si(OH)_4$	<5%
S	Sulphate ion	SO_4^{2-}	>90%*
P	Phosphate ion	PO_4^{3-}	>80%*
Cl	Chloride ion	Cl^-	>90%*
Class 2 – organically associated (covalent or ionic bonding with tissue)			
Ca	Calcium pectate	macromolecule	0.8–2.6%
Mg	Chlorophyll, magnesium pectate	$C_{55}H_{72}MgN_4O_5$, macromolecule–	8–35%
Mn	Various organic structures	Mn^{2+}, Mn^{3+}, Mn^{4+}	>90%*
Fe	Organic complex, organic sulphates	Fe^{3+}, Fe^{2+}	>80%*
S	Sulpholipids, amino acids, proteins	SO_4^{2-}, S	–
P	Nucleic acids	PO_4^{3-}	–
Class 3 – precipitated (pure compound, amorphous or crystalline)			
Ca	Calcium oxalate	$CaC_2O_4 \cdot nH_2O$	30–85%
Fe	Phytoferritin	$(FeO.OH)_8 \cdot (FeO.OPO_3H_2)$	Up to 50% in leaf tissue
P	Phytates	Ca-Mg(-K)-salt of $C_6H_6[OPO(OH)_2]_6$	Up to 50–86% in seeds
Si	Phytolite	$SiO_2 \cdot nH_2O$	

Note: For items marked * no quantities have been reported. The value quoted indicates that the speciation is the dominant one for that element.

Source: [1, 2].

8.2.2 Laboratory characterization techniques for biomass ashes

The most commonly applied techniques for the determination of the ash content and ash composition of coals and other solid fuels in the laboratory involve heating the fuel slowly in air to constant mass at a temperature of 815°C, and subjecting the resultant ash residue to chemical elemental analysis. The ash residue is normally weighed to provide an estimate of the ash content of the fuel, and then analysed for the ten major elements present in coal ashes, i.e. SiO_2, Al_2O_3, Fe_2O_3, CaO, MgO, TiO_2, Na_2O, K_2O, P_2O_5 and SO_3. The elemental concentrations are conventionally expressed as oxides, in their highest oxidation states. The analysis of the ash for its trace element content is also fairly common practice. This is a perfectly reasonable and practical approach for most coals, and many other solid fuels. It has also been applied for biomass ash content determination and chemical analysis on the basis that most practical biomass combustion systems produce ashes, principally in the form of oxides, and at temperatures of 800°C or higher, and that the ten major elements listed above also reflect the principal inorganic constituents of most biomass materials.

For many biomass materials, however, a significant proportion of the inorganic material is volatile at the conventional ashing temperatures for coal, and an ashing temperature of 550°C has been adopted as the standard for ash content determination, to avoid underestimation of the ash content of the fuel, due to loss of a significant portion of the volatile inorganic components (CEN/TS 14775-2004).

For elemental analysis of biomass ash materials, for both major elements and trace elements, wet digestion of the raw biomass rather than of the laboratory-prepared ash is preferred, prior to elemental analysis using inductively coupled plasma–atomic emission spectrometry (ICP-AES) techniques, again principally because of concerns about the potential for the loss of volatile inorganic components during the ashing process (CEN/TS 15290-2006, CEN/TS 14297-2006).

Ash elemental analysis data for a number of clean biomass materials which are important as fuels for combustion processes on both domestic and industrial scales are given in Table 8.2. It is clear from the data presented in this table that the ash content of the majority of the biomass materials listed, with the exception of the poultry litter, are around 5–7 per cent on a dry basis, or lower.

Clean wood materials, either in chip or in pellet form, are utilized as a fuel for a wide range of combustors for small domestic stoves and boilers, for commercial and industrial scale boilers and for co-firing in large coal-fired utility boilers. They have a relatively low ash content, less than 2 per cent, and this is largely dependent on the level of bark in the fuel. Clean white wood materials have a very low ash content, generally less than 1 per cent.

Straw and grass materials and the solid residues from the vegetable oil producing industries are utilized in substantial quantities as fuels for industrial-scale boilers and have ash contents in the 4–7 per cent range. Poultry litter is utilized in significant quantities as an industrial boiler fuel in Britain and there are a number of new projects, particularly in Northern Europe and North America. It has been included as an example of a high ash fuel, with particularly difficult ash behaviour.

It is also clear from the data presented in Table 8.2 that all of the biomass ashes tend to be rich in a fairly similar suite of inorganic species, i.e. the compounds of calcium, potassium, silicon, phosphorus and magnesium. The clean wood materials are particularly rich in Ca, Si and K. In some cases, particularly with waste wood materials, the ashes can have significant levels of some of the heavy metals, particularly lead and zinc. The ash chemistries of the other biomass materials vary significantly but all are dominated by Ca, K, Si or P compounds.

One of the key properties of fuel ash materials in combustion systems is their behaviour at elevated temperatures and, in particular, their fusion behaviour. The sintering and fusion of the ash particles on the grates in stoker-fired combustors, and the sintering, fusion and agglomeration of the ash particles in fluidized bed-fired combustors, are important processes. The fusion behaviour of the ashes is also an important factor in determining the propensities of the fuels to form fused or partly fused slag deposits on the furnace wall surfaces in all combustion systems, and may have an influence on the nature of the fouling deposits that can occur on the heat exchanger and other surfaces in boiler plants.

The fusion behaviour of most fuel ashes is a complex phenomenon, which is best described in terms of a melting curve, where the percentage of the ash by mass which is fused is plotted against the temperature. An example of such a curve is reproduced in Figure 8.1. In this example, the melting curves of a model biomass ash system, comprising a mixture of alkali metal salts, have been calculated using a programme called MeltEst,

Table 8.2 *Typical ash elemental analysis data (major elements) for a number of important biomass materials*

Biomass type	Coniferous forestry residue	SRC willow	Cereal straw	Oil seed rape straw	Miscan-thus	Reed canary grass	Olive residue	Palm Kernel	Poultry litter
Sulphur content (wt%, dry ash-free basis)	0.04	0.05	0.1	0.3	0.2	0.2	0.2	0.2	0.5
Chlorine content (wt%, dry ash free basis)	0.01	0.03	0.4	0.5	0.2	0.6	0.2	0.1	–
Ash content (wt%, dry basis)	2	2	5	5	4	6	7	4	13
Elemental analysis (mg/kg, dry basis)									
Al	–	–	50	50	–	–	1500	750	600
Ca	5000	5000	4000	15,000	2000	3500	6000	3000	20,000
Fe	–	–	100	100	100	100	-	900	2500	900
K	2000	3000	10,000	10,000	7000	12,000	23,000	3000	21,000
Mg	800	500	700	700	600	1300	2000	3000	5000
Na	200	–	500	500	–	200	100	200	3000
P	500	800	1000	1000	700	1700	1500	7000	14,000
Si	3000	–	10,000	1000	–	12,000	5000	3000	9000

Note: These data are for relatively clean materials.

which has been developed at Abo Akademi in Finland, based on experimental phase diagrams and thermodynamic data, (see, for instance [3], and the references presented therein).

On the ash melting curves, two key temperatures are commonly identified and used to describe the behaviour of the ashes:

1 The T_{15} temperature, at which 15 per cent of the ash material is molten. This is considered to be the temperature at which the surfaces of the ash particles or slag deposits begin to become sticky and receptive to incoming particles; and
2 The T_{70}, temperature at which 70 per cent of the ash is molten. This is the level at which the outer surface of an ash deposit on a vertical boiler tube will begin to flow.

The standard ash fusion test, which has been applied for the characterization of the fusion behaviour of coal ashes for many decades, is based on the determination of three or four key temperatures on the melting curve. This procedure has been developed, and has been

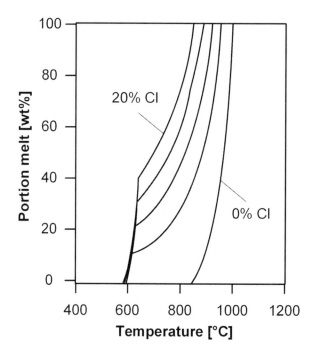

Figure 8.1 *Calculated melting curves for salt mixtures with K/Na molar ratio 90/10, SO_4/CO_3 molar ratio 80/20 and Cl varying between 0 and 20 per cent of the total alkali*

Source: [3].

widely applied, for alumino-silicate, coal ash systems, which have complex melting behaviour and tend to produce relatively viscous melts. The results of the application of this test to biomass ashes, which are not alumino-silicate systems in the main, and many of which do not behave in the same way, can provide useful information, but should always be treated with caution.

The use of **chemical fractionation techniques**, which were developed originally for the characterization of the inorganic components of coals, has also been applied to the characterization of biomass materials. The procedure involves the treatment of a small sample of the fuel by a standardized leaching process, with a series of progressively more severe chemical reagents, for example:

Water → ammonium acetate solution → hydrochloric acid solution

Four fractions are obtained:

1 water leachable components, principally the alkali metals salts, sulphur and chlorine compounds;
2 buffer solution leachable components, generally organically associated;
3 acid leachable components, generally carbonates and sulphates; and
4 residue, principally silicates and species insoluble in mineral acids.

It is generally considered that the water and acetate-leachable elements are those that are more readily released into the vapour phase, and form the very finest aerosol fraction of the ash generated by combustion of the biomass. The acid soluble and residue fractions are not considered to be released into the vapour phase during the combustion process, and tend to be represented predominantly in the coarser fractions of the ash.

The results of chemical fractionation analysis of a number of fuels, including woodchips, forestry residues, bark, wheat straw, peat and bituminous coal are presented in Figure 8.2. These results are particularly interesting in that they illustrate the differences in the leaching behaviour of biomass materials and of more familiar fuels, i.e. bituminous coal and peat.

The results of the chemical fractionation tests can be summarized as follows:

- For the bituminous coal, very little of the inorganic material in the fuel, and principally only the sodium and the chloride, was in the water- and acetate-soluble fraction.
- For the peat, there was a higher level of water- and acetate-soluble material, with some of the calcium in the acetate soluble category.
- The biomass materials, in general, had significantly higher levels of water- and acetate-soluble material, with the majority of the potassium and chloride and some of the calcium and magnesium being in this category. In the case of the wheat straw and the forest residue 1 samples, a significant part of the silica was also considered to be water/acetate soluble.

For the purposes of making simple comparisons between different fuels, a summary of the results of the chemical fractionation test work can be prepared, and an example from the same source is presented in Figure 8.3. In this case, the distribution of the ash-forming material in the fuels, expressed as g/kg, dry fuel, between the water/acetate soluble and the HCl/insoluble residue fractions, are presented. The differences between the fuels are clear. The peat and coal have the higher ash contents, but only a relatively small proportion of the mineral material is in the water and acetate soluble fractions and is considered to contribute to the formation of the fine ash/aerosol material. In the case of the biomass materials, the total mineral matter contents are lower, but a much higher proportion of the mineral material is considered to contribute to the formation of the fine ash/aerosol fraction.

Clearly, the chemical fractionation techniques can provide useful information about the nature and potential behaviour of a wide range of ashes in combustions systems. It is also clear, however, that the technique has distinct technical limitations and that the results should be interpreted very carefully.

Mineralogical analysis and microscopic techniques are increasingly employed for the characterization of mineral materials in fuels, fuel ashes and deposits. X-ray diffraction is commonly used for the identification of the major crystalline phases in these materials. Computer controlled scanning electron microscopy (CCSEM) with energy dispersive X-ray spectrometry is a particularly powerful analytical technique for the examination of the microstructure and chemistry of ashes and deposits.

These techniques are most commonly employed for fundamental research on the characteristics of ashes, and for investigative work on the microstructural characteristics of ashes and deposits.

A number of **slagging and fouling indices** are available for the assessment of the propensity of fuel ashes to form deposits in combustors or furnaces. A detailed description

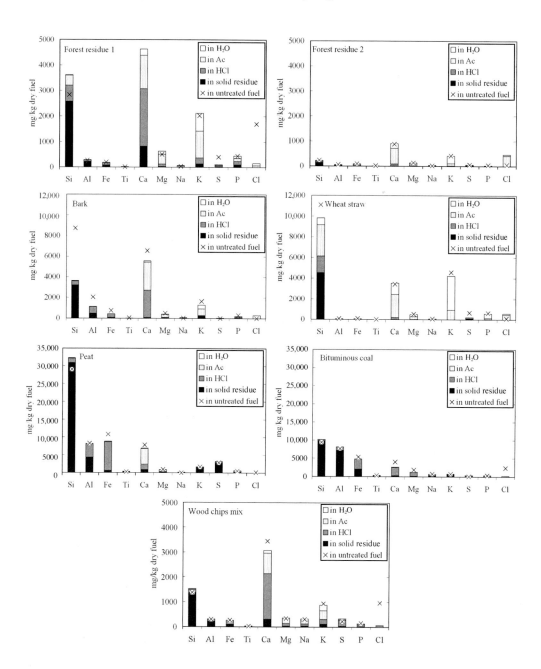

Figure 8.2 *Basic chemical fractionation data for a range of solid fuels*
Source: [4].

of the technical basis and use of these indices is presented in [7]. These are based either on the fuel ash content and the ash chemical composition, or on the results of laboratory tests that can be performed on small samples of the fuel. In the main, these indices have been

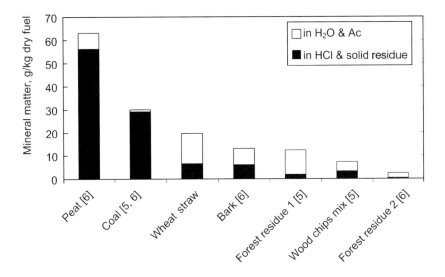

Figure 8.3 *Summary of the chemical fractionation data for a range of solid fuels*
Source: [5, 6].

developed for the assessment of coals, and are applied, with appropriate modifications, to other solid fuels, including wastes and biomass materials, and to the mixed ashes produced by the co-firing of biomass materials with coal.

The majority of the coal **slagging indices** are concerned with the fusion behaviour of the ash, and the traditional indices are based either on the results of the ash fusion test or on the chemical composition of the ash, commonly the ratio of the acidic metal oxides, (SiO_2 and Al_2O_3) to the basic oxides (Fe_2O_3, CaO, MgO, Na_2O and K_2O). These indices provide a general assessment of the ash fusion temperature, which is then employed to rank the ash in terms of its propensity to form fused slag deposits. Despite the technical limitations of both of these approaches, these are still used widely in the industry for fuel specification, boiler design and plant operational purposes.

A number of more sophisticated approaches to the assessment of the slagging propensity, for instance based on the use of phase diagrams of the appropriate alumino-silicate systems, or the use of mineralogical analysis data derived from characterization of the fuel using scanning electron microscopes, have been developed; however, these have enjoyed only limited use within the industry.

Since the majority of the coal ash slagging indices are based on the assessment of the fusion behaviour of alumino-silicate coal ashes, the application to biomass ash systems, which are chemically very different, can be problematic, and great care should be applied when interpreting the conventional slagging index values for biomass.

When considering the potential slagging behaviour of the mixed ashes from the co-firing of biomass with coal, it is clear that, apart from SiO_2, all of the significant components of most biomass ashes, and principally the alkali and alkaline earth metals, are powerful fluxes for alumino-silicate systems. It is expected, therefore, that the introduction of biomass will result in a reduction in the fusion temperatures, and hence an increase in the slagging potential. This will, of course, depend on the level of fluxing agents in the coal and on the co-firing ratio. The effect will be more dramatic when biomass is co-fired with coals with high fusion temperature ashes, i.e. the effect of adding fluxing elements to a coal ash with low levels of the fluxing elements is much greater than to a coal with lower ash fusion temperatures and higher levels of fluxing elements.

For the co-firing of biomass at relatively low levels, the mixed ash is still predominantly an alumino-silicate system and the normal coal slagging assessment methods can be applied, with some confidence.

The **fouling indices** for coal ashes are, in the main, based on the sodium content of the fuel. The deposition of the sodium compounds by a volatilization/condensation mechanism is considered to be the principal driving force for convective pass fouling in coal boilers. The potassium in coal ash is principally present as a constituent of the clay minerals, and is not considered to be available for release by volatilization in the flame.

For most biomass materials, potassium tends to be the dominant alkali metal and, as is the case for the sodium, is generally in a form that is available for release by volatilization. The fouling indices for biomass materials tend, therefore, to be based on the total alkali content of the fuel [8].

Overall, therefore, a suite of ash characterization techniques and ranking methods are available for biomass materials. The majority of these techniques were originally developed and applied for the characterization of coals and other conventional solid fuels and, partly because they are already familiar within the industry, have been adapted for use with biomass materials. In general, the application of these techniques to biomass materials has been technically successful, provided that the appropriate level of care, particularly in the interpretation of the results for equipment design and operational purposes, has been applied.

8.3 High temperature behaviour of inorganic constituents of biomass in combustion systems

8.3.1 Introduction

A general schematic illustration of the processes undergone by the inorganic fraction of solid fuel particles at elevated temperatures in combustion systems is presented in Figure 8.4. As stated above, the inorganic species in biomass materials and other fuels are generally distributed within the fuel particles in three forms:

1. atomically dispersed within the fuel organic matter;
2. in grains within the fuel particles; and
3. as extraneous material.

As described in Figure 8.4, there are inorganic constituents of biomass materials that are volatile at combustion temperatures, particularly some of the alkali metal (K and Na) compounds, the phosphates and some of the heavy metals species. These species can be released into the combustion gases in the form of a fume or as a condensable vapour. Other inorganic species, principally the compounds of calcium and silicon, can be released from the burning fuel particle as very finely divided, sub-micron particles.

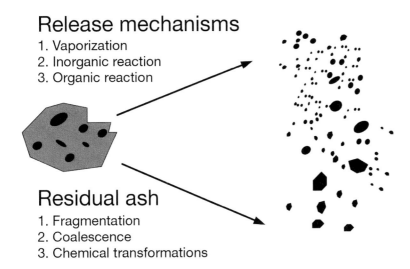

Figure 8.4 *Fate of inorganic material in solid fuels during combustion processes*

Source: [9].

The balance of the inorganic fraction of the biomass will undergo a number of chemical and physical transformations, depending on the chemical composition, combustion temperature and residence time, to form the larger, residual ash particles.

The key mineral transformations and chemical reactions that may occur at elevated temperatures include [10]:

- the fusion or partial fusion of quartz and silica particles and, at high temperatures, chemical interactions with other ash components, principally to form alkali and alkaline earth metal silicates;
- the fusion or partial fusion of alumino-silicates;
- the decomposition of carbonates, oxalates, chlorides, etc. and other inorganic salts;
- the volatilization of alkali metals and some heavy metals;
- particle fragmentation by thermal shock and the rapid release of gaseous species from particles; and
- the coalescence of intra-particle mineral particles.

The specific details of the 'release mechanisms' and 'residual ash formation processes' depend largely on the type of combustor, i.e. on the flame temperatures and the residence times at elevated temperatures.

As stated in Chapter 5, for most solid fuels, including biomass materials, there are three general modes of combustion:

1   grate combustors, which are generally employed for domestic, and small and medium-sized industrial/commercial applications;
2   fluidized bed combustors, of the bubbling bed and circulating types, which are commonly employed for medium-sized industrial/commercial applications; and

3 pulverized fuel combustors, which are employed in the main for large industrial and utility applications. These are rarely used for 100 per cent biomass firing, and the interest is principally in the co-firing of biomass in large pulverized coal-fired boilers.

The basic characteristics and applications of these combustor types are described in Chapter 5. The mode of combustion, however, is relevant to the ash behaviour at elevated temperatures since the combustion conditions tend to be significantly different. The majority of the laboratory and industrial-scale experimental work on these processes has recognized this, and is aimed specifically at one or the other of these types of combustor.

8.3.2 Grate-fired combustors

In grate-fired systems, the fuel particles are distributed over a moving or static grate to form a fuel bed, with some of the combustion air being supplied from underneath. The normal intention is to retain the majority of the ash on the grate to be removed, either manually or mechanically, to an ash pit, although a significant quantity of the ash will be released from the bed as fly-ash particles entrained in the combustion gases or in the form of vapours and fine fumes.

The results of important recent experimental work on the formation of aerosols and coarser fly-ash particles from the grate combustion of biomass materials can be summarized in the reaction scheme represented in Figure 8.5. This reaction scheme provides a good overview of the principal ash and aerosol formation processes occurring with wood fuels, but the key features apply to the grate combustion of most biomass materials. The general description of these processes applies particularly to domestic and small industrial biomass combustors, although the details of the grate design and the air supply arrangements may vary significantly.

The reaction scheme shows the fly ash and aerosol formation processes that are considered to occur on and above the bed. The diagram shows the fuel bed on the grate, with the under-grate wind-box. In this case, the fuel is introduced on the right hand side of the grate, and is first dried and then ignited by radiation from the furnace as it passes along the grate from right to left. The combustible volatile components of the dried fuel are then released into the volume of the furnace, and a flame is generated above the fuel bed. When the fuel has released all of the combustible volatile material, the flame dies down and the fuel char burns out over the last section of the grate. The bed depth decreases as the fuel is progressively combusted, leaving a bed of the coarser ash particles on the grate.

The maximum fuel bed temperatures that apply in grate combustion of biomass materials are generally of the order of 1000–1200°C, and the overall residence times on the grate are relatively long, commonly of the order of several minutes.

Depending on the ash chemical composition and the local bed temperatures, a degree of sintering or fusion of the bed ash particles may occur. If the degree of fusion of the ash is excessive, relatively large ash agglomerates may form. This can interfere with the distribution of air through the fuel, and may affect burnout of the char. The bed ash composition is usually fairly similar to that of the laboratory-prepared fuel ash, although it can sometimes be depleted in the more volatile inorganic species, principally potassium.

The coarse fly-ash particles are principally small particles of ash, of up to an aerodynamic diameter of 200–500μm or so, which are entrained with the upward flow of under-grate air and the combustion gases, and are carried upwards through the furnace. The

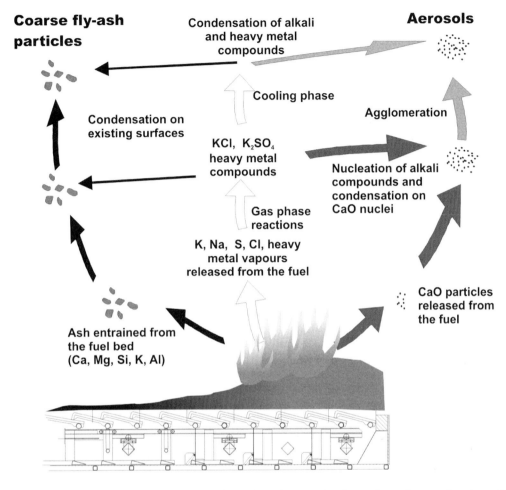

Figure 8.5 *Key processes involved in fly ash and aerosol release from the combustion of wood on a grate*

Source: [11].

chemical composition of the coarse fly ash is normally similar to that of the bed ash. There is commonly condensation of volatile inorganic species, particularly alkali metal sulphates, and some heavy metal compounds, on the surfaces of the coarse fly ash particles.

The fine aerosols, which are generally sub-micron, are generated by the condensation from the vapour phase of the volatile inorganic species, i.e. the alkali metal and some heavy metal compounds, principally as chlorides and sulphates, as the flue gases cool in passing through the boiler. Calcium compounds are not considered to be volatile under these conditions. The calcium content of the fine particulate material tends to increase with increasing particle size, with calcium being present in significant concentration only at particle sizes above around 0.8μm [11].

Obernberger et al [11] report the results of an important investigation of the ash behaviour of three wood-based biomass fuels in conventional grate-fired combustion systems:

1. chemically untreated woodchips, with an ash content of < 0.5 per cent, and an ash composition dominated by K, Ca and Mg, with very low heavy metals content, as described above;
2. bark, which had a much higher ash content, 5–7 per cent, and very much higher Ca content; and
3. waste wood materials with an ash content of around 5–6 per cent, and significant zinc and lead content.

These materials are not untypical of wood-based biomass materials. The authors found that the aerosols generated by the grate combustion of bark had a relatively high calcium content, even at low particle sizes. They suggested that this is due to the release of very finely divided CaO particles from the bark during the combustion process.

The aerosols generated by the combustion of waste wood materials had significant zinc, and to a lesser extent, lead content. Both lead and zinc are considered to be volatile at the temperatures that apply in grate-fired systems.

The authors came to the following general conclusions about fly-ash and aerosol generation from the grate combustion of woody biomass materials:

- The particle size distributions of the fly ashes are commonly bimodal, with the main coarse fly-ash peak in the size range larger than 1μm, with a maximum particle size up to 200–500μm, and with the aerosol peak in the sub-micron range.
- The total fly-ash concentration in the flue gases, is dominated by the generation of the coarse fly ashes by bed ash particle entrainment in the combustion gases. The quantity of fly ash is influenced by the ash content of the fuel and the grate operating parameters.
- The aerosol concentration in the flue gases, and the composition of the aerosol material, are influenced mainly by the chemical composition of the fuel, i.e. by the levels of volatile species (the alkali metals, lead and zinc) and of the very fine Ca-rich particulate material contained within the fuel.

It is clear, therefore, that the recent work of Obernberger et al [11, 12] has provided a very good and detailed description of the key chemical and physical processes involved in the generation of the fly-ash and grate ash materials, and particularly of the fine aerosol particles, from the grate firing of biomass materials.

As stated above, grate-fired systems are designed to cope with a degree of the sintering and partial fusion of the ash on the grate, however the excessive sintering and fusion of the ash material within the fuel bed in grate-fired plants can be a significant practical issue. This can occur fairly rapidly, and is often associated with poor fuel distribution causing areas on the fuel bed which are too deep, have relatively poor air distribution and hence tend to run relatively hot. The formation of large fused ash agglomerates on the bed can further interfere with the air distribution through the fuel bed and this can lead to growth of the agglomerates to form significant 'pancakes' of fused ash on the grate. This can result in poor combustion conditions, and hence reduced combustion efficiencies, with significant levels of unburned fuel in the bottom ash. Excessive levels of fused ash on the grate can also result in difficulties with the removal of the ash from the bed, and in downstream problems in the bottom ash handling systems.

8.3.3 Fluidized bed combustors

The behaviour of biomass ash materials in fluidized bed systems has been fairly extensively studied, particularly in Scandinavia, where this type of combustor is very important industrially, principally for the recovery of energy from the waste products of the pulp and paper industries, see for instance [13, 14].

In fluidized bed combustion systems, the fuel particles are suspended in a fluidizing air stream, along with a relatively coarse-grained bed material. The great majority of the ash leaves the fluidized bed combustors in the form of fly-ash particles, generally of up to around 50–100μm in diameter. This material will also contain fine particles of elutriated bed material, which is commonly quartz, or unreacted limestone, with lime and calcium sulphate/sulphite.

The combustion temperatures that apply in fluidized beds are somewhat lower than those that apply in fixed beds. The bed and freeboard temperatures when burning biomass materials tend to be less than 900°C, and the ash particles tend, in a number of ways, to resemble the ash produced by combustion of the fuel in a laboratory furnace at similar temperatures.

The levels of release of alkali metals by volatilization, and the degree of fusion of the ash, tend to be significantly lower than those that apply at the much higher temperatures in grate combustors, where the ash is exposed to higher temperatures for longer residence times. The inorganic constituents of the biomass and the bed material may also undergo thermal decomposition reactions within in the furnace. The thermal decomposition of calcium carbonate or calcium oxalate to calcium oxide, and the subsequent sulphation of the lime particles, for instance, is a very important process in many fluidized bed systems.

In most fluidized bed combustion systems, the tendency of the fuel ash and bed material particles to sinter and form agglomerates is a key process, the avoidance of which is an important combustor design and operational issue. Excessive agglomeration of the bed particles can lead to poor air distribution and eventually defluidization of the bed. In extreme cases, it may be necessary for the operators to bring the combustor off line to remove and replace the bed material.

The bed ash sintering process involves the formation of particle-to-particle bonds, commonly by a viscous flow sintering mechanism, involving the low melting temperature ash components, initially in the form of necks between the particles. If the conditions are favourable, fairly extensive three-dimensional agglomerates can form.

The key processes responsible for the formation of bonds between the bed particles in biomass-fired fluidized bed combustors are:

- Partial melting of the low melting point ash components to form a liquid phase of low viscosity, which, in turn, forms the necks between the bed particles. This type of sintering mechanism is of key importance for fuels with ashes rich in alkali metals, phosphates and some of the heavy metals.
- In some cases, the solid ash and bed particles, and particularly silicates and alumino-silicate species may be partially soluble in the liquid phase at the sintering

temperature, and this can contribute to the strength of the particle–particle bonding. At higher bed temperatures, it is possible for a silicate melt of high viscosity to form on the surfaces of the bed material. The viscous liquid phase can also flow to form necks between the ash and bed particles.
- It is also possible for chemical reactions to occur at the surfaces of the bed particles, which can increase the strength of inter-particle bonds. For instance, the reaction of lime on the surfaces of the bed particles with SO_2, to form calcium sulphate, is considered responsible for the sintering of bed particles in fluidized bed combustors firing high calcium biomass materials [15].

In operating fluidized bed combustion plants, the key issue is the control over the bed temperatures and the fuel distribution. The formation of agglomerates is often associated with the formation of fuel-rich hot spots in the bed. These can also be formed as a result of slag falls from the upper furnace surfaces or the detachment of pieces of furnace wall refractory material.

Although fluidized bed combustors firing biomass materials operate at relatively low bed temperatures, the release of volatile inorganic species – particularly the alkali metals and some of the heavy metals, into the flue gases does occur, and the fouling of the surfaces of the boiler convective section occurs, as described elsewhere in this chapter, is also a feature of fluidized bed boilers firing biomass materials.

The ash residues from fluidized bed combustors are generally of two basic types:

1. The fly-ash materials carried over from the combustor and captured by the particulate emission abatement equipment. These materials generally comprise the smaller fuel ash particles, and particles of quartz sand or lime/limestone elutriated from the bed. The great majority of the ash discards from the system are of this type.
2. The larger ash particles retained in the furnace, which can be removed periodically through the bed drains. These generally comprise agglomerated fuel ash and bed materials, and slag deposit material detached from the furnace surfaces. When firing biomass materials, the quantities of bed ash discard materials tend to be relatively small.

8.3.4 Pulverized fuel combustion systems

Pulverized fuel combustion systems are generally associated with very large solid fuel boilers for power generation, and it is relatively rare for these to be fuelled with biomass alone, although there are a small number of pulverized wood-fired boilers in operation. It is, however, becoming increasingly common for biomass materials to be co-fired in large coal-fired power station boilers. In the future, it is likely that co-firing with coal will be the preferred means for the utilization of biomass materials for power generation in those countries with significant coal-fired boiler capacity, in preference to smaller, dedicated, biomass-fired power plants.

In pulverized fuel systems, the peak flame temperatures are very high, compared to those in most other combustion systems, commonly around 1600°C, and residence times at these temperatures are relatively short, of the order of a few seconds. The principal concerns in this case are with the behaviour of the inorganic material associated with the biomass when subjected to very high temperatures and the conditions that apply in suspension flames, and with the impact of the co-firing of the biomass on the behaviour of the resultant ash, i.e. with the characteristics of the mixed biomass–coal ashes.

There is a very extensive literature on the characteristics of coal and coal mineral materials and their behaviour in pulverized coal combustion systems. The classic text is a monograph by Erich Raask entitled *Mineral Impurities in Coal Combustion* [7], which presents an excellent summary of the state of knowledge on coal ash characteristics and behaviour in pulverized coal combustors and other systems at that time. The series of conferences on this subject held triennially by the Engineering Foundation for a number of decades also provide an excellent source of technical information.

Literature reports of detailed testwork on pulverized fuel-fired systems firing 100 per cent biomass are relatively rare. One recent and important example [4] describes the results of a programme of trials on an 80MW_{th} down-shot fired boiler in Sweden, firing pulverized wood fuels. Some very interesting data on the fly-ash formation processes occurring in the milled biomass suspension flames, as well as information on the ash deposit formation processes, were obtained.

In this case, detailed analysis of the organic components of the fuels and of the ash materials collected in the electrostatic precipitator were carried out, and samples of the fly-ash particles were taken using a Berner-type low pressure cascade impactor, inserted into the flue gas ductwork at temperatures around 400°C. Chemical analysis of the size fractions, in the range up to 10μm, indicated that the concentrations of the volatile inorganic elements, i.e. the K, Na, S and Cl, decreased with increasing particle size. The concentrations of the less volatile species, i.e. the Si, Al, Fe, Ca, Mg, Mn, and P, tended to increase with increasing particle size.

The general descriptions in the literature of the flame-imprinted characteristics of biomass fly-ash particles generated by the combustion of pre-milled biomass particles in suspension flames, and by the co-combustion of biomass with coal, have benefited from the extensive literature on coal mineral matter transformations in these systems. The principal inorganic constituents of biomass materials, i.e. the simple inorganic compounds of Si, K, Ca, Mg, Na, P, S and Cl, are also present in coal mineral materials, although there are some important differences in detail.

In general terms, the products of combustion of biomass materials when co-fired with coal in large utility boilers tend to have a higher level of sub-micron fume and vapour, and the fly-ash particles tend to be significantly smaller than those formed by the combustion of pulverized coal.

The key ash-related impacts of the co-firing of biomass with coal are potentially on the slagging and fouling deposit formation potential of the mixed biomass–coal ash, and on the efficiency of the installed particulate collection equipment. As stated above, biomass materials tend to have a relatively low ash content compared to most coals; however, the biomass ash materials tend to be relatively rich in the alkali and alkaline earth metals, and these are effective fluxes for the alumino-silicate coal ashes. In general terms, therefore, the co-firing of biomass with coal will tend to increase both the slagging and fouling propensity of the mixed fuel, depending on the chemistries of the coal and biomass ashes and the co-firing ratio. At low co-firing ratios, at less than about 10 per cent on a mass basis, recent plant experience has indicated that these effects tend to be modest.

8.4 Formation and nature of ash deposits on the surfaces of combustors and boilers firing or co-firing biomass materials

8.4.1 Introduction to ash deposition

Operational problems associated with the deposition and retention of ash materials can and do occur on all of the major gas-side components of combustors and boilers firing or co-firing biomass materials.

The more important occurrences at high temperatures in the combustion system and furnace are associated with:

- The agglomeration of ash particles in the fuel beds of stoker-fired and fluidized bed-fired combustion equipment lead to poor combustion conditions, de-fluidization of fluidized beds, and problems with ash removal and downstream ash handling equipment.
- The deposition of ash materials on burner component and divergent quarl surfaces in large pulverized fuel furnaces can result in interference with burner light-up and operation.
- The build-up of large accumulations of fused and partially fused ashes can interfere with the operation of stokers and fluidized bed combustors and can block ash hopper throats in pulverized fuel-fired furnaces. This can result in the requirement for unplanned outages for off-load cleaning.
- The deposition of fused or partially fused ash deposits on furnace heat exchange surfaces reduces furnace heat absorption, and leads to increased furnace and furnace exit gas temperatures. This can lead to increased ash deposition and high metal temperatures in the convective sections of boilers, and it may be necessary to reduce load or to come off load for manual cleaning.
- The accumulation and subsequent shedding of large ash deposits on upper furnace surfaces can lead to damage to the components of the lower furnace or the combustion system.

These are slag formation processes which occur at relatively high temperatures, generally in excess of around 800°C, on furnace refractory or water wall surfaces in direct receipt of radiation from the flame. They occur relatively rapidly, over a matter of hours, when conditions are favourable and usually involve the deposition and subsequent sintering and fusion of the fly-ash particles.

The accumulation of ash deposits in the convective section of boilers also occurs. These are normally termed fouling deposits, and the more common occurrences are described below:

- The formation of ash deposits on the surfaces of heat exchange banks occurs at flue gas temperatures less than around 1000°C, and is generally a much slower process than slag formation, with significant ash deposits growing over a number of days. The process involves the formation of deposits in which the ash particles are bonded by specific low melting point constituents, principally the alkali metal species. The gas temperatures are too low for significant sintering or fusion of the ash particles. In general terms, as the gas temperatures decrease through the boiler convective section, the deposits tend to be less extensive and physically weaker. This is reflected in the

design of the convective section, i.e. it is possible to reduce the cross pitches of the tube banks as the flue gas temperatures decrease. Convective section fouling is one of the most troublesome ash-related problems associated with the combustion of biomass materials, because of the relatively high alkali metal content, and hence high fouling potential, of many biomass materials. Fouling reduces the heat absorption in the convective banks, and results in increased flue gas temperatures. Increased fouling also increases the gas-side pressure drop across the banks. Uncontrolled fouling will lead to ash bridging between the tubes. This further increases the gas-side pressure drop and can result in the channelling of the flue gas. This, in turn, can result in local overheating of the heat exchange tubes and localized damage by particle impact erosion.
- Ash deposits on heat exchange surfaces at very low flue gas temperatures tend to be relatively weakly bonded. They are commonly initiated by the physical accumulation of ash, often by the gravitational settling of ash material which has been dislodged from upstream primary deposition sites by the action of soot blowers.
- Low temperature fouling and corrosion of air-heater surfaces in larger boilers are also common occurrences. These are complex processes and tend to be very specific to the design and operation of the air-heater.

It is clear from the discussion presented above that the ash deposition processes that occur in biomass combustors and boilers, and when biomass is co-fired with coal, are fairly complex phenomenon, dependent both on the characteristics of the ash and on the design and operation of the combustor or boiler plant. It is instructive, therefore, to try to summarize the major findings described in the significant technical literature, and from the accumulated plant experience, on biomass ash deposition processes.

8.4.2 Slag formation processes

As stated above, slag formation processes are principally associated with the sintering and fusion of ash particles on surfaces within the furnace, at temperatures in excess of 1000°C.

The details of the slag formation mechanisms do depend significantly on the type of combustor and the furnace design, although the basic processes are similar. For slag formation, the dominant process responsible for mass transfer of ash material from the flue gas stream to the surface at the primary deposition site is the inertial impaction of the entrained fly-ash particles. Other mass transfer processes do occur, but they are generally of little importance in most practical situations, except perhaps during the very initial stages, when the deposit is beginning to form on a clean metal surface.

The key issues, therefore, in determining the distribution of slag deposits within a furnace, the rate of slag formation and the nature of the deposits are the adhesion of ash material to, and the removal of material from, the surface at the deposition site over time. The adhesion of particulate material to a surface is dependent on the physical state of both the particles and the surface. For effective particle adhesion to occur, either the particles themselves must be molten or partially molten, or the surface must be receptive, i.e. must be at least partially molten and 'sticky'. As stated above, this is commonly associated with the T_{15} temperature, at which around 15 per cent of the ash particle is fused, and the particle begins to be 'sticky'.

In the initial stages of slag formation, when the metal surfaces are clean and relatively cold, only the partially molten or 'sticky' particles can adhere to the solid surface, and the deposition rate is low. The dominant deposition processes are often the condensation of volatile inorganic species, and the adhesion of that portion of the fly-ash particles arriving at the surface that are at temperatures above T_{15}. The chemical composition of the initial deposit may be very different from that of the bulk fuel ash, and the deposit consists of relatively loosely sintered fly-ash particles, bonded together by low melting point constituents, commonly the alkali metal compounds. This only tends to occur when the boiler is brand new or after an outage during which the furnace surfaces have been very thoroughly cleaned. It is a common observation that there may be a 'period of grace' of several days in duration when little or no slag formation is apparent.

As the deposition proceeds, and the deposit thickness increases, the temperature of the ash material at the outside surface of the deposit also increases. The ash deposit becomes progressively denser and more fused. The surface of the deposit becomes progressively more receptive to the arriving particulate material, and the deposition rate can increase sharply. The chemical composition of the deposit approaches that of the bulk fuel ash.

In some cases, particularly with low fusion temperature ashes and in high temperature regions of the furnace, the deposit can grow to the point where the temperature at the outside of the deposit is in excess of the flow temperature (or T_{70}) for the ash, and a fused, flowing slag is formed. In this condition, the surface is very receptive even to solid particles arriving at the deposition site, and the deposition rate is at a maximum.

The extent to which slag represents an operational issue depends on the fusion behaviour of the ash and the local gas and surface temperatures. These factors generally determine the growth and degree of fusion of the deposits, and whether or not they can be controlled by the online cleaning systems.

The general process description presented above goes some way to explain the distinctly layered structure observed in many furnace slag deposits. The ash material adjacent to the furnace tube surface is generally relatively weakly bonded and has high porosity. This is also the mechanically weakest material and most likely to fracture under the action of soot-blowing or thermal stresses. Slag deposits commonly detach in one piece. The process description also indicates that the deposition rate is often largely controlled by the condition of the deposit surface, and less so by the fly-ash particle flux or the condition of the fly-ash particles.

It is also a common observation that hot spots on the furnace surfaces, often associated with small areas of refractory material, can act as slag initiation points from which slag can grow out into the furnace volume and across the furnace tubes. The refractory surface is not efficiently cooled, and the surface temperature of the refractory or of any ash deposit material attached to it is much higher than that of the adjacent furnace tubes. This means that the surface of the deposit is more receptive than the ash deposit on adjacent tubes. The deposition rate is higher and the nature of the deposit is commonly very different from that on the adjacent tubes, and can act as a point at which the slag can grow in all directions. It is, for this reason, a common observation that the walls of a combustion chamber can be relatively clean except for areas around small patches of refractory material on and around which a fused slag has grown.

It is clear from the description of the slag formation and growth processes presented above that the key fuel-related factor that controls the extent of slag formation and the nature of the slag is the fusion behaviour, both of the individual fly-ash particles and of the surface of the slag. The ash fusion behaviour of biomass materials varies widely, depending on their chemical composition.

In very general terms, three types of biomass ash system have been described by Bryers [16] in terms of their general ash chemical composition and their fusion behaviour, and hence their propensity towards the formation of furnace slag:

1. high silica/high potassium/low calcium ashes, with low fusion temperatures, including many agricultural residues;
2. low silica/low potassium/high calcium ashes, with high fusion temperatures, including most woody materials; and
3. high potassium/high phosphorus ashes, with low fusion temperatures, including most manures, poultry litters and animal wastes.

These properties will be reflected in the slagging index values for individual biomass materials, based either on the ash chemistry or the ash fusion tests described in Section 8.2 above. The fusion properties of the ash will have a key influence on the design of the combustion equipment and the furnace. Furnaces are generally designed with the formation of slag deposits in mind, i.e the furnace geometry should be such that the flames are adequately contained and the gas temperatures local to the walls low enough to avoid excessive slag formation. In boilers with water wall furnaces, it is also important that adequate online cleaning is provided to help maintain the furnace heat absorption.

One of the key functions of the furnace in this regard is to reduce the flue gas temperatures to a level below that at which excessive slag can form, prior to entry of the flue gases into the convective section of the boiler. Failure to do this would result in the formation of fused or partly fused slags in the convective section tube banks, and this is highly undesirable.

8.4.3 Convective section fouling processes

Convective section ash fouling is a lower temperature process, largely driven by the deposition of volatile inorganic species in the ashes, principally the alkali metals and, in some cases, phosphorus compounds, by a volatilization–condensation mechanism. The volatile species will condense on any cooled surface, initiating deposit growth and acting as a bond between the non-volatile ash particles which adhere to the deposits.

On deposition, there is also a tendency for the alkali metal compounds and other inorganic species to sulphate, a reaction with the sulphur oxides in the flue gases. This can add significantly to the mass and volume of the deposits, and may alter the physical properties of the deposits on cooling. Because the deposition occurs predominantly by a condensation mechanism, and involves principally the volatile components of the ash, the deposition efficiency and deposition rate are relatively low compared to slagging processes. The resultant deposits tend to be well adhered to the tube surfaces, and relatively dense and of low porosity.

The physical properties of the ash deposits depend on the flue gas temperature at the deposition site. The deposits in the higher temperature regions of the boiler, and particularly at the entry to the superheater, which is commonly the first packed tube bank encountered by the flue gases, can be partly fused at normal flue gas temperatures. The deposits tend to act as a receptive surface for the other non-volatile ash components, which are present in the flue gas in the form of solid particulate material, and can be tenacious and hard to remove. This is particularly the case if the fouling is extensive and ash bridges have formed between the tubes. On cooling, the deposits are hard and well anchored to the tube bank surfaces. Online cleaning techniques tend to be relatively ineffective if the deposition is extensive. When the boiler is taken offline, the deposits cool to a hard dense solid, which is very difficult to remove.

The deposits formed on boiler surfaces in contact with flue gases at lower temperatures (i.e. below about 600–700°C, depending on the chemistry) tend to be unfused. They are more friable in nature, and are usually easier to remove both on and offline.

For this reason, it is common practice, when designing boilers for the firing of high fouling fuels, e.g. municipal solid wastes and some biomass materials, to have a large radiant furnace section, which acts to reduce the flue gas temperatures at the entry to the first tube bank to temperatures below 700°C. This helps to control the fouling, and specifically helps to avoid the formation of partly fused ash deposits at the entry to the first convective bank.

There are a number of common knock-on problems associated with the formation of excessive levels of ash fouling deposits in the convective sections of boilers. These apply to all boilers fired with solid biomass fuels, particularly if the extent of deposition is sufficient to cause ash bridging between tubes:

- The generation of higher flue gas velocities in the open gas pass channels through the bank, result in increased gas-side erosion wear of the tube surfaces.
- The overuse of sootblowers in an attempt to control the gas-side fouling is a further common cause of boiler tube erosion.
- When there is excessive convective pass fouling, with blockage of a significant number of the flue gas passes in the tube banks, there is a tendency for the generation of high metal temperatures in tubes in the flue gas passes that remain open. This can result in accelerated metal wastage, due to gas-side corrosive attack, and to other thermal damage to the tubes.
- Increased metal wastage of boiler surfaces due to the increased incidence of boiler outages, and hence increased out-of-service corrosion, particularly if water washing of boiler surfaces to remove the ash deposits is applied. The metal wastage rates from this type of aqueous salt solution corrosion can be fairly dramatic if the boiler cannot be drained effectively, or lies wet for a prolonged period after washing.

With increased operating time, there is a tendency for these problems to get progressively worse, i.e. the time between forced outages for cleaning will tend to decrease, and the number of tube leaks and the requirement for significant repair work will tend to increase. This is a fairly common experience with boilers where the extent of convective pass fouling is such that operation on a campaign basis, i.e. with regular outages for manual cleaning, is required.

Fouling deposits on convective pass surfaces at flue gas temperatures lower than around 600°C tend to be very much less strongly bonded, and it is possible to progressively reduce the heat exchange tube cross pitching with decreasing flue gas temperature and, in some instances, to employ extended surface tubing. Ash deposits can and do accumulate in cooler regions of the boiler and the economizer; however, in the main, these are readily controlled provided that the heat exchange bank design is suitable for the purpose and there is adequate online cleaning.

8.4.4 Deposit growth, shedding and online cleaning

In all practical situations, the long-term accumulation pattern of deposits in furnaces and boilers involves competition between processes which tend to add to the deposit and those which remove material from deposits. These processes have been described by a number of authors in the technical literature on fossil fuel and biomass ashes [7, 9, 16]. The key deposit growth processes are:

- Ash particle inertial impaction is the dominant process in high temperature slag formation, and particularly for the larger fly-ash particles. The rate of deposition by impaction is a function of the particle flux, and of the deposition efficiency which, in turn, is dependent on the degree of fusion of the particles and/or the deposit surface.
- The condensation of volatile inorganic species, in vapour or fume form in the flue gases, on cooled surfaces, which is the principal driving mechanism for convective pass fouling, and is of particular importance for biomass materials because of the relatively high levels of volatile species in these fuels.
- Chemical reactions occurring within the deposits, and particularly oxidation, sulphation and chlorination processes.
- Thermophoresis, the transport of small, gas-borne, ash particles to cooled surfaces by the effects of the local gas temperature gradients, is only important for very small, sub-micron particles and particularly during deposit initiation when the local temperature gradients are at a maximum and when the rate of deposition by inertial impaction is low.

Uncontrolled ash deposition, with no shedding or deposit removal would, in most plants firing solid fuels, very quickly result in operational problems, and most solid fuel furnaces and boilers are fitted with on-line cleaning systems of various types. There are also natural ash deposit shedding mechanisms and other processes which are responsible for the reduction in the extent of deposition. The key processes are:

- The principal means of online control of deposition in most furnaces and boilers is the use of the installed soot-blowers. These devices direct a high velocity jet of steam, water or compressed air at the deposits, and employ a combination of mechanical impact and thermal shock to break up and remove them. Sonic soot-blowers can also be deployed, particularly for the removal of relatively weak deposits in the cooler parts of the boiler convective section. In extreme circumstances, where very tenacious and troublesome ash deposits have formed, small explosive charges can be employed to break up the deposit material. The deposit material removed in this way may be carried forward with the flue gases, but can also accumulate elsewhere in the furnace or in the convective pass of the boiler.
- The natural shedding or detachment of deposits also occurs. This can occur when deposits grow too large for the adhesive forces to support them, or due to the effect of thermal expansion differences between the ash deposit and the boiler tube, during shut-downs and boiler load changes. Rapid boiler/combustor load changes can be deliberately used for deposition control. The detachment of large accumulations of slag in this way can, however, result in damage to components or in troublesome accumulations of ash lower in the furnace.

- Heavily fused deposits of low viscosity can drip on to surfaces lower down in the furnace or boiler.
- In the boiler convective section, fly-ash particle impact erosion wear can result in the reduction of the thickness of fouling deposits, particularly on the sides of tubes.

It is clear, therefore, that ash deposition control in biomass combustors and boilers is a fairly complex issue. The key factors are:

- The careful design of the furnace and boiler convective section, which properly recognizes the characteristics and behaviour of the fuel ash, is of prime importance, and the boiler design engineer obviously has a key role in this regard. The incorporation of specific furnace and boiler design features, where appropriate, to minimize ash deposition, and to aid the removal of ash and the avoidance of ash accumulation within the system is also of key importance.
- The correct design, operation and maintenance of the combustion equipment and of the online cleaning systems are important issues. It is also often preferable to maintain the plants at a relatively low level of deposition, rather than to deploy the online cleaning systems only when there is evidence of significant deposition. Intensive cleaning of the furnace and boiler surfaces during outages can be very effective in increasing the operating times between outages. There are specialized online deposition monitoring and soot-blowing control systems that are commercially available and that can assist significantly with the optimization of the soot-blower operations and the control of ash deposition.

8.5 Impact of ash on the flue gas cleaning equipment in biomass-firing systems

8.5.1 Introduction

The cleaning of the flue gases to a particular standard, prior to exhaustion to the atmosphere, is a feature of most industrial biomass combustion systems, worldwide. The larger biomass combustors and boilers generally have fairly tight controls on emission levels, which may depend on whether the biomass fuel is regarded as being a relatively clean fuel or a waste material. The smaller domestic or commercial/industrial systems tend to have less onerous emissions control requirements. Appropriate flue gas cleaning systems, designed to comply with the relevant operating standards for most industrial biomass materials and combustion systems, are available commercially from experienced vendors, and are relatively well proven. A discussion of the regulatory and technical issues associated with the control of the emissions from dedicated biomass combustion systems, and of the other relevant environmental control issues, is presented in Chapter 9.

The main focus in this section, therefore, is on the impacts of the co-firing of biomass materials at increasing co-firing ratios on the flue gas cleaning systems for the larger-scale combustion systems and boilers originally designed for the combustion of conventional fossil fuels, although a number of the technical issues may also apply to dedicated biomass boiler systems.

As stated above, the quantities of biomass co-fired in large coal-fired and other fossil fuel-fired power plant boilers have increased fairly dramatically over the past few years, particularly in Northern Europe but also elsewhere in the world. The level of co-firing activity worldwide, and the co-firing ratios at specific plants, are likely to increase further over the next few years.

In modern coal-fired power plant boilers, the principal gaseous and gas-borne emissions control equipment is currently concerned with the control of the following species:

- total particulate emissions control, principally using dry electrostatic precipitators or fabric filters;
- NO_x emissions control, by both primary and secondary measures, with selective catalytic reduction (SCR) being the most commonly applied secondary measure;
- SO_x emissions control principally by wet, dry and semi-dry flue gas desulphurization (FGD) techniques.

All of the flue gas clean-up technologies listed above are well proven industrially for pulverized coal firing, over a wide range of coal qualities and for all relevant scales of operation. There are a number of experienced and specialist vendors for all of the most commonly applied equipment and, generally speaking, the implementation of these emission control measures in large solid fuel-fired boilers has been very successful.

It is interesting to examine the evidence to date on the ash-related impacts of biomass co-firing on the performance of the emissions control equipment. A number of the technical issues described will also be applicable to flue gas cleaning for dedicated biomass plants.

8.5.2 The impact of co-firing on electrostatic precipitators

When considering the effects of biomass co-firing on the performance of the electrostatic precipitators and on particulate emission levels, the principal technical concern is that the fly-ash particles generated from biomass combustion tend to differ chemically from pulverized coal ash, are significantly smaller than those from coal firing, and there is a greater tendency towards the generation of significant levels of sub-micron fumes and vapours. There may, therefore, be a tendency for the particle capture efficiency in electrostatic precipitators to decrease with increasing co-firing ratio and for the total particulate emission levels to increase accordingly.

In most countries, it is necessary to demonstrate to the environmental regulators that the co-firing of biomass materials has no significant additional negative environmental impacts compared to coal firing, and there is a growing body of evidence that, at relatively low biomass co-firing ratios, i.e. less than 10 per cent on a heat input basis, there have been very few incidents of significant increases in the total particulate emission levels that can be attributed to biomass co-firing activities. There are concerns, however, that this may not apply at higher co-firing ratios, and that modifications to the installed particulate collection equipment may be required. However, it should be noted that, to date, the industrial experience of the co-firing of biomass materials at levels above 10 per cent on a heat input basis is limited.

8.5.3 The impact of biomass co-firing on SCR catalysts

The performance of SCR catalysts, and particularly the effective catalyst lifetime, has a significant impact on the economics of SCR systems, and can be influenced by a number of factors, many of which are fuel related [17]:

- thermal degradation by pore sintering;
- ammonia salt condensation inside catalyst pores due to low temperature operation;
- surface blocking of catalyst pores by small ash particles;
- poisoning of catalyst surfaces by volatile inorganic species; and
- fly-ash particle impact erosion of catalyst material.

Early experiences of the use of SCR catalysts in woodchip- and peat-fired boilers in Scandinavia, and of boilers firing animal manures and sludges, indicated that deactivation of the catalysts by sodium and potassium salts, and by phosphorus and silica compounds was a significant issue. The observed deactivation resulted in markedly reduced catalyst lifetimes and increased operating costs [18, 19].

A number of laboratory and plant tests of the impact of biomass co-firing on the fouling and deactivation of catalysts have also been performed over the past few years. Figure 8.6 for instance, presents some of the results of experiments involving the exposure of SCR catalyst materials in a slip-stream reactor on a combustor firing a fuel rich in the alkali and alkaline earth metals. They indicate clearly that there were significant increases in the concentrations of calcium, sulphur and sodium compounds on the surfaces of the catalyst material after prolonged exposure to the flue gas. The results of the test work also indicated that there was significant deactivation of the catalyst.

At Studstrup Power Station in Denmark where, since 2002, cereal straws and other baled biomass materials have been co-fired, at up to 10 per cent on a heat input basis, with coal in a 350MW$_e$ boiler, the impact of the co-firing on a side-stream SCR catalyst block has been studied over a period of up to 5000 hours [21]. The results of this test work have

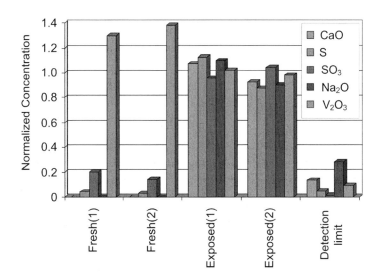

Figure 8.6 *Chemical analysis of key elements on an SCR catalyst surface before and after exposure to a flue gas from an alkali and alkaline earth metal-rich fuel*

Source: [20].

indicated that there was no significant difference in the performance of the catalyst between the straw co-firing test and coal firing alone. Further test work with exposure of the catalysts over periods of time up to 20,000 hours was planned at the station.

The deactivation of SCR catalysts associated with the firing or co-firing of biomass materials has been recognized as being of significant industrial importance, and this has been the subject of a major EC-funded R&D project, CATDEACT, which is co-ordinated by IVD at the University of Stuttgart. The details of the project partners and the work programme are given in the project website, www.eu-projects.de.

A programme of test work using a side-stream, pilot-scale catalyst block was also carried out at Borssele Power Station in the Netherlands in 2004 [17]. The boiler (425MW$_e$) was co-firing a fairly wide range of biomass materials with coal at up to around 10 per cent on a mass basis. The initial catalyst exposure test was over a period of 1000 hours, and no significant decrease in catalyst activity, compared to that of a baseline test firing coal and refinery off-gas only, was detected. No evidence of significant increases in the levels of deposition of arsenic, sodium or potassium were detected, although there was a 30–40 per cent increase in the P_2O_5 concentration.

Overall, it would appear that the increased deactivation rates of SCR catalysts due to alkali metal- and/or phosphate-driven fouling is a significant technical issue when firing or co-firing biomass materials with high levels of these species. For dedicated biomass combustion plants this is a significant technical issue. At the biomass co-firing levels that currently apply in large power plants, i.e. generally lower than 10 per cent on a heat input basis, there have not, as yet, been any significant operational problems, and the results of a number of side-stream tests at low co-firing ratios on operating plants have been relatively encouraging in this regard.

In the event of significant catalyst deactivation, it is possible to water wash the catalyst blocks to remove soluble alkali metal and other salts, and hence recover the catalyst activity and catalyst life. At Avedøre Power Station in Denmark, where co-firing of wood pellets is done with heavy fuel oil and natural gas, a water washing system for the SCR catalyst blocks has been in successful commercial operation for a number of years [22].

8.5.4 The impact of biomass co-firing on FGD plants

The great majority of clean biomass materials of industrial importance have sulphur contents that are significantly lower than those in most coals and, in the great majority of cases, they also have similar or lower chlorine levels. The impact of biomass co-firing, therefore, in the great majority of cases is to reduce the acid gas abatement duty of the installed FGD system, and hence reduce the lime/limestone usage and plant operating costs.

The great majority of biomass materials also have significantly lower levels of most of the key trace element and heavy metal species than most coals, and the duties of the waste-water treatment plants are reduced. This is not the case, of course, for a number of the biomass-based waste materials.

The evidence to date from coal-fired power plants which have been co-firing clean biomass materials, albeit at relatively low co-firing ratios, has been that co-firing has had no significant negative impacts on the operation and performance of the FGD plants.

8.6 Impact of biomass ash on boiler tube corrosion, and on erosive and abrasive wear of boiler components.

8.6.1 Technical background to gas-side corrosion processes in boilers

The corrosion processes that occur on the gas-side surfaces of boiler tubes are very complex and, since they occur at high temperatures and in very aggressive conditions, they are extremely difficult to study. Nevertheless, because of their great importance to the designers and operators of boiler plants, these processes have been the subject of a great deal of technical work, at laboratory, test rig and plant scale, over many decades. There is a substantial technical literature on this subject, particularly for fossil fuel-fired boilers [23], and increasingly for biomass boilers and for boilers co-firing biomass with fossil fuels, as the industrial importance of biomass utilization for power generation and CHP applications has increased. Gas-side corrosion processes have generally been of less importance for domestic and small commercial/industrial biomass boilers, which tend to operate at much lower steam temperatures.

In general terms, it has been found that the gas-side corrosion rates of boiler tubes are controlled by a number of factors:

- the tube material;
- the flue gas and metal temperatures, and the heat flux at the tube metal surface;
- the chemical composition of the ash deposit material at the metal–deposit interface;
- the chemical composition of the flue gases; and
- the operating regime of the plant.

For most biomass boilers, it has been found that, provided the combustion equipment and furnace are well-designed and operating reasonably well, the principal concern has been with the high temperature corrosion rates that apply in the superheater section of the boilers. The concerns are principally associated with the final stage superheaters, with the leading elements and steam outlet legs being subject to the most aggressive attack, principally because of the relatively high metal and gas temperatures that apply in these locations.

As stated previously, the majority of biomass materials have the following key chemical characteristics, which have an influence on the high temperature corrosion processes:

- Biomass ashes tend to be relatively rich in alkali metals, particularly potassium compounds, and in some cases, phosphates, which tend to form deposits on the surfaces of the superheater tubes, via a volatilization/condensation mechanism.
- Most biomass materials have relatively low total sulphur contents, generally less than 0.5 per cent;
- The chlorine content of biomass materials varies significantly, but can be up to around 1 per cent or so in some cases.

As described in Section 8.4 above, the ash deposits that form on the superheater tube surfaces, tend to be rich in potassium salts, principally sulphates, chlorides and phosphates, depending on the fuel composition and the gas and metal temperatures. The biomass ash deposits, therefore, tend to have relatively high potassium contents and relatively high chloride to sulphate ratios. This can have a significant impact on corrosion behaviour, particularly at high metal temperatures on superheater surfaces. In general, it is necessary, therefore, to design dedicated biomass boilers with final steam temperatures that are significantly lower than those that apply in large coal-fired boilers.

A number of very good technical reviews of the chemistry of biomass ash deposits and the corrosion processes in biomass boilers have been published in recent years (see for instance [24, 25] and the references cited therein). There has also been increasing interest over the past few years in the impact of co-firing biomass materials on boiler corrosion processes (see, for instance [26] for a very useful, recent overview).

8.6.2 Corrosion mechanisms

It has been fairly common industrial experience with biomass boilers, that significant corrosion rates of the conventional superheater materials have occurred at metal temperatures in excess of 450–500°C and under the ash chemistries described above. A number of potential corrosion mechanisms are possible under these conditions, and have been described, including:

- corrosion processes involving the reactions between the metal/metal oxides and gaseous chlorine species, i.e. Cl_2 and HCl;
- solid phase reactions involving alkali metal chlorides; and
- corrosion reactions involving molten alkali metal and other chlorides.

The experience from the operation of biomass boilers, particularly in Scandinavia, is that the most severe corrosion is commonly associated with ash deposits containing alkali metal chlorides on high temperature superheater surfaces. The deposits can be responsible for significant metal wastage rates at temperatures below the melting temperature of KCl, and are considered to be associated with the presence of low melting eutectic mixtures.

The melting and eutectic temperatures of the relevant species are listed in Table 8.3. It is clear from the data presented in the table that a number of the key binary salt mixtures, and particularly the KCl–$FeCl_2$ and $NaCl$–$FeCl_2$ systems have low temperature eutectics, in the range 340–390°C, i.e. local liquid phases can form within the deposits as the KCl or NaCl in the deposited material reacts with the metal or the metal oxide scale.

Another reaction considered to be responsible for enhanced corrosion processes in biomass boilers involves the sulphation of the deposited alkali chlorides in contact with SO_2/SO_3 in the flue gases, with the release of HCl local to the metal surface/scale/deposit interface, according to the following reaction scheme:

$$2KCl_{(s)} + SO_{2(g)} + 0.5O_{2(g)} + H_2O_{(g)} \rightarrow K_2SO_{4(s)} + 2HCl_{(g)}$$

The gaseous HCl can then diffuse to the metal/scale surface and react to form metal chlorides. These mechanisms are considered to apply when firing biomass fuels with low sulphur contents and significant chlorine contents, i.e. where the ash deposit chemistry is dominated by the alkali metal chlorides.

Riedl et al [27] described a potential corrosion mechanism, 'active oxidation', that they considered may be responsible for the accelerated corrosion observed in small, hot water, fire tube boilers firing bark, woodchips and sawdust in Austria. They described the enrichment of alkali metal chlorides on the heat exchange surfaces, by a condensation

Table 8.3 *Melting/eutectic temperatures for pure compounds and binary mixtures relevant to chloride-rich, biomass ash deposits and associated corrosion reactions*

System	Melting/eutectic temperature (°C)	Composition at the eutectic point (mole % alkali)
NaCl	801	–
KCl	772	–
$FeCl_2$	677	–
$CrCl_2$	845	–
NaCl–$FeCl_2$	370–374	c. 56
KCl–$FeCl_2$	340–393	45.8–91.8
NaCl–$CrCl_2$	437	53.7
KCl–$CrCl_2$	462–457	36–70

Source: [24].

mechanism, as described in Section 8.4. They further suggested that the chlorides can react with the SO_2/SO_3 in the flue gases to form sulphates, and the subsequent release of gaseous chlorine, according to the following reaction schemes:

$$2NaCl + SO_2 + O_2 = Na_2SO_4 + Cl_2$$
$$2KCl + SO_2 = K_2SO_4 + Cl_2$$

One of the results of the sulphation reaction is the release of gaseous Cl_2 within the deposit and local to the metal surface, and this can react with the low alloy steel to form $FeCl_2$. The $FeCl_2$ has significant vapour pressure under the relevant conditions, and can diffuse through the corrosion product/deposit layer and react with the excess oxygen in the flue gases, according to the following reactions:

$$3FeCl_2 + 2O_2 = Fe_3O_4 + 3Cl_2$$
$$2FeCl_2 + 1.5O_2 = Fe_2O_3 + 2Cl_2$$
$$FeCl_2 + O_2 + Fe_3O_4 = 2Fe_2O_3 + Cl_2$$

The authors considered that these reactions are responsible for the regeneration of the gaseous chlorine local to the metal surface, and the resultant severe corrosion. The basic principles of the 'active oxidation' mechanisms are described graphically in Figure 8.7. This provides an illustration of the key chemical reactions that are considered to be occurring within the tube metal/corrosion scale/ash deposit layers of the interactions with the flue gases.

8.6.3 Plant experience and boiler probing trials with biomass firing and biomass–coal co-firing

A number of plant-based corrosion probing trials in biomass boilers have been carried out over the past few years. These trials involve the exposure of small coupons of the test materials to the flue gases from biomass combustion over relatively long periods of time.

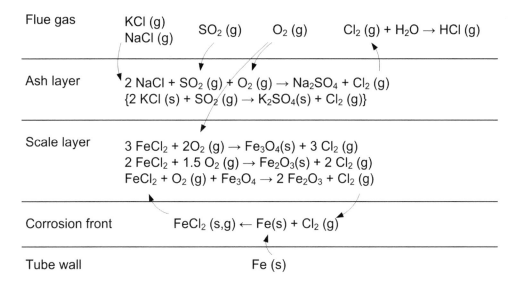

Figure 8.7 *Principles of the proposed 'active oxidation' corrosion mechanism*

Source: [27].

They are relatively expensive to perform; however, along with carefully recorded plant experience of corrosion processes, they represent probably the best means obtaining corrosion rate data under reasonably realistic conditions.

A very important programme of plant-based corrosion test work involving the insertion of corrosion probes into the flue gas stream was done in the three small straw-fired power plants, Masnedo, Rudkobing and Ensted in Denmark [25, 28, 29]. The stations were firing cereal straws with ash levels typically around 5–7 per cent, chlorine levels typically around 0.3–0.5 per cent, sulphur contents less than 0.2 per cent and potassium contents up to around 2 per cent on a dry basis.

Three types of metal test piece exposure trials were performed:

1 the exposure of test metal rings on water-/air-cooled probes;
2 the exposure of candidate tube materials in a test superheater; and
3 the exposure of test tube sections inserted into an existing superheater.

Both ferritic and austenitic superheater alloys were exposed to the flue gas atmospheres at metal temperatures in the range 450–620°C. It was found that the measured corrosion rates for the test materials increased with increasing metal temperature, from less than 50 nm/h at 470°C to values in excess of 1000 nm/h at temperatures in excess of 600°C.

The basic findings on the corrosion processes and deposit morphologies were summarized and this has been reproduced in Table 8.4. These results indicate that there were significant differences in the deposit morphologies and corrosion chemistry at different temperatures. At relatively low temperatures, up to around 520°C, there was the formation of the normal, stable protective oxide layer. At the higher temperatures, no stable protective oxide layer was formed, and rapid metal wastage by selective corrosion and grain boundary attack was observed. All of the alloys tested gave fairly similar corrosion rates, although there appeared to be a shallow optimum in corrosion resistance for alloys with chromium contents in the range 15–18 per cent.

Table 8.4 *Summary of the findings on corrosion and deposit morphologies on superheater surfaces in straw-fired boilers*

Plant	Steam temperature (°C)	Metal surface temperature (°C)	Corrosion morphology	Deposit
Masnedo	520		Protective oxide formation with grain boundary attack – oxide has a tendency to spallation.	KCl with K_2SO_4 adjacent to the surface oxide. Iron oxide dendrites in both.
	450–570	460–615	Selective corrosion and grain boundary attack at the higher temperatures. Protective oxide at the lower temperatures.	KCl
		Average 580	Selective corrosion and grain boundary attack.	
Rudkoping	450		Formation of a protective oxide.	Predominantly KCl, minor amounts of sulphates and silicates or K_2SO_4 with some KCl. Iron oxide dendrites in both.
		520–650	Selective corrosion.	
Ensted	447–470		Selective corrosion.	K_2SO_4 with minor amounts of KCl. Iron oxide dendrites.

Source: [28].

It is clear, therefore, that the enrichment of alkali metal chlorides at the metal/oxide/ash deposit interface can lead to increased rates of metal wastage due to gas-side corrosion, particularly on high temperature surfaces in the superheater and reheater sections of the boiler. In a number of cases, rapid metal loss, leading to significant corrosion-related operational problems has resulted.

This subject has been studied in some detail, particularly in Denmark, where there has been particular interest in the co-firing of cereal straws with coal in large power boilers for some time [28–31]. This is of particular technical interest because of the high final steam temperatures, and hence high superheater metal temperatures, in coal-fired utility boilers, and the understandable concerns about the risks associated with biomass co-firing.

Important and relevant information on these processes has been derived from plant experience and test work on the 80MW$_{th}$ CFB boiler at Grenaa in Jutland, which was designed for the firing of up to a 50:50 mixture of straw with coal, on a heat input basis.

This experience has indicated that the corrosion rates of superheater tubes when co-firing straw were of the order of 5–25 times greater than those measured when the boiler was firing coal alone.

The history of the superheater fouling and corrosion at Grenaa is very interesting and the findings may be instructive in helping to identify the key plant operational and other factors that can affect the corrosion process. For the first six months of operation, the unit was never run above 80 per cent load, and no particular problems with the formation of superheater fouling deposits were reported. Increasing the load to 100 per cent resulted in significant increases in the flue gas temperatures in the cyclone from around 850°C to around 900–1000°C. Serious problems with excessive fouling on the surfaces in the cyclones and the superheaters were reported. After 18 months of operation, it was found that corrosive damage to the superheater elements, which were operating at a final steam temperature of 505°C, was so serious that replacement of the final superheater elements was necessary. Selective chloride corrosion was considered to be responsible for the accelerated metal loss. Examination of the superheater tubes and of test specimens suggested that the chromium had been selectively removed from the metal surface and along grain boundaries. The protective oxide layers on the metal surface were irregular, porous, cracked and partly exfoliated. In many cases, significant enrichment of chloride was detected at the corrosion front in the degraded metal zone.

Considerable efforts were made to provide a description of the key factors responsible for the accelerated superheater tube wastage. One of the most important observations was that the great majority of the potassium in the superheater tube fouling deposits, and particularly at the metal/oxide/ash deposit interface, was in the form of potassium chloride, and it was considered that this was the principal agent of the selective chlorine corrosion.

A number of corrosion mechanisms have been suggested to describe the corrosion process at Grenaa in detail. The mechanism offered by Hansen et al **[32]** involves sulphation of the solid phase KCl followed by the release of chlorine gas, or possibly HCl, adjacent to the metal surface. The evidence for this type of mechanism lies in the results of the detailed microstructural examination of the corroded metal specimens. The chlorine selective corrosion involves the removal of chromium and to some extent iron from the metal grains and particularly at the grain boundaries. The chromium and iron formed oxides which are found at the degraded metal surface and between the degraded metal grains. In many cases, the degraded metal layer is porous, with no included metal oxides. No solid or liquid ash or flue gas components have been found between the grains in the degraded metal except for chlorine, which is associated with the chromium and iron. Hansen et al were of the view that this makes a liquid phase corrosion mechanism involving fused ash components unlikely, and that a mechanism involving a gas phase corrodent, probably chlorine gas or hydrogen chloride, is more plausible.

Significant modifications were made in response to the accelerated superheater corrosion and other operational problems at Grenaa. These included:

- the installation of additional heat transfer surface to reduce furnace temperatures;
- a switch to a coal of lower sulphur content; and
- a switch to a limestone of higher quality.

These modifications had the effect of providing significant reductions in the extent of fouling deposition in the boiler convective section and in the rate of superheater metal wastage.

The results of the corrosion test work performed during the straw/coal co-firing trials, at up to 20 per cent straw on a heat input basis, in a 150MW$_e$ pulverized coal boiler at Studstrup Power Station are also relevant in this context. The results were very different from those obtained from the CFB boiler at Grenaa. At Studstrup, the impact of co-firing straw, at a co-firing ratio of 10 per cent, on the measured corrosion rates were modest, and the evidence indicated a very different corrosion mechanism from that observed at Grenaa [28].

Examination of the fly-ash material indicated that the great majority of the potassium was associated with the alumino-silicates in the coal ash. It was considered that the potassium species released into the vapour phase in the flame reacted with the other ash constituents, and that little or no KCl was present at the metal/oxide ash deposit interface. The chlorine released from combustion of the straw was present in the flue gases in the form of HCl, and possibly Cl_2. It was found that the dominant corrosion mechanism relevant to the superheater tubes, at straw co-firing levels up to 10 per cent, was oxidation with perhaps some alkali sulphate melt corrosion, i.e. the normal corrosion mechanisms associated with coal firing.

It was considered that the key difference between the CFB at Grenaa and the pulverized fuel combustion at Studstrup is that the flame temperatures in pulverized fuel systems are considerably higher, and that this permits the interaction between the released potassium species and the other ash components. Furnace gas temperatures in CFB systems are too low for substantial interaction of this type to occur. It should also be noted that the straw/coal co-firing ratio at Grenaa was significantly higher than at Studstrup and that the differences observed in the behaviour of the potassium and chlorine may also be related to the co-firing ratio.

It is clear, therefore, that the co-firing of biomass materials with coal introduces a risk of accelerated corrosion rates of high temperature boiler tubes, and that the risk is associated with the increased levels of available alkali metal species and chlorine released into the boiler flue gases. Under unfavourable circumstances, as for instance at Grenaa, the impact of the accelerated metal wastage can be dramatic and there can be significant impacts on the operation and integrity of the plant. These risks do appear to be lower in pulverized fuel-fired boilers with biomass fuels with lower ash, potassium and chlorine contents and at low co-firing ratios.

The situation in boilers where biomass materials are co-fired with pulverized coal (see for instance, [29] and the references cited therein) has been summarized in the following terms:

- The coal ash deposits tend to be dominated by alkali metal sulphates, and particularly sodium sulphates. The co-firing of some biomass materials may introduce increased levels of chloride into the system. In most cases that have been of industrial interest, or that have been studied to date, the co-firing ratios have been relatively low and the chemistry of the ash deposit has been dominated by the alkali metal sulphates.
- There is also evidence of reaction at very high temperatures within pulverized coal furnaces of volatilized potassium species with silicate and alumino-silicate coal ash particles, to form potassium silicates and alumino-silicates, and reducing the availability of the potassium for the formation of chlorides and sulphates local to the tube surfaces.

Overall, therefore, there is evidence from both laboratory- and plant-scale experimental work, and plant experience, particularly at Studstrup Power Station in Denmark, where cereal straws and other baled materials are co-fired with coal at a co-firing ratio up to around 10 per cent on a heat input basis [21, 33], that the co-firing of biomass materials at low co-firing ratios has only had a very modest impact on the gas-side corrosion rates of high temperature boiler surfaces.

8.6.4 Preventive and remedial measures for fire-side corrosion

The general experience with the operation of boiler plants firing a wide range of clean biomass materials has been that, at final steam temperatures in the range 500–540°C, unacceptably high rates of metal wastage of superheater elements can occur. In boilers firing contaminated biomass materials and a wide variety of waste materials, with high alkali metal contents, significant chlorine contents and low sulphur contents, significant corrosion of the superheater tubes can occur at even lower final steam temperatures.

There are a number of potential approaches to these problems:

- control of the final steam temperatures, at the boiler design stage, to levels at which the corrosion rates are acceptable, for the fuel being fired and the superheater materials employed;
- the use of fire-side additives to modify the flue gas and ash deposit chemistry; and
- the selection of more resistant alloys for construction of the final superheaters.

The conventional approach to the control of superheater corrosion is by selection of the appropriate combination of tube materials and final steam temperatures for the fuel being fired. For instance, in modern incineration plants for municipal solid wastes, where the flue gases and ash deposits are extremely aggressive, it is common practice to limit the final steam temperatures to around 400°C, and to protect leading tubes in the final superheater against corrosion and erosion processes with SiC sleeving. Modern biomass-fired boiler plants commonly have final steam temperatures in the range 450–540°C, depending principally on the characteristics of the fuel and the materials selected for the construction of the final superheater elements. These are design decisions taken by the boiler supplier, based on previous experience and the best technical information available.

A number of plant-based studies, which involve the exposure of test coupons prepared from the candidate superheater materials to the flue gases from biomass combustion at controlled metal temperatures and for prolonged periods of time, have been carried out. The results of a significant programme of plant-based corrosion test work in small straw-fired power plants, i.e. Masnedo, Rudkobing and Ensted, in Denmark, has been described in Section 8.6.3.

The power stations were firing cereal straws and both ferritic and austenitic superheater alloys were exposed to the flue gas atmosphere at metal temperatures in the range 450–620°C. It was found that the measured corrosion rates for the test materials increased with increasing metal temperature, from around <0.05mm/1000h at 470°C to values in excess of 1mm/1000h at temperatures above 600°C. Clearly, one of the key parameters controlling the corrosion rate is the tube metal temperature.

They also found that all of the alloys tested gave fairly similar corrosion rates at any given temperature, although there appeared to be a shallow optimum in corrosion resistance for alloys with chromium contents in the range 15–18 per cent. These results are not untypical for boilers firing biomass materials, where the corrosion process is driven by the presence of alkali metal chlorides at the metal/corrosion product/deposit interface.

The use of fire-side additives to modify the chemistry of the ash deposits may be of some benefit. This is common practice, for instance, in oil-fired boilers, where magnesia-based additives are fairly commonly employed to reduce the rates of metal wastage associated with the relatively aggressive vanadium oxides and sulphates in the oil ash deposits.

Vattenfall has recently developed a fire-side additive, which has been of some value in reducing the active chloride concentration in biomass-fired boilers or possibly waste incineration plants, and which may have wider application [34]. They reported the use of a proprietary liquid fire-side additive, ChlorOut, which was effective in removing KCl from the flue gases, but had only a small effect on the SO_2 concentration and on the pH of the flue gas condensate. The results of 1000-hour corrosion tests in a 100MW$_{th}$ BFB boiler firing demolition wood, forestry residues and coal in Sweden indicated that there was a significant reduction in the measured corrosion rates when the additive was applied. Vattenfall are currently applying the ChlorOut system, comprising the liquid additive and delivery system, with an in-furnace alkali chloride measurement system, for use in biomass boilers and waste incineration plants.

As stated above, the conventional responses to the excessive metal wastage of boiler tubes due to high temperature corrosion include replacement of the affected tubing with a more resistant material or protection of the tubing with a more corrosion-resistant coating or sleeving, or a weld overlay. The selection of the appropriate materials for these applications is obviously a key issue and this has been the subject of significant laboratory experimentation and boiler probing work.

The key results of an important programme of work which involved the insertion of corrosion probes, with a range of relevant conventional superheater materials, into the flue gas stream of a straw-fired boiler, and the measurement of the metal wastage rates are summarized in Figure 8.8. This shows the measured corrosion rates in mm/1000h as a function of the metal temperature. The major trend is clear, with the corrosion rates increasing sharply with increasing metal temperature, as expected. All of the superheater materials tested, which have chromium contents in the range 11.7–18.4 per cent, show the same trends with temperature. In very general terms, the authors were of the view that the corrosion resistance of the materials tested, with chromium contents in the range quoted above, were fairly similar.

Looking at the absolute corrosion rate values measured during these tests, it is clear that at metal temperatures around 460°C the measured corrosion rates for all of the materials were less than 0.05mm/1000h, the equivalent of around 0.4mm p.a. This is a relatively high corrosion rate, and is probably a result of the relatively short exposure times for the probing trials. At metal temperatures above about 500°C, the measured corrosion rates for all of the test materials were in the range 0.1–0.3mm/1000h, or of the order of 0.8–2.4mm p.a. These are very high corrosion rates, fairly severe if they were to apply to the boiler tubes.

Overall, therefore, it is clear from both laboratory and plant test work, and from plant experience, that the high temperature corrosion of superheater tubes is a significant concern in biomass boilers, and particularly in boilers firing fuels with significant chlorine contents. The key to the avoidance of excessive rates of metal wastage is the selection, at the design stage, of the correct combination of the final steam temperature and the tube material for

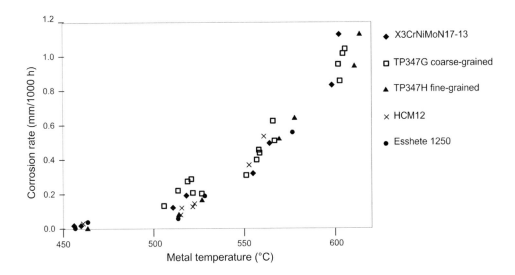

Figure 8.8 *Corrosion rates of boiler tube material specimens exposed to flue gases from straw combustion, plotted against the metal temperature*

Source: [25].

the particular application. It is also clear that the use of fuel additives can be of value in some cases, where problems arise on operating plants.

8.6.5 Erosion and abrasion of boiler components and other equipment

The erosion and abrasion of boiler components and other equipment in solid fuel-fired plants, at all scales of operation, is associated predominantly with the presence in the fuels and ashes of hard mineral particles, and particularly those that are harder than the steels and refractory materials employed for the construction of the interior surfaces in the boiler. The only mineral species that is commonly found in clean biomass materials in significant levels, and which is in this category, is quartz, and only high quartz biomass materials or those contaminated with significant levels of tramp materials are expected to present problems with erosion and abrasion in the fuel handling and firing equipment.

The fly ashes from biomass combustion tend, in the main, to be in very fine particulate form and are relatively soft, and are not considered to be particularly abrasive or erosive in nature to boiler tubes and components of the ash handling systems.

In general terms, the majority of the biomass materials under consideration have relatively low ash contents and, for this reason, erosion and abrasion processes tend, on the whole, to be less important than they are in coal-fired plants. There are, however, one or two specific areas where erosion and abrasion can be significant issues:

- the utilization of some biomass materials, such as rice husks, which have particularly high quartz content can give rise to abrasion problems in the fuel and ash handling systems and the erosive wear of boiler components;

- the formation of fused ash material, particularly the bottom ashes from grate-fired systems, can cause erosion and abrasion damage to the components of both mechanical and pneumatic ash handling systems; and
- in biomass boiler systems that suffer severe convective section fouling problems, excessive rates of particle impact erosive wear of boiler tubes and erosive wear associated with the regular use of convective pass soot-blowers, in an attempt to control the impact of the fouling, are common occurrences.

In general terms, however, the experience has been that the ash abrasion and erosion problems associated with the utilization of the great majority of biomass materials are significantly less important than those experienced when firing more conventional solid fuels.

8.7 Biomass ash utilization and disposal

With the increase in the utilization of biomass materials for the production of heat and power, and for co-firing with fossil fuels in large power plants, the optimization of biomass ash utilization for beneficial purposes, and the minimization of any negative impacts of ash disposal, are becoming of increasing environmental and commercial importance. For the sustainable utilization of biomass in dedicated domestic and industrial/commercial combustion plants, the most beneficial approach involves the return of the ashes to the field or the forest. The ashes normally have some value as a relatively low grade fertilizer, based principally on the lime, potassium and phosphorus content of the ash.

For the small-scale utilization of clean, untreated biomass materials, recycling of the ashes is fairly widely practised worldwide. In the more industrialized countries, where the biomass may contain significant levels of inorganic pollutants, for ashes from combustion plants burning treated biomass materials, and for large plants burning or co-firing biomass, there may be restrictions on the quantities of ash that can be utilized in this way, or specific limits on the concentrations of particular trace element/heavy metal species in ashes that can be returned to the soil. The levels of the inorganic pollutants in the ashes are dependent entirely on the composition of the fuel being fired.

There are also some concerns about the potential for ashes to contain significant levels of organic pollutant species, i.e. polychlorinated dibenzodioxins/furans (PCDDs and PCDF) and polyaromatic hydrocarbons (PAH). The concerns are particularly with the fly-ashes and boiler deposit materials from domestic and industrial/commercial grate and fluidized bed boilers, and particularly in cases where the combustion conditions are relatively poor and the unburned char levels in the fly ashes are high. There are generally fewer concerns with the bottom ashes, which have usually been subjected to higher temperatures for longer residence times. Since the organic pollutants are principally a product of poor combustion conditions, the preferred means of controlling the levels of organic pollutants in the ashes, in most cases, is to specify a maximum unburned carbon level in the fly ashes to be utilized for a particular purpose.

If the ash cannot be recycled to the place of origin of the biomass or cannot be utilized as a fertilizer, for the reasons described above, there are a number of other potential beneficial uses [35]:

- the bottom ashes from grate-fired and fluidized bed combustors may be utilized as road construction or landscaping materials;
- fly ashes may be used as a component of cement blends and mortars; and
- fly ashes may also be ulitized as a component of lightweight aggregates.

The main technical and commercial barriers to biomass ash utilization for the purposes listed above are that they tend to be available in relatively small quantities compared to other competitive materials, and it is considered that the fly-ash quality from biomass combustion processes tends to be relatively inconsistent.

When co-firing biomass materials with coal in large coal-fired boilers, mixed biomass ash/coal ash discard streams are generated, and the technical and commercial issues associated with ash utilization/disposal from large utility boilers are somewhat different from those for the smaller domestic and industrial/commercial biomass combustion plants.

A number of the more important environmental impacts of the operation of the large coal-fired power plants, which may be employed for co-firing biomass, are associated with the very large quantities of solid materials discarded from these processes. These are, principally:

- the furnace bottom ash residues;
- the boiler ashes from the boiler hoppers and the fly ashes collected in the particulate emission control equipment; and
- the solid residues of FGD processes.

In the main, these materials are produced and handled separately. There are, however, a small number of plants that produce a mixed coal ash/FGD residue discard material. The quality of the ash residues may be affected by the co-firing of biomass. The quality of the FGD residue material is not normally significantly affected by co-firing.

Very large quantities of these materials are produced. Annual statistics on this are collected by the European Coal Combustion Products Association (ECOBA),* in Essen, Germany, by the American Coal Ash Association (ACAA)† in North America, and by other national organizations.

As stated above, the boiler ash discards are produced from a number of outlets, including:

- fly ashes from the electrostatic precipitator or bag filter, and boiler ashes from the ash hoppers under the boiler economizer and air preheater; and
- bottom ashes and slags from the furnace ash hoppers and slag taps.

On average, around 80 per cent of the total ash discard material from large pulverized fuel-fired boiler plants is in the form of **fly ashes** from the particle collection equipment, and the economizer and air preheater hoppers. This is a fine powder with the great majority of the particles being less than 50μm in diameter.

---

* European Coal Combustion Products Association (www.ecoba.com) publishes useful information on the solid discards from coal plants.

† The American Coal Ash Association (www.acaa-usa.org) provides up-to-date statistics on coal ash production and utilization in the US.

The overall level of utilization of fly ashes in Europe is around 50 per cent; however, the rate in different countries varies markedly. In Holland, Germany, Belgium, Italy and Denmark utilization rates are in excess of 80 per cent, and significant advances have been made in the reclamation of old fly-ash dumps. In other countries, utilization rates are much lower. In Britain, for instance, which produces around 10 million tonnes per annum, the utilization level is less than 50 per cent. Overall, there is a trend towards the increased utilization of fly ashes, particularly for higher value uses, as a constituent of construction materials.

For some of the higher value uses of boiler fly ashes, such as for cement and concrete applications, the quality of the fly-ash is controlled by a standard specification or classification systems. A very useful listing of the international standards and specifications of boiler ashes is given at www.iflyash.com.

These standards usually specify the loss on ignition, the SO_3 content and the fineness, and sometimes include a specified level of pozzolanic behaviour for the ash, i.e. the tendency to set hard when mixed with lime and water and allowed to dry. In most countries, specialist commercial companies are involved in the treatment, storage and sale of boiler ashes, principally to the manufacturers of construction materials and the civil engineering industry.

On average, around 20 per cent or so of the total ash discards from pulverized coal-fired boilers is in the form of **bottom ashes and slags**, which have been retained within the furnace and are collected from the furnace ash hopper. In the main, this material comprises fused and partly fused furnace slags and ash deposit material dislodged from the upper furnace. It is very variable in nature, and can range in size from small particles less than 1mm in diameter to large lumps of fused slag up to 1m or so in size. Chemically, the boiler bottom ashes tend to be similar to the fly ashes, however they tend to have relatively low unburned carbon contents and low SO_3 contents.

The co-firing of biomass materials with coal will result in the production of mixed ashes, i.e. the fly-ashes, and the bottom ashes will comprise of the chemical constituents of both the coal and the biomass ashes. The chemical and physical properties of the mixed ash discards will clearly depend on the ash contents and compositions of the coal and the biomass ashes, and on the co-firing ratio. The technical and environmental aspects of the utilization and disposal of these mixed ash discards will require careful consideration.

For the higher value uses, particularly of the fly ashes, compliance with the standard specifications and classifications will be a requirement, and it should be noted that these standards were not specifically prepared with biomass co-firing in mind. It is sometimes the case that the standards have been poorly drafted, and they may be over-prescriptive. Standards in a number of European countries have been redrafted recently to permit the ashes from biomass co-firing to be utilized in concrete production [36]; see also Section 7.6.15.

8.8 References

1 KORBEE, R., KIEL, J. H. A., ZEVENHOVEN, M., SKRIFVARS, B.-J., JENSEN, P. A. and FRANDSEN, F. J. (2001) 'Investigation of biomass inorganic matter by advanced fuel analysis and conversion experiments', in *Power Production in the 21st Century: Impacts of Fuel and Operations*, Snowbird, Utah

2 MARSCHNER, H. (1997) *Mineral Nutrition of Higher Plants*, Academic Press, London

3   BACKMAN, R., SKRIFVARS, B.-J. and YRJAS, P. (2005) 'The influence of aerosol particles on the melting behaviour of ash deposits in biomass-fired boilers', *Aerosols in Biomass Combustion*, Graz, March
4   SKRIFVARS, B.-J., LAUREN, T., HUPA, M., KORBEE, R. and LJUNG, P. (2004) 'Ash behaviour in a pulverised wood fired boiler – a case study', *Fuel*, vol 83, pp1371–1379
5   SKRIFVARS, B.-J., BLOMQUIST, J.-P., HUPA, M. and BACKMAN, R. (1998) 'Predicting the ash behavior during biomass combustion in FBC conditions by combining advanced fuel analyses with thermodynamic multi-component equilibrium calculations', presented at the 15th Annual International Pittsburgh Coal Conference, Pittsburgh, PA, US, September
6   ZEVENHOVEN, M., SKRIFVARS, B.-J., YRJAS, P., HUPA, M., NUUTINEN, L. and LAITINEN, R. (2001) 'Searching for improved characterisation of ash forming matter in biomass', in *Proceedings of the 16th ASME International FBC Conference*, Reno, Nevada, US, May
7   RAASK, E. (1985) *Mineral Impurities in Coal Combustion: Behavior, Problems, and Remedial Measures*, Hemisphere Publishing, New York
8   MILES, T. R., MILES, T. R. jnr., BAXTER, L. L., BRYERS, R. W., JENKINS, B. M. and ODEN, L. L. (1995) *Alkali Deposits Found in Biomass Power Plant: A Preliminary Study of their Extent and Nature*. National Renewable Energy Laboratory, Golden, Co. US
9   BAXTER, L. L. (1993) 'Ash deposition during biomass and coal combustion: A mechanistic approach', *Biomass and Bioenergy*, vol 4, pp85–102
10 BAXTER, L. L., MILES, T. R., MILES, T. R., jnr., JENKINS, B. M. MILNE, T., DAYTON, D., BRYERS, R. W. and ODEN, L .L. (1998) 'The behaviour of inorganic material in biomass-fired power boilers: Field and laboratory experiences', *Fuel Processing Technology*, vol 54, pp47–78
11 OBERNBERGER, I., BRUNNER, T. and JOLLER, M., (2001) 'Characterisation and formation of aerosols and fly ashes from fixed-bed biomass combustion', in Nussbaumer, T. (ed.), *Aerosols from Biomass Combustion*, International Seminar, 27 June 2001 in Zurich, organized on behalf of International Energy Agency (IEA) Bioenergy Task 32: Biomass Combustion and Cofiring and the Swiss Federal Office of Energy, Verenum, Switzerland
12 OBERNBERGER, I. and BRUNNER, T. (2005) 'Fly ash and aerosol formation in biomass combustion processes – an introduction', *Aerosols in Biomass Combustion*, vol 6 in Thermal Biomass Utilization Series, BIOS Bioenergy Systeme, Graz, Austria
13 VALMARI, T. (2000) 'Potassium behaviour during combustion of wood in circulating fluidised bed power plants', VTT Publication 414
14 BAIN, R. L., OVEREND, R. P. and CRAIG, K. R. (1998) 'Biomass-fired power generation', *Fuel Processing Technology*, vol 54, pp1–16
15 SKRIFVARS, B.-J., SFIRIS, G., BACKMAN, R., WIDEGRIN-DAFGARD, K. and HUP, A. M. (1996) 'Ash behaviour in a CFB boiler during combustion of Salix', in *Proceedings of the IEA Conference on Biomass Utilisation*, Banff
16 BRYERS, R. (1996) 'Fireside slagging, fouling and high temperature corrosion of heat transfer surfaces due to impurities in steam raising fuels', *Prog. Energy Combustion Science*, vol 22, pp29–120
17 BILL, A., DECKER, R. and LARSSON, A.-C. (2005) 'SCR for biomass derived fuels: Operating experience and performance results from a slipstream at the Borssele Power Plant', presented at Power-Gen International, Las Vegas, December
18 AHONEN, M. (1996) 'Long-time experience in catalytic flue gas cleaning and catalytic NO_x reduction in biofueled boilers', *VTT Symposium*, vol 163
19 BECK, J., UNTERBERGER, S., SCHEFFKNECHT, G., JENSEN, A., ZHENG, Y, and JOHNSSON, J. E. (2005) 'The effect of biomass co-combustion on De-NO_x catalysts', presented at the Second Internatioal Conference on Clean Coal Technologies for our Future, Cagliari, May
20 BAXTER, L. L. and KOPPEJAN, J. (2005) 'Biomass-coal co-combustion: Opportunity for affordable renewable energy', available on the IEA Bioenergy Task 32 website www.ieabcc.nl
21 OVERGAARD, P., SANDER, B., JUNKER, H., FRIBORG, K. and LARSEN, O. H. (2004) 'Two years' operational experience and further development of full-scale co-firing of straw', presented at the 2nd World Conference on Biomass for Energy, Industry and Climate Protection, Rome, May
22 OTTOSEN, P. (2005) 'Impact of co-combustion of wood pellets at Avedore Power Plant fired with

heavy fuel oil and natural gas', presented at IEA Bioenergy Workshop 2, Co-utilisation of biomass with fossil fuels, Copenhagen, May, and available on www.ieabioenergy.com, click on 'workshops'

23 RAASK, E. (1988) *Erosion Wear in Coal Utilisation*. Hemisphere
24 NEILSEN, H. P., FRANDSEN, F. J., DAM-JOHANSEN, K. and BAXTER, L. L. (2000) 'The implications of chlorine-associated corrosion on the operation of biomass-fired boilers' *Prog. Energy and Combustion Science*, vol 26, pp283–298
25 HENRIKSEN, N., MONTGOMERY, M. and LARSEN, O. H. (2002) 'High temperature corrosion in biomass-fired boilers', *VDI-Berichte*, No. 1680
26 BAXTER, L. L., TREE, D., FONSECA, F. and LUCAS, W. (2003) 'Biomass combustion and co-firing issues overview: Alkali deposits, fly ash, NO_x/SCR impacts', presented at the International Conference on Co-utilisation of Domestic Fuels, Gainesville, FL, February
27 RIEDL, R., DAHL, J., OBERNBERGER, I. and NARODOSLAWSKY, M. (1999) 'Corrosion in fire tube boilers of biomass combustion plants', *Proceedings of the China International Corrosion Control Conference*, Beijing
28 MONTGOMERY, M., KARLSSON, A. and LARSEN, O. H. (2002) 'Field test corrosion experiments in Denmark with biomass fuels. Part 1: Straw firing', *Materials and Corrosion*, vol 53, pp121–131
29 MONTGOMERY, M. and LARSEN, O. H. (2002) 'Field test corrosion experiments in Denmark with biomass fuels. Part 2: Co-firing of straw and coal', *Materials and Corrosion*, vol 53, pp185–194
30 HENRIKSEN, N. and LARSEN, O. H. (1997) 'Corrosion in ultrasupercritical boilers for straw combustion'. *Materials at High Temperature*, vol 14, pp227–236
31 HENRIKSEN, N. and LARSEN, O. H. (1996) 'Fouling and corrosion in straw and coal-straw fired USC plants', in *Conference on Biomass Energy Environment, European Bioenergy Conference*, pp192–197
32 HANSEN, L. A., NIELSEN, H. P., FRANDSEN, F. J., DAM-JOHANSEN, K., HORLYCK, S. and KARLSSON, A. (2000) 'Influence of deposit formation on corrosion in a straw-fired boiler', *Fuel Processing and Technology*, vol 64, pp189–209
33 FRIBORG, K., OVERGAARD, P., SANDER, B., JUNKER, H., LARSEN, O. H., LARSEN, E. and WIECK-HANSEN, K. (2005) 'Full-scale co-firing of straw, experience and perspectives', presented at the Second International Conference on Clean Coal Technologies for our Future, Cagliari, May
34 HENDERSON, P., SZAKALOS, P., PETTERSSON, R., ANDERSSON, C. and HOGBERG, J. (2006) 'Reducing superheater corrosion in wood-fired boilers', *Materials and Corrosion*, vol 57, pp128–134
35 PELS, J. R., DE NIE, D. S. and KIEL, J. H. A. (2005) 'Utilisation of ashes from biomass combustion and gasification', 14th European Biomass Conference and Exhibition, Paris
36 SPLIETHOFF, H., UNTERBERGER, S. and HEIN, K. R. G. (2001) 'Status of co-combustion of coal and biomass in Europe', presented at the Sixth International Conference on Technology and Combustion for a Clean Environment, Oporto, July

9

Environmental Aspects of Biomass Combustion

9.1 Introduction

Biomass combustion influences the environment mainly through emissions to the atmosphere. Depending on the emission component, it influences both the local, regional and global environment. The local environment is affected mainly by particle emissions and other components caused by incomplete combustion. The regional environment is affected by acid precipitation originating mainly from NO_x and SO_2 emissions, while the global environment is affected by emissions of direct or indirect greenhouse gases and through ozone depletion.

The amount of pollutants emitted to the atmosphere from different types of biomass combustion applications is highly dependent on the combustion technology implemented, the fuel properties, the combustion process conditions, and the primary and secondary emission reduction measures that have been implemented.

Due to the wide diversity of biomass fuels, with highly varying elemental composition, moisture content, density and thermochemical behaviour, many different types of biomass combustion applications have been developed. These biomass combustion applications cover a wide range, from small-scale units for room heating to large-scale power plants. Hence, the air pollution control technologies must be selected with care, and economy will always be a limiting factor. However, implementation of more stringent emission limits forces the development of low emission biomass combustion applications forward.

In general, emission reduction measures developed for combustion of fossil fuels can also be applied for biomass combustion applications. All biomass combustion applications benefit from an optimized combustion process, which reduces emissions from incomplete combustion. However, to reduce the emissions further or to reduce emissions from complete combustion, secondary measures must usually be applied. For small-scale biomass combustion applications, secondary emission reduction measures are often not cost-effective, and the emission regulations are therefore usually not as strict as for large-scale biomass combustion applications.

For biomass combustion applications using virgin biomass as fuel, emission regulations are usually applied for emissions from incomplete combustion, such as some kinds of particles and CO. In some cases emission regulations are also applied for NO_x, which to some extent can also be reduced by primary emission reduction measures. However, in special cases it may be necessary to apply secondary NO_x emission reduction measures. Emissions of SO_2 are usually not significant for wood combustion applications due to the low sulphur content in wood. However, for biomass fuels such as miscanthus, grass and straw, emissions of SO_2 may be significant and SO_2 emission reduction measures must be applied.

In this chapter, the emission reduction measures applicable for biomass combustion applications will be presented, including both emissions from incomplete and complete combustion. However, to give the reader an overview of the environmental aspects of biomass combustion as well, the various emission components are first presented and their sources and impact on climate, environment and health are indicated. Then an overview of typical emission levels from various biomass combustion applications is given. These emission levels will be very dependent on the emission reduction measures implemented and are only guiding values reflecting today's typical biomass combustion applications, not the emission reduction potential. Following the presentation of primary and secondary emission reduction measures, emission limits for biomass combustion applications in selected IEA member countries will be presented as examples of where the emission limits stand today. However, establishment of emission limits is a dynamic process and emission limits are often adjusted according to the available technology and economic considerations. Finally, the environmental impacts of emissions from biomass combustion applications are analysed and compared with the environmental impacts of combustion of fossil fuels.

Various literature sources dealing with different aspects of air pollution exist. For further reading on air pollution in general, the following references are recommended: [1–7].

9.2 Environmental impacts of biomass combustion

In this section, the environmental impacts of emissions from biomass combustion applications are analysed and compared with the environmental impacts of combustion of fossil fuels. Furthermore, in Section 9.3 options for ash disposal and utilization will be discussed.

The environmental impacts of air pollution from most biomass combustion applications today are far from negligible. However, compared to fossil fuel combustion applications there are several advantages.

First of all, biomass is a renewable fuel, and is considered as being CO_2-neutral with respect to the greenhouse gas balance. However, this is only true if we are able to achieve very low levels of emissions from incomplete combustion [8], and if we do not include our use of fossil fuels in harvesting and transportation of biomass fuels, and our use of electricity produced from fossil fuels.

To evaluate the real environmental impacts of biomass combustion, a life cycle analysis (LCA) should ideally be carried out. This type of evaluation includes the various stages of the life cycle of the biomass, from procurement of the fuel, transportation, storage and conversion to the discharge and handling of ashes. The construction, operation, maintenance and decommissioning of the energy converting technology should also be included in the assessment according to the LCA method. The exchanges to the surrounding environment in terms of emissions to air, soil and water are then inventoried for each stage in the life cycle. These exchanges are added up to indicate environmental impacts such as global warming, acidification and ozone depletion in the environmental impact assessment.

In [9], an LCA study is presented in which the procurement and conversion of biomass and fossil fuels are compared. The study includes emissions of CO_2 (expressed as CO_2 equivalents), NO_x and SO_x. It concludes that procurement of biomass fuels, in general, consumes less energy than procurement of fossil fuels. Furthermore, it is concluded that fossil fuels cause the highest emissions for each of the parameters, CO_2, NO_x and SO_x.

Many of the activities related to the procurement of fuels involve the use of fossil fuels. Therefore, it is not fully correct to define biomass fuels as CO_2-neutral when considering the complete life cycle. However, from the results in [9] it can be concluded that the gap between the CO_2 impact from biomass and fossil fuels grows bigger when procurement is included in the considerations. The gap is even larger in the case of favourite biomass fuels.

Other aspects of the procurement should ideally be included in the assessment as well, such as, for example, the use of fertilizers and the release of N_2O during the growth of the biomass. N_2O is a very potent greenhouse gas and can therefore, even in small amounts, change the picture. The handling of the ashes should also be included, whether it is reused/recycled or must be sent to landfill due to a high content of heavy metals such as cadmium.

The increasing concentration of greenhouse gases in the atmosphere is the main incentive for the substantial increase in the use of biomass for heat and power production in IEA countries today. As a renewable fuel, biomass will be available for heat and power production in substantial amounts after the fossil fuel resources have diminished. Reserves of oil and natural gas are fast decreasing, while the reserves of coal will be available for much longer. However, as can be seen in Table 9.1, coal is the least attractive fossil fuel for heat and power production due to its high emission levels.

Table 9.1 *Emissions from a typical 2000MW fossil fuel power station using coal, oil or natural gas*

	Coal, conventional without flue gas desulphurization [kt/year]	Oil, conventional [kt/year]	Gas, combined cycle [kt/year]
Carbon dioxide	11,000	9000	6000
Airborne particulates	7	3	Negligible
Sulphur dioxide	150*	170	Negligible
Nitrogen oxides	45	32	10
Carbon monoxide	2.5	3.6	0.270
Hydrocarbons	0.750	0.260	0.180
Hydrochloric acid	5–20	Negligible	Negligible
Solid waste and ash	840	Negligible	Negligible
Ionizing radiation (Bq)	10^{11}	10^9	10^{12}

Note: * Coal without desulphurization is not a state-of-the-art technology today, so this figure does not represent the present best practice.
Source: [10].

NO_x and SO_x emissions from biomass combustion applications are, in general, low compared to coal and oil combustion. In most cases, the NO_x emission level can be significantly reduced by the use of primary emission reduction measures, and can be further reduced by implementing secondary emission reduction measures. Emissions of SO_x can easily be reduced by secondary emission reduction measures, and in some cases also by primary emission reduction measures. However, for many biomass combustion applications, secondary emission reduction measures for NO_x and SO_x are not cost-effective due to the fact that biomass combustion applications are usually much smaller than fossil fuel combustion applications.

The main disadvantage of biomass combustion applications, especially small-scale applications such as wood-stoves, fireplaces, and wood log boilers, is their high level of emissions from incomplete combustion compared to fossil fuel combustion applications. Many of these small-scale biomass combustion units are based on natural draught and are also operated as batch or semi-continuous systems. In addition, combustion process control systems are usually not cost-effective.

One of the main advantages of biomass combustion for heat and power production is the utilization of a renewable fuel source, which in most cases is locally available, and which in many cases has a very low or even negative alternative value. Emission reduction measures are known and are available for all harmful emission components; it merely depends on emission limits and cost-effectiveness whether the emission reduction measures are implemented or not. By increasing the size of the biomass combustion applications, improved emission reduction possibilities will become cost-effective. However, local availability of the biomass fuel and transportation costs will usually be a limiting factor for the size of a biomass combustion application. The possibility of co-firing biomass and fossil fuels is promising, both with respect to effective biomass utilization for heat and power production and for achieving low emission levels.

In short, biomass is an environmentally sustainable fuel for heat and power production that is important today and which will increase in importance in the years to come. Technology to reduce emissions and improve combustion efficiency is being developed continuously. Thus the negative environmental impacts of biomass combustion that exist today will be reduced in future plants.

9.2.1 Emission components and their main influencing factors

Emissions from biomass combustion applications can be divided into two major groups: emissions from complete combustion and emissions from incomplete combustion. It should be noted that particle emissions can be a result of both complete and incomplete combustion. They are treated in the two subsections below.

9.2.1.1 Emissions from complete combustion

The following components are emitted to the atmosphere as a result of complete combustion in biomass combustion applications.

Carbon dioxide (CO_2)

CO_2 is a major combustion product from all biomass fuels, originating from the carbon content in the fuel. However, CO_2 emissions from biomass combustion are regarded as being CO_2-neutral with respect to the greenhouse gas effect and this is considered to be the main environmental benefit of biomass combustion.

Nitrogen oxides (NO_x)

NO_x emissions from biomass combustion applications are mainly a result of complete oxidation of fuel nitrogen (see Section 2.3.2.3), in contrast to fossil fuel combustion applications where nitrogen in the air also contributes, to some extent, to the NO_x emission level. NO_x is formed both in the gas phase combustion and in char combustion. The main nitrogen oxide emitted is nitric oxide, NO, which is converted to NO_2 in the atmosphere.

The possible gas phase reaction mechanisms for NO_x formation in biomass combustion applications are:

1. **The fuel NO_x mechanism [11–13]:** Fuel nitrogen is converted to NO (> 90 per cent) and NO_2 (< 10 per cent) through a series of elementary reaction steps called the fuel NO_x mechanism. Important primary nitrogen-containing components are NH_3 and HCN. However, significant amounts of NO and N_2 may also be found in the pyrolysis gas. If sufficient O_2 is available, NH_3 and HCN will mainly be converted to NO through different reaction routes. However, in fuel-rich conditions NO will react with NH_3 and HCN, forming N_2. This is utilized as a primary NO_x reduction measure. By optimizing the primary excess air ratio, temperature and residence time, a maximum conversion of NH_3 and HCN to N_2 can be achieved (see Section 9.2.4.7).
2. **The thermal NO_x mechanism [14, 15]:** Nitrogen in the air starts to react with O radicals and forms NO at temperatures above approximately 1300°C. The amount of NO formed increases with increasing temperature, O_2 concentration and residence time. However, in biomass combustion applications the combustion temperatures are, in general, lower than 1300°C. Thermal NO_x is a post-flame problem, meaning that the major formation of thermal NO_x occurs in the post-flame gases, after the main combustion, due to its dependence on O_2 concentration and residence time.
3. **The prompt NO_x mechanism [11, 16]:** Nitrogen in the air may also react with CH, mainly, forming HCN, which then follows the reaction steps of the fuel NO_x mechanism. The prompt NO_x mechanism is less temperature-dependent and much faster than the thermal NO_x mechanism. However, it is only important in fuel-rich conditions and is very dependent on the CH concentration. The prompt NO_x mechanism has not been found to be of significant importance in biomass combustion applications, in contrast to fossil fuel combustion applications.

Additionally, fuel nitrogen is retained in the char and is largely oxidized to NO in the char combustion phase, but may subsequently be reduced to N_2 by a fast heterogeneous reaction with the char [17]. The amount of fuel nitrogen retained in the char relative to the amount of fuel nitrogen released in the devolatilization phase is determined in part by the thermal exposure of the fuel [18].

The fuel NO_x emissions increase with increasing nitrogen content in the fuel, excess air ratio and combustion temperature, up to a point where all fuel nitrogen intermediates have been converted to either NO_x, N_2O or N_2. However, the fraction of fuel nitrogen converted to NO_x decreases with increasing nitrogen content in the fuel, as can be seen in Figure 9.1. This has also been shown by other investigators [19–21]. Figure 9.2 illustrates the relative importance of the fuel, thermal and prompt NO_x mechanisms as a function of combustion temperature. NO_x emissions may be reduced by both primary and secondary emission reduction measures. Further references: [8, 22–46].

Nitrous oxide (N_2O)

N_2O emissions are a result of complete oxidation of fuel nitrogen. However, the N_2O emission levels measured in various biomass combustion applications are very low. This is a result of several influencing factors, which will not be discussed here. Even though the N_2O emission levels from biomass combustion are very low, they do contribute to some degree to the greenhouse gas effect, because of the high global warming potential (GWP) factor of

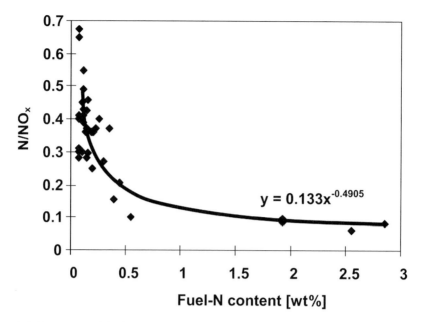

Figure 9.1 *Fraction of fuel nitrogen converted to NO_x for various wood fuels in various wood combustion applications, as a function of fuel nitrogen content, together with a trend line*

Source: [31].

N_2O, and to ozone depletion in the stratosphere. N_2O emissions may be reduced by primary emission reduction measures. Further references: **[21, 33, 46–53]**.

Sulphur oxides (SO_x)

Sulphur oxides (SO_x) are a result of complete oxidation of fuel sulphur. It is mainly SO_2 (> 95 per cent) that is formed, however, some SO_3 (<5 per cent) may be formed at lower temperatures. The fuel sulphur (see Section 2.3.2.3) will not be completely converted to SO_x; a significant fraction will remain in the ashes while a minor fraction is emitted as a salt (K_2SO_4) or as H_2S at lower temperatures. Measurements at two district heating plants in Denmark **[54]** using straw as fuel showed that 57–65 per cent of the sulphur was released into the flue gas, while the remainder was bound in the ashes. Houmøller and Evald **[55]** reported nine closed sulphur balances from several full-scale measurements in Denmark, four straw-fired units, three woodchip-fired units and two pellet-fired units. They found similar results. SO_2 emissions may be reduced by primary measures such as lime or limestone injection or by secondary measures. Further references: **[33, 51, 56–61]**.

Hydrogen chloride (HCl)

Part of the chlorine content in the fuel will be released as HCl. The chlorine content of wood is very low (see Section 2.3.2.3). However, significant amounts of HCl may be formed from biomass fuels containing higher amounts of chlorine, such as miscanthus, grass and straw. The fuel chlorine will not be completely converted to HCl; the main fraction is retained in salts (KCl, NaCl) by reaction with K and Na, while traces are emitted as dioxins and organic

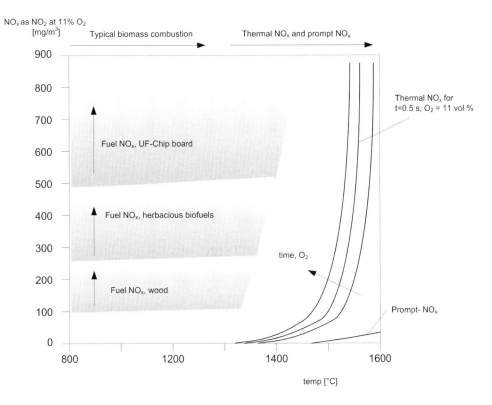

Figure 9.2 *Fuel-NO_x emission levels as a function of temperature and fuel type (fuel-N content), and comparison with thermal and prompt NO_x formation*

Source: [32].

chlorine components. HCl emissions may be reduced by washing of the fuel, which is utilized to some extent for straw due to its high chlorine content, and by secondary emission reduction measures. Further references: [60–63].

Particles

Particle emissions originate from several sources. Among these are fly-ash, which is a result of entrainment of ash particles in the flue gas, and aerosols. Other types of particle emissions are mentioned in the next section on emissions from incomplete combustion.

Fly-ash consists of coarse fly-ashes (particles with a diameter larger than 1μm) and aerosols (particles with a diameter less than 1μm). While coarse fly-ashes result from entrainment of ash and fuel particles from the fuel bed, aerosols are formed from compounds (e.g. salts like KCl, NaCl, K_2SO_4). Aerosols are a result of reactions between K or Na and Cl or S, released during combustion from the burning fuel to the gas phase, and subsequently from sub-micron particles by nucleation and condensation processes. Since the main constituents of aerosols are easily volatile elements including heavy metals, and due to the fact that the compounds of these elements are characterized by comparatively low melting temperatures, aerosols can cause severe problems concerning deposit formation in boilers and increase the risk for boiler corrosion, as well as cause emission problems [64]. See section 8.3.2 for a more detailed discussion on aerosols in grate furnaces.

Secondary particle emission reduction measures – electrostatic precipitators (ESPs) and baghouse filters – are utilized to reduce the particle emission level in large-scale biomass combustion applications. Since such filters are only economically affordable for medium- to large-scale applications, small-scale biomass combustion units have been identified as a significant source of fine particles, which can cause health problems as immissions.

By optimal design of combustion chambers, coarse fly-ash particles entrained in the flue gas may to some degree be prevented from leaving the combustion chamber, instead falling down to the bottom of the combustion chamber to be removed as bottom ash.

Heavy metals

All virgin biomass fuels contain heavy metals to some degree (the most important are Cu, Pb, Cd, Hg). These will remain in the ash or evaporate, and also attach to the surface of particles emitted to the atmosphere or be contained inside fly-ash particles. Contaminated biomass fuels, such as impregnated or painted wood may contain significantly higher levels of heavy metals. One example is the presence of Cr and As in CCA (chromium-cadmium-arsenic) impregnated wood. Heavy-metal emissions can be reduced by secondary emission reduction measures.

9.2.1.2 Emissions from incomplete combustion

Emissions caused by incomplete combustion are mainly a result of either:

- inadequate mixing of combustion air and fuel in the combustion chamber, which produces local fuel-rich combustion zones;
- an overall lack of available oxygen;
- too low combustion temperatures;
- too short residence times; or
- too low radical concentrations, in special cases, for example in the final stage of the combustion process (the char combustion phase) in a batch combustion process.

These variables are all linked together through the reaction rate expressions for the elementary combustion reactions. However, in cases in which sufficient oxygen is available, temperature is the most important variable due to its exponential influence on the reaction rates. An optimization of these variables will, in general, contribute to reduced levels of all emissions from incomplete combustion.

The following components are emitted to the atmosphere as a result of incomplete combustion in biomass combustion applications:

Carbon monoxide (CO)

Conversion of fuel carbon to CO_2 takes places through several elementary steps, and through several different reaction paths. CO is the most important final intermediate. It is oxidized to CO_2 if oxygen is available. The rate at which CO is oxidized to CO_2 depends primarily on temperature. CO can be regarded as a good indicator of the combustion quality. Large-scale biomass combustion applications usually have better opportunities for optimization of the combustion process than small-scale biomass combustion applications. Hence, CO emission levels are usually lower for large-scale biomass combustion applications. Figure 9.3 shows the CO emission level as a function of excess air ratio for various biomass combustion applications, while Figure 9.4 shows the CO emission level as a function of combustion temperature. For a given system, CO emission is lowest at a specific excess air ratio: higher excess air ratios will result in a decreased combustion temperature while lower excess air ratios will result in inadequate mixing conditions. In addition, sufficient residence time is important to achieve low CO emission levels, mainly because CO is generally a later intermediate than hydrocarbons.

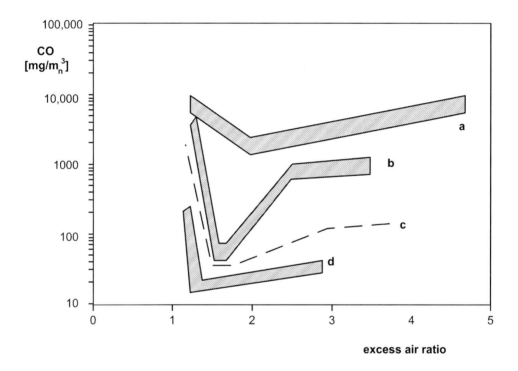

Figure 9.3 *CO emissions in mg/m_n^3 as a function of excess air ratio λ*

(a) a simple, manually charged wood boiler;
(b) a down-draught wood log boiler;
(c) an automatic furnace with combustion technology as of 1990;
(d) an automatic furnace with enhanced combustion technology as of 1995. Automatic furnaces with appropriate combustion process control can be operated under optimum conditions

Source: [65].

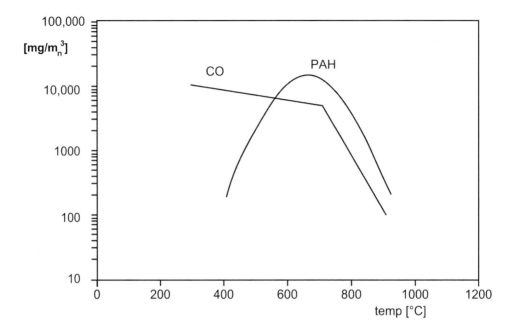

Figure 9.4 CO and PAH emissions in mg/m$_n^3$ as a function of combustion temperature
Source: [66].

Methane (CH$_4$)

CH$_4$ is usually mentioned separately from the other hydrocarbons since it is a direct greenhouse gas. In biomass combustion applications it is an important intermediate in the conversion of fuel carbon to CO$_2$ and fuel hydrogen to H$_2$O. As for CO, emissions of CH$_4$ are a result of too low combustion temperatures, too short residence times, or lack of available oxygen. Hydrocarbons are, in general, earlier intermediates than CO, which means they have lower emission levels.

Non-Methane Volatile Organic Components (NMVOC)

This group includes all hydrocarbons except CH$_4$, PAH (polycyclic aromatic hydrocarbons) and other heavy hydrocarbons which condense and form particle emissions. They are all intermediates in the conversion of fuel carbon to CO$_2$ and fuel hydrogen to H$_2$O. As for CO, emissions of NMVOC are a result of too low combustion temperatures, too short residence times, or lack of available oxygen.

Polycyclic Aromatic Hydrocarbons (PAH)

PAH are usually mentioned separately from other hydrocarbons due to their carcinogenic effects. They are all intermediates in the conversion of fuel carbon to CO$_2$ and fuel hydrogen to H$_2$O. As for CO, emissions of PAH are a result of too low combustion temperatures, too short residence times or lack of available oxygen. A comparison of the combustion temperature influence on the PAH and CO emission level is shown in Figure 9.4. Further references: [67–75].

Particles

Particle emissions from incomplete combustion can be found as soot, char or condensed heavy hydrocarbons (tar). Soot consists mainly of carbon, and is a result of a local lack of available oxygen in the flame zone and/or local flame extinction. Char particles may be entrained in the flue gas due to their very low specific density, especially at high flue gas flow rates. Condensed heavy hydrocarbons are an important, and in some cases the main, contributor to the total particle emission level in small-scale biomass combustion applications such as wood-stoves and fireplaces. As for CO, emissions of particles may be a result of combustion temperatures that are too low, residence times that are too short or a lack of available oxygen. However, due to the diversity of particle emission components, reducing particle emission levels by primary measures is not as straightforward as it is for CO, except for particles consisting of condensed heavy hydrocarbons. Measures to reduce particle emissions follow the same principles as mentioned in the section on particles formed from complete combustion.

Polychlorinated dioxins and furans (PCDD/PCDF = PCDD/F)

Polychlorinated dioxins and furans are a group of highly toxic components. They are found to be a consequence of the de novo synthesis in the temperature window between 180°C and 500°C [32]. Carbon, chlorine, catalysts (Cu) and oxygen are necessary for the formation of PCDD/F. PCDD/F can be formed in very small amounts from all biomass fuels containing chlorine (see Section 2.3.2.3). The emissions of PCDD/F are highly dependent on the conditions under which combustion and flue gas cooling take place; therefore, wide variations are found in practice. Although herbaceous biomass fuels have high chlorine contents, their PCDD/F emissions are usually very low. This may be explained by their high alkali content, which leads to the formation of salts (KCl, NaCl) and thus to a lower level of gaseous chlorine for the de novo synthesis. Because of the many factors influencing PCDD/F formation, wide variations may appear even within the same biomass combustion installation. In general, the PCDD/F emission level from biomass combustion applications using virgin wood as fuel is well below the health risk limit. PCDD/F emissions can be reduced by primary and secondary emission reduction measures. Further references: [34, 63, 71–82].

Ammonia (NH_3)

Small amounts of NH_3 may be emitted as a result of incomplete conversion of NH_3, formed from pyrolysis/gasification, to oxidized nitrogen-containing components. This occurs in special cases in which the combustion temperature is very low. Additionally, secondary NO_x reduction measures utilizing NH_3 injection may contribute to the NH_3 emission level due to NH_3 slippage. NH_3 emissions can be reduced by general primary emission reduction measures for emissions from incomplete combustion, and by optimizing the NH_3 injection process.

(Ground level) Ozone (O_3)

O_3 is a secondary combustion product formed from photochemical atmospheric reactions including CO, CH_4, NMVOC and NO_x. It is a direct greenhouse gas and also influences the local and regional environment. As such, it is a highly unwanted by-product of biomass combustion applications. O_3 emissions can be reduced indirectly by reducing emissions from incomplete combustion, and by primary and secondary NO_x emission reduction measures.

Table 9.2 shows the pollutants from biomass combustion and their impacts on climate, environment, and health.

Table 9.2 *Pollutants from biomass combustion and their impacts on climate, environment and health*

Component	Biomass sources	Climate, environmental and health impacts
Carbon dioxide (CO_2)	Major combustion product from all biomass fuels.	**Climate:** Direct greenhouse gas. However, CO_2 emissions from biomass combustion are regarded as being CO_2-neutral with respect to the greenhouse gas effect[1] since biomass is a renewable fuel.
Carbon monoxide (CO)	Incomplete combustion of all biomass fuels.	**Climate:** Indirect greenhouse gas through O_3 formation. **Health:** Reduced oxygen uptake especially influences people with asthma, and embryos. Suffocation in extreme cases.
Methane (CH_4)	Incomplete combustion of all biomass fuels.	**Climate:** Direct greenhouse gas. Indirect greenhouse gas through O_3 formation.
Non Methane Volatile Organic Components (NMVOC)	Incomplete combustion of all biomass fuels.	**Climate:** Indirect greenhouse gas through O_3 formation. **Health:** Negative effect on the human respiratory system.
Polycyclic Aromatic Hydrocarbons (PAH)	Incomplete combustion of all biomass fuels.	**Environment:** Smog formation. **Health:** Carcinogenic effects.
Particles	Soot, char and condensed heavy hydrocarbons (tar) from incomplete combustion of all biomass fuels. Fly-ash and salts.	**Climate and environment:** Reversed greenhouse effect through aerosol formation. Indirect effects of heavy-metal concentrations in deposited particles. **Health:** Negative effect on the human respiratory system. Carcinogenic effects.
Nitrogen oxides (NO_x (NO, NO_2))	Minor combustion product from all biomass fuels containing nitrogen. Additional NO_x may be formed from nitrogen in the air under certain conditions.	**Climate and environment:** Indirect greenhouse gas through O_3 formation. Reversed greenhouse gas effect through aerosol formation. Acid precipitation. Vegetation damage. Smog formation. Corrosion and material damage. **Health:** Negative effect on the human respiratory system. NO_2 is toxic.
Nitrous oxide (N_2O)	Minor combustion product from all biomass fuels containing nitrogen.	**Climate:** Direct greenhouse gas. **Health:** Indirect effect through O_3 depletion in the stratosphere.[2]

Table 9.2 *continued*

Component	Biomass sources	Climate, environmental and health impacts
Ammonia (NH_3)	Small amounts may be emitted as a result of incomplete conversion of NH_3, formed from pyrolysis/gasification, to oxidized nitrogen-containing components. Secondary NO_x reduction measures by NH_3 injection (SNCR, SCR).	**Environment:** Acid precipitation. Vegetation damage. Corrosion and material damage. **Health:** Negative effect on the human respiratory system.
Sulphur oxides (SO_x (SO_2, SO_3))	Minor combustion product from all biomass fuels containing sulphur.	**Climate and environment:** Reversed greenhouse gas effect through aerosol formation. Acid precipitation.[3] Vegetation damage. Smog formation. Corrosion and material damage. **Health:** Negative effect on the human respiratory system, asthmatic effect.
Heavy metals	All biomass fuels contain heavy metals to some degree, which will remain in the ash or evaporate.	**Health:** Accumulate in the food chain. Some are toxic and some have carcinogenic effects.
(Ground level) Ozone (O_3)	Secondary combustion product formed from atmospheric reactions including CO, CH_4, NMVOC and NO_x.	**Climate and environment:** Direct greenhouse gas. Vegetation damage. Smog formation. Material damage. **Health:** Indirect effect through O_3 depletion in the stratosphere.[2] Negative effect on the human respiratory system, asthmatic effect.
Hydrogen chloride (HCl)	Minor combustion product from all biomass fuels containing chlorine.	**Environment:** Acid precipitation. Vegetation damage. Corrosion and material damage. **Health:** Negative effect on the human respiratory system. Toxic.
Dioxins and furans (PCDD/PCDF)	Small amounts may be emitted as a result of reactions including carbon, chlorine, and oxygen in the presence of catalysts (Cu).	**Health:** Highly toxic. Liver damage. Central nervous system damage. Reduced immunity defence. Accumulate in the food chain.

Notes

[1] **The greenhouse gas effect:** The natural greenhouse gas effect keeps the Earth's mean temperature at about 15°C. Without the greenhouse gas effect, the Earth's mean temperature would be −18°C. Anthropogenic sources of greenhouse gases are generally believed to contribute to an increasing greenhouse gas effect, causing the Earth's mean temperature to slowly increase. From 1750 to 1994, the concentration of the three most important greenhouse gases – CO_2, CH_4, and N_2O – increased by 30, 145 and 15 per cent, respectively (contributing to an increasing greenhouse gas effect). However, particles, SO_2 and NO_x contribute, to some degree, to a reversed greenhouse gas effect caused by aerosol formation.

[2] **Depletion of the ozone layer:** The atmospheric ozone layer is found in the stratosphere, 10–40km above ground level. Ozone absorbs ultraviolet radiation from the sun and prevents damaging radiation from reaching the Earth's surface. Ozone in the stratosphere may be reduced by reactions with NO, where NO may be formed from N_2O in a first reaction step. Ground level ozone, however, is a pollutant and a greenhouse gas.

[3] **Acid precipitation:** Emissions of NO_x, SO_x and HCl result in acid precipitation through the formation of acids.

9.2.2 Measuring emissions from biomass combustion

A brief description follows on some of the main principles for sampling and measuring emissions from biomass combustion. For details and for descriptions of methods for other compounds, please refer to national and international standards from, e.g., US Environmental Protection Agency (US-EPA) and Verein Deutscher Ingenieure (VDI).

9.2.2.1 Measuring particulate emissions
Sampling of particulate matter (PM) in flue gases is carried out using an isokinetic probe. The basic principle is measurement of gas velocity in the duct, and a corresponding suction in the sampling nozzle to ensure that the same velocity is maintained in the nozzle as in the gas stream (isokinetism).

9.2.2.2 Sampling of gases
The method is generally extractive, i.e. a sample of the flue gas is extracted from the flue gas stream for analysis in external equipment. Before analysis, the gas is cleaned of particulate matter, and the moisture is removed.

9.2.2.3 Analysing gas compounds
To measure for CO and CO_2 a non-dispersive infrared (NDIR) gas analyser is often used. O_2 is typically analysed using a zirconium oxide analyser. NO_x are usually measured using a chemiluminiscense analyser.

9.2.3 Emissions data

Emissions from biomass combustion can, in general, be divided into emissions that are mainly influenced by combustion technology and process conditions, and emissions that are mainly influenced by fuel properties.

The amount of pollutants emitted to the atmosphere from various types of biomass combustion applications is highly dependent on the combustion technology implemented, the fuel properties, the combustion process conditions, and the primary and secondary emission reduction measures that have been implemented. When available, such data generally refer to a single fuel-technology combination. In order to obtain an objective view of emission levels from various biomass combustion applications, it is necessary to collect emission data from a wide range of fuel–technology combinations. However, due to the many parameters involved, the data can still only give an indication of typical emission levels.

One problem in trying to compare emission levels from different biomass combustion applications is that many different denominators are used. Usually the basic data needed to recalculate the emission level into other denominators are not given, and assumptions must be made. Procedures and formulae for converting emission levels between different denominators are given in Chapter 2. Below, selected emissions data are presented.

9.2.3.1 Domestic applications
Domestic applications include wood-stoves, fireplaces, fireplace inserts, wood log boilers, heat-storing stoves and pellet stoves. Any of these may in principle be equipped with a catalytic converter. Skreiberg [83] investigated and compared the emission levels from an advanced down-draught staged-air wood-stove, a traditional wood-stove and a wood-stove

equipped with a catalytic converter. The results showed that the traditional stove had significantly higher levels of emissions from incomplete combustion than the staged-air wood-stove and the catalytic stove. The catalytic stove emitted significantly less CO than the staged-air wood-stove, but the difference for C_xH_y and particles was less significant. However, wide variations in the emission levels from the same stove were reported, depending on operating conditions. In many cases, the levels of emissions from incomplete combustion may be many times higher than reported at nominal load, usually increasing exponentially with decreasing load relative to nominal load [84].

In 1994 the Technical University of Munich performed an extensive measuring programme on emission levels from domestic wood applications [85]: wood-stoves, fireplace inserts, heat-storing stoves, pellet stoves and catalytic wood-stoves. The applied test method was DIN 18891/18895. Calculated arithmetic average values of the reported values are shown in Table 9.3.

Table 9.3 *Arithmetic average emission levels in mg/m^3_0 at 13 per cent O_2 from small-scale biomass combustion applications*

	Load [kW]	Excess air ratio	CO [mg/m³]	C_xH_y [mg/m³]	Particles [mg/m³]	NO_x [mg/m³]	Temp [°C]	Efficiency [%]
Wood-stoves	9.33	2.43	4986	581	130	118	307	70
Fireplace inserts	14.07	2.87	3326	373	50	118	283	74
Heat-storing stoves	13.31	2.53	2756	264	54	147	224	78
Pellet stoves	8.97	3.00	313	8	32	104	132	83
Catalytic wood-stoves	6.00		938					

Note: The term m^3_0 designates volume at standard reference condition; pressure 101.3kPa and temperature 273K.

The emission levels of pellet stoves are very low and can be compared to those of oil burners. This is to be expected since pellet stoves are operated as a continuous combustion process, with combustion process control possibilities. The CO emission level for the catalytic wood-stoves is lower than for the other units, except for the pellet stove, which is also to be expected (see Section 4.4.4). The heat-storing stoves have lower emission levels than the wood-stoves and fireplace inserts. Heat-storing stoves usually operate at a high load, which increases the possibilities for reduced emission levels. The fireplace inserts have lower emission levels than the wood-stoves, which is somewhat surprising. However, they operate at significantly higher loads than the wood-stoves. The NO_x emission levels are highly influenced by the fuel nitrogen content, and no specific conclusions can be drawn from this table.

Table 9.3 illustrates the main trends that can be expected; however, one must be careful interpreting these results and emission values from domestic biomass combustion applications in general. Also, measured emission levels are highly influenced by the test method (forced or natural draught). The influence of standards, test procedures and calculation procedures on the emission level of a catalytic wood-stove tested in nine IEA countries, according to their national standards and test procedures, was investigated as a part of the

IEA Bioenergy, Task X – Conversion, Combustion activity. The results were published in the journal *Biomass and Bioenergy* [86].

Skreiberg and Saanum [87] inventoried emission levels of various biomass combustion applications, both domestic and industrial, and the results are summarized in Table 9.4.

Table 9.4 *Arithmetic average emission values from wood combustion applications*

	NO_x as NO_2 [mg/MJ]	Particles [mg/MJ]	Tar [mg/MJ]	CO [mg/MJ]	UHC as CH_4 [mg/MJ]	VOC [mg/MJ]	PAH [µg/MJ]
Cyclone furnaces	333	59	n.m.	38	n.m.	2.1	n.m.
Fluidized bed boilers	170	2	n.m.	0	1	n.m.	4
Pulverized fuel burners	69	86	n.m.	164	8	n.m.	22
Grate plants	111	122	n.m.	1846	67	n.m.	4040
Stoker burners	98	59	n.m.	457	4	n.m.	9
Wood boilers	101	n.m.	499	4975	1330	n.m.	30
Modern wood-stoves	58	98	66	1730	200	n.m.	26
Traditional wood-stoves	29	1921	1842	6956	1750	671	3445
Fireplaces	n.m.	6053	4211	6716	n.m.	520	105

Note: The data were collected from investigations in various IEA countries (Norway, Switzerland, Finland, the UK and Denmark). n.m. = not measured. UHC = unburned hydrocarbons.

9.2.3.2 Industrial applications

Nussbaumer and Hustad [88] present typical ranges of emission levels from various automatic wood furnaces (understoker furnaces, grate firings and dust firings), as can be seen in Tables 9.5 and 9.6.

Table 9.5 *Emissions mainly influenced by combustion technology and process conditions: comparison between poor and high standard furnace design*

Emissions at 11% O_2	Poor standard	High standard
Excess air ratio, λ	2–4	1.5–2
CO (mg/m³₀)	1000–5000	20–250
C_xH_y (mg/m³₀)	100–500	< 10
PAH (mg/m³₀)	0.1–10	< 0.01
Particles, after cyclone (mg/m³₀)	150–500	50–150 *

Note: * except for dust firings, usually > 150.

Table 9.7 presents the emissions of wood-burning installations used in the Netherlands with a capacity of 30–320kW$_{th}$ [89, 90]. As these installations use the same fuel (clean woodchips), the technologies can be compared. It can be noted that these installations are relatively efficient and have low particle emissions. In all cases, particle emissions are well below the Dutch emission guidelines. The table further indicates that automatic operation and combustion process control drastically reduce the emission of CO and C_xH_y.

Table 9.6 *Emissions mainly influenced by fuel properties: comparison between various fuel types (typical values)*

Emissions at 11% O_2	Fuel type	Typical data
NO_x (mg/m³₀)	Native wood (soft wood) Native wood (hard wood) Straw, grass, miscanthus, chip boards Urban waste wood and demolition wood	100–200 150–250 300–800 400–600
HCl (mg/m³₀)	Native wood Urban waste wood and demolition wood, Straw, grass, miscanthus, chipboards (NH_4Cl)	< 5 raw gas: 100–1000 with HCl absorption: < 20
Particles (mg/m³₀)	Native wood Straw, grass, miscanthus, chipboards Urban waste wood and demolition wood	after cyclone: 50–150 after cyclone: 150–1000 after bag- or electrostatic filter: < 10
Σ Pb, Zn, Cd, Cu (mg/m³₀)	Native wood Urban waste wood and demolition wood	< 1 raw gas: 20–100 after bag- or electrostatic filter: < 5
PCDD/F (ng TE/m³₀)	Native wood Urban waste wood and demolition wood	typical: < 0.1 range: 0.01–0.5 typical: 2 range: 0.1–20

Note: TE = toxicity equivalent.

Table 9.7 *Emissions from small industrial woodchip combustion applications in the Netherlands*

Manual/automatic operation	Combustion principle	Draught control	Capacity [kW]	CO [mg/m³₀]	C_xH_y [mg/m³₀]	NO_x [mg/m³₀]	Particles [mg/m³₀]	Efficiency [%]
Manually operated	Horizontal grate	Natural, uncontrolled	36	2390	124	156	21	85.0
		Forced, uncontrolled	34.6 30	3450 656	130 21	172 182	29 34	83.5 90.0
Automatically operated	Understoker	Forced, controlled	~40 320	66 31	2.5 3.7	118.7 186	80 51	85.4 89.1

Note: All emission values at 11% O_2.
Source: [91].

Obernberger inventoried emission data from various publications [92], representing capacities ranging from 0.5 to 10MW$_{th}$. The fuels used were particle board, woodchips, medium density fibreboard (MDF) and bark. Table 9.8 shows the results.

Table 9.8 *Emissions from industrial wood-fired installations, using particle board, woodchips, MDF and bark*

Component	Emission	Number of observations
CO	125–2000	25
C_xH_y	5.0–12.5	25
PAH	0.0006–0.06	unknown
Benzopyrene	0.000005–0.001	4
NO_x (as NO_2)	162–337	22
Particles	37–312	29
SO_2	19–75	17
Cl	10	12
F	0.25	unknown

Note: mg/m^3 at 11% O_2 dry.

Source: [92].

It can be seen from the Table 9.8 that the emissions of SO_2, Cl and F are relatively low. This is due to the low content of these elements in the fuel. Also the emissions of C_xH_y, PAH and Benzopyrene are low; these can be further reduced through optimization of the combustion process. Emissions of CO were found to be relatively high, particularly for old combustion installations. CO emissions can be reduced by avoiding intermittent boiler operation and/or through improved combustion process control (see Section 9.2.4.6). NO_x emissions can be further reduced by applying/optimizing staged combustion, see Section 9.2.4.7. From a study of the individual measurements carried out in the same study, the following was concluded:

- The emissions usually decrease as the size of the combustion installation increases, due to improved process control possibilities and efficient flue gas cleaning facilities. Combustion installations exceeding $4MW_{th}$ often have electrostatic filters or flue gas condensation units in addition to a cyclone, which is usually installed in smaller installations. For smaller combustion installations, such investments are usually not economically viable. However, in Scandinavian installations, economic flue gas condensation has been achieved with boiler installations below $1MW_{th}$.
- NO_x emissions are an exception. The fuel NO_x emissions increase with increasing nitrogen content in the fuel, excess air ratio and combustion temperature, up to a point where all fuel nitrogen intermediates have been converted to either NO_x, N_2O or N_2. At low combustion temperatures, the temperature influence is more important than the influence of excess air ratio, resulting in lower NO_x emissions for smaller combustion installations.

9.2.3.3 CHP plants

Detailed information on emissions from biomass CHP plants are rarely found in the literature. One study, [93], summarizes some emissions from straw-fired CHP plants in Denmark.

Table 9.9 *Emissions from straw-fired CHP plants*

Component	Observation	Emission
SO_2 [mg/m_n^3] at 6% O_2	Estimated annual mean	150
HCl [mg/m_n^3] at 6% O_2	Estimated annual mean	100
CO [mg/m_n^3] at 10% O_2	Average measured in 4 plants	114
PAH [µg/m_n^3] at 10% O_2	Average measured in 4 plants	1.7
PCDD/DF [pg I-TEQ/m_n^3] at 10% O_2	Average measured in 4 plants	2.5

Note: I-TEQ = international toxic equivalent quantity.
Source: [93].

Further information on the environmental impact of biomass CHP plants, including information on water consumption and ash production, are found in the comprehensive study [94].

9.2.4 Primary emission reduction measures

Reduction of harmful emissions through flue gases and effluents can be obtained by either avoiding creation of such substances (primary measures) or removing the substances from the flue gas (secondary measures). In general we distinguish these two basic methods by stating that a primary measure is a modification of the combustion process (e.g. limestone injection in the furnace), and a secondary measure takes places after the combustion process (e.g. ammonia injection in the flue gas channel).

One might even expand the concept of primary measures to include emission reduction by decreasing fuel consumption through increasing heat network overall efficiency. The topic is outside the scope of this book; however, the impact on emission reduction from, for example, thermal insulation of heat network pipes, insulation of buildings, energy efficiency measures or energy management systems in any building should not be overlooked.

In this section, measures for reducing primarily emissions from incomplete combustion and NO_x are presented. However, SO_x can also be reduced by primary measures, such as lime or limestone injection. N_2O emissions may be reduced by primary measures, but this is not a straightforward process, and these measures will not be presented here due to the low emission levels of N_2O from biomass combustion applications. PCDD/F emissions are a result of incomplete combustion. However, emission reduction measures for PCDD/F deviate from measures related to other emissions from incomplete combustion. It should be mentioned that the main primary emission reduction measures to avoid PCDD/F are complete burnout of the fly ash and an operation of the combustion at low excess air ratio and under stable conditions.

Primary emission reduction measures aim to prevent or reduce the formation of emissions and/or a reduction of emissions within the combustion chamber. Several possible measures exist. The measures we will discuss here are:

- modification of the fuel composition;
- modification of the moisture content of the fuel;
- modification of the particle size of the fuel;

- selection of the type of combustion equipment;
- improved construction of the combustion application;
- combustion process control optimization;
- staged-air combustion;
- staged fuel combustion and reburning; and
- catalytic converters.

In practice, these measures are often interrelated.

9.2.4.1 Modification of the fuel composition

Decreasing the amount of those elements in the fuel that contribute to harmful emissions or operational problems can be carried out to some extent. Several methods have been developed for reducing the sulphur and nitrogen content in natural gas and oil. Methods also exist for coal, but these are not cost-effective. Virgin biomass fuels are solid fuels, like coal, with limited possibilities for decreasing the amount of specific elements in the fuel. However, by choosing the fuel specification, one can influence the fuel composition and hence the emissions, and for the case of straw, washing the fuel has been shown to be effective.

The K, Na, S, Cl, Zn and Pb concentrations in the fuel determine the mass of aerosol emissions as well as their chemical composition [95]. Obviously untreated wood performs better than treated wood, and the origin of the fuel also has an influence on aerosol emission.

Washing of straw has been shown to reduce the amount of chlorine and potassium significantly. Washing can either be performed by leaving the straw on the field for some time after the harvest, exposing it to rain or by controlled washing. Leaching experiments with barley straw performed in Denmark [54] showed that after 150mm of rain, the chloride content had dropped from 0.49 per cent to below 0.05 per cent and the potassium content had dropped from 1.18 per cent to 0.22 per cent.

Controlled washing of straw has also been carried out in Denmark, by boiling the straw at 160°C and by washing the straw at 50–60°C. The latter is considered the most economic option. So far, straw washing has only been tested at small plants. The energy losses caused by washing, drying and leaching of organic matter amount to approximately 8 per cent of the calorific value of the straw. This is offset, however, by the prolonged life of the boilers, because corrosion problems are avoided. Washing of straw is also expected to have advantages for the subsequent application of the fly ash, since straw ash that does not contain alkaline salts and other impurities may be used as a filler in building materials in Denmark. As chlorine is needed for the formation of dioxins and furans, washing straw will also contribute to reduced dioxin and furan emission levels.

9.2.4.2 Modification of the moisture content of the fuel

The moisture content in biomass can vary widely. Wood for energy purposes for instance, may vary in moisture content from approximately 10 to 60 per cent of water by weight. The first value represents wood residuals from the wood industry where drying has been applied, while fresh wood from the forest may contain up to 60 per cent water.

High moisture content in the fuel makes it difficult to achieve a sufficiently high temperature in the combustion chamber. Often a temperature above approximately 850°C is desired to ensure a sufficiently low level of CO. If such a high temperature is not reached, incomplete combustion occurs with high emissions as a result.

In general, unless waste heat from another process can be accessed at a very low cost, the cost of artificial drying is too high to make the drying process itself economically feasible.

The design of the combustion chamber is of great importance when biomass with a high moisture content is used. Improvement of the combustion quality can be reached by use of a high amount of ceramic linings and insulation of the combustion chamber. This measure, together with a high preheating temperature of the combustion air, may make it possible to utilize fuel with a high moisture content in an environmentally acceptable manner.

Even in cases where improvement of the combustion process is achieved, a certain decrease in the boiler efficiency must be accepted. This is due to the fact that the amount of moisture in the fuel leads to a higher flow of flue gas, including water vapour, from the boiler. This represents an energy loss.

But combustion of wood with high moisture content can be advantageous if combined with a flue gas condensing system, and provided there is a sufficiently low heat sink. Condensation of the water vapour in the flue gas raises the overall efficiency to such an extent that improvement of the overall economy of plant operation may be achieved.

9.2.4.3 Modification of the particle size of the fuel

The fuel particle size is very relevant for the combustion technology selection process. The fuel size in biomass combustion applications may vary from whole wood logs to fine sawdust.

In small-scale biomass combustion applications, such as wood-stoves, fireplaces and wood log boilers, the fuel generally consists of wood logs with bark, of varying size. Hence, the fuel size has little influence on the selection of combustion technology within this segment of the biomass combustion market. However, by various arrangements of the wood logs inside the combustion chamber, increasing or decreasing the overall active surface area, it is possible to influence the combustion process to some degree. Also, by combining large and small wood logs and pieces of bark, the combustion process can be influenced.

In large-scale biomass combustion applications with automatic fuel feeding, the fuel size is more decisive. If the fuel consists of both very small and very large pieces, a shredder or chipper can be used to reduce the particle size of the largest particles. In this way, a more homogeneous particle size is obtained. Hence, a wider range of technology options can be used. However, particle size reduction is only attractive if the benefits outweigh the additional investment and energy costs.

9.2.4.4 Selection of the type of combustion equipment

When selecting the combustion technology for a biomass combustion application there are several aspects to be considered, both with respect to the combustion process and to primary and possibly secondary emission reduction measures. Also, the heat/power capacity of the application usually limits the choice of combustion technology, either due to technological or economic considerations.

First of all, fuel characteristics such as fuel composition, moisture content and particle size are important. For wood fuels, only the nitrogen content may limit the choice of combustion technology, if there are NO_x emission limits to be met. The moisture content, however, will be very decisive for wood fuels such as woodchips and bark if drying of the fuel prior to combustion is not an option.

For other types of biomass fuels, additional fuel constituents, such as ash, chlorine, potassium and sulphur may influence the combustion process in such a way that certain preferences should be made when selecting the combustion technology, e.g. for straw combustion.

Secondary emission reduction measures can be bought and fitted to most biomass combustion applications, depending on the emission limits to be met. However, substantial emission reduction can be achieved for emissions from incomplete combustion and NO_x emissions by selection of the best possible combustion technology for a given fuel and by optimizing the combustion process, including primary NO_x reduction measures. This may remove the need for secondary emission reduction measures other than for particle removal.

9.2.4.5 Improved construction of the combustion application

In order to obtain optimal combustion, with minimal emissions from incomplete combustion, one has to achieve:

- sufficiently high combustion temperatures;
- sufficiently long residence times; and
- optimal mixing of fuel gases and air, also with changing heat and/or power output.

These factors are partly determined by the combustion technology and the design of the furnace, and partly by the combustion process operation. Recently, a number of combustion process control systems have been developed for the optimization of combustion processes. These will be mentioned in the next section.

9.2.4.6 Combustion process control optimization

A process controller aims to govern selected process parameters according to a predefined scheme. The primary aim of a process control device in a biomass combustion application is to adjust the heat production according to the heat demand. In addition to this, the process control device can be programmed for simultaneous optimization of the combustion process with respect to minimizing emissions and maximizing thermal efficiency.

For biomass combustion, typical process parameters that can be used as process control parameters are the CO, C_xH_y and O_2 concentrations in the flue gas, as well as combustion chamber temperatures and boiler temperature. Process variables that can be directly adjusted to achieve the targets for the aforementioned process parameters are typically the amount of fuel fed into the furnace, and the amount of primary and secondary combustion air supplied.

Minimizing emissions

The combustion quality can be modified by adjusting the amounts of fuel and primary and secondary air, based on measured concentrations of CO, C_xH_y, O_2 and the combustion chamber temperature.

In the case of direct process control, CO and C_xH_y are measured continuously and the governing variables are adjusted to obtain minimum emissions. Because of process fluctuations, the concentrations of CO and C_xH_y often remain high.

In the case of indirect process control, the ideal excess air ratio (λ) is first established for all expected process conditions (boiler load, fuel moisture content, etc.) to obtain minimum emissions. Then the measured value for O_2 is used as a process control parameter. Control of λ ensures a stable combustion process, but as the actual process parameters often deviate from expected values, emissions are not always minimized in practice.

Direct and indirect process control can also be combined to obtain a stable combustion process with minimized emissions.

Controlling heat output

In addition to minimizing emissions, there is a need to control the heat output from the furnace or boiler. This control can be based on using measured temperature difference and mass flow of boiler water. However, the control of the boiler water temperature is more commonly used. The relation between fuel input and the input of primary and secondary air is established after the boiler is installed. Using this relation, these parameters can be adjusted to keep the boiler water temperature at a given value.

Modification of an existing biomass boiler

Existing boilers can often be modified to successfully achieve reduced emission levels, higher thermal efficiency and improved control of heat output. An example is given below.

For a 500kW$_{th}$ Nolting underscrew feeder wood combustion plant with cyclone, TNO installed a λ sensor in order to control the combustion process and heat output. Also, flue gas recirculation was applied and the combustion chamber was modified. Table 9.10 shows that as a result of the optimization, the efficiency increased while the emissions were effectively reduced [35].

Table 9.10 *Effect of optimization on emissions and efficiency*

Property	Before optimization			After optimization		
	1	2	3	1	2	3
CO [mg/m$_o^3$]	3516	4439	4327	82	313	103
C$_x$H$_y$ [mg/m$_o^3$]	262	303	269	2	28	2
NO$_x$ [mg/m$_o^3$]	772	722	764	652	872	706
Dust [mg/m$_o^3$]	219	235	214	99	157	106
Flue gas temperature [°C]	163	164	158	109	162	132
Flue gas losses [%]	17	–	17	7	–	8
Losses due to incomplete combustion [%]	1.5	–	2.0	0.1	–	0.1
Overall efficiency [%]	81	–	81	93	–	92

Note: Mix at 11% O$_2$, dry.

This example illustrates that it is possible to significantly reduce boiler emissions and increase efficiency with relatively simple process modifications.

9.2.4.7 Staged-air combustion

Staged-air combustion is widely applied in biomass combustion applications, also in small-scale applications. However, the possibilities for an accurate control of the combustion air are usually limited in small-scale applications, which may result in higher emission levels. Staged-air combustion makes a simultaneous reduction of both emissions from incomplete combustion and NO$_x$ possible through a separation of devolatilization and gas phase combustion. This results in improved mixing of fuel gas and combustion air. In the first stage, primary air is added for devolatilization of the volatile fraction of the fuel, resulting in a fuel gas consisting mainly of CO, H$_2$, C$_x$H$_y$, H$_2$O, CO$_2$ and N$_2$. For NO$_x$ emission reduction, the fuel gas content of NH_3, HCN and NO is also of particular interest. In the second stage, sufficient secondary air is supplied to ensure a good burnout and low emission levels from incomplete combustion.

An improved mixing of fuel gas and secondary air reduces the amount of secondary air needed, resulting in higher flame temperatures, and also a lower overall excess air ratio. Hence, emissions from incomplete combustion are reduced by a temperature increase, which speeds up the elementary reaction rates, and by improved mixing, which reduces the residence time needed for mixing the fuel gas and the secondary combustion air. However, this does not mean that the NO_x emission level is automatically reduced as well. An efficient reduction of both emissions from incomplete combustion and NO_x emissions can only be achieved by optimization of the primary excess air ratio.

As mentioned before, fuel nitrogen is converted to NO (> 90 per cent) and NO_2 (< 10 per cent) through a series of elementary reaction steps, called the fuel NO_x mechanism. Important primary nitrogen-containing components are NH_3 and HCN. However, significant amounts of NO and N_2 may also be found in the pyrolysis gas. If sufficient O_2 is available, NH_3 and HCN will mainly be converted to NO through different reaction routes. However, in fuel-rich conditions NO will react with NH_3 and HCN, forming N_2. This mechanism is utilized as a primary NO_x reduction measure. By optimizing the primary excess air ratio, temperature and residence time a maximum conversion of NH_3 and HCN to N_2 can be achieved.

Figure 9.5 shows the ratio between the total fixed nitrogen (TFN) amount (includes all nitrogen-containing species except N_2) and the fuel nitrogen content (TFN/Fuel-N) as a function of primary excess air ratio for chemical kinetic calculations in two ideal flow reactors, utilizing a detailed elementary reaction scheme. A typical pyrolysis gas composition from wood was used as fuel gas composition. As can be seen, the choice of flow reactor highly influences the reduction potential and illustrates the importance of mixing effects. The plug flow reactor (PFR) and the perfectly stirred reactor (PSR) can be regarded as the two extremes among mixing models. The reduction potential is also influenced by the choice of nitrogen-containing species in the fuel gas. The reduction potential for NH_3 as nitrogen-containing species is, in general, higher than it is for HCN, as the conversion of HCN to N_2 requires a longer and more complex reaction path. Temperature, residence time and initial fuel nitrogen content also influence the reduction potential.

In a 25kW test reactor [37] with fixed bed up-draught gasification and separate reduction chamber, followed by gas phase combustion, it has been shown that an NO_x reduction of up to 50–75 per cent can be achieved by staged-air combustion. The percentage increases with increasing fuel-N content, at optimum conditions:

- residence time in the reduction chamber ≈ 0.5s (> 0.3s);
- reduction chamber temperature ≈ 1100–1200°C; and
- primary excess air ratio ≈ 0.7.

The NO_x emission level as a function of primary excess air ratio is shown in Figure 9.6 for the 25kW test reactor, with two-stage combustion and a reduction chamber temperature of 1150°C. An additional reduction can be achieved by staging the primary air. If the use of a separate reduction chamber is not applicable, then the NO_x emission reduction potential decreases, but it can still be significant if efficient air staging is applied in the combustion chamber. To ensure a constant primary excess air ratio and reduction chamber temperature,

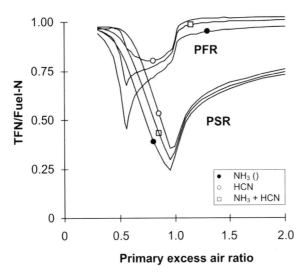

Figure 9.5 *TFN/fuel-N ratio for NH_3, HCN, and a mixture of 50 per cent NH_3 and 50 per cent HCN as a function of primary excess air ratio at a constant temperature (1173K), residence time (100ms), and fuel-N content (1000ppm) for two ideal flow reactors, a PFR and a PSR*

Source: [36].

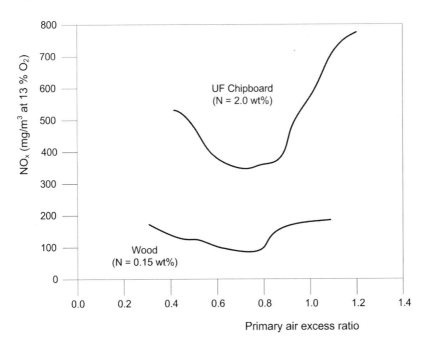

Figure 9.6 *NO_x emission level as a function of primary excess air ratio for the 25kW test reactor*

Source: [37].

a suitable combustion process design and accurate combustion process control are necessary. Further references: [38–43].

9.2.4.8 Staged fuel combustion and reburning

Staged fuel combustion and reburning are other possible methods for NO_x reduction in biomass combustion applications. The primary fuel is combusted with an excess air ratio above 1, and no significant NO_x reduction occurs. A secondary fuel is then injected into the flue gas after the primary combustion zone, without additional air supply. A sub-stoichiometric reducing atmosphere is created in which NO_x formed in the primary zone may be reduced, by reactions with NH_3 and HCN formed from the secondary fuel (if the secondary fuel contains nitrogen), in a similar manner as for staged-air combustion (see Figure 9.7). Additionally, NO is converted back to HCN by reactions with HCCO and CH_i radicals (i = 0–3) formed from the secondary fuel. This is called reburning. Under typical reburning conditions, HCCO appears to be the most effective radical for removing NO [44]. Finally, a sufficient amount of air is added after the reducing zone to achieve a good burnout with an overall excess air ratio above 1.

The potential of staged fuel combustion was investigated in an understoker furnace [37], where the secondary fuel was introduced on a second grate above the main fuel bed with an energy input ratio of approximately 70 per cent primary and 30 per cent secondary fuel. A reduction of NO_x by 52–73 per cent was achieved with a temperature in the reduction

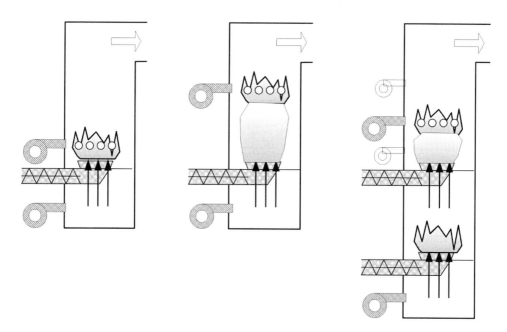

Figure 9.7 *Three principles of combustion*

Note: Diagrams from left to right: conventional combustion, staged-air combustion and staged fuel combustion

Source: [96].

zone of approximately 700°C. Hence, a NO_x reduction potential similar to that of staged-air combustion can be achieved at a significantly lower temperature level in the reduction zone. The optimum overall excess air ratio in the reduction zone is approximately 0.7–0.9. As in the case of staged-air combustion, residence time and initial fuel nitrogen content also influence the reduction potential. The properties of the secondary fuel will also be of importance, especially the fuel-N and the volatile content.

Staged fuel combustion requires automatic feeding of the primary and secondary fuel, and the secondary fuel must be easily adjustable. This limits the use of staged fuel combustion to large-scale biomass combustion applications, since a suitable combustion process design with two fuel-feeding systems as well as an accurate combustion process control is necessary. Natural gas, fuel oil, pyrolysis gas, biomass powder, sawdust or similar fuels can be used as secondary fuels. Further references: **[41, 95, 97]**.

9.2.4.9 Catalytic converters

Catalytic converters are utilized to some degree in small-scale biomass combustion applications such as wood-stoves and wood log boilers. The catalytic converter is usually placed in the flue gas channel after the combustion chamber. A catalytic converter introduces the possibility of heterogeneous reactions. The catalytic surfaces of the catalytic converter provide free surface sites where the reactants adsorb, react and finally desorb as products from the catalytic surface. The activation energy for the reactions at the catalytic surface is in general much lower than it is for the equivalent gas phase reactions. Catalytic converters can therefore efficiently reduce the level of emissions from incomplete combustion in the flue gas at low flue gas temperature levels. If properly ignited, the catalytic converter can oxidize unburned flue gas components at flue gas temperatures down to about 130°C **[86]**.

A catalytic converter consists of a durable, heat-resistant ceramic composition that is extruded into a cellular, or honeycomb, configuration. After extrusion, this ceramic monolith is fired and then covered with a noble-metal catalyst (usually platinum, rhodium or palladium, or combinations of these) or metal oxides. It is of paramount importance that the catalyst has a high thermal stability and a good poison resistance to avoid its deactivation.

The emission reduction efficiency of the catalyst depends on several variables, such as catalyst material, surface area, operation temperature, temperature profile over the catalytic converter, residence time, mass transport limitations, homogeneity of mixing before the catalytic converter, and emission component. Emission reduction efficiencies of 70–93 per cent for CO, 29–77 per cent for CH_4, 80–100 per cent for other hydrocarbons, 43–80 per cent for PAH and 56–60 per cent for tar have been reported for a wood-stove with forced draught **[46]**. A limiting factor for the reduction potential obtainable in small-scale biomass combustion applications is the pressure drop over the catalytic converter. Natural draught is often used in wood-stoves and fireplaces. This limits the possible surface area of the catalytic converter.

A comparison of the emission levels of a wood-stove equipped with a catalytic converter with the emission levels of a traditional wood-stove and with an advanced down-draught staged-air wood-stove **[98]** under similar operating conditions, has shown that the wood-stove with the catalytic converter achieved the lowest CO emission levels by far. However, for hydrocarbons and particles, the difference was smaller, especially compared to the advanced down-draught staged-air wood-stove. In batch combustion units, such as wood-stoves and fireplaces, a significant fraction of the total CO emission level will originate from the char combustion phase due to, usually, too low a temperature for effective gas phase conversion of CO, formed from the heterogeneous char oxidation, to CO_2.

Catalytic converters are not widely applied in small-scale wood combustion units today. In most IEA Bioenergy member countries, the emission limits can be met without a catalytic converter.

9.2.5 Secondary emission reduction measures

Secondary measures can be applied to remove emissions from the flue gas once it has left the boiler. For virgin wood combustion, particle removal is of particular relevance. For other types of biomass, additional secondary measures may be necessary, depending on the elementary composition and the fuel characteristics of the selected biomass fuel, and the combustion technology.

In this section, emission reduction measures primarily for removal of particles, NO_x and SO_x are presented. Other components that can also be reduced by secondary measures are HCl, heavy metals and PCDD/F; however, secondary emission reduction measures for these components will not be presented in detail. HCl emission levels are reduced in wet throwaway processes applied for SO_x reduction. Furthermore, adsorptives such as activated lignite can be used for a combined extraction of HCl, SO_2 and PCDD/F. PCDD/F emission levels can be reduced by an efficient particle separation at temperatures well below the temperature range of the de novo synthesis. Emissions of heavy metals can be significantly reduced in particle-collecting devices such as bag filters or electrostatic filters.

9.2.5.1 Particle control technologies

Not every particle control technology suits every need. Among the determining factors are the particle's size, required collection efficiency, gas flow size, allowed time between cleanings, the detailed nature of the particles, and the presence of tars in the flue gas. The following rules of thumb [1] may be helpful in selecting particle control technologies for biomass combustion applications:

- Sticky particles (e.g. tars) must be collected in a liquid, as in a scrubber, or in a cyclone, bag filter or an electrostatic filter whose collecting surfaces are continually coated with a film of flowing liquid. There must also be a way to process the contaminated liquid thus produced.
- Particles that adhere well to each other but not to solid surfaces are easy to collect. Those that do the reverse often need special surfaces, e.g. Teflon-coated fibres in filters that release collected particles well during cleaning.
- The electrical properties of the particles are of paramount importance in electrostatic filters, and they are often significant in other control devices where friction-induced electrostatic charges on the particles can aid or hinder collection.
- For non-sticky particles larger than about 5mm, a cyclone separator is probably the only device to consider.
- For particles much smaller than 5mm one normally considered electrostatic filters, bag filters and scrubbers. Each of these can collect particles as small as a fraction of a micron.
- For large flows the pumping cost makes scrubbers very expensive; other devices are preferable.
- Corrosion resistance and dew point must always be considered.

As Figure 9.8 shows, the effectiveness of particle control technologies strongly depends on the particle size. Table 9.11 shows the performance of technologies particularly suitable for collecting larger particles, while Table 9.12 summarizes the performance of different particle control technologies that can be used both for coarse fly ash as well as aerosols. Below, the following particle control technologies will be discussed:

- settling chambers;
- cyclones;
- multicyclones;
- electrostatic filters;
- bag filters;
- scrubbers;
- panel bed filters; and
- rotating particle separators.

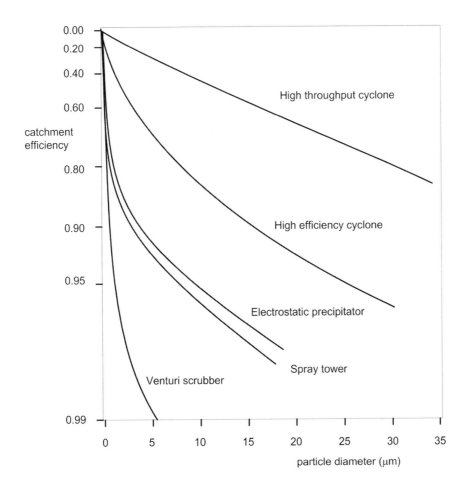

Figure 9.8 *Collection efficiencies for various particle control technologies*

Source: [4].

Table 9.11 *Summary of typical sizes of particles removed by various particle control technologies*

Particle control technology	Particle size (μm)	Efficiency (%)
Settling chambers	> 50	< 50
Cyclones	> 5	< 80
Multicyclones	> 5	< 90
Electrostatic filters	> 1	> 99
Bag filters	> 1	> 99
Spray chambers	> 10	< 80
Impingement scrubbers	> 3	< 80
Cyclone spray chambers	> 3	< 80
Venturi scrubbers	> 0.5	< 99

Table 9.12 *Summary of separation efficiency for different particle control technologies based on full-scale measurements*

Separation technology	Multicyclone	Flue gas cond. unit	Flue gas cond. unit	ESP	ESP	Baghouse filter
Combustion technology	Underfeed stoker	Moving grate	Moving grate	Underfeed stoker	Moving grate	Moving grate
Fuel	Hardwood	Bark	Hardwood	Hardwood	Ind. waste wood	Waste wood
Load [% of nom. capacity]	90	70	50	25	80–100	60–85
Total fly ash (raw gas) [mg/m$_n^3$]	109		113	260	132	433
Total fly ash (clean gas) [mg/m$_n^3$]	74		20	< 10	32	< 2
Coarse fly ash precipitation	**	***	***	***	***	***
Aerosols (raw gas) [mg/m$_n^3$]	22–35	60–110	26–36	90–120	35–58	72–98
Aerosols (clean gas) [mg/m$_n^3$]	22–35	36–55	22–26	< 10	6-8	0.4–0.7
Aerosols precipitation		**	*	**	**	***

Note: Emission data relate to dry flue-gas and 13vol% O_2. *= generally insufficient. **= moderate. ***= generally sufficient.

Source: [99].

Settling chambers

Principle of operation: Particle separation in a settling chamber is based on the principle of gravity (see Figure 9.9). The main disadvantage of this method is the low collection efficiency. However, it is still widely applied because of its ability to extinguish the flame. Typical characteristics of a settling chamber are given in Table 9.13.

Advantages:
- low pressure loss;
- simplicity of design and maintenance;
- high capacity;
- low costs; and
- ability to extinguish the flame.

Disadvantages:
- much space required; and
- low collection efficiency.

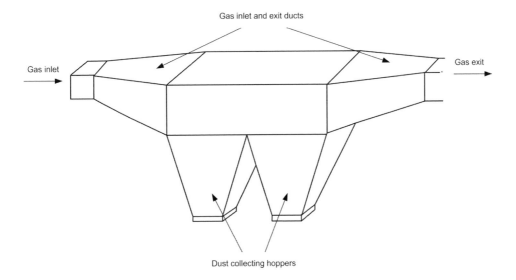

Figure 9.9 *Settling chamber*

Source: [4].

Table 9.13 *Characteristics of a settling chamber*

Principle	Gravity
Separation efficiency	App. 10% for particles < 30μm
	App. 40% for particles < 90μm
Gas velocity	1–3m/s
Pressure drop	< 20Pa
Temperature range	< 1300°C
Pressure range	< 100 bar
Application	First separation step

Cyclones

Principle of operation: Particle separation in a cyclone is based on the principle of gravity in combination with centrifugal forces. Gas and solid particles are exposed to centrifugal forces, which can be created in two ways:

1. gas flows into the cyclone in a tangential direction;
2. gas flows into the cyclone in an axial direction, and is brought into rotation using a fan.

Because of the centrifugal forces, particles hit the wall and slide down into a container. Figure 9.10 illustrates the principle. Cyclones have higher collection efficiency than settling chambers due to the centrifugal force principle.

Advantages:
- simplicity of design and maintenance;
- little floor space required;
- dry continuous disposal of collected; dusts;
- low to moderate pressure loss;
- handles large particles;
- handles high dust loadings;
- temperature independent;
- low costs; and
- ability to extinguish the flame.

Disadvantages:
- much head room required;
- low collection efficiency of small particles;
- sensitive to variable dust loadings and flow rates; and
- tars may condense in the cyclone.

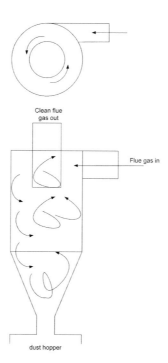

Figure 9.10 *Principle of a cyclone*

Source: [7].

The energy loss is mainly determined by the pressure drop over the cyclone and is about 0.2kWh per 1000m³ of flue gas. Typical characteristics of a cyclone are given in Table 9.14.

Table 9.14 *Characteristics of a cyclone*

Principle	Gravity in combination with centrifugal forces
Separation efficiency	85–95%
Gas velocity	15–25m/s
Pressure drop	60–150Pa
Temperature range	< 1300°C
Pressure range	< 100 bar
Application	First or final particle separation step

Multicyclones

The separation efficiency of a cyclone can be improved by increasing the centrifugal force through reduction of the cyclone diameter. In order to prevent loss of capacity, several cyclones can be used in parallel; this is named a multicyclone, and is illustrated in Figure 9.11. Multicyclones are, however, more complicated, and therefore more expensive. Further, the pressure drop is higher compared to a single cyclone, resulting in higher energy consumption.

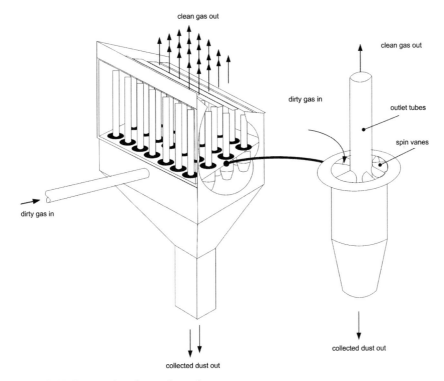

Figure 9.11 *Principle of a multicyclone*

Source: [7].

Electrostatic filters

Principle of operation: The principle of an electrostatic filter is illustrated in Figure 9.12. In an electrostatic filter, the particles are first electrically charged. Then they are exposed to an electric field in which they are attracted to an electrode. Periodically, this electrode is cleaned through vibration, by which the dust falls off the electrode into a collection unit.

In practice, separation can be done in one or two stages. Most of the electrostatic filters found in practice are one-stage filters. In the case of two-stage separation, charging of the particles is first done in a very strong electric field, after which a relatively weak field separates the particles.

Advantages:
- above 99 per cent efficiency obtainable;
- very small particles can be collected;
- particles may be collected wet or dry;
- pressure drops and power requirements are small compared with other high-efficiency collectors;
- maintenance is nominal unless corrosive or adhesive materials are handled;
- few moving parts;
- Can be operated at high temperatures up to 480°C; and
- Applicable for high flue gas flow rates.

Disadvantages:
- relatively high initial costs;
- sensitive to variable particle loadings or flow rates;
- resistivity makes catchment of some materials uneconomical;
- precautions are required to safeguard personnel from high voltage;
- collection efficiencies can deteriorate gradually and imperceptibly; and
- voluminous.

Table 9.15 illustrates some typical characteristics of an electrostatic filter. For typical wood burning applications up to 1.5MW$_{th}$, dust emissions under 50mg/m$^3_0$ can be obtained with electrostatic separation.

Table 9.15 *Characteristics of an electrostatic filter*

Principle	Electrical charging of particles
Separation efficiency	95–99.99%
Gas velocity	0.5–2m/s
Pressure drop	15–30Pa
Temperature range	< 480°C
Pressure range	< 20 bar
Application	Final particle separation step

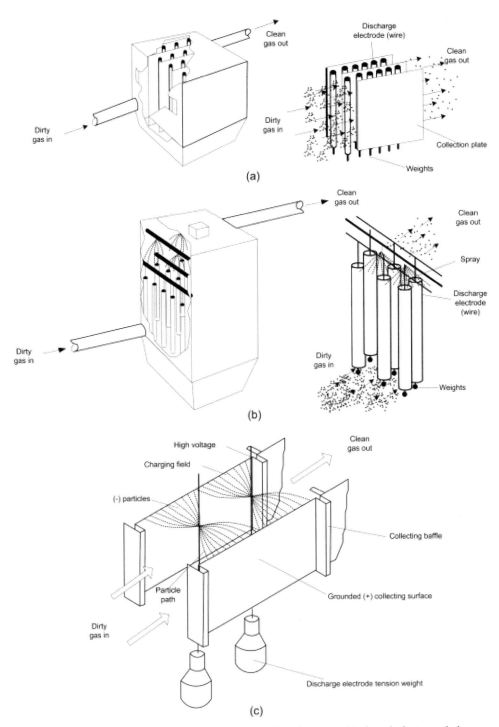

Figure 9.12 *Electrostatic filters: (a) plate type, (b) tube type, (c) detailed view of plate type filter element*

Source: [7].

In the development of electrostatic separators, the following recent trends can be recognized:

- reduced presence of dust particles in the filtered air through further optimization of the geometry of the electrodes and the gas distribution (the distance between the electrodes has been increased to about 800mm, so that the velocity of charged dust particles, and consequently the throughput, can be increased);
- application of advanced, microprocessor-based controllers of high-voltage generators and cleaning mechanisms, according to the filter load and specific dust characteristics (this effectively reduces the energy consumption);
- application of new construction materials that allow operation above temperatures of 480°C;
- application of pressure vessels and airtight isolators that allow operation above 20 bar; and
- application of pulsated electrode voltage, in order to limit any reversed flow of highly charged particles.

Bag filters

Principle of operation: As Figure 9.13 shows, the construction of a bag filter is relatively simple. It consists of a filter or cloth, tightly woven from special fibres and hung up in a closed construction through which flue gas passes. The separation efficiency of bag filters is quite high, even with high flue gas flow rates and high particle content.

Advantages:
- above 99 per cent efficiency obtainable;
- dry collection possible;
- decrease of performance is noticeable; and
- collection of small particles possible.

Disadvantages:
- sensitive to filtering velocity;
- high temperature gases must be cooled;
- affected by relative humidity (condensation);
- susceptibility of fabric to chemical attack;
- voluminous;
- operating temperature limited to about 250°C;
- tars may condense and clog the filter at low operating temperatures; and
- limited lifetime of the cloth (2–3 years).

The first layer of particles in fact improves the filtration efficiency. However, as more particles settle on the cloth, the pressure drop increases. Therefore, periodically the cloth is cleaned by vibration or pressurized air. Cloth filters are usually manufactured in cylindrical shapes. For heavily loaded filters (>100m^3/m^{2}h), flue gas flows inward. For lightly loaded filters (<100m^3/m^{2}h) flue gas flows outward.

Bag filters are usually made of various elements, which can be cleaned in turn with pressurized air. Since the fraction of elements that are cleaned at one moment is rather small compared to the total area, pressure variations over the filter are limited. The filter can be operated with a constant flue gas flow because the pressure drop is relatively low and constant (around 1000–3000Pa).

The operating temperature range is limited to about 250°C; above this, as well as when unburned carbon is present in the fly-ash, there is a significant fire risk. In order to limit the amount of particles settling on the filter and to reduce the chance of fire through sparks, a cyclone can be used. When the operating temperatures used are too low, tars present in the flue gas may condense and clog the cloth.

The materials commonly applied for bag filters (textile, polymers) can resist temperatures up to around 250°C. Recent developments are focused on the improvement of cloth cleaning and the application of materials that can operate at higher temperatures. Examples are glass fibre, special polymers, metal fibre and ceramic fibre. When using metal or ceramic fibre, flue gas temperatures up to 600–800°C can be used. The selection of fibre material is primarily determined by flue gas temperature, more than chemical resistance. Table 9.16 gives an overview of the temperature resistance of applicable bag filter materials.

For bag filters, the following recent developments can be recognized:

- increased collection efficiency through the application of finer cloth fibres, better distribution of dust over the filter area, and use of microprocessor-controlled cleaning devices;
- energy conservation through the use of microprocessor controls that optimize and reduce the use of pressurized air for cleaning the filter; and
- reduction of the filter dimensions through increase of the specific filter area load, and optimization of the geometry of various components.

Table 9.16 *Temperature resistance of bag filter materials*

Material	Maximum operating temperature [°C]	
	Dry gas	Humid gas
Polypropylene	90	90
Polyacrylnitril	130	120
Polyester	150	135
Polyphenylsulphide	180	180
Polyamide	220	180
Polyimide	260	240
Polytetrafluorethylene	250	250
Glass fibre	295	280
Metal (Inconel)	600	550
Ceramic fibre	850	850

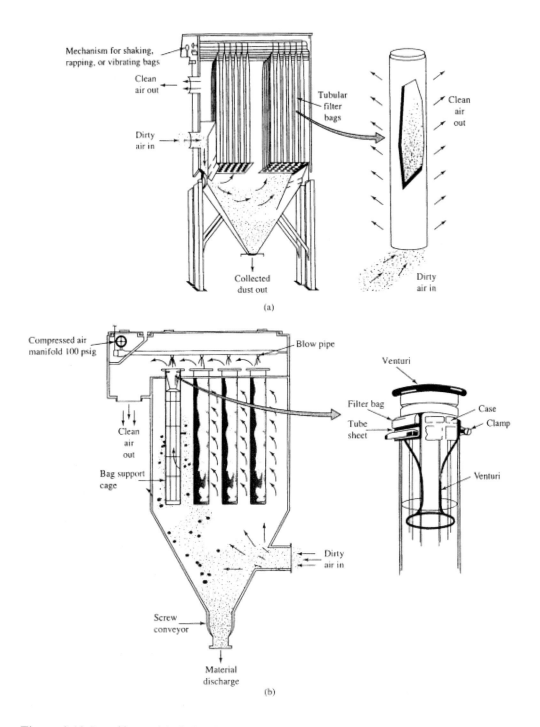

Figure 9.13 *Bag filters: (a) shaker bag filter, (b) pulse-jet bag filter*
Source: [7].

Scrubbers

Principle of operation: In scrubbers, particles are scrubbed out from the flue gas by water droplets of various sizes, depending on the type of scrubber used. The particles are removed by collision and interception between droplets and particles. Upon impact, the particles are wetted and carried by the water droplet, thus effecting removal. The more droplets that are formed, the more efficient the unit will be. Therefore, the droplets must be small. Smaller-diameter spray nozzles will produce smaller droplets but will also result in higher pressure drops, consuming more energy. Since efficiency increases as the droplet size decreases, efficiency increases with increasing pressure drop.

Flue gas scrubbing and condensation are often done in a scrubber-condenser, which contributes to lower emissions of particles and simultaneously higher energy efficiency of the plant.

Figure 9.14 shows various types of scrubbers. Parts (a) and (b) show ordinary spray chambers, a counter-current scrubber and a cross-flow scrubber, respectively. In a counter-current scrubber, flue gas is introduced at the bottom side of the unit and flows upward against the current of the settling of the atomized liquid droplets. In a cross-flow scrubber, flue gas flows across the settling of the atomized spray water droplets. Although two sets of sprays atomize the water in horizontal directions, the settling of the resulting droplets is still downward, across the direction of the flow of the flue gas. Part (c) shows a venturi scrubber and part (d) shows a cyclone spray chamber, which is a combination of an ordinary spray and a cyclone. Several other types of scrubbers exist, such as plate scrubbers, packed-bed scrubbers, baffle scrubbers, impingement-entrainment scrubbers and fluidized bed scrubbers. These will not be presented here.

Advantages:
- simultaneous gas (SO_2, NO_2, HCl) absorption and particle removal;
- ability to cool and clean high-temperature, moisture-laden gases;
- corrosive gases and mists can be recovered and neutralized;
- reduced dust explosion risk; and
- efficiency can be varied.

Disadvantages:
- corrosion, erosion problems;
- added cost of waste-water treatment and reclamation;
- low efficiency on sub-micron particles;
- contamination of effluent stream by liquid entrainment;
- freezing problems in cold weather;
- reduction of buoyancy and plume rise; and
- water vapour contributes to visible plume under some atmospheric conditions.

Panel bed filters

The concept of the panel bed filter was developed by Professor Squires (US) and further developed by SINTEF/NTNU Thermal Energy and Hydropower in Trondheim, Norway. The filter uses sand and other granulates to filter dust from flue gases. Among the most important advantages of this filter in comparison to a dust filter are its temperature resistance as well as its reduced sensitivity to sparks that may cause dust explosions. Figure 9.15 illustrates that the filter consists of layers of fine and course sand, physically separated and enclosed with strips that can open in one direction.

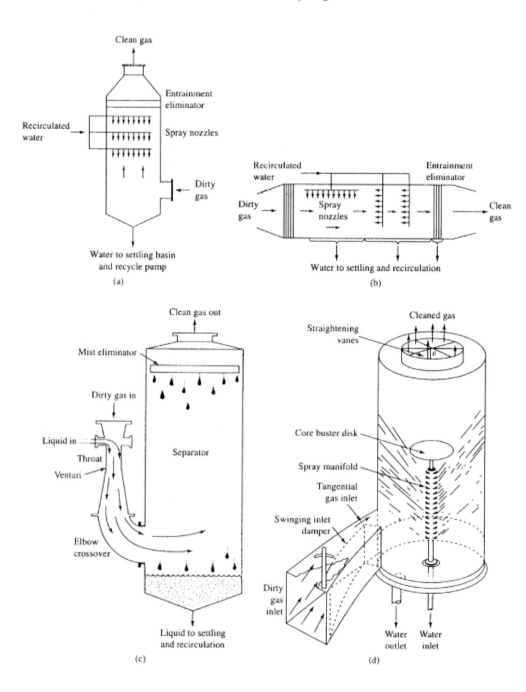

Figure 9.14 *Scrubbers: (a) counter-current spray chamber, (b) cross-flow spray chamber, (c) venturi scrubber, (d) cyclone spray chamber*

Source: [7].

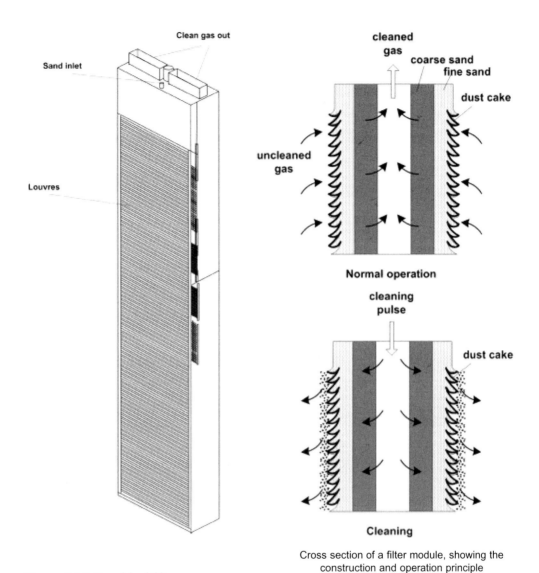

Figure 9.15 *Panel bed filter*

Source: [100].

Both the investment costs and the principle of a panel bed filter are similar to that of a bag filter. At the high-pressure side, a dust cake builds up. Dust in the flue gas that passes this cake first enters a layer of fine sand, then a layer of coarse sand. After a certain time of operation, the pressure drop over the dust cake is increased to a level at which the filter is automatically cleaned. Pulsated pressurized air from the low-pressure side will slightly open the strips on the high-pressure side, thereby removing the dust layer together with a small amount of fine sand. Sand and dust are then separated. The fine sand can be recycled into the filter, while the dust can be removed. The characteristics of a panel bed filter are summarized below:

- temperature resistant to 700°C, and insensitive to fluctuations in temperature and the presence of any hot dust particles;
- resistant against corrosive elements contained in the gas;
- the filter material is relatively cheap and can easily be replaced;
- since the specific filtration capacity is about 6–10 times higher than that of a bag filter, the filter area can be effectively reduced;
- low operating and maintenance costs; and
- capable of removing dust with particle sizes < 1mm, and emission levels below 5mg/m$^3_0$ at 13 per cent O_2 are achievable.

Rotating particle separators

Another recent development is the rotating particle separator (RPS) **[101–103]** (see Figure 9.16), developed in the Netherlands.

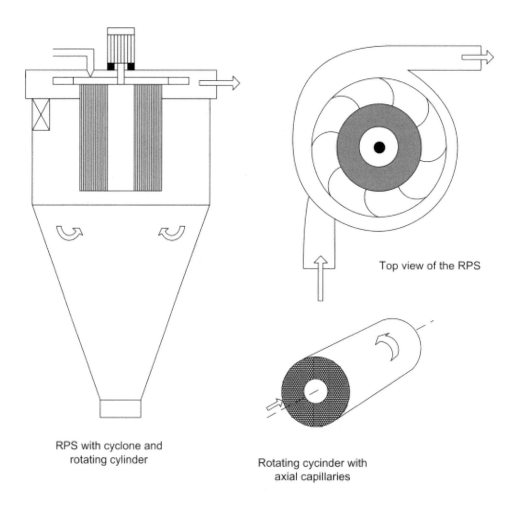

Figure 9.16 *Rotating particle separator*

Source: **[101]**.

The technology can either be used together with a conventional multicyclone or instead of an electrostatic filter. Several field tests are presently being carried out in biomass combustion plants. The separation efficiency depends on the particle size. It is claimed that for particles greater than 1μm, separation efficiencies up to 99.9 per cent can be achieved. The fly-ash concentration can be limited to 50mg/m$^3_0$ at 11 per cent O_2.

In the rotating particle separator, flue gases first enter a circular movement by means of an integrated fan. Here, heavy particles are forced outwards by centrifugal forces and fall down into the ash pit, while the light particles pass through a rotating filter element. This filter element consists of a multitude of axial channels. The liquid and solid particles are pushed against the channel walls and are periodically removed by passing air or water flowing through these channels at high velocity.

9.2.5.2 NO_x control technologies

Nitrogen oxides (NO_x) are often lumped together with sulphur oxides (SO_x) as air pollution control problems, because of the similarities between the two:

- NO_x and SO_x react with water and oxygen in the atmosphere to form nitric and sulphuric acids, respectively. These two are the principal contributors to acid rain. Because the acid rain process removes both NO_x and SO_x from the atmosphere, neither is believed to be increasing in concentration in the global atmosphere.
- NO_x and SO_x undergo atmospheric transformations leading to or contributing to the formation of PM10 (particles of 10μm or less in diameter) in urban areas.
- In high concentrations, NO_x and SO_x are severe respiratory irritants.
- NO_x and SO_x are released to the atmosphere in large quantities from fossil fuel combustion. Coal-fired power plants are the largest emitters. Emissions of NO_x and SO_x from biomass combustion, however, are substantially lower, and are caused by the fuel nitrogen and fuel sulphur content in the fuel, respectively. For SO_x this is due to a much lower sulphur content in biomass compared to coal. For NO_x it is partly due to a generally lower nitrogen content in biomass compared to coal, and partly due to fuel nitrogen being the only NO_x contributor of significance in biomass combustion, while additionally both thermal and prompt NO_x is of importance in coal combustion.

However, focusing on biomass combustion applications, the following major differences can be identified between fuel NO_x and SO_x formation:

- Formation of NO_x in combustion chambers can be greatly reduced by optimizing the combustion process through primary NO_x emission reduction measures, such as staged-air combustion and staged fuel combustion. No such optimization is practically possible for SO_x. However, in special combustion applications such as fluidized bed reactors, lime or limestone injection may be used to convert SO_x to anhydrite ($CaSO_4$), which then can be removed from the flue gas in the form of particles.
- The ultimate fate of sulphur oxides removed by pollution control or fuel-cleaning processes is to be turned into $CaSO_4$, which is an innocuous, low solubility solid, commonly deposited in landfills. There is no correspondingly cheap, innocuous and insoluble salt of nitric acid, so landfilling is not a suitable solution for the NO_x we collect in pollution control devices. The ultimate fate of NO_x is to be converted into molecular nitrogen.

- It is relatively easy to remove SO_2 from combustion gases by dissolving SO_2 in water and causing a reaction with alkali. Aqueous SO_2 quickly forms sulphurous acid, which reacts with the alkali and then is oxidized to sulphate. Collecting nitrogen oxides this way is not nearly as easy because NO, the principal nitrogen oxide present in combustion flue gas has a very low solubility in water. Unlike SO_2, which quickly reacts with water to form acids, NO must undergo a two-step process to form an acid, in which NO first reacts with oxygen to form NO_2, which then reacts with water to form HNO_3. The first reaction is relatively slow. It is fast enough in the atmosphere to lead to the formation of acid precipitation in the several hours or days that the polluted air travels before encountering precipitation. However, it is too slow to remove significant quantities of NO in the few seconds that a flue gas spends in a wet limestone scrubber used for SO_2 control. Some of the NO_2 in the flue gas is removed in such scrubbers, but normally only a small fraction of the total nitrogen oxides is NO_2 (less than 10 per cent).

NO_x emissions can be controlled both by primary emission reduction measures, as shown in Section 9.2.4, and/or by secondary emission reduction measures. The secondary emission reduction measures involve chemical treatment of the flue gas after the combustion chamber, aimed at converting NO_x to N_2.

The secondary NO_x emission reduction measures applicable for NO_x reduction in biomass combustion applications are mainly selective catalytic reduction (SCR) or selective non-catalytic reduction (SNCR). Both utilize injection of a reducing agent, mainly ammonia or urea, to reduce NO_x to N_2, with or without a catalyst, respectively. Additionally, catalytic converters optimized for NO_x reduction can be placed after the combustion chamber in small-scale biomass combustion applications, like catalytic converters optimized for the reduction of emissions from incomplete combustion, which were discussed under primary emission reduction measures. This is not a commonly used NO_x control technology for biomass combustion applications, but it is widely used for NO_x control in motor vehicles.

Selective catalytic reduction (SCR)

SCR reduces NO_x to N_2 by reactions with, usually, ammonia or urea in the presence of a platinum, titanium or vanadium oxide catalyst. The stoichiometric equations for SCR are:

$$4NO + 4NH_3 + O_2 \rightarrow 4N_2 + 6H_2O$$

$$2NO_2 + 4NH_3 + O_2 \rightarrow 3N_2 + 6H_2O$$

SCR operates optimally in a temperature range of 220–270°C [37] using ammonia, and 400–450°C using urea, where a vaporized reducing agent is injected. Approximately an 80 per cent NO_x reduction has been reported for SCR in fossil fuel combustion [2], where it is the most widely used secondary NO_x control technology. However, Nussbaumer [66] reports up to 95 per cent NO_x reduction at 250°C without significant slippage of ammonia in a wood-firing system. For the SCR process, the long-term behaviour of the catalyst can be a problem, as deactivation (the loss of catalytic activity and/or selectivity over time) is likely. Deactivation of the catalyst with time has an impact on the amount of catalyst needed and on the frequency of catalyst replacement in SCR systems, and both have an effect on the operating costs. Three mechanisms for deactivation of SCR catalyst have been identified: fouling (surface deposition); pore condensation (and/or pore blockage) and poisoning. For a given situation one or more of these mechanisms may be occurring.

Figures 9.17 and 9.18 show low dust and high dust wood-firing systems with SCR.

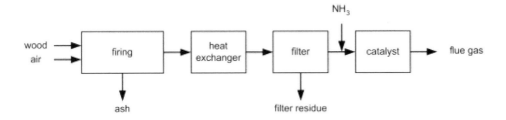

Figure 9.17 *Wood-firing system with NO_x reduction by SCR low dust process*

Source: [53].

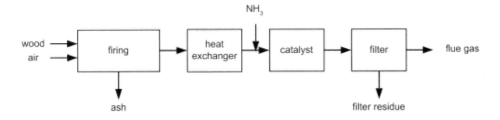

Figure 9.18 *Wood-firing system with NO_x reduction by SCR high dust process*

Source: [53].

Selective non-catalytic reduction (SNCR)

Because of the requirements and demands of catalysts, SNCR processes have been developed that do not require a catalyst for activation of the reaction. Instead, the reaction is run at higher temperatures. In the SNCR process, ammonia or urea is injected into the flue gas at a temperature usually between 850 and 950°C. In a wood-firing system, a temperature between 840 and 920°C proved optimal [37]. Because of the high temperature, this process does not need a catalyst to initiate the reactions. Ammonia is injected at a rate of 1:1 to 2:1 mole ammonia to mole of NO_x reduced. About 60–90 per cent NO_x reduction can be achieved with SNCR. The SNCR process requires an accurate temperature control to achieve optimum NO_x reduction conditions. If the temperature is too high, ammonia is oxidized to NO and if the temperature is too low, ammonia does not react at all and is emitted together with the NO_x. Hence, there exists an optimum temperature window for the SNCR process. Ammonia must be added to the flue gas in a quantity that is proportional to the NO_x content in the flue gas. Good mixing is very important to achieve optimum NO_x reduction conditions. Most SNCR processes have an ammonia slip of about 1–2ppm in the flue gas leaving the stack. Figure 9.19 shows a schematic view of an SNCR process for a wood-firing system.

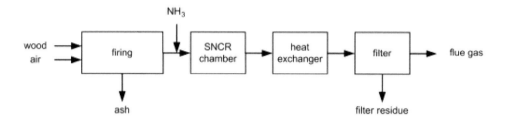

Figure 9.19 *Wood-firing system with NO_x reduction by SNCR*

Source: [53].

In Figure 9.20 the NO_x reduction potential of SCR and SNCR is illustrated as a function of the fuel nitrogen content and compared to the NO_x reduction potential of staged-air combustion with a separate reduction chamber and also to conventional combustion.

9.2.5.3 SO_x control technologies

SO_x emission levels from biomass combustion applications are generally low. Mainly SO_2 (> 95 per cent) is formed. Emissions of SO_2 are usually not significant for wood combustion applications, due to the low sulphur content in wood. However, for biomass fuels such as miscanthus, grass and straw, emissions of SO_2 may be significant and SO_2 emission reduction measures must be applied.

Several measures have been developed for the removal of SO_2 from gases. However, for flue gases from combustion applications, in which the SO_2 concentration is usually below 1000ppm, the common measure is scrubbing the flue gas with water containing finely ground limestone in limestone wet scrubbers, as shown in Figure 9.21. The flue gas, from which the solid fly-ash particles have been removed, passes through a tower where it passes counter-current to a scrubbing slurry containing water and limestone particles. Figure 9.21 shows the scrubber vessel as a spray tower with multiple sprays and a mist eliminator; some other designs use a packing with a very high open area in the tower or specialized bubbler designs.

In the tower, the SO_2 dissolves in the slurry and reacts with limestone, producing CO_2, which enters the gas stream, and $CaSO_3$. The latter is almost entirely oxidized to $CaSO_4$, partly by the excess oxygen in the flue gases in the tower, partly in the effluent hold tank, or the thickener, or in an additional oxidizing vessel, sparged with air. The slurry is recirculated from the hold tank. A side-stream is taken to a thickener (also called a settler) and a filter to remove solids; ultimately the captured SO_2 must leave the system as $CaSO_4$ or $CaSO_3$ in this waste solid. Finely ground fresh limestone is added to the effluent hold tank. The scrubber operates at or near the adiabatic saturation temperature of the entering

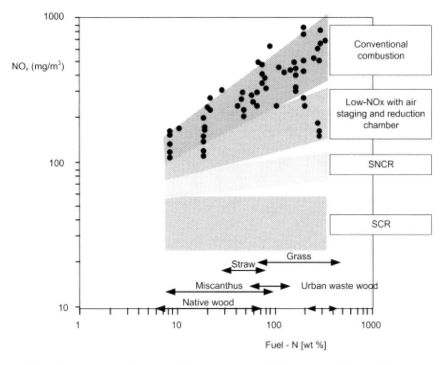

Figure 9.20 *Comparison of NO_x reduction potential for various NO_x reduction measures versus fuel-nitrogen*

Source: [32].

flue gas, which is about 52°C. The cleaned flue gas is normally reheated to about 80°C to restore plume buoyancy and to prevent acid corrosion of the ducts and stack downstream of the reheater, and then released to the stack. The moisture content of the waste slurry is reduced by thickening and then vacuum filtration. The filter cake is often mixed with dry fly-ash from the same plant to make a solid waste stream that has a still lower moisture content. The resultant mixture is easier to handle and store as it goes to its ultimate destination, landfill.

The wet limestone scrubber is a wet throwaway process. However, dry throwaway processes have also been developed to avoid the solids handling and wet sludge disposal difficulties that come with wet throwaway processes. Dry throwaway processes experience less corrosion and scaling difficulties and produce a waste product that is much easier to handle and dispose of. All of these systems inject dry alkaline particles (limestone, hydrated lime, sodium carbonate or sodium bicarbonate) into the gas stream, where they react with the gas to remove SO_2. The SO_2-containing particles are then captured in the particle collection device that the plant must have to collect fly-ash. With limestone or lime, high SO_2 collection efficiencies can only be reached by putting a large excess of lime or limestone into the system, thus increasing reagent costs, increasing the load on the particle collector, and increasing the volume of solid wastes to be disposed of. However, if one uses more reactive (and much more expensive) sodium carbonate or sodium bicarbonate, the collection efficiency is much better, mostly because of the much higher reactivity of these salts.

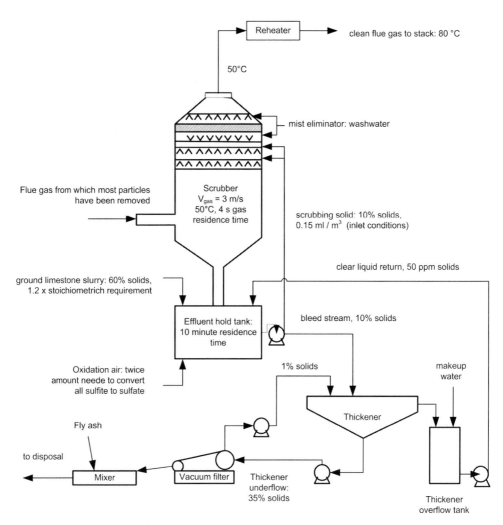

Figure 9.21 *Flow diagram of a typical limestone wet scrubber*

Source: [1].

Wet–dry systems have also been developed, combining some features of the preceding two kinds of systems. The most widely used wet–dry systems are spray dryers.

Simultaneous gas (SO_2, NO_2, HCl) absorption and particle removal can be achieved in scrubbers. However, scrubbers are too expensive for most smaller biomass combustion applications and are not widely utilized. Limestone injection is commonly installed in large-scale fluidized bed reactors burning coal.

9.2.6 Emission limits in selected IEA member countries

For a number of countries currently or previously involved in IEA Bioenergy Task 32, the maximum emissions allowed for biomass combustion plants are listed in this section. The overview was initially produced by Heikki Oravainen of VTT [104] and has been revised for this second edition of the book. The figures, which generally apply to new plants, vary significantly from country to country. This has a large influence on the technology implemented in the individual country. The various units used make comparison difficult; therefore standardization is desirable. The term m^3_0 used in this section designates volume at standard reference condition; pressure 101.3kPa and temperature of 273K.

9.2.6.1 European Union

In EU member countries, emission legislation has been derived from the guidelines for Large Combustion Plants (LCP) (2001/80/EG, PB L 309) as well as the Waste Incineration Directive (WID) (2000/76/EG, PB L 332). Guideline 2001/80/EG applies for biomass, defined as products from agriculture and forestry, vegetable waste from agriculture, forestry and from the food production industry, untreated wood waste and cork waste). Guideline 2000/76/EG applies to the incineration of all other types of waste. The emission guidelines are listed in Table 9.17.

Table 9.17 *Overview of LCP and WID (data in mg/m³)*

	LCP (clean biomass, at least 50MW$_{th}$)		Waste Incineration Directive	
	Co-firing (6% O$_2$)	Stand-alone (6% O$_2$)	Co-firing (6% O$_2$)	Stand-alone (11% O$_2$)
NO$_x$ (measured as NO$_2$)	200*	200–400	200*	200
SO$_2$	200*	200	50*	50
Dust	30*	30–50	10*	10
Cd + Tl			0.05	0.05
Hg			0.05	0.05
Sum heavy metals			0.5	0.5
HCl			10*	10
HF			1*	1
Dioxins/furans (ng/m³)			0.1	0.1
VOC			10*	10
CO			50*	50

Note: * Mixing rule applies for composition of total flue gas.

9.2.6.2 Austria

Table 9.18 *Emission limits for residential heating boilers fired with solid biomass fuels with a nominal heat output of up to 300kW*

Feeding system	CO [mg/MJ$_{NCV}$]	VOC [mg/MJ$_{NCV}$]	NO$_X$ [mg/MJ$_{NCV}$]	Particles [mg/MJ$_{NCV}$]
Manual	1100	80	150	60
Automatic	500	40	150	60

Table 9.19 Emission limits for steam boilers (STB) and hot water boilers (HWB) in industrial and commercial plants

Nominal fuel capacity [MW]	CO [mg/m³₀] STB	CO [mg/m³₀] HWB	VOC [mgC/m³₀] STB	VOC [mgC/m³₀] HWB	NO_x [mg NO_2/m³₀] STB	NO_x [mg NO_2/m³₀] HWB	SO_2 [mg/m³₀] STB	SO_2 [mg/m³₀] HWB	Particles [mg/m³₀] STB	Particles [mg/m³₀] HWB
Solid fuels without wood, lignite in brackets if differing (6% O_2)										
< 0.35	–	1000	–	–	–	–	1	–	150	150
0.35–1	–	1000	–	–	–	400	1	–	150	150
1–2	–	150	–	–	2	400	1	–	150	150
2–5	250	150	–	–	2	400	1	–	150	50
5–10	250	150	–	–	2	400	1	–	50	50
10–50	250	150	–	–	2	350	1000 (2000)	400	50	50
50–150	250	150	–	–	600	100	1000	200	50	50
150–300	250	150	–	–	450	100	200 (600)	200	50	50
300–500	250	150	–	–	300	100	200 (600)	200	50	50
> 500	250	150	–	–	200	100	200 (600)	200	50	50
Natural wood (13% O_2)[4]										
< 0.15	–	800	–	50	–	250/300/500[3]	–	–	–	150
0.15–0.35	–	800	150	50	–	250/300/500[3]	–	–	150	150
0.35–0.5	–	250	150	20	–	250/300/500[3]	–	–	150	150
0.5–1	–	250	100	20	–	250/300/500[3]	–	–	150	150
1–2	–	250	50	20	–	250/300/500[3]	–	–	150	150
2–5	250	250	50	20	–	250/300/500[3]	–	–	120	50
5–10	250	100	50	20	–	250/300/350[3]	–	–	50	50
10–50	250	100	50	20	–	200/200/350[3]	–	–	50	50
50–300	250	100	50	20	300	200/200/350[3]	–	–	50	50
> 300	250	100	50	20	200	200/200/350[3]	–	–	50	50

Notes: [1] lignite or lignite briquettes must have a S content below 1%; [2] as low as possible, fulfilled, if low-NO_x burner, FBC, flue gas recirculation or staged combustion is applied; [3] natural wood/beech, oak, natural bark, brushwood, tree cones/wood residues treated with halogen- and heavy-metal free agents; [4] if PCDD/F formation is possible due to the fuel composition in plants with a nominal fuel capacity of more than 10MW PCDD/F in the flue gas must be less than 0.1ng/m³₀

9.2.6.3 Denmark

Recommendations for emission limits are given in the national guideline [105]; actual limits are included in the plants' environmental approval issued by local authorities based on the national guideline.

Fuel input [MW]	CO [ppm]	Particles [mg/m³₀]	NO$_x$ [mg/m³₀]
Wood fuels, like wood pellets, sawdust, woodchips, grain (ref. 10% O_2)			
0.12–1.0	–	300	–
> 1.0	Commonly 500	40 or 100*	–
> 5.0	Commonly 500	40 or 100*	300
Straw (ref. 10 % O_2)			
< 1.0	–	–	–
> 1.0	500	40	–
> 5.0	500	40	300

Note: * Depending on the applied flue gas cleaning method.

9.2.6.4 Finland

General emission limits for indigenous fuels (wood, wood waste, peat, straw) are shown in the table below – local authorities can set tighter requirements for emission limitation.

Heat output [MW]	NO$_x$ [mg/MJ]	SO$_2$ [mg/MJ]	Particles [mg/MJ]
1–5	–	–	200
5–50	–	–	85–4/3 (P-5) *
	NO$_2$ [mg/m$_n^3$ at 6% O_2]	SO$_2$ [mg/m$_n^3$ at 6% O_2]	Particles [mg/m$_n^3$ at 6% O_2]
50–300	400	Biomass: 200 Peat: 400	50
100–300	300	Biomass: 200 Peat: 200	30
> 300	150	Biomass: 200 Peat: 200	30

Note: * P is the capacity in MW. For grate combustion the limit is 200mg/MJ in power range 1–10MW.

Official limits for small-scale wood-burning appliances (up to 300kW) are also expected to be implemented in 2008.

9.2.6.5 Germany

The following table summarizes the current state of legislation in Germany.

Fuel input [MW]	CO [mg/m³₀]	NO$_x$ [mg/m³₀]	SO$_2$ [mg/m³₀]	Particles [mg/m³₀]
Coal (ref. 7% O$_2$) (TA-Luft 5.4.1.2.1)				
1–5	150	fluidized bed: 300 others: 500	fluidized bed: 350 others: 1000 or 1300	50
5–50	150	fluidized bed: 300 others <10MW: 500 others >10MW: 400	fluidized bed: 350 others: 1000 or 1300	20
Peat (ref. 11 % O$_2$) (TA-Luft 5.4.1.2.1)				
1–5	150	fluidized bed: 300 others: 500		50
5–50	150	fluidized bed: 300 others <10MW: 500 others >10MW: 400		20
Straw and similar (ref. 13% O$_2$) (1.BImSchV)				
< 0.1	4000		350*	150
Straw and similar (ref. 11% O$_2$) (TA-Luft 5.4.1.3)				
0.1–1	250	500	350*	50
1–50	250	400	350*	20
Clean wood (ref. 11% O$_2$) (TA-Luft 5.4.1.2.1)				
1–2.5	150	250	350*	100
2.5–5	150	250	350*	50
5–50	150	250	350*	20
Clean wood (ref. 13% O$_2$) (1.BImSchV)				
0.015–0.05	4000	–	–	150
0.05–0.15	2000	–	–	150
0.15–0.5	1000	–	–	150
0.5–1	500	–	–	150
Used wood, low contamination (13% O$_2$) (1.BImSchV)				
0.05–0.1	800			150
> 0.1–0.5	500			150
> 0.5–1	300			150
Used wood, low contamination (11% O$_2$) (TA-Luft 5.4.1.2.1)				
1–5	150	400	350*	50
> 5–50	150	400	350*	20
Used wood, high contamination (11% O$_2$) (17.BImSchV)				
All	50	200	50	10

Note: * Applies above a total mass flow of 1.8kg SO$_2$/h

For selected applications further limitations have to be met for the parameters HC, HCl, PCDD/F and others.

Several limitations have changed recently and further changes are being discussed, particularly for small-scale furnaces below 1000kW. Thus emission levels for the 1.BImSchV applications are being modified in 2007.

9.2.6.6 The Netherlands

In the Netherlands' emission regime, a distinction is made between clean biomass (as defined in the European LCP) and all other types of biomass, regarded as waste. The simplified decision tree is shown in Figure 9.22. The general outline is as follows:

- For combustion or co-firing of clean solid biomass from agricultural, forestry or landscape maintenance operations as defined in the European LCP (such as clean wood, verge grass, composting sieve overflow, etc.), the so-called BEES-A applies.
- For all other solid biomass-containing fuels, the BVA (derived from the European WID) applies. This concerns products such as vegetable, fruit and garden waste, swill, animal manure, mixed industrial wastes, etc.
- BEES-B applies for gaseous fuels produced from anaerobic digestion or thermal gasification of biomass that are combusted in a gas engine.
- An exception is made for industrial combustion of clean waste wood in an installation smaller than 5MW$_{th}$ and owned by the producer of the wood. In such cases, the NER-BR for clean waste wood applies.

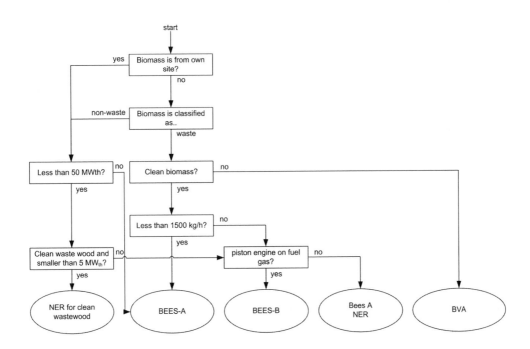

Figure 9.22 *Simplified decision tree for emission limits in the Netherlands*

Source: [106].

Table 9.20 provides an overview of the current emission standards that may apply to bioenergy installations.

Table 9.20 *Overview of most relevant emission limits in the Netherlands for solid biomass combustion in mg/m^3_0*

	BEES-A Clean solid biomass [mg/m^3, 6% O_2 dry]	NER Clean waste wood [mg/m^3, 11% O_2 dry]	BVA, Stand-alone Contaminated solid biomass [mg/m^3, 11% O_2 dry]	BVA, Co-firing Contaminated solid biomass [mg/m^3, 6% O_2 dry]
NO_x	200 (>300MW_{th}) 100 (<300MW_{th})	400[2]	130 (η_{el_eq}>40 %[1], <20MW_{th}) 70 (other cases)	100 (<300MW_{th}) 200 (>300MW_{th})
SO_2	700 (<50MW_{th}) 200 (>50MW_{th})	–	50	200
Dust	20	100 (<0.5MW_{th}) 50 (0.5–1.5MW_{th}) 25 (1.5–5.0MW_{th})	5	20
Cd + Tl		–	0.05	0.015
Hg		–	0.05	0.4mg/kg biomass (<10%$_m$ co-firing) (3.5/%$_m$ + 0.05) mg/kg for larger co-firing
Total heavy metals		–	0.5	0.15
HCl		–	10	Mixing rule
HF		–	1	Mixing rule
Dioxins and furans (ng TEQ)		–	0.1	0.1
VOC		–	10	Mixing rule
CO		250 (1.5–5.0MW_{th})	50	Mixing rule
NH_3		–		

Notes: [1] 1MJ of heat produced is accounted for as 0.47MJ of electricity.
[2] Only if larger than 250kW_{th} and with more than 80% particle board in fuel.

9.2.6.7 Belgium

For the French-speaking part of Belgium, Wallonia, emission limits are defined in the building permit as there are no official general emission limits. The values in the table below are generally followed by the authorities that grant building permits; values refer to plants for clean wood and an O_2 reference of 11 per cent.

Heat output [MW thermal]	CO [mg/m_n^3]		NO_x [mg/m_n^3]		Particles [mg/m_n^3]	
	Existing plant	New plant	Existing plant	New plant	Existing plant	New plant
0.5–1	250	250	500	500	150	150
1–5	250	250	500	250	150	50
5–50	250	250	400	250	50	20

9.2.6.8 Norway

Combustion installations with capacity exceeding about 5MW are given individual permits. The degree of specification of emission limits vary due to fuel type, plant size and issue date of the licence. Newer licences that regulate emissions are more specific and stricter than older ones. The number of issued permits with specific emission limits is not available. Combustion installations covered by the Integrated Pollution Prevention and Control (IPPC) Directive are generally regulated based on best available technology (BAT).

The following emission limits are indications of emission limits in existing combustion installations smaller than 50MW burning biomass and not covered by the IPPC Directive. The emission limits provide guidance for the authority, and emission limits for installations are decided after individual judgements. In some cases higher limits are accepted. Official figures are expected in 2007 when the current round of hearings is finalized. Thus the existing indications of emission limits are provisional; however, they are in actively use already by the Norwegian Pollution Control Authority.

Figures refer to dry gas, a temperature of 273K, a pressure of 103.1kPa and NO_x as NO_2.

Capacity	Dust mg/Nm³	NO_x mg/Nm³	CO mg/Nm³
Existing combustion installations burning biomass (11% O_2)			
0.5–1MW	150/300[1]	–	–
1–5MW	150/300[1]	–	300[2]
5–20MW	100	300	200
20–50MW	50	300	100
New combustion installations burning biomass (11 % O_2)			
0.5–1MW	100	250	150
1–5MW	20	250	100
5–50MW	20	200	100

Notes: [1] waste from corn [2] not waste from corn.

9.2.6.9 Sweden

Generally emission limits are set by the approving authority.

Small-scale equipment

For small-scale domestic equipment, less than 300kW, the emission of organic gaseous compounds (OGC) or carbon monoxide (CO) are regulated. For primary heating sources, i.e. boilers, the limit value is 80, 100 or 150mg OGC/m_n^3 dry gas at 10 per cent O_2 according to the following table. For secondary heating sources other than pellet stoves, the limit value is 0.3vol% CO at 13 per cent O_2. For pellet stoves the limit value is 0.04vol% at 13 per cent O_2.

Nominal capacity, kW	mg OGC per m_n^3 dry gas at 10% O_2
Manual fuel feed	
< 50	150
> 50 < 300	100
Automatic fuel feed	
< 50	100
> 50 < 300	80

Plants between 300kW and 50MW

Emissions of nitrogen oxides are in practice limited by the system for NO_x-fee. The system includes plants larger than 25GWh. The average emission during 2005 was 57mg NO_x/MJ_{fuel}.

For plants of 0.5–10MW there are recommended limits of dust of 100mg/m_n^3 at 13 per cent CO_2 in cities, higher outside cities. In practice, accepted limits today are much less than 100mg/m_n^3 at 13 per cent CO_2.

There are no general limits for CO. Examples of limits permitted are: less than 100mg/m_n^3 for plants larger than 5MW and less than 500mg/m_n^3 for plants smaller than 5MW.

Emissions to air of sulphur are limited to 100mg/MJ. In practice, sulphur is limited by the sulphur tax that covers fuel with sulphur content higher than 0.05wt%.

Plants larger than 50MW

For plants larger than 50MW (supplied power) emissions of SO_2, NO_x and dust are regulated. For solid biofuels in *existing* plants larger than 50MW the limit values for SO_2 are:

Maximum limit value of emission of sulphur dioxide (SO_2) [mg SO_2 per m_n^3 at 6% O_2]		
50–350MW	> 350–500MW	> 500MW
Combustion is only allowed if emission to air is less than 0.19g sulphur per MJ fuel	1000–400 (linear decrease)	400

For solid biofuels in *new* plants larger than 50MW (gas turbines excluded) the limit value for SO_2 is 200mg SO_2 per m_n^3 at 6 per cent O_2.

For solid biofuels in *existing* plants > 50MW the limit values for NO_x are:

Maximum limit value of emission of nitrogen oxides (NO_x) [mg/m_n^3 at 6% O_2]	
50–500MW	> 500MW
600	500

For solid biofuel in *new* plants larger than 50MW (gas turbines excluded) limit values for NO_x are:

Maximum limit value of emission of nitrogen oxides (NO_x) [mg/m_n^3 at 6 % O_2]		
50–100MW	> 100–300MW	> 300MW
400	300	200

For solid biofuels in *existing* plants larger than 50MW the limit values for dust are:

Maximum limit value of emission of dust [mg/m_n^3 at 6% O_2]	
< 500MW	> 500MW
100	50

For solid biofuels in new plants larger than 50MW the limit values for dust are:

Maximum limit value of emission of dust [mg/m$_n^3$ at 6% O_2]	
50–100MW	> 100MW
50	30

9.2.6.10 Switzerland
Emission limit values valid in 2006 are summarized from data in [107]. For installations smaller than 70kW, the emission limit value refers to the type test only and is not subject to emission control in practice.

Heat output	Reference O_2	CO [mg/m$^3_0$]		Gaseous organic substances	NO$_x$	Ammonia and ammonium compounds	Particles
[MW]	[vol%]	natural wood	ind. waste wood	[mg C/m$^3_0$]	[mg NO$_2$/m$^3_0$]	[mg NH$_3$/m$^3_0$]	[mg/m$^3_0$]
0.02–0.07	13	4000	1000	–	–	–	–
0.07–0.2	13	2000	1000	–	–	–	150
0.2–0.5	13	1000	800	–	–	–	150
0.5–1	13	500	500	–	–	–	150
1–5	11	250	250	50	–	30	150
> 5	11	250	250	50	–	30	50

Emission limit values for installations larger than 70 kW, which need to be guaranteed in practice and which are planned to be implemented (provisional future values):

Heat output	Reference O_2	CO	Gaseous organic substances	NO$_x$	Ammonia and ammonium compounds	Particles
	[vol%]	[mg/m$^3_0$]	[mg C/m$^3_0$]	[mg NO$_2$/m$^3_0$]	[mg NH$_3$/m$^3_0$]	[mg/m$^3_0$]
> 70kW	13	500[2]	–	250[1]	–	30
> 350kW	13	500	–	250[1]	–	30
> 600kW	13	500	–	250[1]	–	30
> 1MW	11	250	–	250[1]	–	20
> 10MW	11	150	–	150	30	10

Note: [1] for a mass flow of more than 2.5kg/h; [2] valid from 2009.

Emission limit values for residential wood combustion installations (provisional future values) for type test approval:

All values at 13vol% O_2 Type	Norm	CO [mg/m³] Valid from 1.8.2008	Particles [mg/m³] 1.1.2008	Particles [mg/m³] 1.1.2011
Log wood boiler	EN 300-5, EN 12809	800	60	50
Autom. wood boiler	EN 303-5, EN 12809	400	90	60
Pellet boiler	EN 303-5, EN 12809	300	60	40
Room heater	EN 13240	1500	100	60
Pellet stove	EN 13240, EN 14785	500	50	40
Hearth	EN 12815	3000	110	90
Central heating hearth	EN 12815	3000	150	120
Chimneys	EN 13229	1500	100	60

9.2.6.11 United Kingdom

Integrated Pollution Prevention and Control (IPPC) is a regulatory system implemented in the UK that employs an integrated approach to control the environmental impacts of industrial activities. To gain a permit, operators have to demonstrate that the techniques they are proposing to use, are the BAT for their installation, and meet certain other requirements, taking account of relevant local factors. The essence of BAT is that the techniques selected to protect the environment should achieve an appropriate balance between environmental benefits and the costs incurred by operators. However, whatever the costs involved, no installation may be permitted where its operation would cause significant pollution. Further details are found in [108].

9.3 Options for ash disposal and utilization

9.3.1 Introduction: general, ecological and technological limitations

9.3.1.1 General comments

Due to increasing thermal utilization of biomass, the amounts of residues from the combustion process also increase, given the fact that the ash content of biomass fuels ranges from 0.5wt%(d.b.) for soft wood to 12wt%(d.b.) for some herbaceous biomass fuels (see also Table 2.6). It shows that the quantity of ash is strongly influenced by the bark content in wood fuels. This is a result of higher ash content in bark, on the one hand, and a higher level of mineral impurities (sand, earth, stones) in the bark, on the other hand. Straw, cereals and other herbaceous biomass fuels contain higher ash contents than wood due to their higher uptake of nutrients during their growing periods. Waste wood contains high amounts of mineral and metallic impurities as well as contaminants due to the manufacturing process and the ways it has been used prior to combustion.

Currently, the biomass ash produced is either disposed of or recycled on agricultural fields or forests. Unfortunately, this is carried out without any form of control. Considering the fact that the disposal costs are rising (the actual prices for the disposal of one tonne of wood ash range from €200 to €500), and that biomass ash volumes are increasing, a controllable ash utilization has to be established. A start was made a few years ago, with research activities focusing on the characterization of biomass ashes and their environmentally compatible utilization. This research is still ongoing.

9.3.1.2 Ecological limitations

For a sustainable biomass utilization, it is essential to close the material fluxes and to integrate the biomass ashes within the natural cycles. Therefore the cycle of minerals should be closed as completely as possible:

Soil/nutrients → root/plant → combustion → ash → soil

Previous research has shown that the natural cycle of minerals within the process of energy production from biomass is disturbed by depositions of heavy metals on the forest ecosystem caused by environmental pollution. Therefore, it is not possible to recycle the total amount of ashes produced during the combustion process in most cases. By separating a side-stream rich in heavy metals, preferably in the combustion plant, it should be possible to recycle the major part of the ashes produced (see Figure 9.23).

The main question concerning biomass ash utilization is: *Which ash fractions can be recycled and which cannot?* Therefore, knowledge of the composition and place of origin of various ash fractions from biomass combustion plants is essential.

Sustainable ash utilization also requires that ash from wood or bark combustion is recycled in forest areas, while ash from straw or energy crops is recycled in agricultural areas, but geographic and techno-economic conditions have to be considered.

9.3.1.3 Logistical limitations

To make sustainable biomass ash utilization as practical as possible, ash pre-treatment is necessary in order to offer a useable product to the farmer. The ash must be mixed in a proper way to guarantee the right chemical composition, which makes the installation of a quality assurance system necessary. Furthermore, the ash must be supplied in a way that makes it possible to distribute it with conventional manure spreaders. The ash pre-treatment depends on the size of the combustion plant as well as on the kind of biomass fuel used. Also, storage of ash can be important if it is not possible to directly link ash production and ash utilization.

Besides direct ash utilization on agricultural or forest soils, indirect utilization is also possible, by adding biomass ash to agricultural compost. The indirect ash utilization shows lower dust formation but more manipulation and handling are necessary. For the selection of the right kind of ash utilization, several factors have to be considered, such as the agricultural infrastructure and total costs of a certain ash utilization process. The production of ash granules by hardening the ash with water is a common pre-treatment technology for biomass ashes. It is in use and being tested in Sweden [110]. Ash granulation has the advantages of dust formation reduction during ash manipulation and spreading. Moreover, the leaching and availability of nutrients from the ash is reduced, which makes the ash more soil-like and decreases possible pH-shock effects on plants.

9.3.1.4 Technological and chemical limitations

A detailed analysis of the processes and mechanisms of ash formation as well as a description of the various ash fractions formed in biomass combustion plants are given in Section 2.2.10.

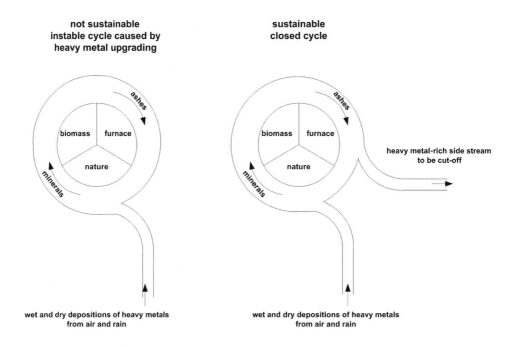

Figure 9.23 *Stable and unstable natural cycles of minerals in recycling of ash from biomass combustion*

9.3.2 Physical and chemical properties of biomass ashes

9.3.2.1 Particle sizes and densities

Table 9.21 shows the standard values for the average mass distribution of various ash fractions related to the total amount of ash for fixed bed combustion plants. The differences are due to different combustion technologies and different particle sizes of the biomass fuels used. The mass distribution depends also on the geometry of the furnace, combustion air inlet, the process control system, and the dust separation technology applied.

The average amount of ash produced depends primarily on the fuel. Fluidized bed combustion systems produce higher amounts of ash than fixed bed units due to the fact that, in addition to the biomass ash, bed material is also discharged from the furnace. The share of bottom ash in the total amount of ash produced in fluidized bed combustion plants is considerably lower than in fixed bed combustion systems and amounts to 20–30 per cent. The remaining 70–80 per cent of the ash produced is fly-ash, which is entrained with the flue gas and separated in the boiler sections and dust precipitators placed behind the boiler sections.

The particle density decreases from the bottom ash to the filter fly-ash (Table 9.22) and can be explained by the decreasing amount of mineral impurities within the ash fraction and the increasing salinity. The bulk density decreases with the fineness of the ash fraction. Ash from straw and cereal combustion shows very low density due to its specific chemical matrix, which, compared to wood or bark ash, contains more salts and less minerals.

Table 9.21 *Distribution of ash fractions for biomass fuels (%)*

Biomass fuel/ash fraction	Bark	Woodchip	Sawdust	Straw and cereal
Bottom ash	65–85	60–90	20–30	80–90
Cyclone fly-ash	10–25	10–30	50–70	2–5
Filter fly-ash	2–10	2–10	10–20	5–15

Explanations: Data for bark and woodchips are from grate furnace or underfeed stokers. Data for sawdust are from underfeed stokers. Data for straw and cereals are from cigar burners. All furnaces are equipped with cyclones, followed by a second and more efficient dust precipitation unit.
Sources: [111–113].

Particle dimensions depend on the particle size of the fuel used, the ash content, the chemical composition of the ash and the amount of mineral impurities within the fuel. The particle size of the bottom ash is influenced by ash sintering while the particle size of the fly-ash depends on the dust separation technology applied as well as on the chemical composition of the biomass fuel (amounts of aerosols formed).

9.3.2.2 Concentrations of plant nutrients

Tables 9.23 and 9.24 show the mean values of plant nutrients in ashes from bark, woodchips, sawdust, straw and cereals. The ashes analysed contain considerable amounts of plant nutrients, which makes ash utilization economically interesting. The only missing plant nutrient in biomass ash is nitrogen. During combustion of the fuel, nitrogen turns to vapour and escapes almost completely with the flue gas.

The Ca concentrations in ashes from bark and woodchip combustion plants are considerably higher than in ashes from straw and cereal combustion plants. For K it is the other way round. The Mg concentrations are similar in the ashes of all biomass fuels investigated. P is upgraded in ashes from cereals, due to the fact that grain contains about four times more P than straw.

Table 9.25 shows the chemical composition of ashes from wood residues and waste wood. The ash composition of wood residues is similar to that of ashes from fresh wood. The ash from waste wood shows lower nutrient concentrations, due to the high amounts of mineral and metal impurities in the fuel.

9.3.2.3 Concentrations of heavy metals

Tables 9.26 and 9.27 show the average heavy-metal concentrations in various ash fractions of wood bark, straw and cereal combustion plants (fixed-bed combustion). Typical concentration ranges of heavy metals in various biomass fuels can be found in Section 2.3.2.4. The concentrations of highly volatile and ecologically relevant heavy metals (Zn, Cd, Pb) increase from the bottom to the filter fly-ash due to the fact that volatile heavy metals turn to the vapour phase during combustion and then condense or react, forming new and very fine fly-ash particles (aerosols), or accumulate on the surface of existing fly-ash particles (see Section 8.3). The smallest ash fraction – the filter fly-ash – considerably exceeds the respective limiting values for utilization of ashes on agricultural fields or in forests for certain heavy metals (see for example, Section 9.4.4). Consequently, this ash fraction has to be disposed of or industrially treated. The

Table 9.22 *Average particle and bulk densities*

Ash fraction	Particle density Average [kg/m³]	Bulk density Mean value [kg/m³]	Bulk density Standard deviation [kg/m³]
Bark combustion (moving grate and underfeed stoker)			
Bottom ash	2600–3000	950	200
Cyclone fly-ash	2400–2700	650	120
Filter fly-ash	2300–2600	350	120
Woodchips combustion (moving grate and underfeed stoker)			
Bottom ash	2600–3000	950	200
Cyclone fly-ash	2400–2700	500	150
Filter fly-ash	2300–2600	–	–
Sawdust combustion (underfeed stoker)			
Bottom ash	2600–3000	650	150
Cyclone fly-ash	2400–2700	300	100
Filter fly-ash	2300–2600	–	–
Straw and cereal incinerators (cigar burner)			
Bottom ash	–	300	80
Cyclone fly-ash	2200	150	60
Filter fly-ash	2200	150	50

Explanations: All results are based on at least five single measurements. All data refer to dry matter. The particle density was determined by a helium pynometer (Autopyknometer 1320, Micromeritics), and the bulk densities following DIN 51705.
Sources: [112, 114, 115].

Table 9.23 *Average concentrations of plant nutrients in ash fractions of bark, woodchip and sawdust combustion plants*

Nutrient	Bottom ash Mean value	Bottom ash St.dev.	Cyclone fly-ash Mean value	Cyclone fly-ash St.dev.	Filter fly-ash Mean value	Filter fly-ash St.dev.
CaO	41.7	8.7	35.2	11.8	32.2	6.9
MgO	6.0	1.2	4.4	0.9	3.6	0.7
K_2O	6.4	2.1	6.8	2.3	14.3	7.2
P_2O_5	2.6	1.0	2.5	0.9	2.8	0.7
Na_2O	0.7	0.2	0.6	0.3	0.8	0.6

Explanations: Concentrations in wt% (d.b.). Ten samples were analysed from each ash fraction and each biomass fuel taken from representative test runs performed over 48 hours in Austrian grate furnaces and underfeed stokers. Type of biomass used: Woodchips and bark from spruce. St.dev.: Standard Deviation.
Sources: [111–113].

Table 9.24 *Average concentrations of plant nutrients in ash fractions of straw and cereal combustion plants*

Nutrient	Bottom ash		Cyclone fly-ash		Filter fly-ash	
	Mean value	St.dev.	Mean value	St.dev.	Mean value	St.dev.
CaO	7.8	7.0	5.9	6.0	1.2	1.0
MgO	4.3	4.2	3.4	3.2	0.7	0.4
K_2O	14.3	14.0	11.6	12.7	48.0	47.0
P_2O_5	2.2	9.6	1.9	7.4	1.1	10.3
Na_2O	0.4	0.5	0.3	0.3	0.5	0.3

Explanations: Concentrations in wt% (d.b.). Two samples were analysed from each ash fraction and each biomass fuel taken from representative test runs performed over 48 hours in Austrian cigar burners. The test runs were performed with winter wheat (straw) and triticale (cereals). St.dev.: Standard Deviation.
Source: [112].

remaining ash fractions (the bottom ash and the cyclone fly ash – usually representing more than 90wt% of the entire ash) should be mixed in a plant-specific ratio – the so-called useable ash – and utilized on agricultural land or in forests as a secondary raw material with fertilizing and liming effects.

Ashes from straw and cereal combustion plants contain 3–20 times lower concentrations of heavy metals than ashes from woody biomass (see Tables 9.26–9.28). In this case, the heavy-metal concentrations of the bottom ash and the cyclone fly-ash remain below the Austrian limiting values for soils (see Table 9.35) and exceed them only for filter fly-ashes. There are several reasons why heavy-metal concentrations are considerably lower in straw and cereal ashes than in wood ashes: lower heavy-metal input by dry and wet depositions on agricultural soils in comparison to forests; higher pH value of the soil (lower heavy-metal mobility); and shorter rotation times.

Table 9.25 *Average concentrations of plant nutrients in ash fractions of combustion plants using wood residues and waste wood*

Nutrient	Bottom ash		Cyclone fly-ash		Filter fly-ash	
	Wood residues	Waste wood	Wood residues	Waste wood	Wood residues	Waste wood
CaO	32.6	31.1	32.3	28.5	–	16.7
MgO	3.0	2.8	3.2	3.0	–	0.5
K_2O	6.6	2.3	7.5	2.7	–	7.5
P_2O	0.9	0.9	1.3	1.4	–	0.4
Na_2O	–	1.1	–	1.1	–	3.3

Explanations: Concentrations in wt% (d.b.). Approximately seven samples were analysed from each ash fraction and each biomass fuel taken from test runs in Switzerland.
Sources: [116–118].

Table 9.26 *Average concentrations of heavy metals in ash fractions of bark, woodchip and sawdust incinerators*

Element	Bottom ash		Cyclone fly-ash		Filter fly-ash	
	Average	St.dev.	Average	St.dev.	Average	St.dev.
Cu	164.6	85.6	143.1	46.7	389.2	246.4
Zn	432.5	305.2	1870.4	598.5	12,980.7	12,195.9
Co	21.0	6.5	19.0	7.3	17.5	5.2
Mo	2.8	0.7	4.2	1.4	13.2	9.8
As	4.1	3.1	6.7	4.3	37.4	41.4
Ni	66.0	13.6	59.6	19.0	63.4	35.4
Cr	325.5	383.0	158.4	61.0	231.3	263.7
Pb	13.6	10.4	57.6	20.5	1053.3	1533.0
Cd	1.2	0.7	21.6	8.1	80.7	59.2
V	43.0	10.0	40.5	16.6	23.6	9.1
Hg	0.01	0.03	0.04	0.05	1.47	2.05

Explanations: Concentrations in mg/kg (d.b.). Ten samples were analysed from each ash fraction and each biomass fuel taken from representative test runs performed over 48 hours in Austrian grate furnaces and underfeed stokers. Type of biomass used: Woodchips and bark from spruce. St.dev.: Standard deviation.

Source: [111–113].

Table 9.27 *Average concentrations of heavy metals in ash fractions of straw and cereal combustion plants*

Element	Bottom ash		Cyclone fly-ash		Filter fly-ash	
	Straw	Cereals	Straw	Cereals	Straw	Cereals
Cu	17.0	47.0	26.0	60.0	44.0	68.0
Zn	75.0	150.0	172.0	450.0	520.0	1950.0
Co	2.0	3.1	1.0	1.6	< 1.0	< 1.0
Mo	< 10.0	< 10.0	< 10.0	10.0	10.0	18.0
As	< 5.0	< 5.0	< 5.0	5.0	22.0	16.2
Ni	4.0	10.5	< 2.5	7.5	< 2.5	< 2.5
Cr	13.5	20.5	17.5	16.5	6.8	5.8
Pb	5.1	4.5	21.5	15.0	80.0	67.5
Cd	0.2	0.2	1.8	1.4	5.2	5.1
V	< 10.0	20.5	< 10.0	16.0	< 10.0	< 10.0
Hg	< 0.1	< 0.1	< 0.1	0.2	0.7	0.1

Explanations: Concentrations in mg/kg (d.b.). Two samples were analysed from each ash fraction and each biomass fuel taken from representative test runs performed over 48 hours in cigar burners. The test runs were performed with winter wheat (straw) and triticale (cereals). St.dev.: Standard deviation.

Source: [112].

Table 9.28 *Average concentrations of heavy metals in ash fractions of residual and waste wood incinerators*

Element	Bottom ash		Cyclone fly-ash		Filter fly-ash	
	Res. w.	Waste w.	Res. w.	Waste w.	Res. w.	Waste w.
Cu	170.0	1234.0	226.0	437.0	–	422.0
Zn	503.0	6914.0	3656.0	15,667.0	–	164,000.0
Co	25.0	21.0	18.0	30.0	–	5.0
Mo	6.6	7.0	10.0	11.0	–	11.0
As	–	17.0	–	59.0	–	104.0
Ni	113.0	179.0	61.0	167.0	–	74.0
Cr	236.0	466.0	212.0	1415.0	–	404.0
Pb	363.0	2144.0	1182.0	8383.0	–	50,000.0
Cd	3.3	20.0	16.0	70.0	–	456.0
V	–	171.0	–	260.0	–	153.0
Hg	< 0.5	< 0.5	< 0.7	0.7	–	< 0.5

Explanations: Concentrations in mg/kg (d.b.). Approximately seven samples were analysed from each ash fraction and each biomass fuel. The test runs were carried out in various Swiss residual and waste wood incinerators. Combustion technology used: grate furnaces or underfeed stokers. Res. w.: wood residues, Waste w.: waste wood.
Source: [117, 119].

Compared to wood and herbaceous biomass ashes, the heavy-metal concentrations in ashes of residual or waste wood incinerators are considerably higher (see Table 9.28) and exceed the respective limiting values for almost all ash fractions produced (see Table 9.35). Ash utilization on soils is therefore not possible if contaminated biomass fuels are used. Mixtures of contaminated and fresh biomass fuels are not recommended for ash utilization either.

9.3.2.4 Organic contaminants and organic carbon

The concentrations of organic contaminants (PCDD/F, PAH) in bottom ashes and cyclone fly-ashes from biomass combustion plants meeting the current technological standards and using chemically untreated biomass are generally low and ecologically harmless (see Table 9.29). In filter fly-ashes, organic contaminants are enriched considerably. Consequently, this ash fraction should be collected separately and should not be used on soils.

Investigation of ashes from bark and wood shows a correlation between the organic carbon content in the cyclone fly-ash and the PAH content. This underlines the importance of a good burnout of fly-ashes [113]. The content of organic carbon (C_{org}) within biomass ashes should be below 5wt% (d.b.) in order to remain within current guidelines for biomass ash utilization on agricultural land and in forests in various countries (e.g. Austria and Denmark). Higher C_{org} concentrations require extra determination of PCDD/F and PAH in the ashes. The comparison of ashes from wood/bark and straw/cereals shows higher PCDD/F values for herbaceous fuels. This is most probably caused by the higher chlorine content in herbaceous fuels, which supports the formation of PCDD/F.

The formation of organic contaminants can be impeded by a high degree of flue gas and ash burnout and by the use of fuels with low Cl content.

Table 9.29 *Organic carbon, chlorine and organic contaminants in biomass ashes*

Ash fraction	$C_{org.}$ [wt% (d.b.)]	Cl⁻ [wt% (d.b.)]	PCDD/F [ng TE/kg d.b.]	PAH [mg/kg d.b.]	B[a]P [μg/kg d.b.]
Bark combustion					
Bottom ash	0.2–0.9	< 0.06	0.3–11.7	1.4–1.8	1.4–39.7
Cyclone fly-ash	0.4–1.1	0.1–0.4	2.2–12.0	2.0–5.9	4.7–8.4
Filter fly-ash	0.6–4.6	0.6–6.0	7.7–12.7	137.0–195.0	900.0–4900.0
Wood chips combustion					
Bottom ash	0.2–1.9	< 0.01	2.4–33.5	1.3–1.7	0.0–5.4
Cyclone fly-ash	0.3–3.1	0.1–0.5	16.3–23.3	27.6–61.0	188.0–880.0
Filter fly-ash	–	–	–	–	–
Sawdust combustion					
Bottom ash	0.2–3.4	< 0.1	1.3–2.1	14.7–21.1	21.0–40.5
Cyclone fly-ash	3.2–15.3	0.1–0.6	1.5–3.7	11.2–150.9	180.0–670.0
Filter fly-ash	–	–	–	–	–
Straw combustion					
Bottom ash	9.0	1.1	2.3	0.1	0.0
Cyclone fly-ash	16.6	13.6	70.8	15.8	17.0
Filter fly-ash	16.1	35.1	353.0	26.0	320.0
Cereal combustion					
Bottom ash	9.4	1.3	22.0	0.3	0.0
Cyclone fly-ash	9.9	5.2	12.2	0.5	0.0
Filter fly-ash	4.9	19.0	56.0	7.3	210.0
Limiting values					
for soils	–	–	5.0 (proposed)	5.0	1000.0 (proposed)
for sewage sludge utilisation	–	–	100.0 (proposed)	20.0	4000.0 (proposed)

Explanations: Guiding values for soils according to ÖNORM L1075. Limiting values for sewage sludge utilisation according to 'Deutsche Klärschlammverordnung'. Results of three test runs for bark, woodchip and sawdust combustion plants (combustion technology used: grate furnace and underfeed stoker) and of two test runs for straw and cereal combustion plants (combustion technology used: cigar burner). TE – toxicity equivalent; PCDD/F – polychlorodibenzo-p-dioxin and dibenzo-furan; PAH – polycyclic aromatic hydrocarbons according to US-EPA (16 compounds); BP – benzopyrene (carcinogenic PAH compound and therefore separately listed).

Source: [111, 112, 113].

Ashes from the combustion of wood residues show a concentration of PCDD/F ranging from 3 to 5ng TE/kg (d.b.) (Content of chlorides in the ash: 0.01–0.6wt% (d.b.)) **[116]**. The PCDD/F concentrations in ashes from waste wood combustion are as follows **[117, 120, 121]**:

	PCDD/F [ng TE/kg (d.b.)]	Chlorides in the ash [wt% (d.b.)]
Bottom ash:	8–14	0.3%
Cyclone fly-ash:	≈ 800	3%
Filter fly-ash:	2650–3800	13.5%

The results show that ashes from waste wood combustion are too contaminated with PCDD/F to be used on soil in an environmentally compatible manner.

9.3.2.5 pH value and electrical conductivity

The pH values of wood ashes normally range from 12 to 13 (see Table 9.30). The respective values for ashes from straw and cereal combustion are lower (10.5–11.5 for bottom ashes and cyclone fly-ashes, and 6.0–9.5 for filter fly-ashes), due to the lower Ca and higher S and Cl concentrations in these ashes.

Table 9.30 *pH value and electrical conductivity of biomass ashes*

	Bottom ash		Cyclone fly-ash		Filter fly-ash	
	pH value in $CaCl_2$	El. cond. [mS/cm]	pH value in $CaCl_2$	El. cond. [mS/cm]	pH value in $CaCl_2$	El. cond. [mS/cm]
Bark	12.7	8.9	12.7	10.8	12.7	35.6
Woodchip/sawdust	12.8	10.2	12.7	13.1	12.6	39.5
Straw	11.4	9.3	10.8	25.8	9.4	49.5
Cereals	10.8	11.4	10.5	21.0	5.9	46.7

Explanations: El. cond.: electrical conductivity.
Source: [111–113].

After the ash is spread on the soil, a rapid conversion of hydroxides into carbonates will occur due to the CO_2 content of the air in soils. Following this, the pH value of the ash will decrease to a neutral value and the pH value of the soil will increase. Parallel to this carbonate formation, the electrical conductivity will also decrease within days to normal values for soils (below 0.75mS/cm). So far, no negative effects of biomass ash applications on soils or plants due to pH shocks have been detected [113].

9.3.2.6 Si, Al, Fe, Mn and carbonate concentrations

Table 9.31 shows the average concentrations of Si, Al, Fe, Mn and carbonate (expressed as CO_2) in mixtures of bottom ash and cyclone fly-ash (so-called 'useable ash') for different kinds of biomass fuel. The ecological impact of recycling these ashes to soils is presented below for the various ash components.

Si is neutral from an ecological point of view; it is insoluble and can improve the structure of soils [122]. Ashes from herbaceous fuels show higher Si concentrations due to the high Si content of the stalk.

The content of Al in the upper layer of soils in, for example, Austria ranges from 15,000 to 60,000mg/kg (d.b.) [123]. Therefore, only ashes from bark show Al concentrations higher than those usually occurring in soils. If the pH values of the soil are above 5, the Al is not soluble and therefore it is neutral in an ecological sense (bound as oxides or hydroxides). If the pH value is below 3.8 (forest soils), the release of Al^{3+} ions will increase and plant damage will occur. Therefore, the Al content of biomass ashes is not a threat to soils with pH values above 3.8. The alkaline effects of biomass ashes will further increase the pH value of the soil and impede the release of Al [113].

Table 9.31 *Si, Al, Fe, Mn and carbonate concentrations in biomass ashes (mixture of bottom ash and cyclone fly-ash) for various fuel types*

Property [wt% (d.b.)]	Mixtures of bottom ash and cyclone fly-ash				
	Bark combustion	Wood chips combustion	Sawdust combustion	Straw combustion	Cereal combustion
SiO_2	26.0	25.0	25.0	54.0	45.0
Al_2O_3	7.1	4.6	2.3	1.8	3.3
Fe_2O_3	3.5	2.3	3.8	0.8	3.2
MnO	1.5	1.7	2.6	0.0	0.0
SO_3	0.6	1.9	2.4	1.2	0.8
CO_2	4.0	3.2	7.9	1.6	1.2

Source: [109].

Fe and Mn are essential nutrients for plants. Due to the low concentration of S in biomass ashes, S can also be evaluated as a nutrient [122].

The concentrations of Fe and Al in the bottom ash and cyclone fly-ash are in the same range. S is highly volatile and therefore concentrates in the cyclone fly-ash. Si and Mn are concentrated in the bottom ash.

The elements in the bottom ash and cyclone fly-ash are mainly available as oxides, but hydroxides, carbonates and sulphates also occur. The amount of carbonates (calculated as CO_2) in the various ash fractions strongly depends on the time the ash was exposed to a moist and CO_2-rich atmosphere (flue gas channel, ambient air) as well as on the furnace temperature. This means that if furnace temperatures are high and residence times of the fly-ash in the flue gas channels are low (full-load operation of the combustion plant), the carbon content of the ashes will be lower. The second influencing factor for the formation of carbonates is the concentration of Ca and Mg in the ash. This is why the carbonate content is low in straw combustion plants, in spite of relatively low furnace temperatures (Si forms oxides; K forms oxides, chlorides or sulphates).

9.3.2.7 Leaching behaviour

Figure 9.24 shows the leaching behaviour of biomass ashes. Approximately 30 per cent of K from ashes of straw and cereal incinerators is water soluble. The solubility of K in these ashes is more than twice as high as that of ashes from bark and woodchip combustion. P is partly water soluble in straw and cereal ashes, and almost insoluble in woodchip and bark ashes. Since K and P are the most important nutrients in biomass ashes, the short-term availability of nutrients in straw and cereal ashes should be higher than in ashes from bark and woodchip combustion plants.

Cl is almost completely soluble in water. The same applies for S in straw and cereal ashes, while S in ashes from wood fuels is only partly soluble in water (about 20 per cent).

Concerning the leaching behaviour of heavy metals, bottom ash and cyclone fly-ash are ecologically harmless as fertilizing and liming agents for soils. If biomass ashes are disposed of, the relevant limiting values for heavy metals in leachates have to be considered [111, 113].

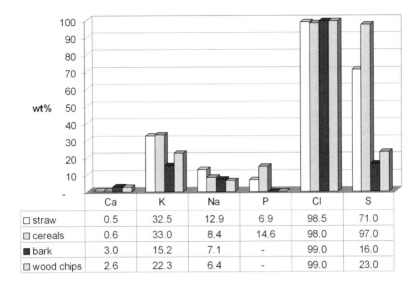

Figure 9.24 *Average water-soluble amounts of elements in mixtures of bottom ash and cyclone fly-ash of biomass fuels*

Source: [111, 113].

9.3.3 Material fluxes of ash-forming elements during combustion of biomass

9.3.3.1 Results of material flux analyses performed

Comprehensive material flux analyses performed in several biomass combustion plants for ash-forming elements reveal the behaviour of these elements and their recovery in the various ash fractions. Figures 9.25 and 9.26 show the average distributions of elements among the various ash fractions for woodchip and bark combustion in fixed bed units. The distribution of elements is similar in fixed bed straw and cereals combustion plants.

The following principles generally apply to biomass combustion plants. They can be derived from the results of the material flux analyses:

- Most of the environmentally relevant heavy metals in biomass ashes (Zn, Pb, Cd) are contained in the filter fly-ash fraction, not in the bottom-ash fraction. This is due to the high volatility of these elements. The filter fly-ash fraction has the highest adsorption potential for these elements.
- From an environmental point of view, the two most important heavy metals in wood and bark ashes are first, Cd and, second, Zn. In modern biomass combustion plants it is possible to retain 30–60wt% (w/w) of the total Cd input and 25–50wt% (w/w) of the total Zn input in the filter fly-ash, which forms only 2–15wt% (w/w) of the total amount of wood ash produced.
- About 85–95 per cent of the plant nutrients in biomass ashes (Ca, Mg, K, P) occur in the so-called 'useable ash' mixture of bottom ash and cyclone ash.

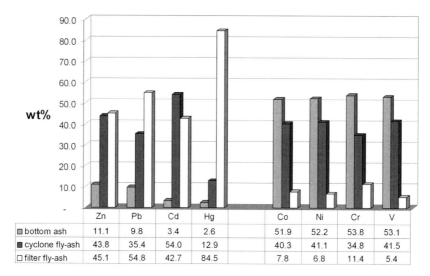

Figure 9.25 *Average distribution of heavy metals among the ash fractions of bark and woodchip combustion plants*

Explanations: Average values of 11 test runs carried out in a biomass district heating plant equipped with moving grate technology, multicyclone and flue-gas condensation unit (nominal boiler capacity: 4MW$_{th}$). Biofuels used: bark for five test runs and woodchips for six test runs. The tests were carried out at full and partial load.

Source: **[109, 111]**.

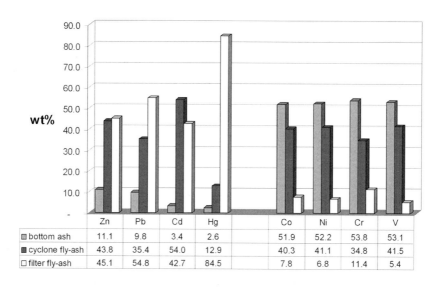

Figure 9.26 *Average distribution of nutrients among the ash fractions of bark and woodchip combustion plants*

Explanations: see Figure 9.25.

Source: **[109, 111]**.

9.3.4 Conclusions

The ideal natural cycle of minerals in ash from the combustion of chemically untreated biomass is disturbed by the deposition of heavy metals on plants and soils **[124, 125]**. Therefore, the following recommendations should be taken into consideration for an ecological utilization of ash from the combustion of chemically untreated biomass (no waste wood or wood residues):

- In order to keep the natural cycle of minerals stable, a side-stream – as small as possible and enriched in heavy metals – should be separated from the process.
- The disturbance of the cycle of nutrients by the separation of this side-stream should be as small as possible.

Applying these recommendations to the results of ash analyses and material flux calculations leads to the following conclusions about sustainable utilization of ashes from chemically untreated biomass fuels (see Figure 9.27).

The following measures can be carried out in order to prevent the heavy metals from accumulating in the ash cycle:

- From the ash fractions produced in biomass combustion plants (bottom ash, cyclone fly-ash and filter fly-ash), a mixture of bottom ash and cyclone fly-ash in a plant-specific ratio (= useable ash) is normally suitable as a secondary raw material with fertilizing and liming effects for agriculture and forestry (closing the mineral cycle to a great extent).
- The filter fly-ash is the smallest ash fraction and is enriched with heavy metals. Therefore, it should be separately collected and disposed of in order to stabilize the natural cycle of minerals.
- In order to remove heavy metals from the ash cycle without excluding minerals, it is advisable to install a multicyclone before a filter fly-ash precipitator. The minerals contained in the ash fraction of the multicyclone can thus be recycled, while the heavy metals contained in the fly-ash precipitator are disposed of.
- Further development of biomass combustion technology should focus on improving the heavy-metal fractionation potential by appropriate primary measures, in order to further upgrade the volatile heavy metals in the filter fly-ash. By this approach, the heavy-metal concentrations in the useable ash should be minimized in order to guarantee stable and sustainable biomass utilization in the long run.

The utilization of ashes from the combustion of waste wood on agricultural land or in forests is not possible without special ash pre-treatment, because of the high concentrations of heavy metals and other contaminants in waste wood ashes. This contamination is due to manufacturing processes and the previous utilization of the wood (e.g. coatings). During waste wood combustion a certain fractionation of heavy metals is achieved but, due to the fact that waste wood also contains high concentrations of non-volatile heavy metals (like Cr and Ni), or heavy metals that are enclosed in other metallic and mineral matrices, the contamination of the bottom ash remains significant. It follows that ashes from waste wood combustion can only be utilized after heavy-metal separation by secondary measures (e.g. acid leaching or thermal treatment). Otherwise, appropriate disposal of the ashes is necessary.

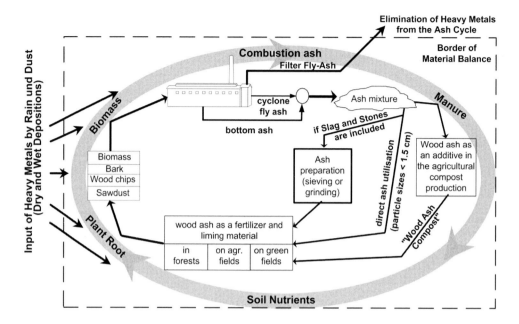

Figure 9.27 *Environmentally compatible cycle economy for ashes from biomass combustion plants with possibilities of ash-conditioning*

The composition of ashes from the combustion of wood residues is in between that of ash from virgin wood and ash from waste wood. The kind of ash utilization possible depends on the kind of contamination and has to be evaluated on a case-by-case basis.

9.4 Treatment technologies and logistics for biomass ashes

9.4.1 Combustion technology

There are two different ways of influencing the composition and characteristics of ashes from biomass combustion (see Table 9.32). The primary and secondary measures mentioned here aim at maximizing the amount of ash that can be utilized in an ecological way.

Fractionated heavy metal separation technologies as a primary measure influence the combustion and fly-ash precipitation process in order to concentrate heavy metals in the filter fly-ash. In this way, expensive and complex secondary measures can be avoided.

Ash treatment aimed at separating heavy metals can be of interest for ash fractions that are not useable on soils. Various methods are possible, as listed in Table 9.32 **[109, 126]**. These methods are also of interest for ashes from waste wood combustion plants. In practice, ash treatment technologies will be applied if the overall costs are cheaper than disposal costs. The heavy metals separated can be used as raw materials (e.g. in metallurgical processes).

Table 9.32 *Possibilities for influencing the composition and characteristics of ashes*

Primary measures during combustion Combustion and precipitation technology	Secondary measures Post-combustion treatment
Good ash burnout	Chemical ash treatment
Production of low amounts of fly ash	Biochemical ash treatment
Controlled fly-ash precipitation	Thermal ash treatment
Controlled ash precipitation temperature	Combined processes

9.4.2 Downstream processes: ash pre-treatment and utilization

Table 9.34 summarizes the most important conditions for a sound closed-cycle economy with biomass ashes. It is advisable to set down the procedures necessary for controlled and sustainable ash utilization within guidelines or legal provisions (see Section 9.4.3).

Table 9.34 shows an overview of requirements for the smooth transfer of the ash from the biomass combustion plant to the farmer (ash pre-treatment, downstream process technology) and from the farmer to nature (ash utilization and spreading).

In selecting an appropriate downstream process technology for a biomass combustion plant, the plant capacity and the kind of fuel used must be taken into consideration. The various points mentioned concerning the selection of an ash utilization technology should also be considered.

Requirements of downstream process technology

The following requirements must be met to supply the farmer with a useful product:

- Bottom ash and cyclone fly-ash have to be mixed in the plant-specific ratios. The filter fly-ash must be disposed of or treated prior to utilization.
- The ash must be suitable for fertilizer spreaders, i.e. free from slag, stones or particles larger than 15–20mm.
- Impurities, like metal particles, have to be separated.

Table 9.33 *Conditions for a controlled and ecologically friendly ash utilization*

- Exclusive use of ashes from chemically untreated biomass fuels.
- Periodic analysis of the useable ash fraction for nutrients and heavy metals.
- Periodic analyses of soils concerning their suitability for ashes from biomass combustion.
- Determination of:
  - the properties that need to be analysed;
  - the limiting values for pollutants; and
  - appropriate analysis methods.
- Regulations for the amounts of ash used in forests and on agricultural land.
- Obligatory recording of ash quantities produced by the plant operator.
- Recycling the ash on the same kind of soil that the fuel was taken from (e.g. wood ash on forest land, straw ash on agricultural land).

Table 9.34 *Logistics of a smooth-running closed-cycle economy for biomass ashes*

Downstream process technology	• Mixing of the ash fractions appropriate for utilization on soils in the proper ratio. • Supplying the ash in a way that makes the use of conventional fertilizer spreaders possible. • Prevention of dust emissions during ash manipulation. • Possibility for ash storage at the biomass combustion plant.
Ash utilization and spreading technologies	• Direct or indirect ash utilization (depends on the regional agricultural infrastructure). • Selection of a suitable spreading technology (depends on the biomass ash and the on-site conditions). • Selection of the proper period for ash spreading.

Source: [127].

To fulfil these requirements several process steps, like screening, metal separation and wrapping, must be integrated into the ash manipulation chain at the combustion plant. Biomass combustion plants with a nominal thermal capacity below 1MW are usually equipped with a manual ash removal system only. Due to the low amounts of ash produced and the fineness of the ash particles (in these plants usually only woodchips and sawdust are used), no special pre-treatment is necessary.

For biomass combustion plants with higher thermal capacities, mechanical and automatic ash removal installations are recommended. If the bark content of the fuel exceeds 30 per cent, special ash pre-treatment processes are necessary (sieving, milling). Bark usually shows a high content of mineral impurities (sand, stones) which causes local ash melting on the grate. Also, combustion plants for herbaceous biomass fuels need ash pre-treatment units due to the low ash-melting point of these fuels. Furthermore, the separation of metal parts (e.g. nails) by a magnetic separator situated in the ash conveying channel is generally recommended for wood fuels.

Requirements of ash spreading

The technology used for ash spreading should allow time-saving, cost-effective, smooth and low-dust operation.

For ash spreading on agricultural fields and meadows, fertilizer spreaders with dust protection, as applied for liming, are suitable. Of the existing fertilizer spreaders without dust protection, only the screw-type fertilizer spreader is suitable. This type can also handle wet and lumped ash as agglomerated ash particles are comminuted by the screw conveyors.

Ash utilization in forests requires equipment that can blow the ash from the road into the forest. On horizontal fields, ash can be blown up to 50m with fertilizer spreaders. Practical tests with a small-scale fertilizer spreader (capacity $2m^3$) caused no damage to young trees directly exposed to the ash stream [128]. The possibility of increased wear-out of the equipment because of slag and stones must be considered.

Biomass ashes can also be used as an additive in agricultural compost production. This indirect ash utilization has the advantage of low dust emissions. Moreover, it eliminates large slag particles: these are comminuted or separated during the composting process. As a guide value, the amount of biomass ash should not exceed 5% vol of the overall volume of all materials used for combusting **[129]**.

9.4.3 Examples of guidelines for the utilization of biomass ashes in Austria

Austria is the first European country to introduce clear legislation for the utilization of biomass ashes. These regulations are briefly outlined below.

'Wood ash' is defined as ash from the thermal utilization of chemically untreated woody biomass such as woodchips, bark and sawdust. The wider term 'biomass ash' also includes ashes from the thermal utilization of straw, cereals, hay and other agricultural residues.

Bottom ash, cyclone fly-ash and filter fly-ash from biomass combustion plants are legally considered industrial waste and not hazardous waste. The utilization of waste materials makes them a secondary raw material, provided the process is ecologically friendly and meaningful. Moreover, ashes from biomass combustion are not regarded as fertilizers as their chemical composition varies too much. The utilization of biomass ashes as a secondary raw material must be regulated.

On the basis of comprehensive research results, the Austrian Ministry for Agriculture and Forestry has worked out two guidelines **[130, 131]** for the proper utilization of biomass ashes on agricultural fields and in forests. These guidelines regulate:

- which kind of biomass ash can be used as a fertilizing and liming agent for agricultural and forest soils (the guidelines cover the kind of ash fractions as well as the concentration of heavy metals in the ash);
- how and when ash spreading is possible (spreader technology, climatic and weather conditions);
- the maximum amounts of ash to be applied; and
- the demands on the composition of the soil (soil type, chemical analysis).

The use of filter fly-ash on soils is forbidden. This ash fraction must be treated as industrial waste. Disposal of filter fly-ash without treatment is only possible in landfills equipped for leachates of eluate class III according to ÖNORM S 2072 **[121, 132]**.

In Denmark, regulations on the use of ashes from combustion of biomass have been effective since January 2000. In Sweden and Germany, guidelines for appropriate biomass ash utilization on soils are under development.

9.4.4 Recommended procedure and quantitative limits for biomass ash-to-soil recycling in Austria

The following recommendations are based on comprehensive ash analysis, field tests and material balances concerning the 'Ash-Soil – Ground Water – Plant' system for agricultural land, grassland and forests in Austria **[112, 113]**. These recommendations are also incorporated in the Austrian guidelines for the use of biomass ashes on soils (see Section 9.4.3).

From the ash fractions produced in biomass combustion plants (bottom ash, cyclone fly-ash and filter fly-ash), a mixture of bottom ash and cyclone fly-ash in the plant-specific ratio (useable ash) is usually applied as a secondary raw material with fertilizing and liming effects on soils.

To avoid ecological incompatibilities and damage, the following procedure should be followed:

1. Only ashes from the combustion of chemically untreated biomass fuels are allowed to be utilized.
2. The filter fly-ash and the cyclone fly-ash should be collected separately. The filter fly-ash has to be disposed of or industrially utilized (after an appropriate treatment).
3. The useable ash of a biomass combustion plant has to be analysed for its content of nutrients and its content of ecologically relevant heavy metals before it is used on soils. Ash analysis should be repeated at regular intervals.
4. The following quantitative limits are recommended for the utilization of useable ash from bark, woodchips and sawdust combustion on soils [130, 131]:
   - 1000kg ha.per year on agricultural land,
   - 750kg ha.per year year on grassnland,
   - 3000kg/ha once in 50 years in forests.
5. The amounts of useable ash from straw and cereal combustion should be determined according to the nutrient demand of the plants and the soil. Due to the low heavy-metal concentrations of these ashes, there should be no environmental limitations on their utilization on soils as long as the filter fly-ash fraction is separated.

The Austrian limits for utilization of biomass ashes on agricultural land and in forests as well as the guide values for soils are shown in Table 9.35.

The limits on the amounts of ash used for agricultural land and grassland are based on the heavy-metal content of the ash. As long as the heavy-metal content of the ash used is lower than the limiting value in the guideline, the amount of ash can be increased [131].

The limit for the amount of ash to be used in forests is based on the content of Cd in the ash. This limit guarantees that the input of Cd into the forest due to ash recycling does not exceed the output of Cd from forest due to wood harvesting [124, 125, 130].

Ashes from the combustion of bark or wood should be used in forests or in short-rotation stands, while ashes from the combustion of straw or cereals should be used on agricultural land.

Further considerations for the use of wood ash in forests

Due to the high content of Ca and Mg in wood and bark ash, the effects of using wood ash as a fertilizer are similar to fertilizing the forest with lime (similar ratio $CaO/CaCO_3$, high pH value of wood ash). The larger particle size of wood ash lowers its aggressiveness in comparison to lime. Therefore, wood ash is recommended for forest soils where an increase of the pH value is desirable.

Further considerations for the use of biomass ash on agricultural land

In general, the amount of ash used for fertilizing depends on the specific cultivation, the soil and the additional fertilizers used, and should be calculated annually by nutrient balance. The average amounts of nutrients in biomass ashes are listed in Section 9.3.2.2.

Concerning the availability of K, biomass ashes are similar to industrial fertilizers. Therefore, biomass ashes are suitable for plants which are sensitive to Cl (trees, bushes, several kinds of vegetables) and which can stand an increase of the soil pH. Furthermore, biomass ash can be used for annual energy crops substituting fertilizers with high Cl content in order to decrease the Cl uptake of the plants [113, 133].

Table 9.35 *Limiting values for concentrations of heavy metals in biomass ashes used on agricultural land and in forests and guiding values for soils according to existing Austrian regulations*

Element	Limiting values forest	Limiting values agriculture*	Guide values for soils
Cu	250	250	100
Zn	1500	1000	300
Ni	100	100	60
Cr	250	250	100
Pb	100	250	100
Cd	8	5	1

Explanations: Guiding values for soils – according to ÖNORM L1075. Concentrations in mg/kg d.b. * according to quality class I.
Source: Limiting values [130, 131].

As for P, wood and bark ashes can only keep the P level in soils stable. The concentration as well as the availability of P is higher in straw and cereal ashes. If soils lack P, industrial fertilizers should be used [113, 133].

9.5 Waste-water handling from flue gas condensation

As described in Section 5.6, flue gas condensation is a heat recovery process by which the overall plant efficiency is increased. The condensation process also involves a precipitation of fly-ash. The contact between fly-ash, flue gas and water implies partial dissolution of the solid compounds in the ash and absorption of gaseous components in the condensate. The remainder of the fly-ash is precipitated as condensate sludge.

Flue gas condensation is mainly of interest when burning biomass with high moisture content, such as forest woodchips (as opposed to straw). This section therefore refers strictly to condensate from wood-burning plants.

The condensate from the process contains several harmful elements, of which the heavy metals pose a particular threat to the environment. Most of the heavy metals, however, are found in the condensate sludge, which can be separated from the condensate more or less effectively, depending on the method used. Knowledge of the composition of the condensate and the sludge, as well as the parameters of significance for the composition, is therefore of great importance in order to be able to treat the condensate and sludge properly before discharging/disposing of it.

Amounts of sludge and condensate

An investigation at three biomass-fired heating plants in Austria gave an indication of the amounts of condensate and sludge that can be expected from flue gas condensation (the figures will depend on the type and operation of the boiler and flue gas condensing unit) [109]:

Sludge: 0.01–0.3kg dry matter per MWh of heat produced in the boiler. The amount of sludge also depends on the load (full/partial) during the

Condensate: 150–500 litres per MWh of heat produced in the boiler. The amount of condensate depends on the moisture content of the fuel, the temperature of the flue gas leaving the condenser, and the content of oxygen in the flue gas (the ratio of excess air in the combustion process).

operation, since the efficiency of other particle separation devices prior to the condensation process (such as multicyclone, etc.) can, in turn, depend on the load.

A general relation between the amount of condensate and the amount of sludge does not exist, since these amounts depend on various factors, described above. Furthermore, the amount and the particle size distribution of the fly-ash in the flue gas depend on the construction and operation mode of the boiler and the type of fuel.

Content of heavy metals

Most of the heavy metals are bound to the particles in the sludge. By regulating the pH value to 7.5–8.5 and performing an efficient filtration of the condensate it can be possible to discharge the condensate to the recipient **[109]** (depending on the limit values of the country in question, of course). The sludge on the other hand must be disposed of or used for industrial purposes. It can definitely not be brought back to the soil due to its high content of heavy metals, in particular Zn, Cd and Pb.

Examples of levels of Cd, Pb and Zn content in the condensate after separating the sludge and condensate are shown in Table 9.36. The measurements were taken at three Austrian and two Danish plants. As can be seen from this table, the measured contents of heavy metals in the condensate from the five plants vary widely. This has partly to do with the fact that the efficiency of the separation of condensate and sludge is relevant to the amount and the particle size distribution of the solid matter leaving with the condensate. The amount of heavy metals remaining in the condensate is also influenced by the efficiency of the separation, since the smallest particles have the highest concentration of heavy metals (due to a relatively bigger surface). The retention time of the condensate in the separation unit decreases as the plant load increases and thus the efficiency of the separation process decreases as well **[109]**.

Table 9.36 *Contents of Cd, Pb and Zn in the condensate after separating the sludge and condensate, measured at three Austrian and two Danish plants*

	Average values from 15 tests at 3 Austrian plants	Average values from 3 analyses at the Danish Ørum plant	Average values from 3 analyses at the Danish Uldum plant
Cd [µg/L]	5–160	3	39
Pb [µg/L]	10–100	13	16
Zn [µg/L]	630–5310	84	9816

Source: **[109, 134]**.

Three methods of separating the condensate and sludge

Three commonly used methods for separating sludge from condensate can be outlined:

1. a sedimentation tank;
2. a filter vessel with wood dust; and
3. a belt filter.

The principles are shown in Figures 9.28–9.30.

In Figure 9.28, the condensate with sludge is led to the tank, where the sludge settles in the bottom. From there it is removed. The cleaned condensate is led out through an overflow. The small particles are easily stirred. Hence this cleaning method depends highly on unstirred conditions existing in the tank.

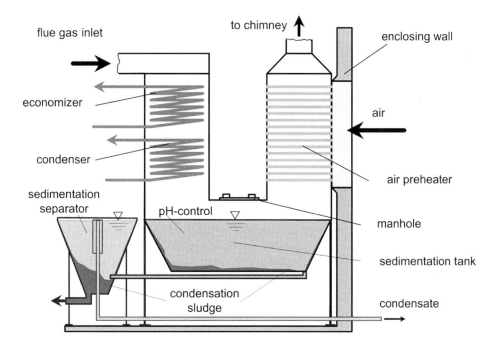

Figure 9.28 *Separation of condensate and sludge using a combination of sedimentation tank and sedimentation separator*

Source: [109].

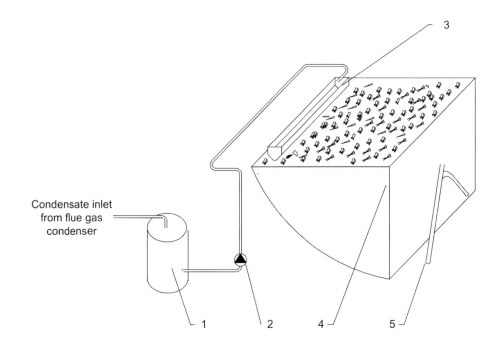

Figure 9.29 *Separation of condensate and sludge using a filter with wood dust*

Notes: 1 Plastic vessel with a float-controlled switch; 2 Condensate pump; 3 Inlet manifold; 4 Steel vessel with wood dust; 5 Outlet of cleaned condensate.
Source: [134].

In Figure 9.29, the condensate with sludge is led from the plastic container into the filter with wood dust in batches. A float-controlled switch activates the condensate pump when a certain level in the plastic vessel is reached. The condensate with sludge is led through the filter. Due to the resistance of the wood dust, the condensate leaves the filter in a continuous flow. The retained particles colour the wood dust black and indicate when it must be replaced.

At a Danish plant in the town of Uldum, the flue gas condensation unit has a capacity of approximately 460kW. The condensate is cleaned in a filter with wood dust. The plastic container can hold 200L and the vessel with wood dust is 4m^3. The wood dust particles are 3–5mm in length [134].

The processes in the belt filter in Figure 9.30 require a pH value of 6.5 [134]. The pH value can be adjusted by adding NaOH to the condensate. The belt filter is a conveyor mounted with an endless filter cloth. The condensate with sludge is led through the cloth, which retains the particles. The particles leave the liquid due to the movement of the conveyor and a scraper removes the particles from the cloth, after which they are collected in a container.

The belt filter is placed in a vessel, where chemicals are added to the condensate in order to increase the efficiency of the filter. A sodium aluminate is usually added to precipitate solid matter in the condensate, and a flocculating agent is also added.

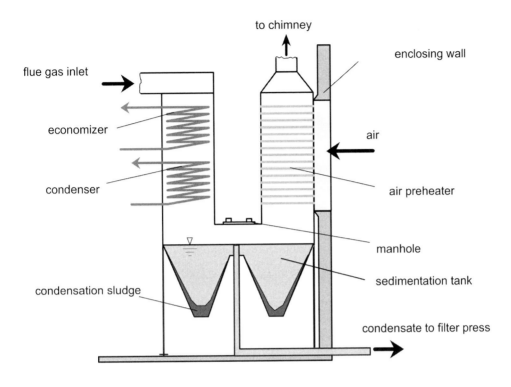
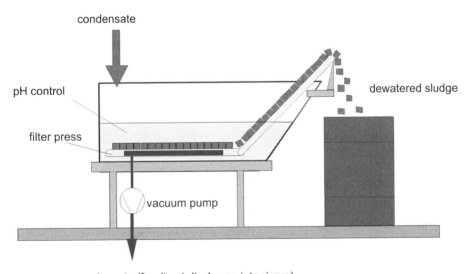

Figure 9.30 *Separation of condensate and sludge using a combination of sedimentation tank and filter press*

Source: [109].

9.6 References

1. DE NEVERS, N. (1995) *Air Pollution Control Engineering*, McGraw-Hill, Inc.
2. ROBERTS ALLEY, E. (1998) *Air Quality Control Handbook*, McGraw-Hill, Inc.
3. CHEREMISINOFF, P. N. (1993) *Air Pollution Control and Design for Industry*, Marcel Dekker, Inc.
4. FLAGAN, R. C. and SEINFELD, J. H. (1988) *Fundamentals of Air Pollution Engineering*, Prentice-Hall, Inc.
5. BOUBEL, R. W., FOX, D. L. TURNER, D. B. and STERN, A. C. (1994) *Fundamentals of Air Pollution*, Academic Press, Inc.
6. CORBITT, R. A. (1999) *Standard Handbook of Environmental Engineering*, McGraw-Hill, Inc.
7. SINCERO, A. P. and SINCERO, G. A. (1996) *Environmental Engineering – A Design Approach*, Prentice-Hall, Inc.
8. KARLSVIK, E., HUSTAD, J. E., SKREIBERG, Ø. and SØNJU, O. K. 'Greenhouse gas and NO_x emissions from wood-stoves', in *Energy, Combustion and the Environment*, Gordon and Breach, vol 2, no A, pp539–550
9. FLYVER CHRISTIANSEN, H., DANISH ENERGY AGENCY, FOCK M. W., CENTER FOR BIOMASS TECHNOLOGY at dk-TEKNIK ENERGY and ENVIRONMENT (2000) 'LCA of Procurement and Conversion of Biomass and Fossil Fuels – used for Energy Production in Denmark 1997', poster presented at the First World Conference and Exhibition on Biomass for Energy and Industry, 5–9 June, Sevilla
10. ABB (1994) *Emissions and the Global Environment – Statistics and Scenarios*,
11. MILLER, J. A. and BOWMAN, C. T. (1989) 'Mechanism and modelling of nitrogen chemistry in combustion', *Prog. Energy Combustion Science*, vol 15, pp287–338
12. GARDINER, W. C. (ed.) (2000) *Gas-Phase Combustion Chemistry*, 2nd edn. Springer-Verlag
13. BOWMAN, C. T. (1992) 'Control of combustion-generated nitrogen oxide emissions: Technology driven by regulation', in *Proceedings of the 24th International Symposium on Combustion*, The Combustion Institute, Pittsburgh, PA, pp859–878
14. ZELDOVICH, Y. B., SADOVNIKOV, P. Y. and FRANK-KAMENETSKII, D. A. (1947) *Oxidation of Nitrogen in Combustion*, Academy of Sciences of the USSR, Moscow
15. ZELDOVICH, Y. B., BARENBLATT, G. I., LIBROVICH, V. B. and MAKHVILADZE, G. M. (1985) *The Mathematical Theory of Combustion and Explosions*, Consultants Bureau, New York, pp30–36
16. FENIMORE, C. P. (1971) 'Formation of nitric oxide in premixed hydrocarbon flames', in *Proceedings of the Thirteenth Symposium (International) on Combustion*, The Combustion Institute, Pittsburgh, pp373–380
17. JENSEN, L. S., JANNERUP, H. E., GLARBORG, P., JENSEN, A. and DAM-JOHANSEN, K. (2000) 'Experimental investigation of NO from pulverised char combustion', *Proc. Combust. Inst.*, vol 28
18. JOHNSSON, J. E. (1994) 'Formation and reduction of nitrogen oxides in fluidised bed combustion', *Fuel*, vol 73, no 9
19. WINTER, F., MAGOR, W. and HOFBAUER, H. (1999) 'NO and N_2O formation and destruction in biomass grate combustors: A comprehensive study – Practical implications', in *Proceedings of the 5th International Conference on Technologies and Combustion for a Clean Environment (Clean Air V)*, 12–15 July 1999, Instituto Superior Tecnico, Lisbon, Portugal, pp983–989
20. WEISSINGER, A. and OBERNBERGER, I. (1999) 'NO_x reduction by primary measures on a travelling-grate furnace for biomass fuels and waste wood', in *Proceedings of the 4th Biomass Conference of the Americas*, Pergamon, pp1417–1424
21. LOFFLER, G., WINTER, F. and HOFBAUER, H. (2000) 'Volatile nitrogen conversion to NO and N_2O during biomass combustion – An extensive parametric modelling study', presented at Progress in Thermochemical Biomass Conversion, 17–22 September, Tyrol, Austria
22. KILPINEN, P. (1992) Kinetic modelling of gas-phase nitrogen reactions in advanced combustion processes, Report 92–7, PhD thesis, Abo Akademy University, Åbo, Finland
23. KELLER, R. (1994) Primärmaßnahmen zur NO_x-Minderung bei der Holzverbrennung mit dem Schwerpunkt der Luftstufung, research report no. 18 (1994), Laboratorium für Energiesysteme, ETH Zürich

24 NUSSBAUMER, T. (1989) *Schadstoffbildung bei der Verbrennung von Holz*, Forschungsbericht Nr. 6, Laboratorium für Energiesysteme, ETH Zürich
25 MARUTZKY, R. (1991) *Erkenntnisse zur Schadstoffbildung bei der Verbrennung von Holz und Spanplatten*, WKI-Bericht Nr. 26, Wilhelm-Klauditz-Institut, Braunschweig, Germany
26 SALZMANN, R. and NUSSBAUMER, T. (1995) *Zweistufige Verbrennung mit Reduktionskammer und Brennstoffstufung als Primärmaßnahmen zur Stickoxidminderung bei Holzfeuerungen*, research report, Laboratorium für Energiesysteme, ETH Zürich
27 NUSSBAUMER, T. (1997) 'Primär- und Sekundärmaßnahmen zur NO_x-Minderung bei Biomassefeuerungen', in *Thermische Biomassenutzung - Technik und Realisierung*, VDI Bericht 1319 VDI Verlag GmbH, Düsseldorf
28 BRUNNER, T. and OBERNBERGER, I. (1996) 'New technologies for NOx-reduction and ash utilisationin biomass combustion plants – JOULE THERMIE 95 Demonstration Project', in *Proceedings of the 9th European Bioenergy Conference*, Volume 2, Elsevier Science Ltd, Oxford, UK
29 WEISSINGER, A. and OBERNBERGER, I. (1999) 'NO_x reduction by primary measures on a travelling – grate furnace for biomass fuels and waste wood', in *Proceedings of the 4th Biomass Conference of the Americas*, September, Oakland (California), US, Elsevier Science Ltd., Oxford, UK, pp1417–1425
30 OLANDERS, B. and GUNNERS, N. E. (1994) 'Some aspects of the formation of nitric-oxide during the combustion of biomass fuels in a laboratory furnace', *Biomass and Bioenergy*, vol 6, pp443–451
31 SKREIBERG, Ø., HUSTAD, J. E. and KARLSVIK, E. (1997) 'Empirical NO_x modelling and experimental results from wood-stove combustion', in *Developments in Thermochemical Biomass Conversion*, Blackie Academic & Professional, pp1462–1476
32 NUSSBAUMER, T. (1998) 'Furnace design and combustion control to reduce emissions and avoid ash slagging', International Energy Agency, *Biomass Combustion Activity*, Final Report of the Triennium 1995–1997, Report No. IEA Bioenergy: T13: *Combustion*, no. 03
33 ZHANG, J., SMITH, K. R., MA, Y., YE, S., JIANG, F., QI, W., LIU, P., KHALIL, M. A., RASMUSSEN, R. A. and Thorneloe, S. A. (2000) 'Greenhouse gases and other airborne pollutants from household stoves in China: A database for emission factors', *Atmospheric Environment*, vol 34, pp4537–4549
34 SCHATOWITZ, B., BRANDT, G., GAFNER, F., SCHLUMPF, E., BUHLER, R., HASLER, P. and NUSSBAUMER, T. (1994) 'Dioxin emissions from wood combustion', *Chemosphere*, vol 29, pp2005–2013
35 SULILATU, W. F. (1997) *Onderzoek naar de haalbaarheid van de toepassing van l- regelsystemen bij bestaande houtverbrandingsinstallaties*, TNO-rapport October
36 SKREIBERG, Ø., GLARBORG, P., JENSEN, A. and DAM-JOHANSEN, K. (1997) 'Kinetic NO_x modelling and experimental results from single wood particle combustion', *Fuel*, vol 76, no 7, pp671–682
37 NUSSBAUMER, T. (1997) Primary and secondary measures for the reduction of nitric oxide emissions from biomass combustion', in *Developments in Thermochemical Biomass Conversion*, Blackie Academic & Professional, pp1447–1461
38 HILL, S. C. and SMOOT, L. D. (2000) 'Modeling of nitrogen oxides formation and destruction in combustion systems', *Progress in Energy and Combustion Science*, vol 26, pp417–458
39 ZABETTA, E. C., KILPINEN, P., HUPA, M., STAHL, K., LEPPALAHTI, J., CANNON, M. and NIEMINEN, J. (2000) 'Kinetic modeling study on the potential of staged combustion in gas turbines for the reduction of nitrogen oxide emissions from biomass IGCC plants', *Energy and Fuels*, vol 14, pp751–761
40 PURVIS, M. R., TADULAN, E. L. and TARIQ, A. S. (2000) 'NO_x control by air staging in a small biomass fuelled underfeed stoker', *International Journal of Energy Research*, vol 24, pp917–933
41 PADINGER, R. et al (1999) *Reduction of Nitrogen Oxide Emissions from Wood Chip Grate Furnaces*. Final report for the EU-JOULE III project JOR3-CT96-0059, Joanneum Research, Graz, Austria
42 NUSSBAUMER, T. (1997) 'Primary measures for the reduction of nitric oxides in wood-fired furnaces. 1. Formation processes of nitric oxides', *Brennstoff-Warme-Kraft*, vol 49, no 46

43 KELLER, R. (1994) 'NO_x reduction on the primary side with the help of air staging on wood burning', *Brennstoff-Warme-Kraft*, vol 46, pp483–488
44 GLARBORG, P., ALZUETA, M. U., DAM-JOHANSEN, K. and MILLER, J. A. (1998) 'Kinetic modeling of hydrocarbon/nitric oxide interactions in a flow reactor', *Combustion Flame*, vol 115, pp1–27
45 SMOOT, L. D., HILL, S. C. and XU, H. (1998) 'NO_x control through reburning', *Progress In Energy and Combustion Science*, vol 24, pp385–408
46 HUSTAD, J. E., SKREIBERG, Ø. and SØNJU, O. K. (1995) 'Biomass combustion research and utilisation in IEA countries', *Biomass and Bioenergy*, vol 9, nos 1–5, pp235–255
47 WINTER, F., WARTHA, C. and HOFBAUER, H. (1999) 'NO and N_2O formation during the combustion of wood, straw, malt waste and peat', *Bioresource Technology*, vol 70, pp39–49
48 GUSTAVSSON, L. and LECKNER, B. (1995) 'Abatement of N_2O emissions from circulating fluidised bed combustion through afterburning', *Industrial & Engineering Chemistry Research*, vol 34, pp1419–1427
49 AMAND, L. E. and LECKNER, B. (1994) 'Reduction of N_2O in a circulating fluidised bed combustor', *Fuel*, vol 73, pp1389–1397
50 AMAND, L. E. and LECKNER, B. (1993) 'Formation of N_2O in a circulating fluidised bed combustor', *Energy and Fuels*, vol 7, pp1097–1107
51 LECKNER, B. and KARLSSON, M. (1993) 'Gaseous emissions from circulating fluidised bed combustion of wood', *Biomass and Bioenergy*, vol 4, pp379–389
52 HULGAARD, T. and DAM-JOHANSEN, K. (1992) 'Nitrous-oxide sampling, analysis, and emission measurements from various combustion systems', *Environmental Progress*, vol 11, pp302–309
53 MANN, M. D., COLLINGS, M. E. and BOTROS, P. E. (1992) 'Nitrous-oxide emissions in fluidised bed combustion – fundamental chemistry and combustion testing', *Progress In Energy And Combustion Science*, vol 18, pp447–461
54 CENTRE FOR BIOMASS TECHNOLOGY (1998) *Straw for Energy Production, Technology - Environment - Economy*, 2nd edn, www.videncenter.dk/
55 HOUMØLLER, S. and EVALD, A. (1999) 'Sulphur Balances for Biofuel Combustion Systems', presented at the 4th Biomass Conference of the Americas
56 HESCHEL, W., RWEYEMAMU, L., SCHEIBNER, T. and MEYER, B. (1999) 'Abatement of emissions in small-scale combustors through utilisation of blended pellet fuels', *Fuel Processing Technology*, vol 61, pp223–242
57 SPLIETHOFF, H. and HEIN, K. R. (1998) 'Effect of co-combustion of biomass on emissions in pulverised fuel furnaces', *Fuel Processing Technology*, vol 54, pp189–205
58 PEDERSEN, L. S., NIELSEN, H. P., KIIL, S., HANSEN, L. A., DAM-JOHANSEN, K., KILDSIG, F., CHRISTENSEN, J. and JESPERSEN, P. (1996) 'Full-scale co-firing of straw and coal', *Fuel*, vol 75, pp1584–1590
59 NORDIN, A. (1995) 'Optimisation of sulfur retention in ash when co-combusting high-sulfur fuels and biomass fuels in a small pilot-scale fluidised bed', *Fuel*, vol 74, pp615–622
60 DAYTON, D. C., JENKINS, B. M., TURN, S. Q., BAKKER, R. R., WILLIAMS, R. B., BELLEOUDRY, D. and HILL, L. M. (1999) 'Release of inorganic constituents from leached biomass during thermal conversion', *Energy and Fuels*, vol 13, pp860–870
61 LAWRENCE, A. D., BU, J. and GOKULAKRISHNAN, P. (1999) 'The interactions between SO_2, NO_x, HCl and Ca in a bench-scale fluidised combustor', *Journal of the Institute of Energy*, vol 72, pp34–40
62 NIELSEN, H. P., FRANDSEN, F. J., DAM-JOHANSEN, K. and BAXTER, L. L. (2000) 'The implications of chlorine-associated corrosion on the operation of biomass-fired boilers', *Progress In Energy and Combustion Science*, pp26, pp283–298
63 KAUFMANN, H. and NUSSBAUMER, T. (1999) 'Formation and behavior of chlorine compounds during biomass combustion', *Gefahrstoffe Reinhaltung Der Luft*, vol 59, pp267–272
64 OBERNBERGER, I. and BRUNNER, T. (eds), (2005) *Aerosols in Biomass Combustion – Formation, Characterisation, Behaviour, Analysis, Emissions, Health Effects*, volume 6 of Thermal Biomass Utilization series, BIOS Bioenergiesysteme, Graz, Austria

65   NUSSBAUMER, T. (1998) *Furnace Design and Combustion Control to Reduce Emissions and Avoid Ash Slagging*, International Energy Agency, Biomass Combustion Activity, Final Report of the Triennium 1995–1997, Report No. IEA Bioenergy: T13: Combustion, no 03
66   NUSSBAUMER, T. (1994) *Emissions from Biomass Combustion*, IEA Biomass Agreement, Task X – Biomass Utilization, Activity 1: Combustion, Final Report of the Triennium 1992–1994
67   MCDONALD, R. D., ZIELINSKA, B., FUJITA, E. M., SAGEBIEL, J. C., CHOW, J. C. anD WATSON, J. G. (2000) 'Fine particle and gaseous emission rates from residential wood combustion', *Environmental Science and Technology*, vol 34, pp2080–2091
68   OANH, N. T., REUTERGARDH, L. B. and DUNG, N. T. (1999) 'Emission of polycyclic aromatic hydrocarbons and particulate matter from domestic combustion of selected fuels', *Environmental Science and Technology*, vol 33, pp2703–2709
69   MARBACH, G. and BAUMBACH, G. (1998) 'Organic compounds in particulate emissions from wood combustion and in airborne particles', *Gefahrstoffe Reinhaltung Der Luft*, vol 58, pp257–261
70   BARREFORS, G. and PETERSSON, G. (1995) 'Volatile hydrocarbons from domestic wood burning', *Chemosphere*, vol 30, pp1551–1556
71   LAUNHARDT, T. and THOMA, H. (2000) 'Investigation on organic pollutants from a domestic heating system using various solid biofuels', *Chemosphere*, vol 40, pp1149–1157
72   WUNDERLI, S., ZENNEGG, M., DOLEZAL, I. S., GUJER, E., MOSER, U. WOLFENSBERGER, M., HASLER, P., NOGER, D., STUDER, C. and KARLAGANIS, G. (2000) 'Determination of polychlorinated dibenzo-p-dioxins and dibenzo-furans in solid residues from wood combustion by HRGC/HRMS', *Chemosphere*, vol 40, pp641–649
73   FIEDLER, H. (1998) 'Thermal formation of PCDD/PCDF: A survey', *Environmental Engineering Science*, vol 15, pp49–58
74   NUSSBAUMER, T. and HASLER, P. (1997) 'Formation and reduction of polychlorinated dioxins and furans in biomass combustion', in *Developments in Thermochemical Biomass Conversion*, Blackie Academic & Professional, pp1492–1508
75   SALTHAMMER, T., KLIPP, H., PEEK, R. D. and MARUTZKY, R. (1995) 'Formation of polychlorinated dibenzo-p-dioxins (PCDD) and polychlorinated dibenzofurans (PCDF) during the combustion of impregnated wood', *Chemosphere*, vol 30, pp2051–2060
76   PFEIFFER, F., STRUSCHKA, M., BAUMBACH, G., HAGENMAIER, H. and HEIN, K. R. (2000) 'PCDD/PCDF emissions from small firing systems in households', *Chemosphere*, vol 40, pp225–232
77   CHAGGER, H. K., KENDALL, A., MCDONALD, A., POURKASHANIAN, M. and WILLIAMS, A. (1998) 'Formation of dioxins and other semi-volatile organic compounds in biomass combustion', *Applied Energy*, vol 60, pp101–114
78   LAUNHARDT, T., STREHLER, A., DUMLERGRADL, R., THOMA, H. and VIERLE, O. (1998) 'PCDD/F and PAH emission from house heating systems', *Chemosphere*, vol 37, pp2013–2020
79   ANDERSSON, P. and MARKLUND, S. (1998) 'Emissions of organic compounds from biofuel combustion and influence of different control parameters using a laboratory scale incinerator', *Chemosphere*, vol 36, pp1429–1443
80   ZUBERBUHLER, U., ANGERER, M., BAUMBACH, G., VATTER, J. and HAGENMAIER, H. (1996) 'Flue gas emission measurements at small-scale and industrial-scale wood firings within the scope of the 1st Amendment to the Federal German Pollution Control Act', *Gefahrstoffe Reinhaltung Der Luft*, vol 56, pp415–418
81   VESTERINEN, R. and FLYKTMANN, M. (1996) 'Organic emissions from co combustion of RDF with wood chips and milled peat in a bubbling fluidised bed boiler', *Chemosphere*, vol 32, pp681–689
82   STRECKER, M. and MARUTZKY, R. (1994) 'PCDD and PCDF formation during incineration of untreated and treated wood and particleboards', *Holz Als Roh-Und Werkstoff*, vol 52, pp33–38
83   SKREIBERG, Ø. (1997) Theoretical and Experimental Studies on Emissions from Wood Combustion, PhD thesis, Norwegian University of Science and Technology, ITEV-report 97:03
84   KARLSVIK, E., HUSTAD, J. E. and SØNJU, O. K. (1993) 'Emissions from wood-stoves and fireplaces', in *Advances in Thermochemical Biomass Conversion*, Blackie Academic & Professional, pp690–707

85 STREHLER, A. (1994) *Emissionsverhalten von Feurungsanlagen fur feste Brennstoffe*, Technische Universitat Munchen-Weihenstephan, Bayerische Landesanstalt fur Landtechnik, Freising
86 SKREIBERG, Ø., KARLSVIK, E., HUSTAD, J. E. and SØNJU, O. K. (1997) 'Round robin test of a wood-stove: The influence of standards, test procedures and calculation procedures on the emission level', *Biomass and Bioenergy*, vol 12, no 6, pp439–452
87 SKREIBERG, Ø. and SAANUM, Ø. (1994) *Comparison of Emission Levels of Different Air Pollution Components from Various Biomass Combustion Installations in the IEA Countries*, Supplemental report to IEA report *Emissions from Biomass Combustion*, IEA Biomass Agreement, Task X – Biomass Utilization, Activty 1: Combustion, Final Report of the Triennium 1992–1994
88 NUSSBAUMER, T. and HUSTAD, J. E. (1997) 'Overview of Biomass Combustion', in *Developments in Thermochemical Biomass Conversion*, Blackie Academic & Professional, pp1229–1243
89 PRÜFBERICHT BUNDENANSTALT FUR LANDENTECHNIK, Wieselberg/Erlauf Prot.Nr.033/94, 32/94,024/95, 025/94, 035/95, 023/93
90 WILDEBURGER, J. M. F. (1996) Staatlich befugter und beeideter Zivilingenieure für Technische Chemie, Polten (Aus), Prüfberichte PZ 1304/ZJW 920
91 SULILATU, W. F. et al (1992) *Kleinschalige verbranding van schoon resthout in Nederland*, NOVEM rapport November 1992
92 OBERNBERGER, I. (1997) 'Stand und Entwicklung der Verbrennungstechnik', in *VDI Berichte 1319, Thermische Biomassenutzung, Technik und Realisierung*, April, VDI Verlag GmbH, Düsseldorf, Germany
93 SANDER, B. et al (2000) 'Emissions, corrosion and alkali chemistry in straw-fired combined heat and power plants', in *1st World Conference on Biomass for Energy and Industry*, Sevilla, Spain, pp775–778
94 EVALD, A. and WITT, J. (2006) *Biomass CHP Best Practice Guide, Performance Comparison and Recommendations for Future CHP Systems Utilising Biomass Fuels*, FORCE Technology, Denmark
95 BRUNNER, T., JÖLLER, M. and OBERNBERGER, I. (2004) 'Aerosol formation in fixed-bed biomass furnaces – results from measurements and modelling', in *International Conference Science in Thermal and Chemical Biomass Conversion*, Victoria, Canada, September
96 NUSSBAUMER, T. (2000) 'Partikel- und stickoxidemissionen, hausheizungen sowie stromerzeugung als aktuelle themen der holzenergie', *Holz-Zentralblatt*, vol 126, no 144, pp1996–1998
97 NUSSBAUMER, T. and SALZMANN, R. (2000) 'Fuel staging for NO_x reduction in automatic wood furnaces', Presented at Progress in Thermochemical Biomass Conversion, 17–22 September, Tyrol, Austria
98 SKREIBERG, Ø., (1994) 'Advanced techniques for wood log combustion', *Proceedings of COMETT Expert Workshop on Biomass Combustion*, Austria, May
99 OBERNBERGER, I. (2005) 'Thermochemical Biomass Conversion', lecture, Department for Mechanical Engineering, Process Technology Section, Technical University Eindhoven, the Netherlands
100 RISNES, H. and SØNJU, O. K. (2000) 'Evaluation of a novel granular bed filtration system for high temperature applications', presented at Progress in Thermochemical Biomass Conversion, 17–22 September, Tyrol, Austria
101 BROUWERS, B. (1996) 'Rotational particle separator: A new method for separating fine particles and mists from gases' *Chemical Engineering and Technology*, vol 19, pp1–10
102 BROUWERS, J. J. (1997) 'Particle collection efficiency of the rotational particle separator', *Powder Technology*, vol 92, pp89–99
103 HASLER, P. H., NUSSBAUMER, T. H., SCHAFFNER, H. P. and BROUWERS, J. J. H. (1998) 'Reduction of aerosol particles in flue gases from biomass combustion with a rotational particle separator (RPS)', Biomass for Energy and industry, 10th European Conference and Technology Exhibition, June 8–11, Würzburg, Germany
104 ORAVAINEN, H. (1998) *Emission Limits for Biomass Fired Boilers in some IEA Member Countries*, VTT Energy, Finland, January
105 MILJØSTYRELSEN (2001) *Luftvejledningen, Begrænsning af luftforurening fra virksomheder, Miljø- og Energiministeriet, København*, national guideline for emission limits issued by the Danish Environmental Agency, Copenhagen

106 INFOMIL (2006) Wetswegwijzer bio-energie, available at www.infomil.nl
107 JANSEN, U. (2006) 'Aktionsplan Feinstaub des Bundes im Bereich Holzfeuerungen und verschärfte Emissionsgrenzwerte', in *Feinstaubminderung und Stromerzeugung im Rahmen der zukünftigen Energieversorgung, 9. Holzenergie-Symposium*, 20 October, ETH Zürich, Zürich, Verenum, Zürich und Bundesamt für Energie, Bern
108 ENVIRONMENT AGENCY (2002) *IPPC Sector Guidance Note Combustion Activities*, Environment Agency, UK
109 OBERNBERGER, I. (1997) 'Nutzung fester Biomasse', in *Verbrennungsanlagen unter besonderer Berücksichtigung des Verhaltens aschebildender Elemente*, volume 1 of Thermal Biomass Utilization series, BIOS, Graz, Austria, dbv-Verlag der Technischen Universität Graz, Graz, Austria
110 LUNDBORG, A. (1998) 'Ecological and economical evaluation of biomass ash utilisation – the Swedish approach;, in *Ashes and Particulate Emissions from Biomass Combustion*, volume 3 of Thermal Biomass Utilization series, BIOS, Graz, Austria, dbv-Verlag
111 OBERNBERGER, I., BIEDERMANN, F. and KOHLBACH, W. (1995) *FRACTIO – Fraktionierte Schwermetallabscheidung in Biomasseheizwerken*, annual report, Institute of Chemical Engineering, Graz University of Technology, Austria
112 OBERNBERGER, I., WIDMANN W., WURST F. and WÖRGETTER M. (1995) *Beurteilung der Umweltverträglichkeit des Einsatzes von Einjahresganzpflanzen und Stroh zur Fernwärmeerzeugung*, annual report for the research project, Institute of Chemical Engineering, Graz University of Technology, Austria
113 RUCKENBAUER, P., OBERNBERGER, I. and HOLZNER H. (1996) *Erforschung der Verwendungsmöglichkeiten von Aschen aus Hackgut- und Rindenfeuerungen*, final report of phase 2 of a research project of the same name, Institute for Plant Production and Plant Breeding, BOKU Wien, Vienna, Austria
114 OBERNBERGER, I., PÖLT P. and PANHOLZER, F. (1995) 'Charakterisierung von Holzasche aus Biomasseheizwerken, Teil II: Auftretende Verunreinigungen, Schütt- und Teilchendichten, Korngrößen und Oberflächenbeschaffenheit der einzelnen Aschefraktionen', in *Umweltwissenschaften und Schadstoff-Forschung – Zeitschrift für Umweltchemie und Ökotoxikologie*, Verbandes für Geoökologie in Deutschland (VGöD), Germany, vol 7, no 1, pp15–24
115 OBERNBERGER, I. (1996) *Prüfung, Beurteilung und Optimierung des Einsatzes von rotierenden Partikelabscheidern für Biomassefeuerungen*, Forschungsbericht I-B2-1997, Ingenieurbüro BIOS, Graz, Austria
116 TOBLER, H. and NOGER, N. (1993) *Brennstoff und Holzverbrennungsrückstände von Altholzfeuerungen: 1. Teilbericht zum Projekt HARVE*, EMPA St. Gallen, Bundesamt für Umwelt, Wald und Landschaft, Bern, Switzerland
117 NOGER, D., FELBER, H. and PLETSCHER, E. (1995) *Holzasche und Rückstände, deren Verwertung oder Entsorgung*, draft final report for the project HARVE, EMPA St. Gallen, Bundesamt für Umwelt, Wald und Landschaft, Bern, Switzerland
118 NOGER, D., FELBER, H. and PLETSCHER, E. (1994) *Zusatzanalysen zum Projekt HARVE*, research report no. 22′032 C, EMPA St. Gallen, Bundesamt für Umwelt, Wald und Landschaft, Bern, Switzerland
119 TOBLER, H. and NOGER, D. (1993) *Brennstoff und Holzverbrennungsrückstände von Altholzfeuerungen*, 1st report for the project HARVE, EMPA St. Gallen, Bundesamt für Umwelt, Wald und Landschaft, Bern, Switzerland
120 HASLER, P. (1994) *Rückstände aus der Altholzverbrennung, Charakterisierung und Entsorgungsmöglichkeiten*, report for DIANE 8: Forschungsprogramm Energie aus Altholz und Papier, Bundesamt für Energiewirtschaft, Bern, Switzerland
121 OBERNBERGER, I., PANHOLZER, F. and ARICH, A. (1996) *System- und pH-Wert-abhängige Schwermetalllöslichkeit im Kondensatwasser von Biomasseheizwerken*, final report for a research project of the same name for the Ministry for Science and the State Government of Salzburg; Institute for Chemical Engineering, Graz University of Technology, Austria
122 SCHEFFER/SCHACHTSCHABEL (1992) *Lehrbuch der Bodenkunde*, 13th edn, ENKE Verlag Stuttgart, Germany

123 STEIERMÄRKISCHE LANDESREGIERUNG (1991) *Steiermärkischer Bodenschutzbericht*, State Government of Styria, Graz, Austria
124 NARODOSLAWSKY, M. and OBERNBERGER, I. (1996) 'From waste to raw material – the way of cadmium and other heavy metals from biomass to wood ash', in *Journal of Hazardous Materials*, vol 50/2–3, pp157–168
125 OBERNBERGER, I. (1994) Sekundärrohstoff Holzasche – Nachhaltiges Wirtschaften im Zuge der Energiegewinnung aus Biomasse, Ph.D. thesis, Institute for Chemical Engineering, Graz University of Technology, Austria
126 DAHL, J. and OBERNBERGER, I. (1998) 'Thermodynamic and experimental investigations on the possibilities of heavy metal recovery from contaminated biomass ashes by thermal treatment', in *Proceedings of the 10th European Bioenergy Conference*, June, Würzburg, Germany, CARMEN, Rimpar, Germany
127 OBERNBERGER, I. (1995) 'Logistik der Aschenaufbereitung und Aschenverwertung', in *Proceedings of the conference 'Logistik bei der Nutzung biogener Festbrennstoffe'*, May, Stuttgart, Germany; Bundesministerium für Ernährung, Landwirtschaft und Forsten, Bonn, Germany
128 OBERNBERGER, I. and NARODOSLAWSKY, M. (1993) *Aschenaustrags- und Aufbereitungsanlagen für Biomasseheizwerke*, final report for a research project for the Landesenergieverein Steiermark; Institute for Chemical Engineering, Graz University of Technology, Austria
129 NARODOSLAWSKY, M. and OBERNBERGER, I. (1995) *Verwendung von Holzaschen zur Kompostierung*, final report for the research project Nr. 4159 of the 'Jubiläumsfonds der Österreichischen Nationalbank', Institute for Chemical Engineering, Graz University of Technology, Austria
130 BUNDESMINISTERIUM FÜR LAND- UND FORSTWIRTSCHAFT (1997) *Der sachgerechte Einsatz von Pflanzenaschen im Wald*, guideline, Ministry for Agriculture and Forestry, Vienna, Austria
131 BUNDESMINISTERIUM FÜR LAND- UND FORSTWIRTSCHAFT (1998) *Der sachgerechte Einsatz von Pflanzenaschen im Acker- und Grünland, Richtlinie*, Ministry for Agriculture and Forestry, Vienna, Austria
132 OBERNBERGER, I. (1995) 'Aschenbehandlung und Verwertung bei Frisch- und Restholzfeuerungen', in *Handbook for the VDI-Seminar 'Alt- und Restholz - Thermische Nutzung und Entsorgung, Anlagenplanung, Energienutzung'*, October, Salzburg; VDI-Bildungswerk, Düsseldorf, Germany
133 HOLZNER, H. and RUCKENBAUER, P. (1994) 'Pflanzenbauliche Aspekte einer Holzascheausbringung auf Acker- und Grünland', in *Proceedings of the International Symposium 'Sekundärrohstoff Holzasche'*, September, Graz; Institute for Chemical Engineering, Graz University of Technology, Austria
134 EVALD, A. (2001) 'Tungmetaller i aske fra anlæg fyret med halm og træ', dk-Teknik Energy & Environment, Denmark

10

Policies

10.1 Introduction

Bioenergy currently accounts for 10.6 per cent or about 45EJ of global energy demand [1], which is larger than any other renewable energy option. Although the vast majority (39EJ) is the use of biomass in traditional wood-stoves in developing countries (which remains rather constant over the years), the real growth is currently observed in biomass-based electricity generation systems, currently accounting for approximately 6EJ. Over 40GW_e of biomass-based electricity production capacity is installed worldwide [2], producing approximately 260TWh of electricity (derived from [3]).

There is an increased recognition in both developing and developed countries that energy from biomass can contribute to various economical, environmental and social policy objectives. Due to differences in the local socio-economic, environmental and cultural situation, different countries have progressed in different ways.

In the majority of predictions and scenarios available for the short and medium term (up to roughly 2010–2015), bioenergy is expected to grow further in most countries. Biomass combustion and particularly the co-firing of biomass with coal are mentioned as key options for heat and power generation. In the longer term, advanced gasification-based power cycles are expected to gain market share.

In order to harness the benefits of bioenergy, it is important that appropriate conversion technologies are developed and put into practice. This chapter describes some of the key markets of biomass combustion based energy systems, as well as policy arguments and policy measures that can support efficient market introduction.

10.2 Global expansion of biomass combustion

The largest markets for biomass combustion systems currently exist in North America and Europe, totalling about two-thirds of current biomass electricity production. Other important regions in the world are Latin America (particularly Brazil) and Asia [3, 4].

Europe and North America are expected to grow in installed capacity by an average 500 and 250MW_e per year, respectively, but relative growth rates will be highest in Asian and Latin American economies (see Figure 10.1). By 2030 biomass-fuelled electricity production is projected to triple and provide 2 per cent of world total requirements, 4 per cent in OECD Europe, as a result of government policies to promote renewables.

At least up to 2015–2020, it can be expected that mainstream biomass power technologies will be based on direct combustion and co-combustion using steam cycles. As the global capacity of coal-fired power stations increases further, increased synergy will be found between biomass and coal-based power generation.

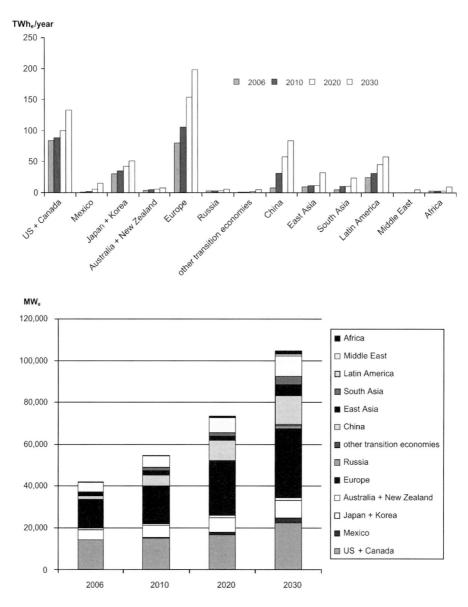

Figure 10.1 *Expected growth in biomass power generation (top) and installed capacity (bottom)*

Source: Figures derived from [3].

The expected growth in biomass combustion systems will take place in dedicated biomass combustion systems using agricultural and process residues, as well as various biomass wastes. On an industrial scale, biomass combustion-based power generation using both grate fired and fluidized bed boilers will continue to dominate the market, as these concepts are already reliable and cost-effective for various fuels and are continuing to improve further. As the financial feasibility of steam cycle-based biomass power systems has a minimum limit of approximately 1MW$_e$, there will be room for innovative but reliable CHP concepts in the medium term, such as organic Rankine cycle (ORC) and Stirling engines, as they are reliable, do not require a pressurized boiler and require little or no user involvement.

Options for using biomass in synergy with coal will become even more important than they are today. Coal retains the largest market share of the world's electricity generation (roughly 40 per cent, [5, 6]) and is expected to grow by 2.2 per cent per year, from 1119GW$_e$ in 2003 to 1997GW$_e$ in 2030 [7]. In 2003, non-OECD economies on the whole relied on coal for roughly 43 per cent of generation, slightly more than the OECD economies. Coal reserves in the US, China, India and Australia are among the largest in the world, and China and the US in particular lead the world in coal-fired capacity additions, adding 546GW$_e$ and 154GW$_e$ up to 2030, respectively [6]. Since coal-based power generation results in the highest release of CO_2 per kWh, there will be political pressure to enable biomass co-firing and carbon capture in these plants.

For biomass fuels that are clean and brittle, direct co-firing in existing pulverized coal-fired power stations will remain the cheapest option for biomass power. There will also be a place for large-scale fluidized bed gasification systems for biomass that is less brittle and/or contaminated, as this technology can provide a clean gas that can be co-fired in coal-fired and potentially also natural gas-fired power stations. Biomass fuels containing challenging components (e.g. Cl) can be burned in a separate boiler, providing steam of medium conditions to even very advanced, ultra-supercritical power plants.

10.2.1 Trends in selected OECD member countries

In **the US**, biomass currently delivers some 3 per cent to the primary energy mix. An estimated 10GW$_e$ of biomass power generation is in operation, of which about 60 per cent is derived from forest products (wood residues, black liquor, etc.) and agricultural residues, 35 per cent from municipal solid waste, and the remainder from other options such as landfill gas (extrapolated from [8]). The mission of the Department of Energy's (DOE's) Biomass Power Program that was launched in 1998 was to encourage and assist industry in the development and validation of renewable, biomass-based electricity generation systems so that bioenergy could become a major cost-competitive contributor to power supplies in both domestic and international markets through the integration of efficient biomass power production with dedicated feedstocks from sustainable farms and forests.

US, DOE expects that in the short term, biomass power generation will mainly take place by biomass co-firing with coal (see Figure 10.2). A vast number of demonstrations were done, co-firing several types of biomass with various coals in different coal-fired power plants. Regretfully, these demonstrations have rarely led to any significant commercial activity, as a result of a lack of financial incentives.

In **Australia**, biomass fuels are slowly finding a place alongside fossil fuels because of their considerable potential as a renewable energy source. Among various technologies for power production from biomass, co-firing of biomass/coal blends in existing pulverized fuel boilers (e.g. 500–1000MW per boiler) has attracted much attention in Australia. This is primarily due to the fact that co-firing in large-scale coal systems increases the conversion efficiency of biomass to power by about a factor of 2, compared to direct combustion

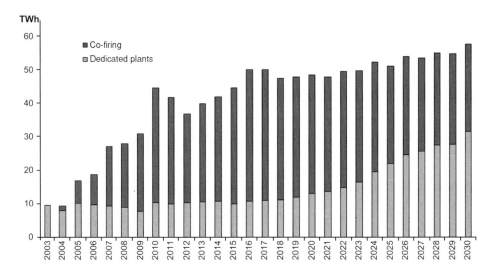

Figure 10.2 *Projected electricity production from biomass in the US*
Source: [8].

in purposed-built biomass power plants. In addition, co-firing provides the most economic outlet for biomass utilization as it uses existing plants and, therefore, investment costs are low.

However, there are certain operational problems associated with co-firing that can be broadly classified into handling and processing issues, combustion-related issues and ash deposition issues (e.g. slagging and fouling). Much of the R&D activities in Australia are focused on finding solutions to these problems. Programmes are being carried out in academic institutions, research organizations such as the Commonwealth Scientific and Industrial Research Organisation (CSIRO) and industry.

The leading academic institution in the area of biomass utilization technology is the University of Newcastle, which through its Co-operative Research Centre (CRC) for Black Coal Utilisation has a number of active projects on biomass co-firing, biomass gasification and co-utilization of gasified biomass and coal. The co-firing project studies the burnout properties of coal/biomass blends and investigates the major parameters that cause the blend properties to be different from those of coal and biomass. In particular, the role of the biomass particle size on the overall burnout properties of the blend is being investigated. The project group is also studying the mechanism of ash deposition when co-firing coal/biomass blends, and examining various methods and techniques for minimizing the impact of fouling and slagging under such conditions. Furthermore, the project is aimed at developing suitable procedures for processing blends of Australian coal and biomass in typical pulverized fuel power stations.

Among industrial organizations, Delta Electricity, Macquarie Generation and CS Energy Ltd, have been more active in implementing the concept of co-firing technology. Delta Electricity is undertaking collaborative projects with the University of Newcastle on a variety of issues related to co-combustion technology. The CSIRO Division of Energy Technology recently initiated a series of projects addressing issues related to biomass combustion. A database of biomass fuel characteristics, including combustion and gasification, is being developed for Australian biomass materials. CSIRO's Divisions of Energy Technology and Forestry and Forest Products are working to assess the potential of plantation crops for bioenergy in Australia, both in large-scale and smaller distributed energy plants.

In 1997 the **European Commission** launched the White Paper **[9]** in which targets for renewable energy were set, including 828PJ_e of biomass-based electricity generation in 2010 and a contribution of biomass in the order of one-third of the amount of renewable electricity in 2010. The first target implies a 10-fold increase compared to 1995, or an average annual growth rate of 17 per cent. These goals were thought to be achievable by expedient development of advanced energy conversion technologies and by implementing incentive-driven policies. Over the years 1995–2004, biomass-based power generation in EU25 countries did achieve impressive annual average growth rates of 10.2 per cent **[10]**, but much more needs to be done. This is certainly a challenge, given that cheaper biomass resources are increasingly being utilized.

Basically, every single European country has specific ambitions for the growth of renewable energy in general, and bioenergy in particular. In the last 10–15 years, average annual growth rates for heat, electricity and transportation fuels from biomass in Europe were already impressive with 2 per cent, 9 per cent and 23 per cent, respectively.

Most of the dedicated biomass power systems in Europe are located in Finland, Sweden, Denmark and Austria (district heating systems, pulp and paper industry, and forestry industries), where biomass energy already accounts for 10–20 per cent of respective primary energy supply. Since 2001, biomass power generation has also grown rapidly in Germany through guaranteed feed-in rates for 20 years. Biomass power is also growing rapidly in the UK and the Netherlands through co-firing, done by large utility companies. In other countries, development is lagging far behind and substantial developments will be required over the next two decades to meet the EU's proposed targets. The progress being made in a number of European countries is shown below.

In 2003, **Austria** derived 21 per cent of its energy from renewable resources while biomass contributed roughly 12 per cent (168PJ) of the primary energy demand. Of the latter, 60 per cent is used for heating applications, 21 per cent for process heat, 11 per cent for CHP and 8 per cent for district heating. Thus, at present biomass is mostly used for heating applications. The Austrian State Governments have worked out appropriate feed-in tariffs for green electricity as well as appropriate regulations for electricity distribution via the grid, which are relatively high. As a result, various innovative CHP concepts are demonstrated here. In the short term, Austria will also explore ways and means to introduce biomass gasification CHP systems as an alternative to the widely deployed combustion-based district heating operations.

In **Denmark**, the 1996 Energy 21 program is the basis for Danish energy policies. The use of renewable energy is expected to increase by 1 per cent per year until 2030. At this rate renewable energy is expected to contribute 12–14 per cent of the primary energy demand (or 100PJ) by 2005 and 35 per cent (or 235PJ) by the year 2030. In Denmark, biomass is the largest renewable energy resource. Although co-firing with biomass is widely practised to cut back on coal consumption, the population distribution and the demand for heat and electricity present a compelling case for building small-scale biomass CHP plants. To assist in the deployment of renewable energy, legislative measures have been put in place to guarantee a 'green' electricity sale price of about €0.08/kWh.

Biomass is also the largest renewable resource in **Finland**. In 2001, Finland derived 20 per cent of its energy (approximately, 280PJ) from wood fuels. Government policies are designed to raise the contribution of renewable energy to 30 per cent by 2010. The opportunities for increased biomass utilization exist in supplying the thermal energy needs of industry and CHP. To reach this goal, government policies are now focused on implementing measures to increase the supply of low-cost biomass. Finland, with a population of approximately 5 million, a large forest products industry, and the absence of a nationwide natural gas distribution system, has come to rely on biomass-based CHP plants.

Germany's policies are directed towards ensuring energy security by focusing on coal and nuclear energy, while aggressively pursuing the use of renewable energy to cut back on fossil fuels. Germany's commitment to international greenhouse gas (GHG) treaties and to secure energy supplies is based on recovering significant quantities of energy from biomass and renewable wastes. Biomass combustion and gasification technologies are an essential part of Germany's plans to convert biomass and waste for both the poly-generation of power, heat and cold, and the production of synthesis gas for biofuels and hydrogen-rich gases for fuel cells. In 2001 a system of guaranteed electricity prices for 20 years came into force, with attractive feed-in rates. This has boosted the application of biomass power generation.

In **the Netherlands**, the contribution of biomass to primary energy supply is approximately 2 per cent. This comes predominantly from electricity production by co-firing biomass in coal-fired power plants, enforced through a covenant between the coal power sector and the government and made financially feasible through a premium that should cover additional costs. Although the goal of 9 per cent renewable electricity in 2010 is expected to be met, achieving the ambitious longer-term goals, namely obtaining 10 per cent all primary energy by 2020 from renewable sources and 30 per cent of total energy from biomass in 2040 will require importing large amounts of biomass from abroad. The long-term R&D programme has been set up by the Dutch government to generate the necessary innovations to facilitate energy transition and to attain the 2040 goals. These include the production of clean fuel or synthesis gas from fossil fuels and biomass energy generation and efficient use of energy.

Sweden: The previous Energy Policy Act of 1997 has been replaced with the new Energy Policy Act of 2002, to promote the efficient use of energy with innovative co-generation applications. From 2003, the trading of renewable energy (RE) certificates has been the main driving force for the exploitation and use of biomass in Sweden. The RE quota of 7 per cent in 2003 will increase to 17 per cent in 2010. The target is to build 10TWh of additional RE capacity by 2010. In 2002, biomass contributed a total of 50TWh towards 600TWh of annual national energy demand. The majority of biomass, along with peat, was employed for district heating. In 2005, the Prime Minister of Sweden established a commission, involving industrial, economic and institutional stakeholders from both the private and public sector, with the objective of establishing policies and strategies for the complete elimination of fossil fuels by 2020. The commission's recommendations are supportive of exploiting the abundant biomass resources of Sweden to offset imports.

The UK government made provision in the Utilities Act 2000 for an obligation to be placed on licensed electricity suppliers to supply specified amounts of electricity from renewable sources. The obligation started at 3 per cent in 2002/2003 rising to 10 per cent by 2010, and has been extended recently to 15 per cent by 2015. The other key element of the scheme is that eligible generators of renewable energy receive a Renewable Obligation Certificate (ROC) for each MWh of renewable electricity generated. Electricity suppliers can either present certificates to cover their obligation, or they can pay a 'buy-out' price of £30 per MWh. All proceeds from the 'buy-out' payments are recycled to the generators in proportion to the number of ROCs presented. ROCs can be traded freely, and the price is dependent on the ratio of ROCs generated to 'buy out' payments. The Obligation, therefore, provides incentives for major electricity supply companies and others to invest in renewable electricity through a market-based mechanism which is consistent with the liberalized electricity sector in Britain. The price to the customer is capped by the 'buy-out price'. Since the ROC price is not dependent on the technology, the least cost, lowest risk technologies are deployed and, in principle, the government/regulator is not involved in the selection of technologies or of individual projects, beyond the establishment of the rules under which the Obligation is to operate.

Since the introduction of the Renewables Obligation, there have been significant increases in electricity generation from landfill gas, onshore wind, and biomass co-firing in coal-fired power plants. As the costs of co-firing of biomass in coal-fired power plants are significantly lower than those of establishing new, dedicated biomass-to-energy plants, it was anticipated that co-firing could provide a boost to the biomass supply infrastructure in Britain, and particularly to provide secure markets in the short to medium term for energy crop materials. In practice, however, most of the biomass is imported from overseas. In order to give other renewable options a chance, limits were placed both on the total quantity of eligible ROCs from co-firing, and on the timescales over which co-firing is eligible under the Obligation. One of the results of the introduction of the Renewables Obligation in April 2002 has been a dramatic increase in biomass co-firing involving all of the large coal-fired power plants in Britain. As of 2006, the total cumulative power generation by biomass co-firing is in excess of 4 million MWh. It is clear that, from a standing start, the electricity supply industry in Britain has responded relatively rapidly to the financial incentives represented by the introduction of the Renewables Obligation.

10.2.2 Trends in selected non-OECD member countries

In **Asian** and **Latin American** countries, there is still a large potential for dedicated biomass-based power generation systems based on agro-processing and forest industries. Retrofits of obsolete CHP plants that are still in operation in sugar mills provide a low risk opportunity for investment, often with high returns as well as positive local environmental impacts. Green field projects based on field residues pose additional challenges on fuel collection infrastructure, which may be complicated in some cases.

In **China**, power generation from biomass is only just beginning. Great opportunities exist in co-firing of biomass fuels, such as agricultural or forestry residues in both existing and still-to-be-built pulverized coal-fired systems and fluidized bed combustion systems for high-ash coals. In 2006 an IEA action was launched to ensure that China considers biomass as a (secondary) fuel in these plants.

10.2.3 Relevant policy issues

In contrast to many other energy technologies, biomass combustion links to many policy areas, such as climate, energy, agriculture and waste policies. The availability and use of biomass resources are often intertwined with various major sectors of the economy: agriculture, forestry, food processing, paper and pulp, building materials and, of course, the energy sector in the widest sense. Seen from the positive side, this gives bioenergy many opportunities to generate multiple benefits apart from energy generation.

By using local biomass resources, a country's fuel portfolio is diversified and dependency on (often imported) fossil fuels is reduced. This has a positive effect on foreign exchange bills and can also reduce the dependency on fuel import from a limited number of politically unstable countries.

Except for a limited amount of indirect greenhouse gas emission related to fuel pretreatment and supply, biomass combustion is considered carbon-neutral and can thereby contribute to a reduction in overall greenhouse gas emissions.

Due to the typically lower amount of sulphur and low combustion temperatures, combustion of biomass can result in lower emissions of SO_2 and NO_x and can thereby lead to a reduction in acid rain.

Use can be made of local process resources that may otherwise cause environmental hazards due to uncontrolled dumping or burning. For example, insufficiently maintained forests increase the risk of forest fires as, for exampe, happened in Portugal in 2005. In many developing countries, agricultural process residues are left unused or even dumped in lagoons, often causing environmental hazards.

The supply of biomass from local resources may create significant employment opportunities and thereby support local economies as fuel expenditures remain within the region. In many countries, the market for plantation-grown biofuels to be used for dedicated power generation is not yet mature because of a lack of confidence among entrepreneurs and financial institutions in its feasibility, the absence of realized dedicated biomass power plants or assured purchase contracts, the lack of market support institutions, etc. In such a situation it is as difficult to develop a mature fuel market as it is to establish a biomass power plant in the absence of assured fuel supply. As it provides a relatively cheap method of converting biomass into useful energy, co-firing can be a very effective strategy for large-scale infrastructure development for biomass fuels. It can then help to create a fully fledged market for biomass fuels in which sellers and buyers can easily trade biofuels. This is particularly true if the combustion equipment used has a wide fuel acceptance range. Worldwide, much experience has been gained with co-combustion of various fuels. However, further research and development are being done to optimize co-firing applications and to reduce the negative effects of co-firing, such as boiler fouling and corrosion [11–16].

In many developing countries in particular, the demand for electricity is rising steeply (e.g. in countries such as China and India by approximately 4.5 per cent per year [3]). This requires huge amounts of public investment, typically in coal-based power generation. Independent private power generation using underutilized biomass residues can make a significant local impact, reduce public expenditure and open up new markets for existing equipment suppliers. In the Indian sugar sector alone, there is a potential of roughly $3GW_e$ to upgrade outdated bagasse-based co-generation facilities to up-to-date condensing steam turbine systems, with typical payback periods of three years [17].

The downside of the many linkages and interests is that the implementation of bioenergy projects is often very complex. Together with certainty over fuel availability and price over time, steady and supporting government policy frameworks in all the mentioned areas (not only energy) are required to support actual investments in bioenergy systems. Also, pricing structures need to be adapted to internalize externalities sufficiently and make these options financially feasible on the long run [18].

10.3 Financial support instruments

Governments can take different courses of action to promote the deployment of biomass combustion based systems. In order to make them cost-effective and accepted by the market, policy actions need to be designed specifically to local conditions, addressing particular local barriers such as limited awareness or financial performance.

As a result of wide variations in biomass sources, combustion processes and sizes applied, the costs of a biomass combustion systems may vary widely from case to case. A range of different financial support systems is operational in different countries. The most important financial support systems currently maintained are feed-in tariffs, premiums, green certificates, investment subsidies and tax credits. An explanation of each is given below.

10.3.1 Fixed feed-in tariffs and fixed premiums

These systems exist in various European countries and are characterized by a specific premium or total price, normally set for a period of several years, that domestic producers of green electricity receive. The additional costs of these schemes are either paid by suppliers in proportion to their total sales volume and are passed through to the power consumers, charged directly to buyers of green electricity or paid by national governments using environmental taxes on conventional electricity. Fixed feed-in systems are used, for example, in Austria and Germany. Fixed-premium systems are used in Denmark, the Netherlands and Spain.

10.3.2 Green certificate systems

A system of green certificate systems currently exists in five EU Member States, as well as Australia. In this case, renewable electricity is sold at conventional power-market prices, but with the right to sell government-issued certificates that guarantee the renewable character of electricity to consumers or producers that are obliged to purchase a certain number of green certificates from renewable electricity producers according to a fixed percentage, or quota, of their total electricity consumption/production.

Since producers/consumers wish to buy these certificates as cheaply as possible, a secondary market of certificates develops where renewable electricity producers compete with one another to sell green certificates.

10.3.3 Tendering

Under a tendering procedure, the state places a series of tenders for the supply of renewable electricity, which is then supplied on a contract basis at the price resulting from the tender. The additional costs generated by the purchase of renewable electricity are passed on to the end-consumer of electricity through a specific energy tax. Pure tendering procedures existed until recently in Ireland and France.

10.3.4 Investment subsidies

In some countries, direct investment subsidies apply for biomass combustion systems. This is the case, for example, in Germany for domestic wood pellet stoves.

10.3.5 Tax deduction

Support systems based only on tax deduction are often applied as an additional policy tool to support renewable energy. In the Netherlands for example, a company investing in a biomass combustion system may deduct an additional 44 per cent of the investment cost from their taxable income.

Table 10.1 *Policies in selected EC countries for promoting bio-electricity production*

Country	Policy
Austria	A federally uniform purchasing and payment obligation for bio-electricity plants (feed-in tariffs).
Belgium	Flanders: quota and tradable certificates. Wallonia: quota and tradable certificates.
Cyprus	Investment subsidies up to 40% of investments. Feed-in tariff: 6.3 €ct/kWh.
Czech Republic	Fixed feed-in tariff (15 years) or green bonus.
Denmark	Fixed feed-in tariff (10–20 years), depending on type of installation and date of commissioning. Tariffs vary from about 5 to 8.1 €ct/kWh, but depend on electricity prices.
Estonia	Fixed feed-in tariff 1.8 times the tariff paid for power from conventional sources (fixed for 7–12 years). No CO_2 charge for companies using biomass internally.
Finland	Fiscal subsidies equivalent to feed-in tariffs of 0.42–0.69 €ct/kWh.
France	Fixed feed-in tariffs (15 years) for installations up to 12MW. Tariffs vary from 4.5 to 5.7 €ct/kWh, depending on biomass/technology combination.
Germany	Fixed feed-in tariffs of 3.9–21.5 €ct/kWh (fixed for 20 years).
Greece	Tax exemptions for investments up to 75%. Investment subsidies up to 40%. Interest subsidy up to 40%. Several other subsidies, and guaranteed feed-in tariff of 90% of existing tariffs for 10 years.
Hungary	Currently in transition from feed-in tariffs to green certificate scheme (quota).
Ireland	Feed-in tariffs. Current level: 7.2 €ct/kWh.
Italy	Quota and tradable certificates.
Latvia	Fixed feed-in tariffs.
Lithuania	Fixed feed-in tariff (fixed for 10 years). Current level: 6.9 €ct/kWh.
Luxembourg	Feed-in tariff of 2.5 €ct/kWh.
Netherlands	Investment tax deduction. Fixed feed-in premiums (fixed for 10 years). Current level: 2.9–9.7 €ct/kWh.
Poland	Quota obligation for producers, but hardly enforced.
Portugal	Fixed feed-in tariffs.
Slovakia	Tax-break and investment subsidy.
Slovenia	Fixed feed-in tariffs. Current level: 10.04 €ct/kWh.
Sweden	Quota and tradable certificates.
UK	Quota and tradable certificates.

Source: [19].

> ### Box 10.1 The dead koala RECs problem
>
> In Australia, Delta Electricity (a utility company) planning to co-fire sawdust originating from forest operations in their coal-fired power plants faced large opposition from a number of NGOs that were afraid that the utility company would use native forest residues in spite of explicit promises that this would not be the case.*
>
> The NGOs started a very strong and vocal media campaign, using posters, advertisements and the internet. The negative impact of this media campaign on the public perception of co-firing has resulted in a price for Renewable Energy Certificates (RECs) for co-firing that are still about 10 per cent lower than for other renewable energy options (this is also known as the 'dead koala RECs problem').
>
> Delta Electricity now increasingly involves communities in the development of new projects. Although winning community support does not guarantee project approval or an absence of appeal from consent authorities, it provides more comfort and a more positive perception to these authorities.

It is difficult to compare the effectiveness of different financial policy instruments; however, in general, it can be stated that investors avoid risk and appreciate long-term stability in financial support mechanisms. In this respect, long-term guaranteed feed-in tariffs provide more clarity to investors than green certificates, of which future prices are yet uncertain. At the same time, tax exemptions and investment subsidies ensure that cash flows are more compatible with traditional energy options, usually characterized by lower initial investment and lower sales prices of the energy produced.

10.4 Other policies that influence the establishment of biomass combustion

The implementation of biomass combustion systems is not solely influenced by financial instruments that support the construction and operation of combustion plants, but also depends on policies for agriculture, forestry and the environment as well as public support.

Efficient coordination among responsible authorities in these areas is essential in smoothing the planning and permission process for new installations. Obtaining public support has proved to be essential for the development of many bioenergy projects. Lack of sufficient public support and full transparency has often resulted in long lead times for obtaining the necessary permits. Arguments that need to be addressed are competition with other biomass-using sectors such as the chipboard and paper industries, and the sustainability of biomass supply (CO_2 balance, emissions and biodiversity effects). It is essential that project developers are fully transparent to external parties from the initial stage of their project.

Within the IEA Bioenergy Agreement, there are two Tasks that particularly address greenhouse gas balances of bioenergy chains and the sustainability of international biomass trade.

---

* As stated in the Commonwealth Regulatory Act of 2000 and the derived NSW Protection of the Environment Operations Act and Regulation, residues from native forest operations are not allowed to be co-fired. This is being controlled with regular audits on fuel supply.

Tightening emission limits on bioenergy plants makes the conversion of biomass to heat and/or power more expensive. In Europe, the Waste Incineration Directive (WID), Large Combustion Plant (LCP) Directive and IPPC Directive all have an important effect on emission restrictions on bioenergy plants.

Environmental policies, on the other hand, might have a positive impact on the market opportunities for biomass combustion systems. An example is the European Landfill Directive (99/31/EC) which aims to prevent, or reduce as far as possible, the negative environmental impacts of landfilling waste. The Directive forces EU member countries to reduce the amount of biodegradable waste going to landfill by 25 per cent in 2006 (taking 1995 as a reference year), by 50 per cent in 2009 and by 65 per cent in 2016.* As a result, landfilling tariffs for combustible waste have increased and it has become more attractive to use various waste streams for energy generation.

10.5 References

1. IEA (2006) *Renewables in Global Energy Supply*, an IEA Fact Sheet, September
2. FAAIJ, A. P. C. et al (2007) *Global Potential for Bioenergy*, paper for the IEA Bioenergy Agreement
3. IEA (2004) *World Energy Outlook*, IEA
4. WESTWOOD, J. (2004) *The World Biomass Report 2004–2013*, Douglas-Westwood
5. IEA COAL (2005) *Life Extension of Coal-fired Power Plants*, PF 05-13,
6. EIA (2006) *System for the Analysis of Global Energy Markets*, Energy Information Administration
7. EIA (2006) *International Energy Outlook 2006*, Energy Information Administration
8. US DOE (2006) Biomass power program, see www1.eere.energy.gov/biomass/
9. EUROPEAN COMMISSION (1997) *Energy for the Future: Renewable Sources of Energy*, White Paper for a Community Strategy and Action Plan, COM(97)599 final,
10. EUROSTAT (2006) 'Gross electricity generation, by fuel used in power-stations', available at http://ec.europa.eu/eurostat
11. SPLIETHOFF, H., SIEGLE, V. and HEIN, K. (1996) 'Zufeuerung von Biomasse in Kohlekraftwerken: Auswirkungen auf Betrieb, Emissionen und Rückstände, Technik und Kosten', in Th. Nussbaumer (ed.), *Feuerungstechnik, Ascheverwertung und Wärme-Kraft-Kopplung*, ENET, Bundesamt für Energie
12. SIPILÄ, K. & KORHONEN, M. (eds.), (1999) *Power Production from Biomass III*, Espoo 14–15 September 1998 VTT, Espoo, Finland
13. NUSSBAUMER, TH., NEUENSCHWANDER, P., HASLER, PH., JENNI, A. and BÜHLER, R. (1998) 'Technical and economic assessment of the technologies for the conversion of wood to heat, electricity and synthetic fuels', in *Biomass for Energy and Industry, 10th European Conference and Technology Exhibition*, 8–11 June, Würzburg (Germany), pp1142–1145
14. BRIDGWATER, A. and BOOCOCK, D. (eds.) (1997) *Developments in Thermochemical Biomass Conversion*, Blackie Academic, London
15. KOPETZ, H., WEBER, T., PALZ, W., CHARTIER, P. and FERRERO, G. (eds.) (1998) *Biomass for Energy and Industry*, 10th European Conference and Technology Exhibition, 8–11 June, Würzburg (Germany)
16. STRAUSS, K. (1992) *Kraftwerkstechnik*, Springer
17. EC-ASEAN (2004) COGEN Programme (Phase 3)
18. FAAIJ, A. P. C. (2006) 'Bio-energy in Europe: Changing technology choices', *Energy Policy*, Department of Science, Technology and Society, Copernicus Institute for Sustainable Development & Innovation, Utrecht University, the Netherlands, vol 34, pp322–342
19. FABER, J., BERGSMA, G. and VROONHOF, J. (2006) *Bio-energy in Europe 2005: Policy Trends and Issues*, March, CE, Delft, the Netherlands

---

* Member States that landfilled more than 80% of their municipal waste in 1995 may postpone each of these targets by a maximum of four years.

11
Research and Development: Needs and Ongoing Activities

This final chapter offers a brief overview of research and development activities that are ongoing or planned, on the basis of the comprehensive information given in this *Handbook of Biomass Combustion and Co-firing* regarding the current state of affairs in this field. Areas in which research and development are being performed are: biomass potentials, fuel pre-treatment technologies, combustion, CHP systems, process control and gas clean-up technologies that can cope with difficult-to-burn feedstock, minimize harmful emissions and increase efficiency. In order to achieve these aims, new and innovative calculation and simulation tools and measurement devices are being used.

Another important target of R&D activities, apart from technological innovation, is the reduction of investment, maintenance and operating costs. However, along with the conditions that can make or break new technologies, the need for research and development also varies per country. Where overlap exists among various countries in research areas, joint research programmes may possibly be executed under the umbrella of the IEA Bioenergy Agreement or other international R&D programmes.

Below are listed a number of important considerations and activities concerning research and development in the area of biomass combustion that are ongoing or foreseen in member states of the European Union and the IEA Bioenergy Agreement, Task 32 'Biomass Combustion and Cofiring'.

11.1 Investigation of potentials of biomass resources

The energy demand has been steadily increasing in recent years worldwide – and this trend is expected to continue. This development has also led to a steady rise in CO_2 emissions. For instance, the European Commission has started initiatives to promote the utilization of renewable energy sources for heat and power generation. The most important of these initiatives are the White Paper (1997), the RES-E Directive (2001), the Commitment on the Green Paper on Security of Energy Supply (June 2002), the Directive on the Energy Performance of Buildings (December 2002), the Directive on Liquid Biofuels (May 2003), the Commitment on Renewable Energies in Europe launched at the Bonn conference (2004) and the Directive on the Promotion of Cogeneration (2004). The combustion of solid biomass for heat, power and combined heat and power (CHP) generation is considered to offer one of the highest potentials for renewable energy utilization and CO_2 emission reduction in the short to medium term. Energy generation from biomass within the EU should be tripled by the year 2010 (based on the year 1995).

These European targets, as well as several targets at national levels, and the increasing utilization of solid biomass (especially woody biomass) for energy generation not only in

Europe but also worldwide raise several questions concerning the availability of sufficient biomass resources and the reliability of fuel supply to meet these goals. Therefore, the investigation of potentials of biomass resources that are available for thermal utilization, becomes more and more relevant. In addition, due to the rising use of solid biomass, efforts must be undertaken to strengthen social and environmental integration along the entire chain from biomass production to provision of energy services to the consumer, because the advantages of the thermal utilization of solid biomass fuels must always be balanced with its disadvantages (e.g. eventual loss of biodiversity, claims on vast areas of land, environmental emissions, hazards and health conditions of workers).

More information: [1–9].

11.2 Development of improved combustion technologies

The development of combustion technologies is still ongoing. The primary aims are to minimize the total costs of heat and/or power production and to maximize safety, ease of operation and efficiency. A future goal is, for instance, to increase the efficiency by higher steam temperatures and pressures and better materials for super heaters. Major goals include the development of new combustion technologies for new biomass fuels (e.g. herbaceous fuels, agricultural waste materials, pellet-fired tiled stoves), and the development of furnaces with a high flexibility regarding biomass fuel quality (multifuel combustion systems).

For medium- and large-scale applications, the use of special biomass fuels, such as energy crops, waste wood and agricultural waste materials is of increasing interest. Annually harvested energy crops (like Salix, miscanthus and grasses) in particular will be on the rise in the coming years. Due to their different chemical compositions (as compared to conventional wood fuels), these biomass fuels require special combustion and flue gas cleaning technologies.

Research activities are ongoing in finding better bed materials for fluidized bed combustion plants and concerning changing the combustion environment with additives.

The development and/or implementation of innovative process control systems (e.g. Fuzzy Logic, model based control strategies) and innovative sensors for biomass combustion technologies in order to improve and stabilize system operation and to further reduce personnel costs is also a key interest.

More information: [10–26].

11.3 Gaseous (especially NO_x) reduction technologies

Biomass combustion systems have reached a high technological level, with low emissions and a high operational performance. However, limiting values for gaseous (especially NO_x) emissions are constantly being driven down by the authorities, which means that major R&D efforts will be required in the future to develop even more advanced systems.

The overall objective for small-, medium- and large-scale combustion units is the reduction of gaseous (especially NO_x) emissions. This can be done with primary measures or combinations of primary and secondary measures. Generally, the decreasing emission limits for NO_x underline the necessity of further developments in this field. These reduction measures further improve the environmental compatibility of thermal biomass utilization, which is one of the most important arguments in the competition with fossil fuel combustion units. The reduction of NO_x emissions is of great importance for small-scale combustion units, as they need simple and affordable solutions. Therefore, technologies well proven in medium- and large-scale combustion units should be simplified and adapted to small-scale applications. An increased utilization of biomass fuels rich in N and ash, such as waste wood and energy crops, necessitates the development and market introduction of efficient emission reduction technologies.

More information: [27–38].

11.4 Ash and aerosol-related problems during biomass combustion including dust (fine particulate) reduction technologies

Ash-related problems in biomass combustion systems form a further topic with a high future R&D demand. These problems cover the areas of particulate formation, deposit formation and corrosion as well as the slagging behaviour of biomass ashes. They are particularly pressing in the combustion of non-wood fuels, such as energy crops, straw, grasses, husks, shells or stones as well as waste wood – due to their high concentrations of alkali metals, S and Cl as well as volatile heavy metals (especially Zn and Pb) in case of waste wood. They influence the design and process control of furnaces and boilers as well as the optimization of dust precipitation units.

Solid ash and soot particles, emitted from biomass combustion installations, are important sources of fine particulate emissions. However, the limiting values for particulate emissions are driven down by the authorities and therefore primary and secondary measures to reduce particulate emissions have to be developed and optimized.

Consequently, mitigation of fine particulate emissions that result from biomass combustion deserves increased attention from research organizations, manufacturers of boilers and particle removal technologies as well as policy makers. The formation and behaviour of fly-ashes and aerosols in biomass combustion units has therefore become an important field of research. Equipment manufacturers need to be encouraged to develop novel, low-cost combustion installations and filtration techniques that result in low particulate emissions, even in small-scale applications. Existing and well-proven dust precipitation technologies that are available for medium- and large-scale applications should be simplified and further developed for their application in small-scale units.

Additionally, some attempts to investigate health risks caused by particulate emissions from biomass combustion systems have already been initiated. This research field is of major relevance within the market competition with fossil fuel-based systems and from a human toxicology point of view and will therefore gain increasing relevance in the coming years. However, due to the high complexity of the problems addressed, interdisciplinary research and a close cooperation of technical and medical sciences is needed to succeed.

Another major focus of R&D activities will be on solving problems concerning deposit formation and corrosion in the heat exchanger sections of large-scale biomass CHP plants. To reduce maintenance and repair costs and to increase the availability of installations, the mechanisms responsible for slagging, fouling and corrosion have to be thoroughly investigated. Further down the road, the search for appropriate primary and secondary measures to prevent deposit formation and corrosion processes as well as to reduce aerosol emissions will also be necessary.

In addition to all this, there are also some open questions regarding the environmentally sound utilization of biomass ashes and the development of appropriate treatment technologies for contaminated biomass ashes. In both directions, the search is on for a closed-cycle economy and the minimization of disposal costs.

More information: [23–26, 39–81].

11.5 Innovative micro-, small- and medium-scale CHP technologies based on biomass combustion

The demand for combined heat and power production (CHP) from biomass is an important new trend. Large-scale biomass CHP systems based on conventional steam turbine cycles are state-of-the-art. Appropriate CHP technologies for small- and medium-scale combustion systems are under development or market introduction, but still require comprehensive R&D in order to reach the demonstration stage or to get further optimized.

In the last few years, several new systems such as the ORC process and the Stirling engine technology have emerged for small- and medium-scale CHP production (10–1000kW$_e$) based on biomass combustion.

The ORC technology for instance, has already been successfully introduced into the market segment aiming at electric capacity ranges between 200kW and 2000kW. However, an optimization potential exists regarding the achievement of an increased electrical efficiency, as well as potential concerning cost reduction by modular design.

For the Stirling engine technology, which is already at the demonstration level, further R&D demand includes fouling and cleaning of high temperature heat exchanger areas, process control and seal development for high pressure systems.

Moreover, some promising technologies for mirco-scale CHP systems have been identified (e.g. thermoelectric generator, Stirling engine). These technologies have to be further developed in order to reach the demonstration level.

In the field of directly or indirectly fired gas turbines utilizing atmospheric as well as pressurized combustion systems, R&D activities are also taking place. These CHP solutions are still in an early stage of development.

R&D efforts will concentrate on the further development and optimization of the CHP technologies until they reach the demonstration and dissemination level.

More information: [82–98].

11.6 Technology development and unsolved problems concerning co-firing of biomass in large-scale power plants

Co-combustion of biomass fuels and fossil fuels (especially coal) is gaining importance worldwide. The R&D demands in this area cover the proper selection and further development of appropriate co-combustion technologies, possibilities of NO_x reduction by fuel staging, problems concerning the deactivation of catalysts, characterization and possible utilization of ashes from co-combustion plants, the formation of striated flows and corrosion problems.

More information: [99–115].

11.7 Fuel pre-treatment technologies

There is an increasing demand for solid biomass fuels individually 'tailored' to needs of the respective application process. This is the driving force behind new upgrading methods or technologies that can be applied either during or immediately after field production (e.g. leaching by rainfall or irrigation) or in a preparatory process prior to energetic use (e.g. stationary leaching, use of additives, compaction). Monitoring of production or separation (fractionation) processes should also focus on fuel properties in order to make optimum use of varying or heterogeneous raw materials. Furthermore, fuel quality aspects are becoming a key target in plant breeding and variety/clone selection. Genetic engineering, although still highly controversial, may also open up new chances for yield and quality improvements of biomass fuels.

Short rotation forestry, which has already gained a certain importance in Scandinavia, is of increasing interest as a possible measure for cleaning industrially degraded land from contaminants (bioremediation) as well as for the utilization of set-aside land. It could also raise the economic competitiveness of this type of biomass fuel. If it does, the thermal conversion process needs to be adapted in order to guarantee an ecologically competitive overall energy production process (ash fractionation, efficient dust precipitation).

Pelletizing technologies for the production of upgraded biomass fuels must be improved in order to lower costs and enhance fuel quality (proper selection of matrices, testing and evaluation of bio-additives for quality improvement and reduction of operating costs, development and testing of pre-treatment technologies for proper conditioning of the raw material). Moreover, future pellet production will have to cope with raw materials beyond the materials currently most commonly used, i.e. wood shavings and sawdust. Due to strongly increasing pellet markets in Europe and worldwide, the production of pellets from, e.g., woodchips, forestry residues, short rotation coppice and different kinds of herbaceous biomass fuels will increase in the future, which makes respective R&D activities necessary.

Fuel-drying technologies directly coupled to the combustion process in order to achieve high overall energy efficiencies are also an interesting development.

More information: **[23, 37, 38, 116–137]**.

11.8 CFD modelling and simulation of thermochemical processes

The application of computational fluid dynamics (CFD) gives a deeper insight into flow-related processes occurring during thermochemical conversion of solid biomass and therefore is a powerful tool for a quicker, less risky and more reliable development of new technologies. The potential of CFD modelling has increased considerably in the last few years. CFD modelling provides the opportunity to calculate reactive flows including species, temperature and residence time distributions as well as multiphase flows (flue gas as well as fuel and fly-ash particles) in biomass furnaces and boilers. Certain units of biomass conversion systems, such as furnaces, heat exchangers and dust precipitators, can be designed and optimized by means of CFD calculations, which are cheaper and less time-consuming than test runs.

Furthermore, various R&D projects are ongoing in order to develop CFD models for solid biomass combustion in entrained flows and packed beds. This will further extend the applicability of this powerful tool to pulverized fuel furnaces as well as to the whole conversion process in spreader stoker, underfeed stoker and grate furnaces.

Concerning the simulation of thermodynamic and chemical processes, there is an ongoing process of improvement of existing programme codes and databases (e.g. reaction mechanisms). Advanced commercial software is also being developed in the fields of chemical reaction kinetics (e.g. models for NO_x and SO_x formation) and thermodynamic equilibrium calculation (e.g. modelling of the behaviour of alkali metals, the formation of low melting compounds as well as the behaviour of heavy metals). Such models deepen our understanding of processes with regard to corrosion, formation of sticky deposits and ash/aerosol formation, thus providing the basis for the development of appropriate technological prevention measures. Research activities will result in models that effectively support the appropriate design of flue gas cleaning systems (for combustion and gasification systems) and can answer special questions of boiler design.

Commercially available databases contain information on the properties of almost all relevant elements and compounds. However, there are still blanks in the information available on thermodynamic and physical properties of certain elements, compounds and especially multicomponent/multiphase systems, which complicate the simulation of chemical reactions and processes at high temperatures as well as simulations for ash-forming elements and kinetically limited and heterogeneous reaction systems (NO_x and SO_x formation) in biomass conversion processes. Consequently, further expansion and improvement of basic data and models is necessary.

Moreover, considerable efforts are currently being made to couple the various CFD models developed and to integrate reaction kinetics and equilibrium modelling into CFD simulations, since this approach provides a powerful opportunity for a spatially resolved simulation and visualization of thermochemical biomass conversion processes. CFD simulations will then not only be flow simulations but rather three-dimensional simulations and visualizations of highly complex and linked physical and chemical processes in thermal biomass conversion plants. This is made possible by the continuously increasing computer performance as well as special new simulation techniques, which allow a significant reduction of calculation time and the application of highly complex models as engineering tools.

In this context, the extension of CFD to NO_x formation including detailed or reduced reaction kinetics as well as deposit and aerosol formation is a major R&D goal in the field of modelling and simulation. Such models give a better understanding of the underlying processes of NO_x formation as well as the formation and composition of coarse fly-ash, aerosols and deposit layers, in order to develop appropriate technological measures.

More information: **[34, 40, 138–184]**.

11.9 References

1 THEK, G., OBERNBERGER, I., HIRTENFELLNER, J. (2005) 'Investigation of woody biomass potentials in selected European countries and their allocation to potential applications for thermal utilisation', in *Proceedings of the 14th European Biomass Conference & Exhibition*, October, Paris, France, ETA-Renewable Energies, Italy, pp96–99

2 VAN DAM, J. et al (2005) 'Biomass production potentials and biofuel trade options of Central and Eastern Europe under different scenarios', in *Proceedings of the 14th European Biomass Conference & Exhibition*, October, Paris, France, ETA-Renewable Energies, Italy, pp100–103

3 FALLOT, A. and GIRARD, P. (2005) 'Spatial assessment of biomass potentials for energy scenarios', in *Proceedings of the 14th European Biomass Conference & Exhibition*, October 2005, Paris, France, ETA-Renewable Energies, Italy, pp104–107

4   KARJALAINEN, T. et al (2005) 'Assessment of energy wood potential in the European Union', in *Proceedings of the 14th European Biomass Conference & Exhibition*, October, Paris, France, ETA-Renewable Energies, Italy, pp112–115
5   BUTTLE, D., McDONNELL, K. and WARD, S. (2005) 'Quantification of energy crops as a biomass resource in Ireland and the UK', in *Proceedings of the 14th European Biomass Conference & Exhibition*, October, Paris, France, ETA-Renewable Energies, Italy, pp116–119
6   HASSELGREN, K. (2005) 'Sustainable and profitable use of waste products and farmland in Europe for production of salix as a fuel resource', in *Proceedings of the 14th European Biomass Conference & Exhibition*, October, Paris, France, ETA-Renewable Energies, Italy, pp120–123
7   GRIGOLATO, S. et al (2005) 'Supply of logging residues for energy production: A GIS-based study in New Zealand and in Italy', in *Proceedings of the 14th European Biomass Conference & Exhibition*, October, Paris, France, ETA-Renewable Energies, Italy, pp132–135
8   STÜLPNAGEL, R., ABEL, H. J. and GREBE, S. (2005) The potential of biomass for energy in Germany – Remarks to possible reserves and to the interactions with the agriculture at present', in *Proceedings of the 14th European Biomass Conference & Exhibition*, October, Paris, France, ETA-Renewable Energies, Italy, pp178–181
9   IEA (2006) *Renewable Energy: RD&D Priorities – Insights from IEA technology programmes*, International Energy Agency (IEA), Paris, France
10 OBERNBERGER, I. (1997) *Nutzung fester Biomasse in Verbrennungsanlagen unter besonderer Berücksichtigung des Verhaltens aschebildender Elemente*, Volume 1 of Thermal Biomass Utilization series, BIOS, Graz, Austria, dbv-Verlag der Technischen Universität Graz, Graz, Austria
11 NUSSBAUMER, T. (1993) *Verbrennung und Vergasung von Energiegras und Feldholz, Jahresbericht 1992 zum gleichnamigen Forschungsprojekt*, Bundesamt für Energiewirtschaft, Bern, Switzerland
12 SWITHENBANK, J., NASSERZADEH, V., GOH, Y. R. and SIDALL, R. (1997) 'Fundamental principles of incinerator design', keynote paper, 1st International Symposium on Incineration and Flue Gas Treatment Technology, University of Sheffield, UK
13 EUROPEAN COMMISSION (1998) *Biomass Conversion Technologies – Achievements and Prospects for Heat and Power Generation*, European Commissions, Belgium
14 OBERNBERGER, I. (1997) 'Decentralised biomass combustion – a technological overview', Keynote lecture, in *Proceedings of the 4th European Conference on Industrial Furnaces and Boilers*, April, Porto, Portugal, INFUB, Rio Tinto, Portugal
15 OBERNBERGER, I. (1996) 'Decentralised biomass combustion – State-of-the-art and future development', keynote lecture at the 9th European Biomass Conference in Copenhagen, *Biomass and Bioenergy*, vol 14, no 1, pp33–56
16 MUSIL, B., HOFBAUER, H. and SCHIFFERT, T. (2005) 'Development of pellets fired tiled stoves', presented at the European Conference on Sustainable Energy Systems for Buildings and Regions, 5–8 October, Vienna, Austria
17 MUSIL, B., HOFBAUER, H., and SCHIFFERT, T. (2005) 'Development and analyses of pellets fired tiled stoves', in *Proceedings of the 14th European Biomass Conference & Exhibition*, October, Paris, France, ETA-Renewable Energies, Italy, pp1117–1118
18 HASLINGER, W., PADINGER, R., WÖRGETTER, M. and SPITZER, J. (2004) 'Austrian research and development of novel solid biofuels and innovative small-scale biomass combustion systems', 2nd World Conference and Exhibition on Biomass for Energy, Industry and Climate Protection, 10–14 May, Rome, Italy
19 OBERNBERGER, I. and THEK, G. (2002) 'The current state of Austrian pellet boiler technology', in *Proceedings of the 1st World Conference on Pellets*, September, Stockholm, Sweden, Swedish Bioenergy Association, Stockholm, Sweden, pp45–48,
20 JOHANSSON, L. S., LECKNER, B., GUSTAVSSON, L., COOPER, D., TULLIN, C., POTTER, A. and BERNTSEN, M. (2005) 'Particle emissions from residential biofuel boilers and stoves – old and modern techniques', in *Proceedings of the International Seminar 'Aerosols in Biomass Combustion'*, March, Graz, Austria, Volume 6 of Thermal Biomass Utilization series, BIOS BIOENERGIESYSTEME GmbH, Graz, Austria, pp145–150

21 GOOD, J., NUSSBAUMER, T., DELCARTE, J. and SCHENKEL, Y. (2004) 'Methods for efficiency determination for biomass heating plants and influence of operation mode plant on efficiency', in *Proceedings of the 2nd World Conference and Exhibition on Biomass for Energy, Industry and Climate Protection*, May, Rome, Italy, Volume II, ETA-Florence, Italy pp1431–1434

22 MERCKX, B., GOOD, J. and DELCARTE, J. (2005) 'Evaluation of uncertainties of a wood chips boiler yield', in *Proceedings of the 14th European Biomass Conference and Exhibition*, October, Paris, France, ETA-Renewable Energies, Italy, pp1347–1348

23 ROBINSON, A. L., BUCKLEY, S. G., YANG, N. and BAXTER, L. L. (2000) 'Experimental Measurements of the Thermal Conductivity of Ash Deposits: Part 1. Measurement Technique', submitted to *Energy and Fuels*, Sandia Report 2000-8599, Sandia National Laboratories, US

24 OBERNBERGER, I., WIDMANN, W., WURST, F. and WÖRGETTER, M. (1995) *Beurteilung der Umweltverträglichkeit des Einsatzes von Einjahresganzpflanzen und Stroh zur Fernwärmeerzeugung*, Jahresbericht zum gleichnamigen Forschungsprojekt; Institute of Chemical Engineering, Graz University of Technology, Austria

25 National Renewable Energy Laboratory, Sandia National Laboratory, University of California, Foster Wheeler Dev. Corp., US Bureau of Mines, and T. R. MILES (1996) *Alkali Deposits found in Biomass Power Plants*, research report NREL/TP-433-8142 SAND96-8225, vols I and II, National Renewable Energy Laboratory, Oakridge, US

26 SANDER, M. and ANDREN, O. (1997) 'Ash from cereal and rape straw used for heat production: Liming effect and contents of plant nutrients and heavy metals', *Water, Air and Soil Pollution*, vol 93, pp93–108

27 PADINGER, R. et al (1999) *Reduction of Nitrogen Oxide Emissions from Wood Chip Grate Furnaces*, Final report for the EU-JOULE III project JOR3-CT96-0059, Joanneum Research, Graz, Austria

28 KELLER, R. (1994) *Primärmaßnahmen zur NO_x-Minderung bei der Holzverbrennung mit dem Schwerpunkt der Luftstufung*, Forschungsbericht Nr. 18 (1994), Laboratorium für Energiesysteme, ETH Zürich, Switzerland

29 SALZMANN, R. and NUSSBAUMER, T. (1995) *Zweistufige Verbrennung mit Reduktionskammer und Brennstoffstufung als Primärmaßnahmen zur Stickoxidminderung bei Holzfeuerungen*, Forschungsbe-richt, Laboratorium für Energiesysteme, ETH Zürich, Switzerland

30 KILPINEN, P. (1992) Kinetic Modeling of Gas-Phase Nitrogen Reactions in Advanced Combustion Processes, Report 92–7, PhD thesis, Abo Akademy University, Abo, Finland

31 GOOD, J. (1992) *Verbrennungsregelung bei automatischen Holzschnitzelfeuerungen*, Forschungs-breicht Nr. 11, Laboratorium für Energiesysteme, ETH Zürich, Switzerland

32 WEISSINGER, A. and OBERNBERGER, I. (1999) 'NO_x reduction by primary measures on a travelling-grate furnace for biomass fuels and waste wood', in *Proceedings of the 4th Biomass Conference of the Americas*, September, Oakland CA, Elsevier Science Ltd., Oxford, UK, pp1417–1425

33 WEISSINGER, A., OBERNBERGER, I. and SCHARLER, R. (2002) 'NO_x reduction in biomass grate furnaces by primary measures – evaluation by means of lab-scale experiments and chemical kinetic simulation compared with experimental results and CFD calculations of pilot-scale plants', in *Proceedings of the 6th European Conference on Industrial Furnaces and Boilers*, April, Estoril, Portugal, INFUB, Rio Tinto, Portugal

34 WIDMANN, E., SCHARLER, R., STUBENBERGER, G. and OBERNBERGER, I. (2004) 'Release of NO_x precursors from biomass fuel beds and application for CFD-based NO_x postprocessing with detailed chemistry', in *Proceedings of the 2nd World Conference and Exhibition on Biomass for Energy, Industry and Climate Protection*, May, Rome, Italy, Volume II, ETA-Florence, Italy, pp1384–1387

35 WEISSINGER, A., FLECKL, T. and OBERNBERGER, I. (2004) 'Application of FT-IR in-situ absorption spectroscopy for the investigation of the release of gaseous compounds from biomass fuels in a laboratory scale reactor – Part II: Evaluation of the N-release as functions of combustion specific parameters', *Combustion and Flame*, vol 137, pp403–417

36 ZHOU, H., JENSEN, A. D., GLARBOR, G. P. and KAVALIAUSKAS, A. (2005) 'Formation and reduction of nitric oxide in fixed-bed combustion of straw', *Fuel*, vol 85, pp705–716

37 BRUNNER, T. and OBERNBERGER, I. (1996) 'New technologies for NO_x-reduction and ash utilisation in biomass combustion plants – JOULE THERMIE 95 Demonstration Project', in *Proceedings of the 9th European Bioenergy Conference*, Volume 2, Elsevier Science Ltd, Oxford, UK
38 BRUNNER, T., LAUCHER, A. and OBERNBERGER, I. (1997) 'Sägerestholzfeuerung mit Rauchgaskondensation und Ascheaufbereitung', in *VDI Bericht 1319 'Thermische Biomassenutzung – Technik und Realisierung'*, VDI Verlag GmbH, Düsseldorf, Germany
39 ROBINSON, A. L., BUCKLEY, S. G., YANG, N. and BAXTER, L. L. (2000) 'Experimental Measurements of the Thermal Conductivity of Ash Deposits: Part 2. Effects of Sintering and Deposit Microstructure', submitted to *Energy and Fuels*, Sandia Report 2000-8600, Sandia National Laboratories, US
40 FRANDSEN, F., JENSEN, P., JENSEN, A., LIN, W., JOHNSSON, J., NIELSEN, H., ANDERSEN, K. and DAM-JOHANSEN, K. (1999) *Clean and Efficient Utilisation of Biomass for Production of Electricity and Heat*, Report No 9905, Inst. f. Kemiteknik, Technical University Denmark, Lyngby, Denmark
41 ELSAM/ELKRAFT (1996) *Action Plan for Bioenergy*, information letter No.7, ELSAM/ELKRAFT, Copenhagen, Denmark
42 VATTENFALL, S. (1993) *Skogsbränsle*, Slutrapport för Projekt Skogskraft, Vattenfall, Väl-lingby, Sweden
43 NARODOSLAWSKY, M. and OBERNBERGER, I. (1995) 'From waste to raw material – The way of cadmium and other heavy metals from biomass to wood ash', *Journal of Hazardous Materials*, vol 50/2–3, pp157–168
44 OBERNBERGER, I., DAHL, J. and ARICH, A. (1996) *Round Robin on Biomass Fuel and Ash Analysis*, final report, JOULE III project No. JOR3-CT950001 and IEA Bioenergy Agreement, Task XIII, Activity 6 'Integrated Bioenergy Systems', European Commission DG XII, Brussels, Belgium
45 CHRISTENSEN, K. A. (1995) The Formation of Sub-micron Particles from the Combustion of Straw, PhD thesis, Department of Chemical Engineering, Technical University of Denmark, Lyngby, Denmark
46 HALD, P. (1994) Alkali Metals at Combustion and Gasification – Equilibrium Calculations and Gas Phase Measurings, MS thesis, Department of Chemical Engineering, Technical University of Denmark, Lyngby, Denmark
47 SKRIFVARS, B-J., SFIRIS, G., BACKMAN, R., WIDEGREN-DAFGÅRD, K. and HUPA, M. (1996) 'Ash behaviour in a CFB boiler during combustion of Salix', in *Proceedings of the International Conference 'Developments in Thermochemical Biomass Conversion'*, May, Banff, Canada, vol 2, Blackie Academic and Professional, London
48 HUPA, M. and BACKMAN, R. (1983) 'Slagging and fouling during combined burning of bark with oil, coal or peat', in *Fouling of Heat Exchanger Surfaces*, The Engineering Foundation, United Engineering Trustees Inc., New York p419
49 SKRIFVARS, B.-J., HUPA, M., BACKMAN, R. and HILTUNEN, M. (1994) 'Sintering mechanisms of FBC ashes', *Fuel*, vol 73, no. 2, p171
50 BACKMAN, R., HUPA, M. and UPPSTU, E. (1987) 'Fouling and corrosion mechanisms in the recovery boiler superheater area', *Tappi Journal*, vol 70, pp123–127
51 MICHELSEN, H. P., LARSEN, O. H., FRANDSEN, F. and DAM-JOHANSEN, K. (1996) 'Deposition and high temperature corrosion in a 10MW straw fired boiler', in *Proceedings of the Engineering Foundation Conference 'Biomass Usage for Utility and Industrial Power'*, May 96, Utah, Engineering Foundation, New York
52 OBERNBERGER, I., DAHL, J. and BRUNNER, T. (1999) 'Formation, composition and particle size distribution of fly-ashes from biomass combustion plants', in *Proceedings of the 4th Biomass Conference of the Americas*, September, Oakland CA, Elsevier Science Ltd., Oxford, pp1377–1385
53 BRUNNER, T., DAHL, J., OBERNBERGER, I. and PÖLT, P. (2000) 'Chemical and structural analyses of aerosols and fly ashes from biomass combustion units by electron microscopy', in *Proceedings of the 1st World Conference on Biomass for Energy and Industry*, June, Sevilla, Spain
54 OBERNBERGER, I., BRUNNER, T. and JÖLLER, M. (2001) 'Characterisation and formation of aerosols and fly-ashes from fixed-bed biomass combustion', in *Proceedings of the International IEA*

Seminar 'Aerosols in Biomass Combustion', Zurich, Switzerland, IEA Bioenergy Agreement, Task 32 'Biomass Combustion and Cofiring' c/o TNO-MEP, Apeldoorn, the Netherlands, pp69–74

55 BRUNNER, T., OBERNBERGER, I., JÖLLER, M., ARICH, A. and PÖLT, P. (2001) 'Behaviour of ash forming compounds in biomass furnaces – Measurement and analyses of aerosols formed during fixed-bed biomass combustion', in *Proceedings of the International IEA Seminar 'Aerosols in Biomass Combustion'*, Zurich, Switzerland, IEA Bioenergy Agreement, Task 32 'Biomass Combustion and Cofiring' c/o TNO-MEP, Apeldoorn, the Netherlands, pp75–80

56 OBERNBERGER, I. and BRUNNER, T. (2004) 'Depositionen und Korrosion in Biomassefeuerungen', in *Tagungsband zum VDI-Seminar 430504 'Beläge und Korrosion in Großfeuerungsanlagen'*, May, Göttingen, Germany, VDI-Wissensforum GmbH, Düsseldorf, Germany

57 JÖLLER, M. and BRUNNER, T., OBERNBERGER, I. (2005) 'Modeling of aerosol formation in biomass combustion in grate furnaces and comparison with measurements', in *Energy and Fuels*, vol 19, pp311–323

58 JÖLLER, M., BRUNNER, T. and OBERNBERGER, I. (2005) 'Modelling of aerosol formation', in *Proceedings of the International Seminar 'Aerosols in Biomass Combustion'*, March, Graz, Austria, volume 6 of Thermal Biomass Utilization series, BIOS BIOENERGIESYSTEME GmbH, Graz, Austria, pp79–106

59 OBERNBERGER, I. (2005) 'Ash related problems in biomass combustion plants', Inaugural lecture for the appointment as professor for *Thermochemical Biomass Conversion* presented on 20 May 2005 at Technische Universiteit Eindhoven, Netherlands, Technische Universiteit Eindhoven, Netherlands; digital version: www.tue.nl/bib/

60 BIEDERMANN, F. and OBERNBERGER, I. (2005) 'Ash-related problems during biomass combustion and possibilities for a sustainable ash utilisation', in *Proceedings of the International Conference 'World Renewable Energy Congress' (WREC)*, May, Aberdeen, Scotland, Elsevier Ltd., Oxford, UK

61 BRUNNER, T., JÖLLER, M. and OBERNBERGER, I. (2006) 'Aerosol formation in fixed-bed biomass furnaces – results from measurements and modelling', in *Proceedings of the International Conference Science in Thermal and Chemical Biomass Conversion*, September, Victoria, Canada, CPL Press, pp1–20

62 OBERNBERGER, I. (2001) 'Aschen und deren Verwendung', in *Energie aus Biomasse*, Springer Verlag Berlin, Heidelberg, New York, pp412–425

63 ROTHENEDER, E., HANDLER, F. and HOLZNER, H. (2005) 'Assessment of the utilisation of differently processed wood-ashes as fertiliser in agriculture and forestry', *Bioenergy in Wood Industry*, 12–15 September, Jyväskylä, Finland

64 DELCARTE, J., DELCARTE, E., MAESEN, P. and SCHENKEL, Y. (2006) 'Heavy metals, PAH and PCB emissions from short rotation crop combustion', in *Proceedings of the International Conference Science in Thermal and Chemical Biomass Conversion*, September, Victoria, Canada, CPL Press, pp967–974

65 BROUWERS, B. (1996) 'Rotational particle separator: A new method for separating fine particles and mists from gases', *Chemical Engineering and Technology*, vol 19, pp1–10

66 BROUWERS, B. (1995) 'Secondary flows and particle centrifugation in slightly tilted rotating pipes', *Applied Scientific Research*, vol 55, pp95–105

67 OBERNBERGER, I., BRUNNER, T. and BÄRNTHALER, G. (2005) 'Aktuelle Erkenntnisse im Bereich der Feinstaubemissionen bei Pelletsfeuerungen', in *Tagungsband zum 5. Industrieforum Holzenergie*, October, Stuttgart, Germany, Deutscher Energie-Pellet-Verband e.V. and Deutsche Gesellschaft für Sonnenenergie e.V., Germany, pp54–64

68 SCHMATLOCH, V. (2005) 'Exhaust gas aftertreatment for small wood fired appliances – recent progress and field test results', in I. Obernberger and T. Brunner (eds.), *Proceedings of the Work–shop 'Aerosols in Biomass Combustion'*, March, Graz, Austria, Institute of Resource Efficient and Sustainable Systems, Graz University of Technology, pp159–166

69 BERNDTSEN, M. (2005) 'A novel electrostatic precipitator (ESP) for residential combustion', Presented at the Workshop on Recent Innovations in Small Scale Combustion organized by the International Energy Agency Bioenergy Agreement Task 32 'Biomass Combustion and Cofiring', available at http://www.ieabcc.nl (accessed 2 August 2006)

70 BRUNNER, T., BÄRNTHALER, G. and OBERNBERGER, I. (2006) 'Fine particulate emissions from state-of-the-art small-scale Austrian pellet furnaces: Characterisation, formation and possibilities of reduction', in *Proceedings of the 2nd World Conference on Pellets*, May/June 2006, Jönköping, Sweden, pp87–91, Swedish Bioenergy Association, Stockholm, Sweden

71 NUSSBAUMER, T., KLIPPEL, N. and OSER, M. (2005) 'Health relevance of aerosols from biomass combustion in comparison to diesel soot indicated by cytotoxicity tests', in *Proceedings of the International Seminar 'Aerosols in Biomass Combustion'*, March, Graz, Austria, Volume 6 of Thermal Biomass Utilization series, , BIOS BIOENERGIESYSTEME GmbH, Graz, Austria, pp45–54

72 SEGERDAHL, K., PETTERSSON, J., SVENSSON, J. E. et al (2004) 'Is KCl(g) corrosive at temperatures above its dew point? – Influence of KCl(g) on initial stages of the high temperature corrosion of 11% Cr steel at 600 degrees C', *Materials Science Forum*, nos 461–464, pp109–116

73 NORLING, R., NAFARI, A. and NYLUND, A. (2005) 'Erosion-corrosion of Fe- and Ni-based tubes and coatings in a fluidized bed test rig during exposure to HCl- and SO_2-containing atmospheres', *Wear*, vol 258, no 9, pp1379–1383

74 HANSSON, H. C. et al (2006) *Final Report Biofuels, Health and Environment*, Swedish Energy Agency, (see also www.itm.su.se/bhm/)

75 OBERNBERGER, I., BRUNNER, T. and JÖLLER, M. (2003) *Aerosols In Fixed-Bed Biomass Combustion – Formation, Growth, Chemical Composition, Deposition, Precipitation and Separation from Flue Gas*, final report of the EU project BIO-Aerosols (Contract No. ERK6-CT-1999-00003), Institute of Chemical Engineering Fundamentals and Plant Engineering, Graz University of Technology, Graz, Austria

76 MONTGOMERY, M. and LARSEN, O. H. (2002) 'Field test corrosion experiments in Denmark with biomass fuels. Part 2: Co-firing of straw and coal', *Materials and Corrosion*, vol 53, pp185–194

77 FRANDSEN, F. J. (2004) 'Utilizing biomass and waste for power production – a decade of contributing to the understanding, interpretation and analysis of deposits and corrosion products', *Fuel*, vol 84, pp1277–1294

78 FRANDSEN, F. et al (2002) 'Deposit formation and corrosion in the air pre-heater of a straw-fired combined heat and power production boiler', *IFRF Combustion Journal*, Article Number 200204

79 ROBINSON, A. L., JUNKER, H. and BAXTER, L. L. (2002) 'Pilot-scale investigation of the influence of coal-biomass cofiring on ash deposition', *Energy and Fuels*, vol 16, no 2, pp343–355

80 FRANDSEN, F. et al (2003) 'Ash and deposit formation in the biomass co-fired Masnedø combined heat and power production plant', *IFRF Combustion Journal*, Article Number 200304

81 ZBOGAR, A., FRANDSEN, F. J., JENSEN, P. A. and GLARBORG, P. (2005) 'Heat transfer in ash deposits: A modelling tool-box', *Progress in Energy and Combustion Science*, vol 31, nos 5–6, pp371–421

82 PODESSER, E., DERMOUZ, H., LAUER, M. and WENZEL, T. (1996) 'Small scale co-generation in biomass furnaces with a stirling engine', in *Proceedings of the 9th European Bioenergy Conference*, Volume 2, Elsevier Science Ltd., Oxford, UK

83 HEIDELBERGER MOTOR GMBH (1995) *Thermoelektrische Konverter*, company brochure, Heidelberger Motorenwerke G.m.b.H.(ed.), Heidelberg, Germany

84 HIGH SPEED TECH OY LTD. (1996) *Small Scale Electricity Production with BOOST Energy Converters*, company brochure, High Speed Tech Oy Ltd., Tampere, Finland

85 BINI, R. et al (1996) 'Organic Rankine cycle turbogenerators for combined heat and power production from biomass', in *Proceedings of the 3rd Munich Discussion Meeting 1996*; ZAE Bayern, Munich, Germany

86 PIATKOWSKI, R. and KAUDER, K. (1996) 'Schraubenmotor zur Wärmekraftkopplung: Stand der Technik', in *Tagungsband zum 4. Holzenergie-Symposium*, October, Zürich; Bundesamt für Energiewirtschaft, Bern, Switzerland

87 OBERNBERGER, I. and HAMMERSCHMID, A. (1999) *Dezentrale Biomasse-Kraft-Wärme-Kopplungstechnologien – Potential, technische und wirtschaftliche Bewertung, Einsatzgebiete*, volume 4 of Thermische Biomassenutzung series, dbv-Verlag der Technischen Universität Graz, Graz, Austria

88 CARLSON, H. and BOVIN, J. (2000) 'Biofuel Stirling engines for CHP', in *Proceeding of the 1st World Conference and Exhibitions on Biomass for Energy and Industry*, June, Sevilla, Spain
89 SENGSCHMIED, F. (1995) Ein Beitrag zur Entwicklung einer druckbeaufschlagten Brennkammer für die zweistufige Verbrennung von Holzstaub, PhD thesis, Vienna University of Technology, Vienna, Austria
90 DE RUYCK, J., ALLARD, G. and MANIATIS, K. (1996) 'An externally fired evaporative gas turbine cycle for small scale biomass gasification', in *Proceedings of the International Conference 'Developments in Thermochemical Biomass Conversion'*, May, Blackie, Chapman and Hall Group, Banff, Canada
91 FRANKE, B., BÖRNER, R., BLUM, B. and BIZAJ, B. (2000) 'Konzeptionelle Gestaltung von GuD-Heizkraftwerken mit integrierter Holzvergasung am Beispiel der Projekte Siebenlehn und Elsterwerda', in *Proceedings of the Conference 'Holzvergasung – Teil der Strategie zur CO_2-Minderung'*, April, Fördergesellschaft Erneuerbare Energien e.V., Berlin, Germany
92 MOSER, W., FRIEDL, G. and HASLINGER, W. (2006) 'Small-scale pellet boiler with thermoelectric generator', in *Proceedings of the 2nd World Conference on Pellets*, May/June, Swedish Bioenergy Association, Stockholm, SwedenJönköping, Sweden, pp85–86
93 STANZEL, K. W. (2006) 'Strom und Wärme aus Pellets für Haushalte', in *Proceedings of the European Pellets Forum 2006*, O.Ö. Energiesparverband, Linz, Austria
94 OBERNBERGER, I., HAMMERSCHMID, A. and BINI, R. (2001) 'Biomasse-Kraft-Wärme-Kopplungen auf Basis des ORC-Prozesses – EU-THERMIE-Projekt Admont (A)', in *Tagungsband zur VDI-Tagung 'Thermische Nutzung von fester Biomasse'*, Salzburg, May, VDI Bericht 1588, VDI-Gesellschaft Energietechnik, Düsseldorf, Germany, pp283–302
95 OBERNBERGER, I., THONHOFER, P. and REISENHOFER, E. (2002) 'Description and evaluation of the new 1,000 kWel Organic Rankine Cycle process integrated in the biomass CHP plant in Lienz, Austria', in *Euroheat and Power*, vol 10, pp18–25
96 BIEDERMANN, F., CARLSEN, H., OBERNBERGER, I., SCHÖCH, M. (2004) 'Small-scale CHP plant based on a 75 kWel Hermetic Eight Cylinder Stirling Engine for biomass fuels – development, technology and operating experiences', in *Proceedings of the 2nd World Conference and Exhibition on Biomass for Energy, Industry and Climate Protection*, May, Rome, Italy, Volume II, ETA-Florence, Italy, pp1722–1725,
97 HAMMERSCHMID, A., STALLINGER, A., OBERNBERGER, I. and PIATKOWSKI, R. (2004) 'Demonstration and evaluation of an innovative small-scale biomass CHP module based on a 730 kWel screw-type steam engine', in *Proceedings of the 2nd World Conference and Exhibition on Biomass for Energy, Industry and Climate Protection*, May, Rome, Italy, Volume II, ETA-Florence, Italy, pp1745–1748
98 OBERNBERGER, I., GAIA, M. and BIEDERMANN, F. (2005) 'Biomasse-Kraft-Wärme-Kopplung auf Basis des ORC-Prozesses – Stand der Technik und Möglichkeiten der Prozessoptimierung', in: *Tagungsband zur internat. Konferenz 'Strom und Wärme aus biogenen Festbrennstoffen'*, June, Salzburg, Austria, VDI-Bericht Nr. 1891, VDI-Verlag GmbH Düsseldorf, Germany, pp131–148
99 KOSTAMO, J. A. (2000) 'Co-firing of sawdust in a Coal-fired Utility Boiler', *IFRF Combustion Journal*, Article Number 200001, January
100 BAXTER, L. and ROBINSON, A. (1999) 'Key issues when co-firing biomass with coal', in *Proceedings of the Sixteenth Annual International Pittsburgh Coal Conference*, 11–15 October
101 TILLMAN, D., HUGHES, E. and PLASYNSKI, S. (1999) 'Commercializing biomass-coal firing: The process, status and prospects', in *Proceedings of the Sixteenth Annual International Pittsburgh Coal Conference*, 11–15 October
102 HEIN, K. R. G. and SPLIETHOFF, H. (1999) 'Co-combustion of coal and biomass in pulverised fuel and fluidised bed', in *Proceedings of the Sixteenth Annual International Pittsburgh Coal Conference*, 11–15 October
103 WIECK-HANSEN, K. (1999) 'Co firing coal and straw in PF boilers – performance impact of straw with emphasis on SCR catalyst for $deNO_x$ catalysts', in *Proceedings of the Sixteenth Annual International Pittsburgh Coal Conference*, 11–15 October
104 RASMUSSEN, I. (1996) The ELSAM development of power plants for combined coal and biomass firing, Power-Gen Europe '96

105 VAN LOO, S., BABU, S. and BAXTER, L. (2002) Workshop on biomass cofiring, organized by IEA Bioenergy Task 32 (Biomass Combustion and Cofiring) and Task 33 (Biomass Gasification) as part of the 12th European Conference and Exhibition on Biomass for Energy and Industry, June, Amsterdam, the Netherlands, available at www.ieabcc.nl (accessed 2 August 2006)

106 JUNKER H. et al (2002) 'Agricultural residues for power production', in *Proceedings of the 12th European Biomass Conference*, Volume I, ETA, Florence, Italy, pp552–555

107 BAXTER, L. (2005) 'Biomass-coal co-combustion: Opportunity for affordable renewable energy', *Fuel*, vol 84, pp1295–1302

108 BAXTER L., and KOPPEJAN, J. (2004) 'Biomass-coal co-combustion: Opportunity for affordable renewable energy', *EuroHeat & Power*, vol 1, available at http://www.ieabcc.nl (accessed 2 August 2006)

109 JENSEN, J. P., NIELSEN, K., MONTGOMERY, M. and ANDERSSON, C. (2005) 'Ash chemistry, corrosion and heavy metal emissions from a 800MW wood pellet + oil + natural gas fired power plant', in *Proceedings of the 14th European Biomass Conference & Exhibition*, October, Paris, France, ETA-Renewable Energies, Italy, pp1019–1022

110 OVERGAARD, P. et al (2005) 'Full-scale tests on co-firing of straw in a natural gas-fired boiler', in *Proceedings of the 14th European Biomass Conference & Exhibition*, October, Paris, France, ETA-Renewable Energies, Italy, pp62–67

111 JUNKER, H. et al (2005) 'CFD simulation of coal and straw co-firing', in *Proceedings of the 14th European Biomass Conference & Exhibition*, October, Paris, France, ETA-Renewable Energies, Italy, pp1150–1153

112 BAXTER, L. (2004) 'Biomass cofiring overview', presented at the Workshop 'Biomass cofiring – Current trends and future challenges' organized by the International Energy Agency Bioenergy Agreement Task 32 Biomass Combustion and Cofiring in the framework of the 2nd World Conference and Exhibition on Biomass for Energy, Industry and Climate Protection, Amsterdam, 2005, available at www.ieabcc.nl (accessed 2 August 2006)

113 SANDER, B. (2004) 'Full-scale investigations of straw co-firing', presented at the Workshop 'Biomass cofiring – Current trends and future Challenges' organized by the International Energy Agency Bioenergy Agreement Task 32 Biomass Combustion and Cofiring in the framework of the 2nd World Conference and Exhibition on Biomass for Energy, Industry and Climate Protection, Amsterdam, 2005, available at www.ieabcc.nl (accessed 2 August 2006)

114 MEIJER, R. (2004) 'Biomass cofiring – status in the Netherlands', Presented at the Workshop 'Biomass cofiring – Current trends and future Challenges' organized by the International Energy Agency Bioenergy Agreement Task 32 Biomass Combustion and Cofiring in the framework of the 2nd World Conference and Exhibition on Biomass for Energy, Industry and Climate Protection, Amsterdam, 2005, available at www.ieabcc.nl (accessed 2 August 2006)

115 LIVINGSTON W. R. (2004) 'The current status of biomass co-firing at coal-fired power stations in Britain', presented at the Workshop 'Biomass cofiring – Current trends and future Challenges' organized by the International Energy Agency Bioenergy Agreement Task 32 Biomass Combustion and Cofiring in the framework of the 2nd World Conference and Exhibition on Biomass for Energy, Industry and Climate Protection, Amsterdam, 2005, available at www.ieabcc.nl (accessed 2 August 2006)

116 JIRJIS, R. and LEHTIKANGAS, P. (1992) *Long Term Storage of Mixed Bark-Shavings Fuel*, report No. 230, ISSN 0348-4599, Swedish University of Agricultural Sciences, Uppsala, Sweden

117 NELLIST, M., LAMOND, W., PRINGLE, R. and BURFOOT, D. (1993) *Storage and Drying of Comminuted Forest Residues*, research report ETSU B/W1/00146/REP/A, ETSU, Harwell/Didcot, UK

118 VINTERBAECK, C. J. M. (1997) 'Economics and logistics for upgraded biofuels with experience from the Swedish market', in *Proceedings of the 9th European Bioenergy Conference*, Volume 2, Elsevier Science Ltd, Oxford, UK

119 STYRIAN CHAMBER FOR AGRICULTURE AND FORESTRY, (1997) *Proceedings of the 'International Pellet Workshop'*, February, Graz, Styrian Chamber for Agriculture and Forestry, Graz, Austria

120 SANDER B. (1996) 'Properties of Danish biofuels and the requirements for power production', in *Biomass and Bioenergy*, vol 12, pp177–183

121 LEWANDOWSKY, I. (1996) *Einflußmöglichkeiten der Pflanzenproduktion auf die Brennstoffeigenschaften am Beispiel von Gräsern* volume 6 of Nachwachsende Rohstoffe series, Landwirtschaftsverlag Münster, Germany
122 HARTMANN, H. (1996) 'Characterisation and properties of biomass fuels – optimisation of biomass quality by agricultural management', in *Proceedings of the Biomass Summer School 1996*, Fehring, Austria, Institute for Chemical Engineering, Graz University of Technology, Austria
123 JØRGENSEN, U. (1997) 'Genotypic variation in dry matter accumulation and content of N, K and Cl in Miscanthus in Denmark', in *Biomass and Bioenergy*, vol 12, no 3
124 OBERNBERGER, I., and THEK, G. (2006) *Herstellung und Nutzung von Pellets*, Volume 5 of Thermal Biomass Utilisation series, Institute for Resource Efficient and Sustainable Systems, Graz University of Technology, Graz, Austria (in press)
125 HASLINGER, W., EDER, G. and WÖRGETTER, M. (2004) 'Improvement of straw pellet fuel quality through additives', in proceedings of the *2nd World Conference and Exhibition on Biomass for Energy, Industry and Climate Protection*, 10–14 May, Rome, Italy
126 HASLINGER, W., EDER, G. and WÖRGETTER, M. (2004) 'Straw pellets for small-scale boilers', 1st European Pellets Conference, March, Wels, Austria
127 THEK, G. and OBERNBERGER, I. (2002) 'Wood pellet production costs under Austrian and in comparison to Swedish framework conditions', in *Proceedings of the 1st World Conference on Pellets*, September, Stockholm, Sweden, Swedish Bioenergy Association, Stockholm, Sweden, pp123–128
128 BRUNNER, T., OBERNBERGER, I. and WELLACHER, M. (2004) 'Waste wood processing in order to improve its quality for biomass combustion', in *Proceedings of the 2nd World Conference and Exhibition on Biomass for Energy, Industry and Climate Protection*, May, Rome, Italy, Volume I, ETA-Florence, Italy, pp250–253
129 RINKE, G. (2005) 'Pelletsfertigung aus Waldfrischholz – eine technische Herausforderung', in *Proceedings of the European Pellets Conference 2005*, O. Ö. Energiesparverband, Linz, Austria
130 STOLARSKI, M., SZCZUKOWSKI, S. and TWORKOWSKI, J. (2005) 'Pellets production from short rotation forestry', in *Proceedings of the European Pellets Conference 2005*, O. Ö. Energiesparverband, Linz, Austria
131 VASEN, N. N. (2005) 'Agri-pellets – Perspectives of pellets from agricultural residues', in *Proceedings of the European Pellets Conference 2005*, O. Ö. Energiesparverband, Linz, Austria
132 MARCHAL, D., RYCKMANS, Y. and JOSSART, J. M. (2004) 'Fossil CO_2 emissions and strategies to develop pellet's chain in Belgium', in *Proceedings of European Pellets Conference 2004*, Wels (Austria), 3–4 March, pp346
133 SCHOLZ, V., IDLER, C., DARIES, W. and EGERT, J. (2005) 'Lagerung von Feldholzhackgut, Verluste und Schimmelpilze', *Agrartechnische Forschung 11*, Heft 4, pp100–113
134 BERGMAN, P. C. A., BOERSMA, A. R. and KIEL, J. H. A. (2004) 'Torrefaction for entrained-flow gasification of biomass', in *Proceedings of the 2nd World Conference and Exhibition on Biomass for Energy, Industry and Climate Protection*, May, Rome, Italy, Volume I, ETA-Florence, Italy, pp679–682
135 ZANZI, R. et al (2004) 'Biomass torrefaction'. in *Proceedings of the 2nd World Conference and Exhibition on Biomass for Energy, Industry and Climate Protection*, May, Rome, Italy, Volume I, ETA-Florence, Italy, pp.859–862
136 BERGMAN, P. C. A. and KIEL, J. H. A. (2005) 'Torrefaction for biomass upgrading', in *Proceedings of the 14th European Biomass Conference & Exhibition*, October, Paris, France, ETA-Renewable Energies, Italy, pp.206–209
137 HENRICH, E., DINJUS, E., STAHL, R. and WEIRICH, F. (2005) 'Biomass fast pyrolysis in a twin screw mixer reactor', in *Proceedings of the 14th European Biomass Conference & Exhibition*, October, Paris, France, ETA-Renewable Energies, Italy, pp604–607
138 ZAKARIA, R. et al (2000) 'Fundamental aspects of emissions from the burning bed in a municipal solid waste incinerator', in *Proceedings of the 5th International Conference on Industrial Furnaces and Boilers*, April, Porto, INFUB, Rio Tinto, Portugal
139 HUPA, M., KILPINEN, P., KALLIO, S., KONTTINEN, J., SKRIFVARS, B. J., ZEVENHOVEN, M. and BACKMAN, R. (2000) 'Predicting the performance of different fuels in fluidised bed

combustion', in *Proceedings of the 5th International Conference on Industrial Furnaces and Boilers*, April, Porto, INFUB, Rio Tinto, Portugal

140 DAHL, J. (2000) Chemistry and Behaviour of Environmentally Relevant Heavy Metals in Biomass Combustion and Thermal Ash Treatment Processes, PhD thesis, Institute for Chemical Engineering, Graz University of Technology, Austria

141 BRINK, A., KILPINEN, P., HUPA, M., KJÄLDMAN, L. and JÄÄSKELÄINEN, K. (1995) *Fuel Chemistry for Computational Fluid Dynamics*, 2nd Colloquium on Process Simulation, Espoo, Finland, 6–8 June

142 SCHARLER, R. and OBERNBERGER, I. (2000) 'Numerical modelling of biomass grate furnaces', in *Proceedings of the 5th International Conference of Industrial Furnaces and Boilers*, April, Porto, INFUB, Rio Tinto, Portugal

143 GRISELIN, N., BAI, X. S. and FUCHS, L. (1999) 'Tracking of particles in a biomass furnace', in *Proceedings of the 2nd Olle Lindström Symposium on Renewable Energy, Bioenergy*, Royal Institute of Technology, Stockholm, Sweden

144 MAGEL, H. C. (1997) *Simulation chemischer Reaktionskinetik in turbulenten Flammen mit detaillierten und globalen Mechanismen*, VDI-Fortschrittsberichte, Reihe 6: Energieerzeugung, No 377

145 SCHARLER, R. and OBERNBERGER, I. (2002) 'Deriving guidelines for the design of biomass grate furnaces with CFD analysis – a new Multifuel-Low-NO_x furnace as example', in *Proceedings of the 6th European Conference on Industrial Furnaces and Boilers*, April, Estoril, Portugal, INFUB, Rio Tinto, Portugal

146 SCHMIDT, U., REXROTH, C.-H., SCHARLER, R., CREMER, I. (2004) 'New trends in combustion simulation', in *Proceedings of the 2nd (International) SBF568-Workshop in Collaboration with ERCOFTAC Group SIG 28 (Ed.) and DGLR group 'Propulsion' on 'Trends in Numerical and Physical Modelling of Turbulent Combustion in Gas Turbine Combustors'*, April, Heidelberg, Germany.

147 SCHARLER, R., FORSTNER, M., BRAUN, M., BRUNNER, T. and OBERNBERGER, I. (2004) 'Advanced CFD analysis of large fixed bed biomass boilers with special focus on the convective section', in *Proceedings of the 2nd World Conference and Exhibition on Biomass for Energy, Industry and Climate Protection*, May, Rome, Italy, Volume II, ETA-Florence, Italy, pp1357–1360

148 SCHARLER, R., OBERNBERGER, I., WEISSINGER, A. and SCHMIDT, W. (2005) 'CFD-gestützte Entwicklung von Pellet- und Hackgutfeuerungen für den kleinen und mittleren Leistungsbereich', in *Brennstoff-Wärme-Kraft (BWK)*, vol 57, no 7/8, pp55–58

149 SCHARLER, R., JOELLER, M., HOFMEISTER, G., BRAUN, M., KLEDITZSCH, S. and OBERNBERGER, I. (2005) 'Depositionsmodellierung in Biomasse-befeuerten Kesseln mittels CFD', in *Tagungsband zum 22. Deutschen Flammentag, Braunschweig*, September, VDI-Bericht No 1888VDI-Verlag Düsseldorf, Germany, pp493–503

150 ZAHIROVIC, S., SCHARLER, R. and OBERNBERGER, I. (2006) 'Advanced CFD modelling of pulverised biomass combustion', in *Proceedings of the International Conference on Science in Thermal and Chemical Biomass Conversion*, September, CPL Press, Victoria, Canada, pp267–283

151 SCHARLER, R., WIDMANN, E. and OBERNBERGER, I. (2006) 'CFD modelling of NO_x formation in biomass grate furnaces with detailed chemistry', in *Proceedings of the International Conference on Science in Thermal and Chemical Biomass Conversion*, September, Victoria, Canada, CPL Press, pp284–300

152 SKREIBERG, Ø., BECIDAN, M. and HUSTAD, J. E. (2006) 'Detailed chemical kinetics of NO_x reduction by staged air combustion at moderate temperatures', in *Proceedings of the International Conference on Science in Thermal and Chemical Biomass Conversion*, September, CPL Press, Victoria, Canada, pp40–54

153 YIN, C., ROSENDAHL, L., KAER, S. K. and SORENSEN, H. (2003) 'Modelling of the motion of cylindrical particles in a non-uniform flow', *Chemical Engineering Science*, vol 58, pp2931–2944

154 KAER, S. K., ROSENDAHL, L. A. and BAXTER, L. L. (2006) 'Towards a CFD-based mechanistic deposit formation model for straw-fired boilers', in *Fuel*, vol 85, pp833–848

155 BRYDEN, K. M. (1998) Computational Modeling of Wood Combustion, PhD thesis, Mechanical Engineeering, University of Wisconsin-Madison

156 BRYDEN, K. M. and RAGLAND, K. W. (1996) 'Numerical modelling of a deep fixed-bed combustor', *Energy & Fuels*, vol 10/12, pp269–275

157 WU, Ch. (2006) Fuel-NO_x Formation during Low-grade Fuel Combustion in a Swirling Burner, PhD thesis, Department of Chemical Engineering, Brigham Young University

158 BRINK, A., HUPA, M., KURKELA, E. and SUOMALAINEN, M. (2004) 'Minimizing NO_x emission from a waste derived fuel gasifier gas combustor using CFD combined with detailed chemistry', *Proceedings of 14th IFRF Members Conference*, Noordwijkerhout, the Netherlands

159 GHIRELLI, F. and LECKNER, B. (2004) 'Transport equation for the local residence time of a fluid', *Chemical Engineering Science*, vol 59, pp513–523

160 POPE, S. B. (1997) 'Computationally efficient implementation of combustion chemistry using in situ adaptive tabulation', *Combustion Theory and Modeling*, vol 1, pp41–63

161 GLARBORG, P., JENSEN, A. D. and JOHNSSON, J. E. (2003) 'Fuel nitrogen conversion in solid fuel fired systems', *Progress in Energy and Combustion Science*, vol 29, pp89–113

162 KILPINEN, P., LJUNG, M., BOSTRÖM, S. and HUPA, M. (eds.) (2000) *Final Report of the Tulisija Programme 1997–1999*, (in Finnish), Report No 1-2000, Åbo Akademi University, Tulisija Coordination. 183pp

163 GALGANO, A. and DI BLASI, C. (2003) 'Modeling wood degradation by the unreacted-core-shrinking approximation', *Industrial & Engineering Chemistry Research*, vol 42, pp2101–2111

164 GALGANO, A. and DI BLASI, C. (2004) 'Modeling the propagation of drying and decomposition fronts in wood', *Combustion and Flame*, vol 139, pp16–27

165 KNAUS, H., RICHTER, S., UNTERBERGER, S., SCHNELL, U., MAIER, H. and HEIN, K. R. G. (2000) 'On the application of different turbulence models for the computation of fluid flow and combustion processes in small scale wood heaters', *Experimental Thermal and Fluid Science*, vol 21, pp99–108

166 BRINK, A. (1998) Eddy Break-up Based Models for Industrial Diffusion Flames with Complex Gas Phase Chemistry, PhD thesis, Abo Akademi University, Turku, Finland

167 KÆR S. K. (2001) Numerical Investigation of Ash Deposition in Straw-fired Boilers, PhD thesis, Aalborg University, Aalborg, Denmark

168 MUELLER, C., SKRIFVARS, B. J., BACKMAN, R. and HUPA, M. (2003) 'Ash deposition prediction in biomass fired fluidised bed boilers – combination of CFD and advanced fuel analysis', *Progress in Computational Fluid Dynamics*, vol 3, pp112–120

169 PYYKÖNEN, J. and JOKINIEMI, J. (2003) 'Modelling alkali chloride superheater deposition and its implications', *Fuel Processing Technology*, vol 80, pp225–262

170 DI BLASI, C. (1993) 'Modelling and simulation of combustion processes of charring and non-charring solid fuels', *Progress in Energy and Combustion Science*, vol 19, pp71–104

171 GRÖNLI, M. (1996) A Theoretical and Experimental Study of the Thermal Degradation of Biomass, PhD thesis, NTNU, Trondheim, Norway

172 LINDSJÖ, H., BAI, X. S. and FUCHS, L. (2001) 'Numerical and experimental studies NO_x emissions in biomass furnace', *International Journal on Environmental Combustion Technologies*, vol 2, pp93–113

173 SHIN, D. and CHOI, S. (2000) 'The combustion of simulated waste particle in a fixed bed', *Combustion and Flame*, vol 121, pp167–180

174 PETERS, B. (2002) 'Measurements and application of a discrete particle model (DPM) to simulate combustion of a packed bed of individual fuel particles', *Combustion and Flame*, vol 131, pp132–146

175 BRUCH, C., PETERS, B. and NUSSBAUMER, T. (2003) 'Modelling wood combustion under fixed bed conditions', *Fuel*, vol 82, pp729–738

176 YANG, Y. B., YAMAUCHI, H., NASSERZADEH, V. and SWITHENBANK, J. (2003) 'Effects of fuel devolatilisation on the combustion of wood chips and incineration of simulated municipal solid wastes in a packed bed', *Fuel*, vol 82, pp2205–2221

177 YANG, Y. B., CHANGKOOK, R., KHOR, A., SHARIFI, V. N. and SWITHENBANK, J. (2005) 'Fuel size effect on pinewood combustion in a packed bed', *Fuel*, vol 84, pp2026–2038

178 THUNMAN, H., LECKNER, B., NIKLASSON, F. and JOHNSON, F. (2000) 'Combustion of wood particles – a particle model for eulerian calculations', *Combustion and Flame*, vol. 189, pp30–46

179 THUNMAN, H. (2001) Principles and models of solid fuel combustion, PhD thesis, Chalmers University of Technology, Göteborg, Sweden
180 ELFASAKHANY, A. (2005) Modeling of pulverised wood flames, PhD thesis, Lund Institute of Technology, Lund, Sweden
181 ALBRECHT, B. A., BASTIAANS, R. J. M., VAN OIJEN, J. A. and DE GOEY, L. P. H. (2006) 'NO_x emissions modelling in biomass combustion grate furnaces', in *Proceedings of the 7th European Conference on Industrial Furnaces and Boilers*, April, Porto, Portugal
182 BACKREEDY, R. I., FLETCHER, L. M., JONES, J. M., MA, L., POURKASHANIAN, M. and WILLIAMS, A. (2005) 'Co-firing pulverised coal and biomass: A modelling approach', in *Proceedings of the Combustion Institute*, vol 30, pp2955–2964
183 WILLIAMS, A., POURKASHANIAN, M. and JONES, J. M. (2001) 'Combustion of pulverised coal and biomass', *Progress in Energy and Combustion Science*, vol 27, pp587–610
184 OBERNBERGER, I. and SCHARLER, R. (eds.) (2006) 'CFD modelling of biomass combustion systems', Special issue of the international journal *Progress in Computational Fluid Dynamics*, vol 6, nos 4/5

Annex 1

Mass Balance Equations and Emission Calculation

Input values

X_C	– Mass fraction of carbon in dry fuel
Y_C	– Volume fraction of carbon in dry fuel
Y_H	– Volume fraction of hydrogen in dry fuel
Y_N	– Volume fraction of nitrogen in dry fuel
Y_S	– Volume fraction of sulphur in dry fuel
$X_{H_2O,w}$	– Weight fraction of H_2O in wet fuel
CO_2	– Weighted average CO_2 in dry flue gas (vol%)
CO	– Weighted average CO in dry flue gas (vol%)
NO	– Weighted average NO in dry flue gas (vol%)
NO_2	– Weighted average NO_2 in dry flue gas (vol%)
C_kH_l	– Weighted average C_kH_l in dry flue gas (vol%)
N_2O	– Weighted average N_2O in dry flue gas (vol%)
$m^{F,w}$	– Fuel consumption (kg wet fuel/h)

Constants

$aY_{O_2, air}$	– Volume fraction of O_2 in air
V_{Mole}	– Mole volume (Nm³/kmole)
M_C	– Mole weight for carbon (kg/kmole)
M_H	– Mole weight for hydrogen (kg/kmole)
M_N	– Mole weight for nitrogen (kg/kmole)
M_S	– Mole weight for sulphur (kg/kmole)
M_O	– Mole weight for oxygen (kg/kmole)

Combustion equation

$$a \cdot (Y_C Y_H Y_O Y_N Y_S) + n \cdot \left(\frac{1 - Y_{O_2,Air}}{Y_{O_2,Air}} \cdot N_2 + O_2 \right) \quad \text{(AI.1)}$$

$$\Rightarrow b \cdot CO_2 + c \cdot H_2O + d \cdot O_2 + e \cdot N_2 + f \cdot CO + g \cdot NO + h \cdot NO_2 + i \cdot C_k H_l + j \cdot SO_2 + m \cdot N_2O$$

Dry flue gas balance:

$$e = 100 - b - d - f - g - h - i - j - m \quad \text{(AI.2)}$$

Carbon balance:
$$a = \frac{b + f + k \cdot i}{Y_C} \tag{AI.3}$$

Hydrogen balance:
$$c = a \cdot \frac{Y_H}{2} - \frac{l \cdot i}{2} \tag{AI.4}$$

Nitrogen balance:
$$n = \frac{Y_{O_2,Air}}{1 - Y_{O_2,Air}} \cdot \left(e + \frac{g}{2} + \frac{h}{2} + m - a \cdot \frac{Y_N}{2} \right) \tag{AI.5}$$

Sulphur balance:
$$j = a \cdot Y_S \tag{AI.6}$$

Oxygen balance:
$$a \cdot Y_O + 2 \cdot \frac{Y_{O_2,Air}}{1 - Y_{O_2,Air}} \cdot \left(100 - b - f - \frac{g}{2} - \frac{h}{2} - i - j - a \cdot \frac{Y_N}{2} \right) \tag{AI.7}$$

$$\rightarrow d = \frac{-2 \cdot b - c - f - g - 2 \cdot h - 2 \cdot j - m}{2 \cdot \left(1 + \frac{Y_{O_2,Air}}{1 - Y_{O_2,Air}} \right)}$$

Combustion equation at stoichiometric conditions

$$a_S \cdot (Y_C Y_H Y_O Y_N Y_S) + n_S \cdot \left(\frac{1 - Y_{O_2,Air}}{Y_{O_2,Air}} \cdot N_2 + O_2 \right) \tag{AI.8}$$

$$\Rightarrow b_S \cdot CO_2 + c_S \cdot H_2O + e_S \cdot N_2 + j_S \cdot SO_2$$

Dry flue gas balance:
$$e_S = 100 - b_S - j_S \tag{AI.9}$$

Carbon balance:
$$b_S = a_s \cdot Y_C \tag{AI.10}$$

Hydrogen balance:
$$c_S = a_S \cdot \frac{Y_H}{2} \tag{AI.11}$$

Nitrogen balance:

$$n_S = \frac{Y_{O_2,Air}}{1-Y_{O_2,Air}} \cdot \left(e_S - \frac{a_S \cdot Y_N}{2}\right) \quad (AI.12)$$

Sulphur balance:

$$j_S = a_S \cdot Y_S \quad (AI.13)$$

Oxygen balance:

$$a_S = \frac{2 \cdot \dfrac{Y_{O_2,Air}}{1-Y_{O_2,Air}} \cdot 100}{2 \cdot \dfrac{Y_{O_2,Air}}{1-Y_{O_2,Air}} \cdot \left(Y_C + \dfrac{Y_N}{2} + Y_S\right) + 2 \cdot Y_C + \dfrac{Y_H}{2} + 2 \cdot Y_S - Y_O} \quad (AI.14)$$

Eq. AI.14 calculations

Volume fraction of oxygen in dry fuel:

$$Y_O = 1 - Y_C - Y_H - Y_N - Y_S \quad (AI.15)$$

O_2 in dry flue gas (vol%):

$$O_2 = d \quad (AI.16)$$

N_2 in dry flue gas (vol%):

$$N_2 = e \quad (AI.17)$$

Stoichiometric CO_2 in dry flue gas (vol%):

$$\dot{V}_{Air}\left[\frac{Nm^3}{h}\right] = \frac{\dot{m}_{Air}}{\rho_{Air}} \quad (AI.18)$$

Volume dry flue gas produced (Nm³/h):

$$V_{FG} = \frac{100 \cdot X_C \cdot m_{F,w} \cdot (1 - X_{H_2O,w}) \cdot V_{Mole}}{M_C \cdot (CO_2 + CO + k \cdot C_k H_l)} \quad (AI.19)$$

Volume dry flue gas produced at stoichiometric conditions (Nm³/h):

$$V_{FG,S} = \frac{100 \cdot X_C \cdot m_{F,w} \cdot (1 - X_{H_2O,w}) \cdot V_{Mole}}{M_C \cdot CO_{2,Max}} \quad (AI.20)$$

Volume air added (Nm³/h):

$$V_{Air} = \frac{V_{FG} \cdot n \cdot \left(\frac{1-Y_{O_2,Air}}{Y_{O_2,Air}}+1\right)}{100} \quad (AI.21)$$

Volume air added at stoichiometric conditions (Nm³/h):

$$V_{Air,S} = \frac{V_{FG,S} \cdot n_S \cdot \left(\frac{1-Y_{O_2,Air}}{Y_{O_2,Air}}+1\right)}{100} \quad (AI.22)$$

Excess air ratio:

$$\lambda = \frac{V_{Air}}{V_{Air,S}} \quad (AI.23)$$

Volume of water from hydrogen in fuel (Nm³/h):

$$V_{H_2O,H} = m_{F,w} \cdot (1-X_{H_2O,w}) \cdot \frac{X_H}{2} \cdot \frac{V_{Mole}}{M_H} \cdot \left(1-\frac{i \cdot l}{a \cdot Y_H}\right) \quad (AI.24)$$

Volume of water from water in fuel (Nm³/h):

$$V_{H_2O,E} = X_{H_2O,w} \cdot m_{F,w} \cdot \frac{V_{Mole}}{M_{H_2O}} \quad (AI.25)$$

Total flue gas volume (Nm³/h):

$$V_{FG,w} = V_{FG} + V_{H_2O,H} + V_{H_2O,E} \quad (AI.26)$$

Weight of air added (kg/h):

$$m_{Air} = V_{Air} \cdot \left(Y_{O_2,Air} \cdot \frac{M_{O_2}}{V_{Mole}} + (1-Y_{O_2,Air}) \cdot \frac{M_{N_2}}{V_{Mole}}\right) \quad (AI.27)$$

Weight of total flue gas produced (kg/h):

$$m_{FG,w} = m_{Air} + m_{F,w} \quad (AI.28)$$

Weight of H₂O produced (kg/h):

$$m_{H_2O,H} = V_{H_2O,H} \cdot \frac{M_{H_2O}}{V_{Mole}} \quad (AI.29)$$

Weight of evaporated water (kg/h):

$$m_{H_2O,E} = X_{H_2O,w} \cdot m_{F,w} \quad (AI.30)$$

Volume of CO_2, CO, O_2, N_2, NO, NO_2, C_kH_l, SO_2 and N_2O (Nm³/h):

$$V_i = V_{FG} \cdot \frac{i}{100} \qquad (AI.31)$$

Volume of H_2O (Nm³/h):

$$V_{H_2O} = V_{H_2O,H} + V_{H_2O,E} \qquad (AI.32)$$

Weight of CO_2, CO, O_2, N_2, NO, NO_2, C_kH_l, SO_2, N_2O and H_2O (kg/h):

$$m_i = V_i \cdot \frac{M_i}{V_{Mole}} \qquad (AI.33)$$

Emissions of NO_x as NO_2 equivalents (mg/kg dry fuel):

$$E_{NO_X} = \frac{m_{NO} \cdot \frac{M_{NO_2}}{M_{NO}} + m_{NO_2}}{m_{F,w} \cdot (1 - X_{H_2O,w})} \cdot 10^6 \qquad (AI.34)$$

Emissions of C_kH_l as CH_4 equivalents (mg/kg dry fuel):

$$E_{C_kH_l} = \frac{m_{C_kH_l} \cdot \frac{k \cdot M_{CH_4}}{M_{C_kH_l}}}{m_{F,w} \cdot (1 - X_{H_2O,w})} \cdot 10^6 \qquad (AI.35)$$

Emissions of CO, SO_2 and N_2O (mg/kg dry fuel):

$$E_i = \frac{m_i}{m_{F,w} \cdot (1 - X_{H_2O,w})} \cdot 10^6 \qquad (AI.36)$$

Conversion factor for fuel – N to NO:

$$\frac{NO}{Fuel-N} = \frac{m_{NO} \cdot M_N}{M_{NO} \cdot X_N \cdot m_{F,w} \cdot (1 - X_{H_2O,w})} \qquad (AI.37)$$

Conversion factor for fuel – N to NO_2:

$$\frac{NO_2}{Fuel-N} = \frac{m_{NO_2} \cdot M_N}{M_{NO_2} \cdot X_N \cdot m_{F,w} \cdot (1 - X_{H_2O,w})} \qquad (AI.38)$$

Conversion factor for fuel – N to N_2O:

$$\frac{N_2O}{Fuel-N} = \frac{m_{N_2O} \cdot 2 \cdot M_N}{M_{N_2O} \cdot X_N \cdot m_{F,w} \cdot (1 - X_{H_2O,w})} \qquad (AI.39)$$

Conversion factor for fuel – N to NO_x:

$$\frac{NO_X}{Fuel-N} = \frac{NO}{Fuel-N} + \frac{NO_2}{Fuel-N} \qquad (AI.40)$$

Annex 2

Abbreviations

λ	combustion air ratio
ACAA	American Coal Ash Association
AI	abrasive index
Al	aluminium
As	arsenic
B	boron
BP	benzopyrene
BAT	best available technology
BEES	Besluit Emissie Eisen Stookinstallaties (Dutch emission legislation for large combustion units)
BFB	bubbling fluidized bed
BLA	Besluit Luchtemissie Afvalverbrandingsinstallaties (Dutch decree for airborne emissions from waste incineration units)
Bq	bequerel
C	carbon
Ca	calcium
$CaCO_3$	calcium carbonate
CaO	calcium oxide
CCA	chromium-cadmium-arsenic
CCSD	Collaborative Research Centre for Coal in Sustainable Development
CCSEM	computer controlled scanning electron microscopy
Cd	cadmium
CEN	European Committee for Standardization
CFB	circulating fluidized bed
CFD	computational fluid dynamics
CH_4	methane
CHP	combined heat and power
Cl	chlorine or chloride
Co	cobalt
CO_2	carbon dioxide
$C_{org.}$	organically bound carbon
Cr	chromium
CRC	Co-operative Research Centre (Australia)
CSA	Canadian Standards Association
CSIRO	Commonwealth Scientific and Industrial Research Organisation in Australia
Cu	copper
C_xH_y	unburned hydrocarbons
daf	dry ash-free fuel
d.b.	dry basis

DOE	US Department of Energy
DW	demolition wood
EC	European Commission
ECN	Energy Research Foundation (the Netherlands)
ECOBA	European Coal Combustion Products Association
EDX	energy-dispersive X-ray spectroscopy
EE	DOE's Offices of Energy Efficiency
ElWOG	Austrian Electricity Act
ESP	electrostatic precipitator
EU	European Union
FAO	Food and Agriculture Organization (UN)
FBC	fluidized bed combustion
FE	DOE's Offices of Fossil Energy
Fe	iron
Fe_2O_3	iron (III) oxide
FGD	flue gas desulphurization
FID	flame ionization detector
GCV	gross calorific value
GHG	greenhouse gas
GWP	global warming potential
H	hydrogen
H_2O	water
HCCO	a ketenyl radical
HCl	hydrochloric acid
HCN	hydrogen cyanide
hEN	harmonised European norms
Hg	mercury
HGI	Hard Grove Index
hv-coal	high-volatility coal
IC	inductively coupled plasma mass spectrometry
IFRF	International Flame Research Foundation
IEA	International Energy Agency
IGCC	integrated gasification combined cycle
IPPC	Integrated Pollution Prevention and Control
ISO	International Organization for Standardization
K	potassium
K_2O	potassium oxide
KCl	potassium chloride
KOH	potassium hydroxide
LCA	life cycle analysis
LCP	large combustion plants
LCPD	Large Combustion Plant Directive
LoI	loss of ignition
LP	linear programming
MBS	multifunctional bioenergy systems
MDF	medium density fibreboard
Mg	magnesium

MgO	magnesium oxide
MJ	megajoule
Mn	manganese
MnO	manganese oxide
Mo	molybdenum
MPC	model predictive control
MRET	Mandatory Renewable Energy Target (Australia)
MSP	multifunctional Salix production
MSW	municipal solid waste
N	nitrogen
Na	sodium
Na_2O	sodium oxide
NaCl	sodium chloride
NCV	net calorific value
NDIR	non-dispersive infrared
NER-BR	Nederlandse Emissie Richtlijn-Bijzondere Richtlijn (a special arrangement under the general Dutch emission guideline)
NETL	National Energy Technology Laboratory
NH_3	ammonia
Ni	nickel
Nm^3	normal cubic metre (at 273K, 1 bar)
NMVOC	non-methane volatile organic components
NO_x	nitrogen oxides
NSW	New South Wales
NSW-EPA	NSW Environmental Protection Agency
O	oxygen
OECD	Organisation for Economic Co-operation and Development
OGC	organic gaseous compounds
ORC	organic Rankine cycle
P	phosphorus
P_2O_5	phosphorus pentoxide
PAH	polycyclic aromatic hydrocarbons according to US-EPA (16 compounds)
Pb	lead
PC	pulverized coal
PCB	polychlorinated biphenyls
PCDD	polychlorinated dibenzo-p-dioxins
PCDF	polychlorinated dibenzofurans
PF	pulverized fuel
PFR	plug flow reactor
pg	picograms (10^{-12} grams)
PJ	petajoule (10^{15} joules)
PM	particulate matter
ppm	parts per million
PSR	perfectly stirred reactor
R&D	research and development
RDF	residue derived fuel
RE	renewable energy

RES	renewable energy source
RES-E	renewable energy source electricity
ROC	Renewable Obligation Certificate (UK)
rpm	rotations per minute
RPS	rotating particle separator
S	sulphur
SCR	selective catalytic reduction
SEM	scanning electron microscopic
Si	silicon
SiC	silicon carbide
SiO_2	silicon dioxide
SNCR	selective non-catalytic reduction
SO_x	sulphur oxides
SP	Swedish National Testing and Research Institute
SPF	seasonal performance factor
SRC	short rotation coppice
t	metric tonnes
TE	toxicity equivalent
TEQ	toxic equivalent quantity
TFN	total fixed nitrogen
TGA	thermogravimetric analysis
TQM	total quality management
TSC	two-stage combustion
UHC	unburned hydrocarbons
UN	United Nations
USC	ultra-supercritical
US-EPA	US Environmental Protection Agency
UWW	urban waste wood
V	vanadium
VOC	volatile organic components
waf	wet ash-free fuel
w.b.	wet basis
WHO	World Health Organization
WID	Waste Incineration Directive
wt%	weight per cent
Zn	zinc

Annex 3

European and National Standards or Guidelines for Solid Biofuels and Solid Biofuel Analysis

This annex provides an overview of standards for solid biofuels and biofuel analysis in European countries. It does not claim to be complete and should be updated periodically in order to keep in step with national developments.

Europe (status March 2006)

CEN/TS 14588:2003	Solid biofuels – Terminology, definitions and descriptions
CEN/TS 14774-1:2004	Solid biofuels – Methods for determination of moisture content – Oven dry method – Part 1: Total moisture – Reference method
CEN/TS 14774-2:2004	Solid biofuels – Methods for the determination of moisture content – Oven dry method – Part 2: Total moisture – Simplified method
CEN/TS 14774-3:2004	Solid biofuels – Methods for the determination of moisture content – Oven dry method – Part 3: Moisture in general analysis sample
CEN/TS 14775:2004	Solid biofuels – Method for the determination of ash content
CEN/TS 14778-1:2005	Solid biofuels – Sampling – Part 1: Methods for sampling
CEN/TS 14778-2:2005	Solid biofuels – Sampling – Part 2: Methods for sampling particulate material transported in lorries
CEN/TS 14779:2005	Solid biofuels – Sampling – Methods for preparing sampling plans and sampling certificates
CEN/TS 14780:2005	Solid biofuels – Methods for sample preparation
CEN/TS 14918:2005	Solid Biofuels – Method for the determination of calorific value
CEN/TS 14961:2005	Solid biofuels – Fuel specifications and classes
CEN/TS 15103:2005	Solid biofuels – Methods for the determination of bulk density
CEN/TS 15104:2005	Solid biofuels – Determination of total content of carbon, hydrogen and nitrogen – Instrumental methods
CEN/TS 15105:2005	Solid biofuels – Methods for determination of the water soluble content of chloride, sodium and potassium
CEN/TS 15148:2005	Solid biofuels – Method for the determination of the content of volatile matter
CEN/TS 15149-1:2006	Solid biofuels – Methods for the determination of particle size distribution – Part 1: Oscillating screen method using sieve apertures of 3.15mm and above
CEN/TS 15149-2:2006	Solid biofuels – Methods for the determination of particle size

	distribution – Part 2: Vibrating screen method using sieve apertures of 3.1mm and below
CEN/TS 15149-3:2006	Solid biofuels – Methods for the determination of particle size distribution – Part 3: Rotary screen method
CEN/TS 15150:2005	Solid biofuels – Methods for the determination of particle density
CEN/TS 15210-1:2005	Solid biofuels – Methods for the determination of mechanical durability of pellets and briquettes – Part 1: Pellets
CEN/TS 15210-2:2005	Solid biofuels – Methods for the determination of mechanical durability of pellets and briquettes – Part 2: Briquettes
CEN/TS 15234:2006	Solid biofuels – Fuel quality assurance
CEN/TS 15289:2006	Solid Biofuels – Determination of total content of sulphur and chlorine
CEN/TS 15290:2006	Solid biofuels – Determination of major elements
CEN/TS 15296:2006	Solid biofuels – Calculation of analyses to different bases
CEN/TS 15297:2006	Solid biofuels – Determination of minor elements
prCEN/TS 15370-1 (under approval)	Solid biofuels – Method for the determination of ash melting behaviour – Part 1: Characteristic temperatures method
prCEN/TS 15289 (approved)	Solid biofuels – Determination of total content of sulphur and chlorine
00335031 (under development)	Solid biofuels – Guideline for development and implementation of quality assurance for solid biofuels
00335032 (under development)	Method for the determination of the particle size distribution of disintegrated pellets
00335033 (under development)	Solid biofuels – Guide for a quality assurance system
00335034 (under development)	Solid biofuels – Method for the determination of bridging properties of particulate biofuels

Several published standards have already been adopted by different European countries as national standards (e.g. Austria, Belgium, Denmark, Germany).

Austria

ÖNORM M 7132 (1998)	Energy-economical utilization of wood and bark as fuel – Definitions and properties
ÖNORM M 7133 (1998)	Chipped wood for energetic purposes – Requirements and test specifications
ÖNORM M 7135 (2000)	Compressed wood or compressed bark in natural state – Pellets and briquettes, Requirements and test specifications
ÖNORM M 7136 (2002)	Compressed wood in natural state – Wood pellets – Quality assurance in the field of logistics of transport and storage
ÖNORM M 7137 (2003)	Compressed wood in natural state – Wood pellets – Requirements for storage of pellets at the ultimate consumer

Belgium

NBN CEN/TS 14775	Solid biofuels – Method for the determination of ash content
NBN CEN/TS 14774-1	Solid biofuels – Methods for the determination of moisture content – Oven dry method – Part 1 : Total moisture – Reference method.
NBN CEN/TS 14774-2	Solid biofuels – Methods for the determination of moisture content – Oven dry method – Part 2 : Total moisture – Simplified method
NBN CEN/TS 14774-3	Solid biofuels – Methods for the determination of moisture content – Oven dry method – Part 3 : Moisture in general analysis sample

Four more standards from the European Committee for Standardization (CEN) have not yet been transferred to the Belgian NBN system, but these standards are already applied in Belgium for ISO 17025 certification:

CEN/TS 14961	Solid biofuels – Fuel specifications and classes
CEN/TS 14918	Solid biofuels – Method for the determination of calorific value
CEN/TS 15148	Solid biofuels – Method for the determination of the content of volatile matter
CEN/TS 14780	Solid biofuels – Methods for sample preparation

Denmark

Best practice guidelines for:

- on-site measurements;
- fuel specifications and control measurements; and
- fuel characterization [1–3].

Finland

Quality assurance manuals for solid wood fuels (guidelines for sampling and determination of properties of wood and peat) [4].

Germany

DIN 51731 (1998)	Testing of solid fuels – Compressed untreated wood – Requirements and testing
DIN 51749 (1989)	Testing of solid fuels – Grill charcoal and grill charcoal briquettes – Requirements and test methods

The Netherlands

In the Netherlands, a Best Practice List has been published for biomass fuel and ash analysis [5]. The following norms are stated:

BPM 1-01	Sampling on location, derived from NEN 7300
BPM 2-01	Sample preparation in the laboratory
BPM 3-01	Chemical composition of biomass – Ultimate; C, H, N
BPM 3-02	Chemical composition of biomass – Ultimate; O
BPM 3-03	Chemical composition of biomass – Ultimate; S
BPM 3-04	Chemical composition of biomass – Ultimate; Cl
BPM 3-05	Chemical composition of biomass – Ultimate; F
BPM 3-06	Chemical composition of biomass – Major elements; Al, Si, K, Na, Ca, Mg, Fe, P, Ti
BPM 3-07	Chemical composition of biomass – Minor elements; As, Ba, Be, Cd, Co, Cr, Cu, Hg, Mo, Mn, Ni, Pb, Se, Te, V, Zn
BPM 4-01	Physical and chemical properties of biomass – Proximate; ash content
BPM 4-02	Physical and chemical properties of biomass – Proximate; moisture content
BPM 4-03	Physical and chemical properties of biomass – Proximate; volatile matter
BPM 4-04	Physical and chemical properties of biomass – Proximate; gross calorific value
BPM 4-05	Physical and chemical properties of biomass – Proximate; particle size distribution
BPM 5-01	Composition of process ash from biomass – Major elements; Al, Si, K, Na, Ca, Mg, Fe, P, Ti
BPM 5-02	Composition of process ash from biomass – Minor elements; As, Ba, Be, Cd, Co, Cr, Cu, Hg, Mo, Mn, Ni, Pb, Se, Te, V, Zn
BPM 6-01	Physical and chemical properties of ash from biomass – Leaching characteristics; column test
BPM 6-02	Physical and chemical properties of ash from biomass – Loss on ignition

Norway

NS 4414 (1997)	Firewood for domestic use
NS 3165 (1999)	Biofuel – Cylindrical pellets of pure wood – Classification and requirements
NS 3166 (1999)	Biofuel – Determination of mechanical strength of pellets
NS 3167 (1999)	Biofuel – Determination of moisture content in laboratory samples of pellets
NS 3168 (2000)	Solid biofuels – Briquettes for fuel – Classification and requirements

Poland

Polish standards are available for classification of wood chips (size classes for wood chips, sample preparation, total and analytical moisture, ash content, volatile matter content, total sulphur content, ash sulphur content, C, H and N content, net and gross calorific value, ash fusion temperatures) [7].

Russia

Russian wood fuel specification standards:
GOST 15815-83 (1985): Technological chips. Specifications
GOST 23246-78 (1979): Crushed wood. Terms and definitions
GOST 3243-88 (1990): Firewood. Specifications
Russian peat standards:
GOST 21123-85 (1986): Peat. Terms and definitions
GOST 13674-78 (1979): Peat. Acceptance rules
GOST 11303-75 (1977): Peat and products of its processing. Method of preparation of analysis sample
GOST 10538-87 (1988): Solid fuel. Methods for determination of chemical composition of ash
GOST 18132-72 (1974): Peat briquetts and semibriquetts. Method for determination of mechanical strength
GOST 10650-72 (1974): Peat. Determination of the disintegration degree
GOST 26801-86 (1987): Peat. Method for determination of ash content in deposit
GOST 9963-84 (1986): Peat bricks for heating purposes. Technical requirements
GOST 50902-96 (1997): Fuel peat for pulverised burning. Specifications
GOST 51062-97 (1998): Sod fuel peat for heating purposes. Specifications

Sweden

SS 18 71 06 (1997)   Biofuels and peat – Terminology
SS 18 71 13 (1998)   Biofuels and peat – Sampling
SS 18 71 14 (1992)   Biofuels and peat – Sample preparation
SS 18 71 16 (1995)   Sampling of solid residues from heating plants
SS 18 71 20 (1998)   Biofuels and peat – Fuel pellets – Classification
SS 18 71 23 (1998)   Biofuels and peat – Fuel briquettes – Classification
SS 18 71 70 (1997)   Biofuels and peat – Determination of total moisture content
SS 18 71 71 (1984)   Biofuels – Determination of ash content
SS 18 71 73 (1986)   Biofuels – Calculation of analyses to different basis
SS 18 71 74 (1990)   Biofuels and peat – Determination of size distribution
SS 18 71 75 (1990)   Peat – Determination of mechanical strength of sod peat
SS 18 71 76 (1991)   Solid fuels – Determination of total sulphur content with Eschka and with bomb washing method
SS 18 71 77 (1991)   Solid fuels – Determination of total sulphur using a high temperature tube furnace combustion method (IR detection)
SS 18 71 78 (1990)   Biofuels and peat – Determination of raw bulk density and calculation of dry raw bulk density in a large container
SS 18 71 79 (1990)   Peat – Determination of raw bulk density and calculation of dry raw bulk density in a large container
SS 18 71 80 (1990)   Biofuels and peat – Determination of mechanical strength for pellets
SS 18 71 84 (1990)   Biofuels and peat – Determination of moisture content in the analysis sample
SS 18 71 85 (1995)   Solid fuels – Determination of total chlorine in solid fuels and solid residues by combustion in a calorimetric bomb

SS 18 71 86 (1995)	Solid fuels – Determination of total sulphur in solid residues using a high temperature combustion method with an IR detection procedure
SS 18 71 87 (1995)	Solid fuels – Determination of the amount of unburned material in solid residues from combustion

Switzerland

Guidelines for 'Sorting and Classifying Energy Wood' [6].

SN 166000 (2001)	Testing of solid fuels – Compressed untreated wood, Requirements and testing

United Kingdom

British Biogen, the trade association to the UK bioenergy industry, has developed a system of describing wood fuels (Describing retail wood fuels, published in 2000) [7].

US

ASTM D 871 (1982)	Standard Test method for moisture analysis of particulate wood fuels
ASTM E 897 (1988)	Standard test for volatile matter in the analysis sample of particulate wood fuels
ASTM D 1102	Test method for ash in wood
ASTM D 873 (1982)	Standard test method for bulk density of densified biomass fuels
ASTM D 2898 (1981)	Rain test for pellets

References

1   ANONYMOUS (1987) Standard no. 1 for the determination of fire chip quality in terms of size (Norm nr 1 for bestemmelse af kvaliteten på brændselsflis med hensyn til størrelsesfordeling – in Danish). Danske Skoves Handelsudvalg, 1987, Danish Forest Society, Frederiksberg, Denmark
2   ANONYMOUS (1999) Nye kvalitetsbeskrivelser for brændselsflis (New descriptions of quality for fuel chips), Danish Forest and Landscape Research Institute, Denmark
3   DK-TEKNIK, R. SØ, ELSAM, ELKRAFT (1996) Biomasses brændsels- og fyringskarakteristika. Anbefalede analysemetoder og notater. (Fuel and combustion characteristics of biomass. Recommended analytical methods and notes), Energiministeriets Forskningsudvalg for produktion og fordeling af el og varme, Brændsler og forbrændingsteknik, Denmark
4   ANONYMOUS (1998) Quality assurance manual for solid wood fuels. Publications 7, 27p. Finnish Energy Economy Association, FINBIO, Finland
5   HEEMSKERK, G. C. A. M. et al (1998) *Best Practice List for Biomass Fuel and Ash Analysis*. EWAB-9820, KEMA Power Generation, Arnhem, April
6   Riegger, W. (1996) *Sortieren und Klassieren von Energieholz*, Swiss Association for Wood Energy, Birmensdorf, Switzerland
7   ALAKANGAS Eija et al., 2005: Review of the present status and future prospects of standards and regulations in the bioenergy field, Nordic Innovation Centre (Ed.), Oslo, Norway, ISSN 0283-7234 (available at http://www.nordicinnovation.net)

Annex 4

Members of IEA Bioenergy Task 32: Biomass Combustion and Cofiring

The following persons represent the member countries of Task 32 at the time of printing. The actual contact information can be found on the Task's internet site: www.ieabcc.nl

Country	Contact person	Address
CEC	Erich Nägele	European Commission DG RTD J3 Rue du Champs de Mars 21 B-1049 Brussels tel: +32-2-296-5061 fax: +32-2-299-4991 e-mail: Erich.Naegele@ec.europa.eu
Austria	Ingwald Obernberger	Graz University of Technology Institute for Resource Efficient and Sustainable Systems Infeldgasse 21b A-8010 Graz tel: +43 316 48 1300 12 fax: +43 316 48 1300 4 e-mail: Inwald.Obernberger@tugraz.at
Belgium	Didier Marchal	Département de Génie Rural Centre Wallon de Recherche Agronomiques Chaussée de Namur, 146 B 5030 Gembloux tel: +32 81 627 144 fax: +32 81 615 847 e-mail: marchal@cra.wallonie.be
Canada	Sebnem Madrali	Department of Natural Resources 580 Booth Street Ottawa, Ontario K1A OE4 tel: +1 613 996 3182 fax: +1 613 996 9416 e-mail: smadrali@nrcan.gc.ca

Denmark	Anders Evald	Force Technology Hjortekærsvej 99 DK-2800 Lyngby tel: + 45 72 15 77 50 fax: + 45 72 15 77 01 e-mail: aev@force.dk
Finland	Pasi Vainikka	VTT-Energy PO Box 1603 FIN-40101 Jyväskylä tel: +358 20 722 2514 fax: +358 20 722 2597 e-mail: pasi.vanikka@vtt.fi
Germany	Hans Hartmann	Technologie- und Forderzentrum Dr. Hans Hartmann Schulgasse 18 D-94315 Straubing tel: +49 9421300112 fax: +49 9421 300211 e-mail: hans.hartmann@tfz.bayern.de
Netherlands	Sjaak van Loo (Task Leader)	Procede Group BV PO Box 328 7500 AH Enschede tel: +31 53 489 4355 / 4636 fax: +31 53 489 5399 mob: +31 6 51219646 e-mail: sjaak.vanloo@procede.nl
Netherlands	Jaap Koppejan (Assistant Task Leader)	Procede Biomass BV PO Box 328 7500 AH Enschede tel: +31 53 489 4636 fax: +31 53 489 5399 mob: +31 6 49867956 e-mail: jaap.koppejan@procede.nl
Netherlands	Edward Pfeiffer Coordination Cofiring activities	KEMA PO Box 9035 6800 ET Arnhem tel +31 26 356 6024 e-mail edward.pfeiffer@kema.com

Netherlands	Kees Kwant (Operating Agent)	SenterNovem PO Box 8242 3503 RE Utrecht tel: + 31 30 239 3458 fax: + 31 30 231 6491 e-mail: k.kwant@senternovem.nl
Norway	Øyvind Skreiberg	Department of Energy and Process Engineering Faculty of Engineering Science and Technology NTNU, N-7491 Trondheim tel: +47 69 261831 fax: +47 99 137857 e-mail: Oyvind.Skreiberg@ntnu.no
Sweden	Claes Tullin SP	Swedish National Testing and Research Institute Energy Technology Box 857 S-501 15 Borås tel: +46 33 16 55 55 fax: +46 33 13 19 79 e-mail: claes.tullin@sp.se
Switzerland	Thomas Nussbaumer	Verenum Langmauerstrasse 109 CH - 8006 Zürich tel: Verenum +41 44 377 70 70 tel: direct +41 44 377 70 71 e-mail: thomas.nussbaumer@verenum.ch
United Kingdom	William Livingston	Doosan Babcock Energy Limited Technology Centre High Street RENFREW PA4 8UW tel: + 44 141 886 4141 fax: +44 141 885 3370 email: blivingston@doosanbabcock.com

Index

Page numbers in *italic* refer to Figures, Tables and Boxes. Ranges of page numbers in *italic* signify that one or more illustrations appears on each page in the range, inclusive.

abrasion, boilers 285–286
acid precipitation 1, 291, 303, 386
active oxidation 277–278, *279*
additives 78, 283, 284, 285, 392
adsorptives 318
advantages of biomass applications 294
aerosols 244, 249, 260, 261, *261*, 262, 297–298, 310, 393–394
    and boiler corrosion 44–46, 298
    heavy metals in 36, 46, 297
agricultural residues 54, 203, 222, 381, 385, 386, 392, 393
agriculture
    ash use in 349, 361, *363*, 364–365, 365–366, 366–367, *367*
    sustainable 108
air
    distribution 21
    pre-heating 18
air gasification 10
Al (aluminium) 357–358, *358*
alcohol production 7
alloys, superheaters 279–280, 283, *283*, 283–284, 284–285
aluminium (Al), in ash 357–358, *358*
Amer Power Plant (Netherlands) 218–219, *218–219*
ammonia (NH_3) 9, 301, *303*
anaerobic digestion 7
analysis methods, biomass fuels 50
annual heat output line 169, *170*
annual utilization rate 170, 171
aquastats 122

ash 11
    behaviour 252, 260–262, *261*, 264–265, 266
    elemental analysis data 252, *253*
    elements forming 43–46, *44–45*, 69
    formation 34–38, *35–37*
    fusion behaviour 243, 252, 258
    melting behaviour 44, *45*, 138, 252–254, *254*
    related problems 249, 393–394
    sintering 147, 252, 260, 262, 263–264, 267–269
    sintering temperature 38, 44, *45*
ash agglomerates 249, 260, 262–263, *263*, 264, 266
ash characteristics of biomass fuels 43–46, *44–45*, 249, 250–258, *253–254*, *256–257*
    chemistry 233, 238, 244
ash content of biomass fuels 43, *43*, 83, 232, 238, 244, 249, 348
ash fractions 37–38, *37*, 350, *351*
ash utilization 98, 249, 286–288, 348–362, *350–360*, *362–364*, 363–365
    ashes from co-firing 206, 208, 213, 214, 222, 245–247, *246*
    as concrete/cement additive 214, 245–247, *246*
    in forestry 366
    guidelines/standards 288, 365–367, *367*
    sustainable 46, 158, 361–362, *362*, 394
ashes
    characteristics 250–258, *253–254*, *256–257*
    characterization techniques 251–258, *254*, *256–257*
    composition 276–277, *358*, 362
    deposition *see* deposition
    disposal 71, 206, 222, 288, 293, 361, 394
    as fertilizer 98, 286, 348, 349, 361, 366–367

handling 249, 263, 266, 293
impact on flue gas cleaning systems 272–275
mixed (from co-firing) 222, 243, 249, 257–258, 264–265, 287, 288
pH values 357, *357*
pre-treatment 349, 361, 363–364, *363–364*
properties 350–358, *351–359*
removal 122, 147, 263, 264, 266
treatment technologies 362–367, *363–364, 367*
utilization *see* ash utilization
from waste wood 362
from woody biomass 353, *353–355*
see also fly ash
Asia 3, 379, 381, 385
atmospheric combustion of pulverized biomass 175
Australia 381–383, 387, *389*
co-firing 203, 209–212, *210*, 220–221, *223*, 236, *236*
milling experiments 229–232, *230–231*
Austria 78, 122, 158, 383
ash utilization guidelines 365–367, *367*
co-firing 203, 215–217, 215–219, *215, 217, 217–219*
emissions limits *339, 340*
investment costs 171, *172*
ORC plant 190, *191*
policies 383, 387, *388*
standards 77, *77*, 168–169
Stirling engines 195, 196–197, *196*
transportation 96, *97*
wood classification 69
authorizations to operate 206
automatic combustion systems 113
automatic system operation *see* process control
Avedøre Power Plant (Denmark) 219–220, *220*

back-pressure plants 178, 179, 179–182, *180–181*, 200
bag filters 318, *320*, 325–327, *327–328*
bales 88, 90, *90*, 91, 144–145, *144*
for co-firing 212, 213, 221, 224–225, 226, 228
baling 72–73, *72–73*, 79
bark 40, *40*, 140, 147, 216, 262, 348
storage 79–81, *80*, 84–85, *84*, 86, 88, 155

base load 169, *170*
basic emission conversion calculations 29–31
basic engineering calculations 23–29
BAT (best available technology) 348
batch combustion 8, 13, 14, 20, 21, 31–33, *32*
applications 19, 22, 112–113
bed material 147–150, *151*, 216, 218, 263, 392
Belgium 192, 203, 288, 344
belt conveyors 90–91, *91*, 216
belt dryers 81–83, *82*
belt filters 370, *371*
best available technology (BAT) 348
BFB (bubbling fluidized bed) systems 23, *134*, 135, 147, 148–149, *149–150*, 154
bio-electricity production, policies for 387–389, *388*
BIOBIB database 48
biochemical conversion technologies 7
BIOCOCOMB project 216–217, *217*
bioenergy 1, 379
current status 3–4, *3*
developing countries 2, 3, *3*
system perspectives 96–108, *99–101, 103, 105–107*
technologies 4–5
biofuels 1, 1–2, 4
see also biomass fuels
biological degradation 79, 155, 227–228
biomass 1, *1*, 2
see also biomass fuels
biomass applications, advantages and disadvantages 294
biomass co-firing *see* co-firing
biomass energy 2
biomass fuels
analysis methods 50
ash characteristics 43–46, *44–45*, 250–258, *253–254, 256–257*
ash content 43, *43*, 83, 232, 238, 244, 249, 348
ash-melting behaviour 44, *45*
baling 72–73, *72, 73*, 79
bulk density 39, *39*, *40*, 60, 225, 233
bundling 72–74, *73, 74, 75*
characteristics 38–50, *39–41, 43–45, 47*, 61, 232–233, *234*
chemical composition 13, *14*, 19, 33, 57
chemical fractionation 254–255, *255–256*

for co-firing 203
databases on 48
drying *see* drying
energy content 59–60
guiding values and ranges 46, 47
handling 89–95, *89–94*, 212, 216–217, *217*, 221, *223*, 225–226
imports 223, 385
moisture content *see* moisture content
new 392
pre-processing *see* pre-processing
pre-treatment *see* pre-treatment
price calculations 59–60
standardization 49–50
storage *see* storage
suitable technologies 95, *95*
supply 56, *56*, 58, 60–62, 389, 392
transportation *see* transportation
see also fuel
biomass resources *1*, 2, 391–392
biomass/coal blends 229
boiler full-load operating hours 171
boiler probing trials 278–283, *280*
boilers
    cleaning systems 60, 148
    co-firing 235
    corrosion 200, 206, 238, 249, 267, 270, 276–285, *278*, *280*, *285*, 298
    design 270, 272
    efficiency 235
    erosion 249, 270, 272, 285–286
    metal wastage 249, 270, 279–280, *280*, 281
    modification 313
    size 169, *170*
bomb calorimeters 12
bonded ash deposits 249
bottom ash 36, 37, *37*, 215
breakers for waste wood 71, *71*
briquettes 74, *77–78*, *78*, 79, 119
    peat 119
briquetting 78, *78*, 81, *82*
bubbling fluidized bed *see* BFB
bucket elevators 93, *93*
bulk density
    ash 350
    fuel 39, *39*, *40*, 60, 225, 233
bundling 72–74, *73–75*
burn-back 123

burners 207, 209, 213, 237, 243, 266
    wood pellets 122–125, *123–124*
burnout 213, 215, 235–236, *236*

cadmium (Cd) unloading 98, 101
calorific value 11–14, 167
Canada, standards *131*, 132–133
carbon dioxide *see* CO_2
carbon monoxide *see* CO
carbon sinks 2
carbonate, in ash 357–358, *358*
Carnot process 193
catalysts 11
catalytic converters 17, *114*, 115, 304, 305, *305–306*, 317–318
CCSEM (computer controlled scanning electron microscopy) 255
Cd (cadmium) unloading 98, 101
cement, use of fly-ash in 214
CEN (European Committee for Standardization) 130–132, *131*
central heating systems 11
    pellet burners for 125–126, 6126
cereals 72, 144–145, *144*, 147
certification 124, 129–133
CFB (circulating fluidized bed) combustion 22–23, *134*, 135, 147, 149–150, *151*, *154–155*
    co-firing 207, 216–219, *217*, 218–219, 221
CFD (computational fluid dynamics) modelling 33, *33–34*, 214, 395–396
CH_4 (methane) 208, 300, *302*
chain conveyors 91, *92*
Chalmers Energy Supply Side database 104
char 9–10, 11, 19, 20, 35–36, 41, 301
    from co-firing 219, 235, 236
    combustion 8, 14, 23, 31, *32*, 151, 295
    gasification 9–10, 23
characterization techniques for ashes 251–258, *254*, *256–257*
charcoal 7, 8, 9, 13, 17
chemical fractionation techniques 254–255, *255–256*
China *3*, 381, 385, 386
chippers 64–65, *64*, *65*, *66*
chlorine *see* Cl
CHP (combined heat and power) 69, 108, 177, 182, 187, *195*, 200
    Austria 383

efficiencies 197–199, *197–199*
emissions 308–309, *309*
heat-controlled operation 170
investment 102
ORC 190–191, 381
potential 391
refits of obsolete plants 385
Stirling engines *195*, 381
chunkers 65, *67*
cigar burners 144–145, *144*
circulating fluidized bed *see* CFB
Cl (chlorine) *41*, 42, 57, *57*, 239, *240*, 243, 244
  see also corrosion; HCl
cleaning
  heat exchangers 194
  offline 266, 270
  online 269, 270
climate change, mitigation 97–98
climate impacts of pollutants *302–303*
closed gas turbines 176, *176*, 192, *193*, 201
closed oil cycle 187
closed thermal cycles 175, 175–176, *176*
closed-cycle economy, biomass ashes 363, *364*
CO (carbon monoxide) 9, 23, 33
CO (carbon monoxide) emissions 127, 239, 291, 298–299, *299–300*, *302*, 308, *308*–309
  reduction 308, 317–318
  wood-burning appliances *299*, 305, *305*
CO_2 (carbon dioxide) 9, 31–33, *32*, 294, 298, 300, *302*, 381, 391
  avoidance costs 108, 203
  concern over emissions 209–211
CO_2-neutrality of biomass 23, 30, 292, 293, 294
co-current combustion 137, *137*
co-firing 200, 272
  ash behaviour 264–265, 272–275
  ash deposition *see* deposition
  barriers to 221, 222, 245
  boiler types 105, *106–107*
  and CO_2 reductions 203
  commercial operation 209, 212, 220, 221
  corrosion *see* corrosion
  costs 104, 105, 108, 206, 207, 222
  direct 206, 207, 208–216, *208–210*, *213–215*
  environmental issues 222, 239–247, *240–242*, *246*
  indirect 206, 207, 216–219, *217–219*
investment 203, 206, 207–208
modelling potential 104–107, *105–107*
with natural gas 219–220, *220*
operational experience 203–206, *204–205*, 208–223, *210*, *213–215*, *217–220*, 220–223, 244
operational issues 232–239, *234*, *236*, *238*, 382
parallel 206, 207–208, 219–220, *220*
Polish opportunities 102–108, *103*, *105–107*, 108
progress 222
prospects 272–273, 294, 379, 381, 381–383, *382*, 383, 386
RECs *389*
research and development 207, 394
retrofitting for 205, 221, 237, 244
technical options 206–208, 221
technical risks 203, 205–206, *205*, 243, 244
techno-economic aspects 170, 171
testing 206, 207, 211, 212
UK 385
see also pre-processing
co-firing ratios 237, 245, 258
  impact on plant 209, 211–212, 235, 236, 243, 244, 265, 273
  low 207, 211, 216, 222, 229
  and SO_x 239, *240*
  straw 212, 213, 214
co-generation 179, 181, 182, 184, *185*, 200
coal 175, 211, 293, 381
  ultimate analyses *234*
coal-based plants 102, 273, 379, 381, 386
coal-fired power stations, Australia 209–211
coarse fly-ash 260–261, *261*, 297
combined heat and power *see* CHP
combustion 7, *7*, 8, 11, *316*
  co-firing 237
  efficiency 29, 40
  influencing parameters 4, *4*
  as main bioenergy technology 4–5
  optimizing 79, 291, 308
  process 8, *8*, 14–17, *16–17*
  rate 235–236
  staged-air 8, 313–316, *315*, *316*
  temperature 11, 20, 263, 299, *300*
  variables affecting 11–38
combustion chamber design 311

combustion equipment, selection 311–312
combustion plants 167–171, *169–170*
combustion technologies 4–5, 8, 38, 95, *95*, 134–135, *134*
　development 361, 392
　industrial and district heating systems 154, *154–155*
　see also domestic wood-burning appliances; industrial applications
comminution 60, 61, 62, 64–65, *65*, *66–68*, 221
complete combustion 157
　emissions from 294–298, *296–297*, *302–303*
components of biomass plants 168
compressed hot water 187, *188*
computational fluid dynamics *see* CFD
computer controlled scanning electron microscopy *see* CCSEM
concrete, use of co-firing ashes in 214, 245–247, *246*
condensable vapours 258
condensate 367–370, *368–371*
condensate sludge *158*, 160, 367–370, *368–371*
condensing plants 179, 181, 182, 200
construction 171, 312
contaminants, in steam 185
continuous combustion 8, 31–33
convective section fouling processes 266–267, 269–270
conventional combustion *316*
conversion technologies 7
corrosion 22, 43, 284, *285*
　boilers 249, 267, 270, 276–285, *278*, *280*, *285*, 298
　from co-firing 200, 206, 213, 214, 222, 238, 239, *240*, 243, 278–283, *280*
　gas-side 239, *240*
　heat exchangers 22, 37, 42, 192, 194, 393–394
　mechanisms 277–278, *278–279*, 281
cost functions, minimizing 161
cost optimization, systems modelling by 104–107, *105*
cost-effectiveness 293, 294
costs
　ash treatment/disposal 71, 362
　of biomass systems 387
　co-firing 104, 105, 108, 206, 207, 222

construction costs 171
fuel costs 54, 79, 169, *169*, 171
heat distribution networks 171
heat generation 171
investment costs 79, 148, 169, *169*, 171, *172*, 179, 200
multifunctional energy plantations 101–102, *101*
operating costs 148, 183
ORC 189, 190
pre-treatment of fuels 93, *94*
reduction 391
renewable energy 203
transportation costs 72–73, *73*, 95, 96, *97*, 294
counter-current combustion 137, *137*
crane systems 86, 90, *90*, 213, 225
cross-flow systems 137, *137*
cycle of minerals 349, *350*, 361
cyclone burners 135, 151–152, *153*
cyclone fly ash 37, *37*
cyclones 136, 216, 308, *319*, 322–323, *322–323*
　CFB combustion 23, 149–150, *151*
　multicyclones 37, *320*, 323, *324*, 361
　particle separation 37, *37*, 318, 319, *319*

databases, on biomass fuel characteristics 48
dead koala RECs problem *389*
Delta Electricity 211–212
Denmark 78, 122, 158, 288, 296
　ash utilization 222
　boiler probing trials 278–283, *280*
　co-firing 203, 212–214, *213–214*, 219–220, *220*, 224–225, 274–275, *275*
　emissions limits 340–341
　investment costs 171, *172*
　policies 212, 383, 387, *388*
　Salix fertilizing plans 62, *63*
　Stirling engines 195, 196, *196*
　trends 383
　washing of straw 310
density of fuel 19–20
deposition 22, 249, 266–272, 271, 382, 393–394
　from co-firing 206, 213, 214, 222, 237–238, *238*
　see also corrosion; fouling; slagging
design 21

furnaces 39–40, 269, 311
   techno-economic aspects 167–171, *169–170*
developing countries 2, 3, *3*
devolatilization 8, 9–11, *10*, 236
dioxins *see* PCDD/F
direct co-firing 206, 207, 208–216, *208–210, 213–215*
direct combustion 379
direct investments 102
directly fired gas turbines 175
disadvantages of biomass applications 294
district heating 5, 102, 108
   combustion technologies 154, *154–155*
   energy demand 169
   network costs 171
   technical and economic standards 168–169
   see also CHP
domestic wood-burning appliances 74, 112–113, 121
   emissions 118, 119, 120, 121–122, 304–306, *305–306*
   standards 129–133, *131–132*
   see also fireplace inserts; fireplaces; heat-storing stoves; wood pellet appliances; wood-fired boilers; wood-stoves; woodchips
down-draught boilers 120, *120*
down-draught combustion 117–118, *117*
downstream processes 266
   ash 363–365, *363–364*
draught 8, 20
drum dryers 83
dry throwaway processes 337
dry-matter losses 79, 84, 85, 155, 228
drying (biomass fuels) 11, 78–83, *80–82*, 216, 228
   in fields 11, 57, 61, 72, 79
   to improve efficiency 79, 155, *156*, 395
   to reduce emissions 11, 310–311
   see also field storage
drying (during combustion) 8, *8*, 9
dust 139, 211, 225
dust injection furnaces 152, *153*
dust precipitators 148

EC (European Commission) 102, 103, 203, 221, 383, 391

economic aspects, steam turbines 182–183
economic optimization 171
economizers 170
economy of scale effects 4, 38, 171, 182
efficiencies 200–201, 203, 392, 395
   boilers 235
   calculations 28–29
   closed gas turbines 192
   combustion system 40
   electrical *see* electrical efficiencies
   gasifiers *199*
   improving 155–161, *156*, 392
   internal combustion engines *199*
   moisture content and 40, 79
   ORC 190–191, *199*
   pellet-fired systems 124
   steam engines *199*
   steam piston engines 184, 185
   steam screw engines 187
   steam turbines 177–178, 179, 183, *199*
   Stirling engines 193, 196–197, *199*
   thermal 21, 22, 28–29, *156*, 235
   wood boilers *199*
electric heaters 119, 121
electrical conductivity of ashes 357, *357*
electrical efficiencies 197–199, *198–199*, 200, 201
   condensing plants 179, 181
   Rankine cycle 179–180
   Stirling engines 196–197
electricity
   demand 386
   price 197–198, *198*
electricity generation, growth in biomass use 379, *380*
electricity supply industry 221
electrostatic filters 308, 318, 323–324, *325*
electrostatic precipitators (ESPs) 273, 298, *319, 320*
emergency coolers 179
emission conversion calculations 29–31
emissions 5
   abatement equipment 243–244
   from co-firing 239–242, *240–242*, 294
   from coal-fired plants 273
   from complete combustion 294–298, *296–297, 302–303*
   components 294–303, *296–297, 299–300, 302–303*

data 292, 304–309, *305–309*
from domestic wood-burning appliances 118, 119, 120, 304–306, *305*
dust 139, 211
environmental impacts 292, *302–303*
fossil fuel combustion 292, 293, *293*, 333
fuel properties and 306, *307*
HCl 296–297
hydrocarbons 124, 125, *125*, 126, 127, 300–301
from incomplete combustion *see* incomplete combustion emissions
limits 120, 121, 244, 272, 291, 292, 294, 339–348, *339–348*, 392–393
from low-quality biomass 22
measuring 304
minimizing 22
particles 124, *125*, 291, 297–298, 301, 304, 393
reduction measures 206, 291–292, 294
requirements 129, 130, *131*, 132–133
from wood pellet burners 122, 123–124, *124*, *125*
see also CO; CO_2; NO_x; primary emission reduction measures; secondary emission reduction measures; SO_x
energy, sustainable 102, 108, 203
energy calculations 26–28
energy carriers 7, 8
energy consumption 3
by pre-treatment 93, *94*
energy content of fuels 59–60
energy conversion 1, 7
infrastructure 102, 103, *103*
energy crops 2, 54, 98, 203, 392, 393
energy demand 169, 391
energy density 39, *39*, *40*
energy efficiencies 395
energy loss during storage 79, 80–81
energy systems analysis 97–98, 108
engineering calculations 23–29
engineering process 167–168
entrained ash 34
environmental benefits 1, 98, 294
environmental impacts 203, 206, 292–294, *302–303*
see also emissions
erosion 209, 249, 270, 272, 285–286

ESPs (electrostatic precipitators) 244, 273, 298, *319–320*
estimating state variables 167
EU (European Union) 3, 5, 102, 387
directives 103, 391
emissions limits 339, *339*
policies for promoting bio-electricity 387–389, *388*
power plants database 104
standardization of fuels 49–50
standards 130–132, *131*
targets 383, 391–392
trends 383
Europe
biomass imports 223, 385
co-firing 220–221
energy consumption 3
installed capacity 379, *380*
standards 245
European Committee for Standardization *see* CEN
Europe.biomass imports 223
excess air ratio 18–19, *20*, 31–33, *32*
extraction plants 179, *182*
extraneous inorganic material 250, 258

Fe (iron), in ash 357–358, *358*
feed-in tariffs 383, 387, 389
feeding systems 125, 126, *128*, 129, *144–145*, 145
co-firing 207, 215
fermentation 7
fertilizer, ash as 98, 286, 348, 349, 361, 366–367
fertilizer spreaders 364
fertilizers, for growing biomass fuels 57, 61, 62, *63*, 293
FGD (flue gas desulphurization) 244, 273, 275
field drying 57, 61, 72, 79
field residues 385
field storage 57, 61, 72, 88
filter fly ash *37*, 38
filter presses *371*
filter-ash precipitators 361
filters 218, 244, 318, 370, *370*
bag filters 318, *320*, 325–327, *327–328*
financial instruments 203, 387–389, *388*
fine grinding mills 65, *67*

fine particles, gasification 150
fine wood waste, storage 88, *88*
Finland 61, 158, 383, 384
    bundling forest residues 73–74, *73, 74, 75*
    co-firing 203
    emissions limits 341
    soapstone stoves 116–117, *116*
    underfeed rotating grates 145–146, *145*
fire tube boilers 182
fire-side additives 283, 284, 285
fire-side corrosion, preventive and remedial measures 283–285
fireplace inserts 115, *115*, 116
    emissions data 304, *305*
fireplaces 8, 19, 294, 301
firewood 112
fixed bed combustion 134, *134*, 135, 135–147, *136–147*
fixed grate systems 140
fixed premiums 387
flame stability 237
flexural strength, concrete *246*, 247
flow properties, biomass/coal blends 225–226, *227*
flue gas 175–176, 192, 239, 243
    calculation of flows and compositions 24–26
    cleaning systems 5, 83, 219, 272–275, 392
    cooling 36
    heat exchange 18, 194
    recirculation 137–138, 144, 148, 164
    reducing oxygen content 156–158, *156–157*
    separation of particles and metals 175
flue gas condensation 157, 158–161, *158–160*, 170, 308, *320*
    waste heat used for drying fuels 81–83, *82*
    waste-water handling 367–370, *368–371*
flue gas desulphurization (FGD) 244, 273, 275
fluidized bed combustion 22–23, 134, *134*, 135, 147–150, *149–151*, 259–260, 338
    ashes 252, 263–264, 266, 350
    co-firing 105, *106–107*, 200, 203
    coal 385
    future 380, 392
fluidized bed gasification 381
fly ash 35, 36, *36*, 37–38, *37*, 175, 297, 350–358, *351–359*
    from co-firing 214, 235, 239, 244, 282

fixed bed combustion 135, 260, *261*, 262
fluidized bed combustion 263, 264, 350
heavy metals in 83, 298
pulverized fuel combustion 265
recycling 218, *218–219*
research 393
slag formation 267–268
utilization 286, 287–288
see also ash utilization; ashes
food processing residues 40, 41, 54
forced draught 8, 20
forest residues 54, 58, 60–61, 72–74, *73–75*, 381
forest soils, pH 46
forestry 56, 361, 366
fossil fuels
    for co-firing 203
    dependency 211
    dominance 103, *103*
    emissions 292, 293, *293*, 333
    environmental impacts 203
    for peak load boilers 169, *170*
    replacing 97–98
    reserves 293
fouling 200, 252, 264, 265, 266–267, 281, 393
    avoiding/minimizing 194, 382
    from co-firing 218
fouling indices 256–257, 258
France 387
fresh wood 40, 79
furnaces, corrosion 243
fuel
    composition 13, 19, 33, 310
    consumption 20, 309
    costs 54, 79, 169, *169*, 171
    flexibility 216, 219, 222
    influence on combustion 19–20, 21
    mixtures of 135, 148
    prices 59–60, 197–198, *198*
    properties 8, 306, *307*
    quality 38, 54, 56–60, *56–57, 59*, 69–71, *70*, 209, 392
    supply 54, *54–56*, 55, 56, *59*, 60–62, 392
    see also biomass fuels; co-firing
fuel cells 384
fuel gas 175, 207, *217*
fuel preparation *see* pre-processing; pre-treatment

fuel-feeding systems 8, 21, 89–95, *89–94*, 134, 150–151
fuel-handling systems 89–95, *89–94*, 212, 216–217, *217*, 221, *223*, 225–226
fumes 244, 249, 258, 260, 265
furans *see* PCDD/F
furnaces 164, 243, 392
   design 39–40, 269, 272
fused ashes 266
fusion, ashes 252, 258, 260, 262, 263, 267–269
fuzzy control 120, 166

gas 7, 8
gas combustion 8
gas phase interaction 23
gas phase kinetics 23, 36
gas turbines 11, 175, 220, *220*, 394
gas-side corrosion 239, *240*, 276–277
gaseous reduction technologies 392–393
gasification
   in co-firing 207, 216, 218–219, *218–219*, 221, 222
   in combustion 7, *7*, 8, 9–11, 23, 119, 151
   technologies *199*, 379, 384
GCV (gross calorific value) 11, 11–12, 13, 38, *39–40*, 40, 59
   calculation 11–12
Gelderland Power Station (Netherlands) 208–209, *208–209, 208–210*, 210
genetic engineering 57, 395
geothermal plants 189, 190, 200
Germany 78, 122, 158, 195, 288
   co-firing 203
   emission limits 342
   policies 383, 384, 387, *388*
   standards 77, *77*
   waste wood classification 69
global expansion of biomass combustion 379–386, *380, 382*
global warming 209–211, 295–296
government policies 203, 212, 379, 381–389, *388*, 389–390
grasses 54, 61, *62, 241*, 291, 296, 392
grate combustion 22, 23, *106–107*, 259
   ash behaviour 260–262, *261*
grate furnaces 135, 135–146, *136–145, 154*, 200, 380
green certificate systems 387, *389*

greenhouse gas effect 111, 294, 295–296
greenhouse gas emissions 1, 103, 291, 386
greenhouse gas treaties 384
greenhouse gases 208, 293, 301–302
ground level ozone (O_3) 301–302
growing phase 56, *56–57*, 57–58, 293
guiding values and ranges, biomass fuels 46, 47

hammer mills 65, *68*, 214, *214*, 224, 225, 228–229
handling of biomass fuels 89–95, *89–94*, 212, 216–217, *217*, 221, *223*, 225–226
hardwoods 9, *10*, 19–20
harvesting 57–58, *59*, 61, 62, *62–63*
HCl (hydrogen chloride) 42, 244, 296–297, *303, 307*, 318
health impacts 3, 46, 86, 227, 301, *302–303*, 333, 393
heat distribution 21
heat distribution networks 171, 309
heat exchange 18, 21
heat exchangers 21, 36, 176, 192, *193*, 194, 201
   corrosion 22, 37, 42, 192, 194, 393–394
   deposition on 22, 249, 266
   fouling 252, 266–267
heat generation costs 171
heat output
   control 112, 113, 313
   variable 179, 182
heat price 197–198, *198*
heat production, efficiencies 197–199, *197–199*, 200
heat pumps 198, 198–199
heat recovery systems 158–161, *158–160*, 170, 220
   see also flue gas condensation
heat storage 18
heat storage tanks 112, 120–121
heat transfer mechanisms 17
heat-storing stoves 18, 116–118, *116–117*, 305, *305*
heavy metals 46, 58, 83, 298, *303*, 318
   in aerosols 36, 46, 297
   in ashes 286, 293, 351–353, *354–355*, 358, 359, *360*, 361
   in condensate sludge 160, 368, *368*
   separation 175, 349, *350*, 361, 362, *362*
   volatilization 35

in waste wood ashes 252, 361, 362
herbaceous biomass fuels 40, 61, *62*, 96, 148, 301, 348
    ashes from 238, 353, *353–355*
    baling 72, *72*
    for co-firing 203
    storage 88, 227–228
    see also bales; baling; grasses; miscanthus; straw
heterogeneous combustion 17, 23, 41
heterogeneous fuels 5
horizontally moving grates 142, *142–143*
hot air turbines 192, 201
hot water 10–11, 118, 119, 121
hydraulic piston feeders 91, *144*
hydrocarbon emissions 124, 125, *125*, 126, 127, 300–301
hydrogen chloride *see* HCl

IEA (International Energy Agency) 5
    Bioenergy Agreement 47, 56, 62, 389–390, 391
    Bioenergy Task 32 339
IGCC (integrated gasification combined cycles) 11, 198–199, *199*
impurities 69, 208, 215, 249, 250, 258, 348
incineration plants 283
inclined moving grates 140, *140, 141*
incomplete combustion emissions 11, 14, 20, 31–33, 294, 298–302, *299–300, 302–303*, 309
    reducing 16, 17, 21, 291, 309
India 381, 386
indirect co-firing 206, 207
indirect investments 102
indoor smoke 3
indoor storage 79, *80*, 85, 88, 225
industrial applications
    combustion technologies 154, *154–155*
    emissions data 306–308, *306–308*
    see also fixed bed combustion; fluidized bed combustion; heat recovery systems; process control; pulverized biomass combustion
industrial waste heat 189
infrastructure 102, 103, *103*, 203, 206, 223
inherent ash 34
inherent inorganic material 250, *251*
inorganic material 249, 250, *251*

in ash 286
    high temperature behaviour 258–265, *259, 261*
installed capacity, expected growth 380–381, *380*
insulation 18
integrated gasification combined cycles *see* IGCC
integrated harvesting systems 61
Integrated Pollution Prevention and Control (IPPC) 348
internal combustion engines 9, 11, 175, 193–194, *199*
International Energy Agency *see* IEA
investment 102, 103, 107–108
    subsidies 387, 389
investment costs 79, 169, *169*, 171, *172*, 200
    co-firing 203, 206, 207–208, 221
    fluidized bed plants 148
    ORC plants 189, 190
    Rankine cycle plants 179
    steam turbine plants 183
IPPC (Integrated Pollution Prevention and Control) 348
Ireland 387
iron (Fe), in ash 357–358, *358*
Italy 158, 288

Japan 203

lambda control probes 120
land use 97–98
landfill 208, 293
landfill gas 381
large-scale combustion applications 4, 8, 20, 22–23, 38, 75, 392
    CHP plants 393–394
Latin America 3, 379, *380*, 385
LCA (life cycle analysis) 292–293
leaching 57–58, *58*, 61, 72, 86, 310
    ashes 358, *359*
legislation 5, 211, 221, 222
liberalization of electricity supply industries 221
life cycle analysis (LCA) 292–293
life cycle of biomass fuels 56, *56*
lime/limestone injection 296, 309
limestone wet scrubbers 218, 336–338, *338*

liquefaction 7, 8, 11
liquid biofuels 169, *170*
live steam temperature 180, *181*
load control 163, *164*
lock-in effects 102, 103
lubrication 185

manganese (Mn), in ash 357–358, *358*
market-based mechanisms 5, 385
markets 107–108, 379–381, *380*, 386
material fluxes of ash-forming elements 359, *360*
MBS (multifunctional bioenergy systems) 98–102, *98–101*
medium-scale applications 8, 22–23, 38, 69, 201, 392, 394
melting curves of ash 252–254, *254*
metal wastage 249, 270, 279–280, *280*, 281
methane (CH$_4$) 208, 300, *302*
methane-enriched gas 7
methanol 9, 10
micro-scale CHP technologies 394
microscopic analysis of ash 255
mills 209, 211, 228–232, *230–233*
    hammer mills 65, *68*, 214, *214*, 224, 225, 228–229
mineral impurities 208, 249, 250, 258, 348
mineralogical analysis of ash 255
minerals, cycle of 349, *350*, 361
miscanthus 54, 57, 61, *62*, 72, 291, 296, 392
mixed ashes, from co-firing 222, 243, 249, 257–258, 264–265, 287, 288
mixed fuels
    for co-firing 207
    flow properties 225–226, *227*
mixture of fuels 135, 148
Mn (manganese), in ash 357–358, *358*
model predictive control *see* MPC
model-based process control 164, 165
modelling
    biomass combustion 33–34, *33–34*
    CFD 33, *33–34*, 214, 395–396
modification
    boilers 313
    fuel composition 310
    fuel particle size 219
    moisture content 310–311
moisture content 33, *39*, 135, 163, 164, 228

for co-firing 213, 225, 235
and combustion 9, 11, *12*, 20, 39–40, 163, 164
of different fuels 38, *40*, 78, 145
and efficiency 79, 156
factors influencing 57, 58, *59*
modification 310–311
see also drying
moulds 79, 80, 225
moving grate furnaces 8, 139, 140–142, *140–142*, 161–163, *162*
MPC (model predictive control) systems 154–155, 165–166, *166*, 167
MSP (multifunctional Salix production) 98, 99–102, *101*, *103*, *104*, 108
muffle dust furnaces 151, 152, *152*
multicyclones 37, *320*, 323, *324*, 361
multifuel feed 5
multifunctional bioenergy systems (MBS) 98–102, *98–101*
multifunctional Salix production *see* MSP
municipal solid waste 175, 283, 381
municipal waste water *100*

N (nitrogen) 41–42, *42*
N$_2$O (nitrous oxide) 293, 295–296, *302*, 309
natural draught 8, 19, 20
natural gas
    co-firing with 219–220, *220*
    topping up with 180
NCV (net calorific value) 11, 13, 38, 39, *39–40*, 59–60, 79
    calculation 12
Nepal 3
The Netherlands 94, 275, 288, 306, *306–307*, 383
    co-firing 203, 208–209, *208–210*, 218–219, *218–219*
    emissions limits 343, *343–344*
    policies 384, 387, 388, *388*
(neuro-)fuzzy control 166
new biomass fuels 392
new materials 183
new technologies 4, 5, 103, 185, 222, 294, 391, 392
NH$_3$ (ammonia) 9, 301, *303*
nitrogen (N) 41–42, *42*

nitrogen oxides *see* NO_x
nitrous oxide *see* N_2O
NMVOC (Non-Methane Volatile Organic Components) 300, *302*
North America
  co-firing 220–221
  emissions requirements 130, *131*, 132–133
  energy consumption *3*
  installed capacity 379, *380*
  standards *131*, 132–133
  wood-burning appliances 117, *118*
Norway, emissions limits 345
NO_x (nitrogen oxides) 42, 136, 139, 294–295, *296–297*, *302*
  control technologies 212, 243–244, 333–336, *335–337*, 392–393
  emissions 21, *32*, 33, 148, 216, 292, 293, 308, 386
  emissions reduction measures 206, 216, 222, 291
  from co-firing 200, 212, 239–242, *241–242*
  staged air combustion 313, 314–316, *315*
  wood pellet burners 123–124, *124*, *125*
  wood-burning installations 305, *305*, *306*

O_2 (oxygen)
  concentration 14, 15
  reducing flue gas content 156–158, *156–157*
O_2 (oxygen) gasification 10
O_3 *see* ozone
observers 167
offline cleaning 266, 270
oil 7, 8
  as lubricant in steam 185, 187
oil burners, replacing 122, 125–126
oil-free operation of piston engines 185
olive residues 40, 41, 54
online cleaning 269, 270, 272
open cycles 175
operating conditions 60
operating costs 148, 183
operational modes 197–198, 200
optimization
  co-firing 237
  process control 312–313, *313*
optimized combustion process 291
ORC (organic Rankine cycle) 176, *176*, 189–191, *190–191*, 200, 381, 394
  advantages and disadvantages 200–201
  efficiencies *199*
organic oil 189, 190
organic Rankine cycle *see* ORC
outages 206, 266, 270
outdoor storage 79, *80*, 85, *85*–86, 88, 155
output control 121
over-fire boilers 118–119, *118*
oxygen *see* O_2
ozone (O_3)
  depletion 291, 295–296, 303
  ground level 301–302

PAH (Polycyclic Aromatic Hydrocarbons) 239, 286, 300–301, *300*, 355–357, *356*
panel bed filters 329–332, *331*
parallel co-firing 206, 207–208, 219–220, *220*
part-load operations 189
particle control technologies 192, 312, 318–333, *319–328*, *330–332*, 393
particle density (ash) 350
particle density (biomass) 235
particle size (ash) 35–36, 318, 319, *319–320*, 351
particle size (fuel) 22–23, 38, *39*, 69, 95, *95*, 233–235, 382
  modification 219
particles
  from co-firing 211, 218, 219, 239, 336
  emissions 124, *125*, 291, 297–298, 301, *302*, 304, 393
  in flue gas 175, 192
  gasification of fine particles 150
PCDD/F (polychlorinated dioxins and furans) 42, 239, 296, 301, *303*, 309, 310, 318
  and ash utilization 286, 355–357, *356*
peak load 169, *170*, 220
peat 119, 129
pellet boiler units 122
pellet burners 125–126, *126*
pellet stoves 124, 127–128, *127*, 305, *305*, 387
pelletizing process 75–76, *76–77*, 81, *82*, 228, 395
pellets 74–78, 79, 113, 119, 129, 147
  co-firing 219, 222
  for domestic use 78
  emissions 122, 124, *125*, 305, *305*
  handling and transportation 90, 96

petroleum-derived chemicals 8
pH
    ashes 357, *357*
    forest soils 46
Phyllis database 48
plant breeding 57, 61, 395
plant capacities 200
plant design, techno-economic aspects 167–171, *169–170*
plant dimensioning 169, *170*
plant nutrients, in ashes 351, *352–353*, 359, *360*
plantations 98
    see also MSP (multifunctional Salix production)
pneumatic conveyors 93, *94*
Poland
    biomass potential 107, *107*
    biomass supply estimates *105*
    co-firing 203
    co-firing opportunities 102–108, *103*, *105–107*, 108
    infrastructure 103, *103*
    renewable energy targets 104, 107
policies 203, 212, 245, 379, 381–389, *388*, 389–390
political support 4
pollutants, impacts *302–303*
polychlorinated dioxins and furans see PCDD/F
Polycyclic Aromatic Hydrocarbons see PAH
porosity of fuel 20
potential of biomass resources 391–392
poultry litter 252, *253*
power consumption
    milling 229, *233*
    ORC plants 191, 200–201
power generation 175–176, 200–201, 203, 380
    see also CHP; closed gas turbines; ORC; steam piston engines; steam screw engines; steam turbines; Stirling engines
power generators 206, 211, 221
power output, steam engines 186, *186*
power plants
    efficiencies 197–199, *197–199*
    replacement 102, 103
    retrofitting 205, 221, 237, 244
power-to-heat ratios 197, *197*
pre-mixing of co-firing fuels 207, 211
pre-ovens 128, *128*

pre-processing, for co-firing 208–209, *208–209*, 212, 213–214, *213*, 216, 223, 228–232, *230–232*
pre-treatment
    ash 349, 361, 363–364, *363–364*
    fuels 38, 64–83, *64–68*, *70–78*, *80–82*, 93, *94*, 200, 395
    waste wood 69–71, *70*, *71*, 208–209, *208–209*
preliminary size reduction 223, 224–225
pressurized combustion of pulverized biomass 175
price calculations 59–60
pricing structures 386
primary air 135–136, 151
primary emission reduction measures 304, 309–318, *313*, *315–316*, 361, 392, 393–394
    NO_x 42, 243, 293, 309, 312, *313*, 316–317, 334–336, *335–337*
    particles 301
    SO_x 293, 296, 309, 336–338, *338*
principles of combustion *316*
process control 21, 22, 120, 308
    optimization by 312–313, *313*
    systems 134, *141*, 145, 159, 161, 163–167, *164–165*, 392
process dynamics 161–163, *162*, 164–165
process heat 200
procurement of fuels 292–293
producer gas from biomass gasification 175, 207, *217*
promoting bio-electricity production 387–389, *388*
proven technologies 200
public support 389
pulsating solar biomass drying process 81, *81*
pulverized biomass combustion 20, 22, 134, *134*, 135, 175, 260
    ash behaviour 264–265
    systems 150–152, *151–152*, 155
pulverized coal boilers 203, 206–223, *208–210*, *213–215*, *217–220*
    see also co-firing; pulverized coal combustion
pulverized coal combustion 22–23, 105, *106–107*, 200, 260, 385
    see also co-firing
pyrolysis 7, *7*, 8, *8*, 9, *10*, 11

optimization 10
pyrolysis gas 9

quality control 59–60
quality of fuel 38, 54, 56–60, *56–57*, *59*, 69–71, *70*, 209, 392
quench processes 160

radiation shields 22
Rankine cycle 176, 178, 179–182, *180–181*
RECs (Renewable Energy Certificates, Australia) 389
recuperation 192, *193*
reducing oxygen content in flue gas 156–158, *156–157*
regenerators 195–196, *195*
regulation 22
renewable energy 203, 221, 379
　sources 97–98
　targets 104, 107, 211, 383
Renewable Energy Certificates (RECs, Australia) 389
renewable energy certificates (Sweden) 384
renewable fuel, biomass as 292, 293, 294
Renewable Obligation Certificates (ROCs) 384–385
Renewables Obligation (UK) 385
research and development 50, 192, 201, 382, 384, 386, 391–396
　co-firing 207, 212, 220–221
residence time 11, 14, 15, 16, *17*, 20, 21
residential batch-fired wood-burning appliances 113–122, *114–120*
resources, biomass combustion *1*, 2, 54, 391–392
retrofitting 205, 221, 237, 244
rice husks 285
ROCs (Renewable Obligation Certificates) 384–385
rotating grates 135
　underfeed 145–146, *145*
rotating particle separators 332–333, *333*
rotation periods 46

S (sulphur) *41*, 43
safety requirements 123
St Andra (Austria) 215–216, *215*
Salix 57, 62, *63*, 392

production in Sweden 98, 98–102, *99–101*, 108
sawdust 54, 93, 140, 147, 150, 229
　co-firing 209–212, *210*, 216
　storage 88, *88*
sawmill residues 64
SCR (selective catalytic reduction) 222, 243–244, 334–335, *335*, *337*
　catalysts 214, 243–244, 273, 273–275, *274*
screw conveyors 91, *92*, 145, 216
screw feeders 218
scrubbers 218, 318, *319*, 329, *329*, 336–338, *338*
secondary air 136
secondary burning 112, 113
secondary emission reduction measures 304, 309, 312, 318–338, *319–328*, *330–332*, *335–338*, 392, 393–394
　HCl 297, 318
　NO_x 42, 293, 333–336, *335–337*, 392
　particles 298, 312, 318–333, *319–328*, *330–332*
　PCDD/F 301, 318
　SO_x 218, 293, 296, 336–338, *338*
secondary size reduction 228–232, *230–233*
sedimentation tanks 369, *369*, 371
selective catalytic reduction *see* SCR
selective non-catalytic reduction *see* SNCR
self-heating, stored biomass 79, 84–85, *84*, *86*, 155, 227
sensors 166–167, 392
settling chambers 321, *321*
sewage water *99*, 101
shedding processes 266, 271–272
short rotation coppice 54, 58, 62, *63*, 222, 229
short rotation crops 56
short rotation forestry 395
short-term storage 86–88, *87*
shredded straw 93, 224
Si (silicon), in ash 357–358, *358*
silicon oil 189
silos 88, *88*
sintering 147, 252, 260, 262, 263–264, 267–269
　temperature 38, 44, *45*
size of installation 308
size reduction 223, 224–225, 228–232, *230–233*
slag, shedding 271

slagging 135, 265, 267–269, 393
  minimizing 39, 136, 138, 269, 382
slagging indices 256–258, 269
sliding bar conveyors 86, *87*
small particles, combustion 8
small-scale applications 4, 8, 22, 79, 294
  CHP 187, 394
  steam screw engines *176*, 187–188, *189*
  Stirling engines 1, 195, 201
small-scale combustion appliances, variables influencing 17–22, *18*
SNCR (selective non-catalytic reduction) 216, 243, 334, 335–336, *336–337*
SO$_2$ (sulphur dioxide) emissions 291, *303*, 386
soapstone stoves 116–117, *117*
sod peat 119, 129
soft sensoring 166–167
software-based process control 166–167
softwoods 9, *10*
solar energy 81, *81*, 121, 194
solid biofuels 2, 175
solid fuel kinetics 23
soot 124, 125, *125*, 127, 301, 393
soot-blowing 271, 272
SO$_x$ (sulphur oxides) 296, *303*
  control technologies 206, 212, 216, 244, 336–338, *338*
  emissions 200, 239, *240*, 292, 293, 309
SP (Swedish National Testing and Research Institute) 124, 129–130
Spain 203, 387
species selection, biomass fuel 57
spray towers *319*, 336–337, *338*
spreader stokers 135, 139, *139*, *143*
staged fuel combustion 316–317, *316*
staged-air combustion 8, 21, 136, 313–316, *315–316*
standardization of fuels 49–50
standards 222, 305–306
  ash utilization 288
  Canada *131*, 132–133
  concrete 245
  district heating plants 168–169
  domestic wood-burning appliances 129–133, *131–132*
  EU 130–132, *131*
  pellets and briquettes 77, *77*
  US *131*, 132–133

start-up 148, 150
state variables, estimating 167
steam engines 176, *176*, 199
  see also steam piston engines; steam screw engines
steam piston engines *176*, 183–186, *185–186*, 200
steam plants 190, 379, 394
steam screw engines *176*, 187–188, *189*
steam turbine plants 183, 394
steam turbines 11, 175, 176, *176*, 177–183, *177–178*, *183*, 200, 219–220, *220*
  for co-firing plants 200
  efficiencies 177–178, 179, *199*
Stirling engines 176, *176*, 193–197, *194–196*, 201, 381, 394
  efficiencies *199*
  using biomass 195–197, *195–196*
stoker boilers, co-firing 203
stoker burners 129, *130*
stoker-fired combustion, ash behaviour 252, 266
storage 60, 61, 83–88, *84*, *86*, *87*, 212, 221, 226–228
  bark 79–81, *80*, 84–85, *84*
  indoor 79, *80*, 85, 88, 225
  outdoor 57, 61, 72, 79, *80*, 85, 85–86, 88, 155
  sawdust and fine wood waste 88, *88*
  woodchips 79–81, *80*, 84–85, *84*, 85, 218
stoves 112
  see also heat-storing stoves; pellet stoves; wood-stoves
straw 147, 291, 296, 393
  bales 72, 144–145, *144*
  for co-firing 212–214, *213–214*, 220, 222
  harvesting 57–58, 61, 62
  pre-treatment 57–58, *58*, *58*, 79, *213*, 297, 418
  shredding 93, 224
  storage 213, 228
  washing 297, 310
straw shredders *213*, 214, *214*
Studstrup (Denmark) 212–214, *213–214*, 225
sub-micron particles 258
sub-stoichiometric bed operation 148
subsidies 221, 387, 389

sulphur dioxide *see* SO$_2$
sulphur oxides *see* SO$_x$
sulphur (S) *41*, 43, 239
superheated steam dryers 83
superheaters 180, *180*, 200
    alloys 279–280, 283, 283–284, 284–285
    corrosion and deposition 22, 277, 279–283, *280*, 284–285, *285*
supply chain 55, *55*, 60–62, 392
supply of fuel 54, *54–56*, 55, 56, 60–62, 392
supply phase 56, *56*, 58
sustainable agriculture 108
sustainable biomass supply 389
sustainable energy 102, 108, 203
sustainable utilization of ash 46, 158, 361–362, *362*, 363–365, *363–364*, 394
Sweden 61, 75, 158, 383
    certification 124, 129–130
    emissions limits 345–347
    policies 384, *388*
    Salix production 62, 98, 98–102, *99–101*, 108
    standards 77, *77*
    tiled stoves 116–117
Swedish National Testing and Research Institute (SP) 124, 129–130
Switzerland 158, 190, 347–348
synergy 98, 379, 381
    see also system perspectives
system perspectives, bioenergy 96–108, *99–101*, *103*, *105–107*
systems analysis 97–98, 107–108, *108*
systems modelling by cost optimization 104–107, *105*

tars 9, 11, 127, 218, 219, 301, 318
tax deduction 388
tax exemptions 389
technical risks of co-firing 203, 205–206, *205*, 243, 244
techno-economic aspects of plant design 167–171, *169–170*
technological developments 4, 103, 185, 294, 391, 392
temperature 14, 15, *17*, 18
tendering 387
test protocols 113, 121, 129–133
testing standards 129–133
TGA (thermogravimetric analysis) 9, *10*

thermal behaviour of fuels 19
thermal efficiency 21, 22, 28–29, *156*, 235
thermal power stations 177
thermochemical conversion technologies 7–8, *7*
thermodynamics 23
thermogravimetric analysis (TGA) 9, *10*
thermostats 122, 129
thinnings 54, 61, 65, 96
tiled stoves 116–117
topping up with natural gas 180
total efficiency, calculation 29
trace elements 251, 286
trace metals 207, 239
traditional wood-stoves *2*, 3, 8, 379
tramp materials 208, 215, 250, 348
transparency 389
transportation 95–96, 99
    costs 72–73, *73*, 95, 96, *97*, 294
    distances 96, 200
    forest residues 72–74, *73*, *74*, *75*, 96, *97*
travelling grates *137*, 138–139, *138*, 215–216, *215*
trends in biomass combustion 381–385, *382*
tube-rubber belt conveyors 91, *91*

UK (United Kingdom) 195, 203, 348, 383, 384–385
ultra-supercritical (USC) plants 219–220, *220*
under-fire boilers 119, *119*, 129
underfeed rotating grates 145–146, *145*
underfeed stokers 135, 146–147, *146–147*, 155
understoker boilers 200
urban green waste 229, 231–232, 6231–6233
US (United States)
    co-firing 203
    coal reserves 381
    milling trials 229
    particulate emissions limits 121–122
    standards *131*, 132–133
    trends 381, *382*
USC (ultra-supercritical) plants 219–220, *220*
utilization of ash *see* ash utilization

vaporization 41
vapours 244, 260, 265
variable heat output 179, 182
vibrating grate furnaces 135, 142–144, *143*
volatile content 19, 40–41, *41*

volatile heavy metals 35, 393
volatile inorganic compounds 252
volatile matter 232, 237
volatile organic compounds 83, 239
volatilization 35, 263
vortex flows 136, 146, *147*

walking floors 86, *87*
Wallerawang Power Station (Australia) 209–212, *210, 223*
washing biomass fuels 297, 310
waste combustion systems 5
waste heat, for drying fuels 81–83, *82*
waste water emissions 86
waste wood 54, 64, 298, 348, 392, 393
    co-firing 208–209, *208–209*, 210, 218
    heavy metals in ash 252, 262, 361, 362
    parameters affecting combustion 69, *70*
    pre-treatment 69–71, *70, 71*
    processing plants 69–71, *70*
waste-water treatment 244, 367–370, *368–371*
water cooling 136, 137–138, 144, 148, 151, 164
water heating 10–11, *11*
water treatment 183, 185
water tube boilers 182, 200
wet fuels 9, 145, 161, 211
wet limestone scrubbers 218, 336–338, *338*
wet scrubbers 218
wet throwaway processes 318, 337
wet-dry throwaway processes 338
wheel loaders 86, 89, *89*
willow *see* Salix
wood 2, 9, 40, 79, 112
    ash content 348
    classification (Austria) 69
    co-firing 222
    handling 225–226
    supply chain 55, *55*
    see also forest residues; waste wood; woodchips
wood combustion applications 5
    emissions 306–308, *306, 307*
wood dust filters 370, *370*
wood fire 112
wood logs 8, 119
wood pellet appliances 122–125, *123–125*
wood pellet stoves 124, 127–128, *127*, 305, *305*, 387
wood processing residues 54
wood-fired boilers 9, 20, 118–119, *118–120, 199*, 294
wood-fired heating plants 5
wood-stoves 19, 22, 31–33, *32*, 113–115, *114*
    emissions 20, 294, 301, 304–305, *305–306*
    traditional 2, 3, 8, 379
woodchips
    appliances using 119, 128–129, *129*, 140, 147
    ash behaviour 262
    co-firing 215–216, *215*, 226, *227, 241*
    handling 90, *90*
    storage 79–81, *80*, 84–85, *84*, 85
    transportation 73, *73*, 96
woody biomass
    ashes from 353, *353–355*
    for co-firing 203
    comminution 60, 61, 62, 64–65, *65*, *66–68*
    transportation 96

yield 58, *59*, 61

Zeltweg Power Plant (Austria) 216–217, 216–219, *217, 217–219*
zero-clearance fireplaces 115, 116